TOPOLOGICAL GRAPH THEORY

Jonathan L. Gross
Columbia University
New York, New York

Thomas W. Tucker
Colgate University
Hamilton, New York

DOVER PUBLICATIONS, INC.
Mineola, New York

Dedication

To Ralph H. Fox
To my father, Albert W. Tucker

Copyright

Published in Canada by General Publishing Company, Ltd., 30 Lesmill Road, Don Mills, Toronto, Ontario.

Bibliographical Note

This Dover edition, first published in 2001, is an unabridged republication of the work originally published in 1987 by John Wiley & Sons, New York. A new Preface and Supplementary Bibliography have been specially prepared by the authors for this edition.

Library of Congress Cataloging-in-Publication Data

Gross, Jonathan L.
Topological graph theory / Jonathan L. Gross, Thomas W. Tucker.—Dover ed.
p. cm.
Originally published: New York : Wiley, c1987.
Includes bibliographical references and index.
ISBN 0-486-41741-7 (pbk.)
1. Topological graph theory. 2. Topology. I. Tucker, Thomas W.

QA166.195 .G76 2001
511'.5—dc21

2001017314

Manufactured in the United States of America
Dover Publications, Inc., 31 East 2nd Street, Mineola, N.Y. 11501

Preface to the Dover Edition

When the original edition of this book went to press in 1985, research activity in topological graph theory was still dominated by group-theoretic structures with close connections to classical, low-dimensional algebraic topology: Cayley graphs, regular branched coverings, and group actions on graphs and surfaces. These structures had enabled us to present a unified context for the study of graphs and surface.

Within that context, we were able to describe not only the Ringel–Youngs solution to the Heawood map-coloring problem and the principal results on the genus of a group, but also the incipient programmatic themes of genus distribution and enumeration. Another emerging theme at that time, not fully launched, was algorithmics.

At the time of publication of this Dover reprint, the field is most heavily influenced by the methods developed within a series of more than several dozen papers by Robertson and Seymour, all published since 1983, whose outcome included solution of the Kuratowski problem for general surfaces. Whereas the classical Kuratowski theorem is that there are exactly two obstructions to imbedding a graph in the sphere, namely K_5 and $K_{3,3}$, the general Kuratowski problem is to decide whether the set of obstructions is finite, which Robertson and Seymour settled affirmatively.

Since the Kuratowski problem does not come with any symmetry, this achievement, no less notable than that of Ringel and Youngs, is accomplished with purely combinatorial methods, involving no group-theoretic structure at all. The monograph *Graphs on Surfaces* by Mohar and Thomassen, whose publication is expected to be contemporaneous with that of this new edition of *Topological Graph Theory*, provides the definitive account of developments in this branch of topological graph theory, sometimes called "structural graph theory" by its practitioners, who are referring largely to matroidal structure, not to group theory. Indeed, it is particularly felicitous that this reprint of *Topological Graph Theory* coincides with the publication of *Graphs on Surfaces*, since the two books complement each other so well and overlap so little.

A third monograph of interest, *Graphs of Groups on Surfaces: Interactions and Models* by Arthur White, is also close to completion. It further updates the 1984 edition of the classic text with new material on enumerative and random topological graph theory, as well as graph imbedding models for finite fields and geometries.

The algebraic branch of topological graph theory has flourished in many ways since 1985. It includes substantial work on regular and edge-transitive maps, Cayley maps, enumeration of regular coverings, groups of low symmetric genus, exploration of the non-uniqueness of triangular imbeddings of complete graphs, and lifting of morphisms. The enumeration and structure of the collection of all imbeddings of a given graph is clearly algebraic, as is much of the work on random graph imbeddings. In structural graph theory, there has been extensive work on edge/face width and its connections to planarity, tree width, the existence and structure of noncontractible cycles in imbedded graphs, algorithms and their complexity for various graph imbedding problems, and the flexibility of imbeddings of a graph in a given surface.

For the Dover reprint, we have made no changes in the text. However, we have included a supplement to the original bibliography to assist a reader interested in tracking the more recent trends in the field. We have tended to emphasize references that seem to fit most naturally into the context of our book; for papers more likely to be categorized under structural graph theory, we recommend the extensive bibliography of Mohar and Thomassen's book. A decade can seem like a lifetime in mathematics, but we feel that *Topological Graph Theory* has aged well and that it continues to be useful as a textbook and reference.

December 2000

Jonathan L. Gross
Thomas W. Tucker

Preface

The primitive objective of *topological graph theory* is to draw a graph on a surface so that no two edges cross, an intuitive geometric problem that can be enriched by specifying symmetries or combinatorial side-conditions. To solve the problem and its variants, techniques are adapted from a broad range of mathematics, especially from algebraic topology and group theory, and more recently from enumerative combinatorics and the analysis of algorithms.

Topological graph theory remains a young specialty, in which two decades ago there were but a scant handful of active contributors. Subsequently, however, there have been well over a thousand research papers partially or exclusively relevant to graph placement problems, written by hundreds of different contributors, many with simultaneous mathematical interests elsewhere.

Unmistakably, the turning point was the Ringel–Youngs solution, completed in 1968, of the classical Heawood map-coloring problem. Numerous follow-up investigations analyzed and generalized the methods devised for the Ringel–Youngs theorem, and the resulting techniques were applied to many other graph imbedding problems. Substantial credit is also due to Arthur White, whose lucid study, *Graphs, Groups and Surfaces* (1973), made topological graph theory accessible to newcomers, and the revised edition (1984) continues to serve that purpose well.

The present monograph is intended as a historically sensitive, comprehensive introduction to the foundations and central concerns of topological graph theory, with bridges to frontier topics and to other areas of mathematics, wherever they fit in naturally. It is sufficiently self-contained that it can serve as a first-year graduate text or a self-guided approach for ambitious persons having a background in undergraduate discrete mathematics, in particular, some acquaintance with graphs and with elementary facts about groups and permutation groups. No point-set topology is necessary, beyond what one might see in an honors calculus course.

Although the focus is concentrated on the exposition and application of general topological methods to graph imbeddings, a small amount of effort is invested in glimpses at the frontiers and in interesting digressions. Nonetheless, for the most part we have steered away both from heavy involvement in non-topological methods—as in the proof of the four-color theorem or in solutions to most crossing-number problems—and from complicated topological methods whose cost of mastery has not yet been demonstrated to compensate adequately in imbeddings (or in proofs of non-existence of imbeddings) of interesting graphs. Moreover, an extensive survey of the frontiers would have been incompatible with our desire to stay close to the core, as much as we regret omitting explicit mention of numerous important results, including some of our own.

The first four chapters are basic, beginning with a review of graph theory that stresses whatever is of topological importance. Discussion of imbeddings into surfaces is blended with a complete proof of the classification of closed surfaces. Voltage graphs, which are a combinatorial form of covering spaces, are developed in full generality, acknowledging the role of covering-space constructions as the single most important tools used in the derivation of genus formulas. No prior experience in algebraic topology is needed for our treatment of coverings and group actions on surfaces, and indeed, the specialized approach here might well serve as a preliminary to a general topological treatment.

The foundations thus established are used in a straightforward explanation of the very lengthy proof of the Ringel–Youngs theorem, which includes sufficient detail to enable someone to comprehend the original proof in its entirety. They are also used in an examination of the main results on the genus of a group, in which one studies imbeddings of "Cayley graphs", which are pictorial representations of a group, optimized over all possibilities for the group.

In early 1987, the frontiers of topological graph theory are advancing in numerous different directions, each pursued by a host of researchers, including particularly active schools in France, Italy, the United Kingdom, the United States, and Yugoslavia. An alphabetized partial listing of frontier topics would include algorithms for imbedding problems, covering-space constructions (including wrapped coverings and flows), enumerative analysis of imbedding distributions (sometimes using representation theory of the symmetric group), forbidden subgraph characterizations and their relationship to graph minors, genus of groups, map-theoretic connections, neoclassical imbedding questions, representation of higher-dimensional manifolds by graphs, and VLSI layouts. Moreover, there are some significant one-person and two-person approaches that fall into none of the foregoing categories. A single volume could not hope to treat all of the major developments in detail, even though we are still able to bring the reader within striking distance of most of them.

We are indebted to many colleagues and students who have read parts of our many preliminary drafts, particularly to Arthur White and Ward Klein, whose critical suggestions have greatly helped. Special thanks are due to Mary Forry for typing the final draft.

JONATHAN L. GROSS
THOMAS W. TUCKER

New York, New York
Hamilton, New York
May 1987

Contents

1

Introduction

In beginning the study of topological graph theory, one needs to prepare both for the development of topological operations and for discussion of the relationship between these operations and the other concepts of graph theory. Most of the important topics in graph theory have some connection with the topological viewpoint. Accordingly, this first chapter reads almost like a survey of graph theory, encompassing trees, automorphisms, traversability, factorization, colorings, algorithms, and more.

1.1. REPRESENTATION OF GRAPHS

In topological graph theory, the intuitive viewpoint is that a graph is a network of nodes and curved arcs from some nodes to others or to themselves. Nonetheless, it is a formal convenience to adopt a combinatorial definition, rather than a topological definition. Therefore, we say that a "graph" G consists of a set V_G of vertices and a set E_G of edges. Each edge e has an endpoint set $V_G(e)$ containing either one or two elements of the vertex set V_G. In a formal setting, we call the set $I_G = \{V_G(e) | e \in E_G\}$ the "incidence structure". For simplicity, when G is the only graph under consideration, one omits the subscripts and writes V and E instead of V_G and E_G.

If both the vertex set and the edge set are finite, then the graph G is said to be "finite". With very few exceptions, the graphs discussed here are implicitly considered to be finite.

A graph is called "simplicial" if it has no self-loops or multiple edges, because, in the standard sense of topology, such a graph is a simplicial 1-complex. (In nontopological contexts, the word "simple"[1] is sometimes used in place of "simplicial".)

[1]To a topologist, the word "simple" has pre-emptive meaning. In particular, a "simple" loop is the continuous nonsingular image of a circle, that is, a loop without self-intersections or other singularities. Our general rule is to avoid ambiguity or such conflicts of usage.

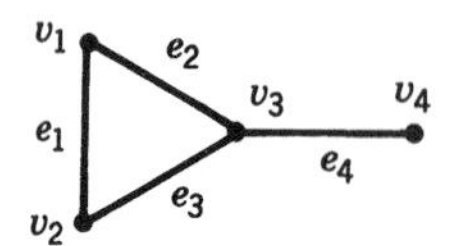

Figure 1.1. The graph of Example 1.1.1.

1.1.1. Drawings

Any finite graph can be geometrically represented by a drawing obtained in the following manner. First, draw a dot for each vertex. Then, for every edge $e \in E$ with two endpoints, draw a line between the dots representing the vertices of $V(e)$, and for every edge with only one vertex, draw a line from the dot representing that vertex to that dot itself. Two examples now illustrate the representation of a graph by a drawing.

Example 1.1.1. *Let* $V = \{v_1, v_2, v_3, v_4\}$, *and let* $E = \{e_1, e_2, e_3, e_4\}$. *Also, let* $V(e_1) = \{v_1, v_2\}$, $V(e_2) = \{v_1, v_3\}$, $V(e_3) = \{v_2, v_3\}$, *and* $V(e_4) = \{v_3, v_4\}$. *A drawing of this graph is shown in Figure* 1.1. *For the sake of clarity, the dots and lines are labeled by the names of the vertices and edges they represent.*

Example 1.1.2. *Let* $V = \{v_1, v_2, v_3\}$, *and let* $E = \{e_1, e_2, e_3, e_4\}$. *Let* $V(e_1) = \{v_1\}$, $V(e_2) = \{v_1, v_2\}$, $V(e_3) = \{v_2, v_3\}$, *and* $V(e_4) = \{v_2, v_3\}$. *Figure* 1.2 *shows two different drawings of this second graph.*

Certain graphs cannot be drawn on a planar surface without line-crossings. Moreover, for the sake of illustration or other convenience, a drawing of a graph sometimes contains more line-crossings than necessary. Fortunately, it is never necessary for the interior of any line to touch any dot. In a drawing of a graph, vertices are represented only by dots, not by line-crossings.

Readers already familiar with graph-theoretic literature may note differences in the present terminology and notation from that appearing elsewhere. For instance, if a graph has an edge with only one endpoint, then Harary (1969) would call it a "pseudograph". Harary's distinction has been widely adopted, because for many purposes, it permits more efficient discussions.

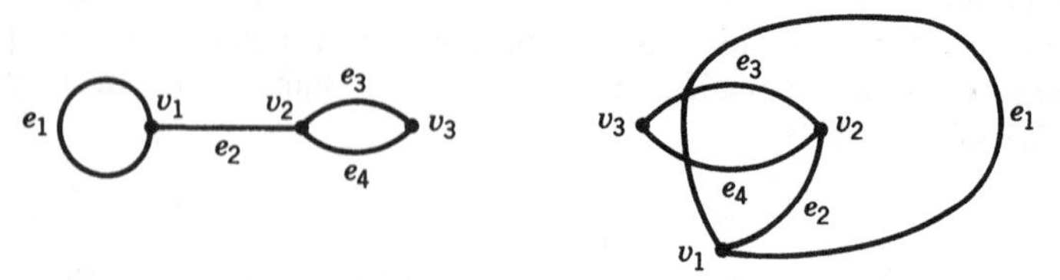

Figure 1.2. Two drawings of the graph of Example 1.1.2, one without line-crossings, the other with crossings.

Certain nonsimplicial graphs, however, such as bouquets of circles (see Section 1.2), are so essential to topological graph theory that it is more efficient to include them here as graphs. In general, the authors have tried to preserve standard terminology and notations except where there is substantive reason in topological graph theory to introduce something new. For the convenience of the reader, a list of all notation used here appears at the end of this book.

1.1.2. Incidence Matrix

The "incidence matrix" for a graph G has $\#V$ rows and $\#E$ columns, where $\#S$ denotes the cardinality of a set S. The entry in row i and column j is 1 if $v_i \in V(e_j)$ and $\#V(e_j) = 2$, 2 if $\{v_i\} = V(e_j)$, and 0 if $v_i \notin V(e_j)$. For example, the graphs of Examples 1.1.1 and 1.1.2 have the following incidence matrices, respectively:

$$\begin{array}{c} \\ v_1 \\ v_2 \\ v_3 \\ v_4 \end{array} \begin{array}{c} \begin{array}{cccc} e_1 & e_2 & e_3 & e_4 \end{array} \\ \begin{bmatrix} 1 & 1 & 0 & 0 \\ 1 & 0 & 1 & 0 \\ 0 & 1 & 1 & 1 \\ 0 & 0 & 0 & 1 \end{bmatrix} \end{array} \qquad \begin{array}{c} \\ v_1 \\ v_2 \\ v_3 \end{array} \begin{array}{c} \begin{array}{cccc} e_1 & e_2 & e_3 & e_4 \end{array} \\ \begin{bmatrix} 2 & 1 & 0 & 0 \\ 0 & 1 & 1 & 1 \\ 0 & 0 & 1 & 1 \end{bmatrix} \end{array}$$

Communicating a graph to a computer by its incidence matrix rather than by its incidence structure may involve a loss of space efficiency, because the zeros occupy storage space not needed for the incidence structure. However, since certain information is recovered more easily from a stored incidence matrix than from a stored incidence structure, there are times when sacrificing the space yields worthwhile gains in operational speed.

1.1.3. Euler's Theorem on Valence Sum

From a geometric viewpoint, the "valence"[2] of a vertex v is the minimum number of line-crossings that occur while one traverses a small circle around the dot representing v in a drawing of the graph. For instance, the valences of v_1, v_2, and v_3 in Example 1.1.2 are 3, 3, and 2, respectively. Roughly speaking, it is sometimes said that the valence of v is the number of edges incident on v, but to get the correct number, one must take care to count an edge twice if v is its only endpoint. Alternatively, one may give a strictly combinatorial definition of the valence of v as the sum of the entries in its row in the incidence matrix.

[2] The synonym "degree" appears to be more frequently used by graph theorists than "valence". However, since "valence" is a self-explanatory abstraction from the terminology of chemistry, and since "degree" is used for polynomials and as the name for a topological invariant of a continuous function, it seems appropriate to use "valence" here for the graph-theoretic concept.

Theorem 1.1.1 (Euler). *The sum of the valences of the vertices of a graph equals twice the number of edges.*

Proof. Every edge contributes exactly 2 to the valence sum. Alternatively, it may be observed that the row sums in the incidence matrix are the valences and that every column sum is 2, corresponding to the contribution of each edge. Of course, the total of the row sums equals the total of the column sums. □

1.1.4. Adjacency Matrix

Two vertices of a graph are called "adjacent" if they are the endpoints of the same edge. They are called "multiply adjacent" if there are two or more edges having them both as endpoints. The "multiplicity of their adjacency" is defined to be the number of edges having them both as endpoints, possibly zero.

A vertex is called "self-adjacent" if there is an edge for which it is the only endpoint. The "multiplicity of its self-adjacency" is defined to be the number of edges for which it is the only endpoint. An edge with only one endpoint is called a "loop". If an edge is not a loop, then it is called a "proper edge".

The "adjacency matrix" for a graph G is square, with $\#V$ rows and columns. The entry in row i and column j is the multiplicity of the adjacency between the vertices v_i and v_j. It follows from this definition that an adjacency matrix is symmetric about the main diagonal. A loop contributes only 1 to a main diagonal entry, so a row or column sum is not necessarily the valence of a vertex. For example, the graphs of Examples 1.1.1 and 1.1.2 have the following adjacency matrices, respectively:

$$\begin{array}{c} \\ v_1 \\ v_2 \\ v_3 \\ v_4 \end{array} \begin{array}{c} \begin{array}{cccc} v_1 & v_2 & v_3 & v_4 \end{array} \\ \begin{bmatrix} 0 & 1 & 1 & 0 \\ 1 & 0 & 1 & 0 \\ 0 & 0 & 1 & 0 \\ 0 & 0 & 1 & 0 \end{bmatrix} \end{array} \qquad \begin{array}{c} \\ v_1 \\ v_2 \\ v_3 \end{array} \begin{array}{c} \begin{array}{ccc} v_1 & v_2 & v_3 \end{array} \\ \begin{bmatrix} 1 & 1 & 0 \\ 1 & 0 & 2 \\ 0 & 2 & 0 \end{bmatrix} \end{array}$$

1.1.5. Directions

A "direction" for an edge e is an onto function $\{\text{BEGIN}, \text{END}\} \to V(e)$. The images of BEGIN and END are called the "initial point" and the "terminal point", respectively. One says that a directed edge goes "from" its initial point "to" its terminal point. In the incidence structure for a list format of a graph, the direction of a proper edge may be indicated by putting the initial point first in its endpoint list.

In topological graph theory, one frequently considers each edge e (even a loop) to have two directions, arbitrarily distinguished as the "plus direction"

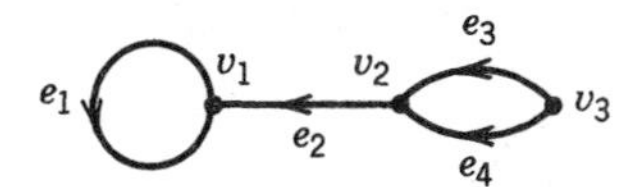

Figure 1.3. Using arrowheads to indicate directions on edges of the graph of Example 1.1.2.

e^+ and the "minus direction" e^-. In a drawing of a graph, an arrowhead is used to show the plus direction, with the initial point behind the arrowhead and the terminal point in front of it. Figure 1.3 shows how the arrowheads are placed on the left-hand drawing of Figure 1.2 if the direction $e_2{}^+$ is from v_2 to v_1 and the directions $e_3{}^+$ and $e_4{}^+$ are both from v_3 to v_2.

An edge e together with either its plus direction or its minus direction is called a "directed edge," and is denoted e^+ or e^-, respectively, thereby slightly abusing the notation.

1.1.6. Graphs, Maps, Isomorphisms

Let G and G' be graphs. A "graph map" $f: G \to G'$ consists of a vertex function $V_G \to V_{G'}$ and an edge function $E_G \to E_{G'}$ such that incidence is preserved, in the sense that for every edge $e \in E_{G'}$ the vertex function maps the endpoints of e [i.e., the set $V_G(e)$] onto the endpoints of the image e. It is customary to denote both the vertex function and the edge function by the same symbol as the graph map, namely, f. Under a graph map, a loop may be the image of a proper edge, but a proper edge may never be the image of a loop.

A graph map $G \to G'$ is called an "isomorphism" if both its vertex function and its edge function are one-to-one and onto. Two graphs are called "isomorphic" if there exists an isomorphism from one to the other.

Example 1.1.3. *Figure* 1.4 *shows two graphs, G and G'. Suppose the function f has the values*

$$f(v_1) = v_1' \qquad f(v_2) = f(v_3) = v_2'$$

$$f(e_1) = f(e_2) = e_1' \qquad f(e_3) = e_2'$$

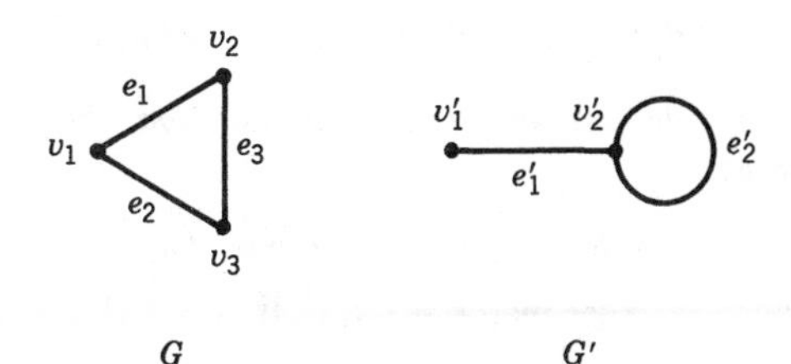

Figure 1.4. The domain G and the range G' of a graph map $f: G \to G'$.

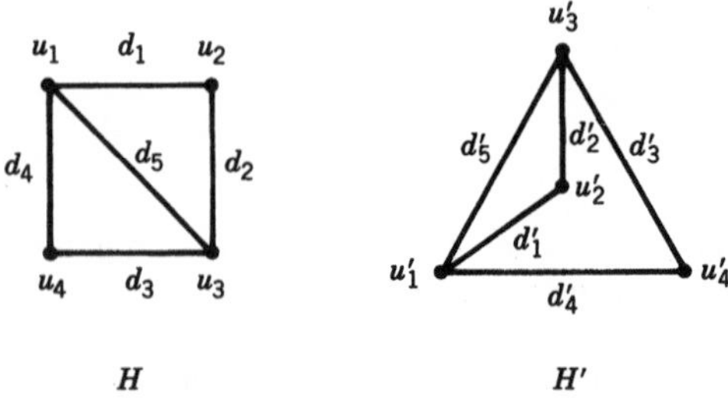

Figure 1.5. Two isomorphic graphs.

The endpoints of the edge e_1 are the vertices v_1 and v_2, which are mapped onto v_1' and v_2', respectively, the endpoints of the edge e_1', which is the image of e_1. The endpoints v_1 and v_3 of the edge e_2 are mapped onto the endpoints v_1' and v_2' of the edge e_1', which is the image of e_2. Both the endpoints v_2 and v_3 of the edge e_3 are mapped onto the single endpoint v_2' of its image e_2'. Thus the vertex map f and the edge map f together form a graph map $f\colon G \to G'$. However, f is not an isomorphism.

Example 1.1.4. *Figure 1.5 shows two isomorphic graphs, H and H'. One isomorphism $g\colon H \to H'$ is given by the rule*

$$g(u_i) = u_i' \quad \text{for } i = 1,\dots,4$$
$$g(d_i) = d_i' \quad \text{for } i = 1,\dots,5$$

1.1.7. Automorphisms

An isomorphism from a graph to itself is called an “automorphism”. Under the operation of composition, the family of all automorphisms of a graph forms a group, called “the automorphism group of the graph” and denoted Aut(G).

Example 1.1.5. *Consider a graph map $h\colon H \to H$ on the left-hand graph H of Figure 1.5, in which h has the values*

$$\begin{array}{lllll} h(u_1) = u_3 & h(u_2) = u_2 & h(u_3) = u_1 & h(u_4) = u_4 & \\ h(d_1) = d_2 & h(d_2) = d_1 & h(d_3) = d_4 & h(d_4) = d_3 & h(d_5) = d_5 \end{array}$$

Alternatively, using disjoint permutation-cycle notation, the graph map h may be written as follows:

$$\text{vertex map } h = (u_1\, u_3)(u_2)(u_4)$$
$$\text{edge map } h = (d_1\, d_2)(d_3\, d_4)(d_5)$$

Then h is an automorphism.

1.1.8. Exercises

1. Draw the graph with vertices v_1, v_2, and v_3, with edges e_1, e_2, e_3, e_4, and e_5, and with endpoint sets $V(e_1) = \{v_1, v_2\}$, $V(e_2) = \{v_2, v_3\}$, $V(e_3) = \{v_1, v_3\}$, $V(e_4) = \{v_1, v_3\}$, and $V(e_5) = \{v_3\}$.
2. Write the incidence structures for the graphs of Figure 1.4.
3. What is the incidence matrix for the graph of Exercise 1?
4. What is the adjacency matrix for the graph of Exercise 1?
5. The "eigenvalues" of a graph are defined to be the eigenvalues of its adjacency matrix. What are the eigenvalues of the graph in Figure 1.1?
6. The "valence sequence" of a graph is the list of all the valences of its vertices (usually in increasing order), so that if there are n vertices, there are n numbers in the valence sequence. Construct two graphs that have the same valence sequence, but that are not isomorphic.
7. (a) Prove that there is no graph map from the graph G' of Figure 1.4 to the graph G of that figure. (b) Prove that G and G' are not isomorphic.
8. Draw two nonisomorphic graphs with four vertices and four edges each and no loops or multiple adjacencies.
9. Prove that if the graph map $f: G \to G'$ is an isomorphism, then the inverse map $f^{-1}: G' \to G$ is also an isomorphism.
10. Find another isomorphism $H \to H'$ for the graphs of Figure 1.5 besides the one given in Example 1.1.4.
11. In permutation notation, list all the automorphisms of the graph H of Figure 1.5.
12. Draw a graph with two vertices and no automorphisms except for the identity.
13. Draw a graph with six vertices, no loops, and no automorphisms except the identity.
14. Identify the automorphism group of the graph of Figure 1.2.
15. Draw a graph whose automorphism group is isomorphic to the cyclic group $\mathscr{Z}_3$.
16. Prove that every graph has an even number of vertices of odd valence.
17. Two graphs are called "co-spectral" if they have the same eigenvalues (see Exercise 5) with the same multiplicities. Draw two co-spectral, connected, 6-vertex, 7-edge graphs.

1.2. SOME IMPORTANT CLASSES OF GRAPHS

Numerous classes of graphs, arising in algebra or geometry as well as in combinatorics or topology, are of special interest in topological graph theory. In this section we examine a few such classes.

1.2.1. Walks, Paths, and Cycles; Connectedness

A "walk" in a graph is the combinatorial analog of a continuous image of a closed line segment, which may cross itself or retrace upon itself, backwards or forwards, arbitrarily often. Precisely, if u and v are vertices of a graph G, then a "walk of length n from u to v" is an alternating sequence of vertices and directed edges,

$$W = v_0, e_1^{\sigma_1}, v_1, e_2^{\sigma_2}, \ldots, e_n^{\sigma_n}, v_n \quad (\text{each } \sigma_i = + \text{ or } -)$$

whose initial vertex v_0 is u, whose final vertex v_n is v, and for $i = 1, \ldots, n$, the directed edge $e_1^{\sigma_1}$ runs from the vertex v_{i-1} to the vertex v_i. If $u \neq v$, then W is called an "open walk". If $u = v$, then W is called a "closed walk".

An open walk is called a "path" if its vertices are distinct. Thus, a path is the combinatorial analog of a homeomorphic image of a closed line segment. The standard path with n vertices is called the "n-path" and is denoted P_n. A closed walk is called a "cycle" if every pair of vertices except its starting and stopping vertex are distinct. Thus, a cycle is the combinatorial counterpart to the homeomorphic image of a circle. The standard cycle with n vertices is called the "n-cycle" and is denoted C_n.

In order to relieve the cumbersomeness of the notation for a walk, path, or cycle, one commonly omits the vertices from the sequence specified in the definition, since there is no difficulty in inferring them from the resulting sequence of directed edges. Moreover, when for $i = 1, \ldots, n-1$ the edges e_i and e_{i+1} have only one endpoint in common, one usually also omits the signs, and writes

$$W = e_1, e_2, \ldots, e_n$$

There are many different notations for a walk, path, or cycle, and the sensible practice in a particular instance is to use the least cumbersome notation that makes sense, even if it is not the same as the notation used in a previous instance. Thus, fluency has priority over consistency.

A graph is called "connected" if for every pair of vertices u and v, there is a path from u to v. Specifying plus directions on some of the edges does not affect the connectedness of a graph, since the direction on an edge in a path (or walk) is permitted to be the opposite of the plus direction. Thus, connectedness in the present combinatorial sense agrees exactly with the underlying topological sense.

1.2.2. Trees

A "tree" is a connected graph with no cycles, as illustrated in Figure 1.6. It is one of the most important kinds of graphs in both applications and theory.

Theorem 1.2.1. *Let u and v be any vertices of a tree T. Then there is a unique path in T from u to v.*

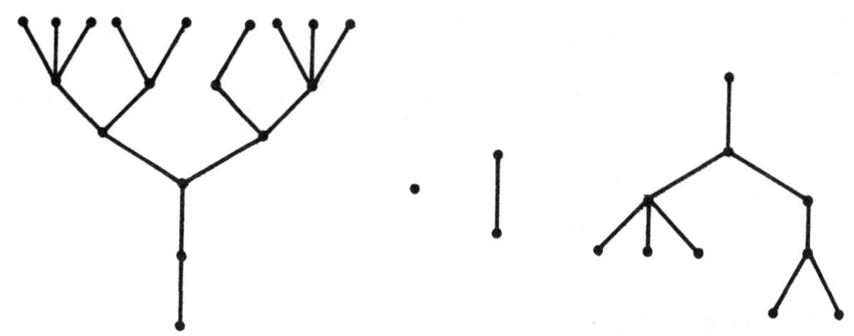

Figure 1.6. Four trees.

Proof. Since the tree T is connected, there exists a path P from u to v. Suppose that P' is another path from u to v. Then let w be the first vertex on P and P', as we proceed from u to v, whose successor in the path P is not its successor in the path P'. Let w' be the next vertex on P after w that also lies on the path P'. Then the path segment from w to w' in P combines with the path segment from w to w' in P' to form a cycle in T, which contradicts the fact that trees have no cycles. □

Theorem 1.2.2. *Let T be a tree. Then $\#V_T = \#E_T + 1$.*

Proof. If $\#V_T = 1$, then $\#E_T = 0$; otherwise there would be a loop, which yields a cycle. By way of induction, assume the proposition is true for trees with p or fewer vertices, and let T have $p+1$ vertices. It follows from Theorem 1.2.1 that removing an edge e from T disconnects it into two smaller trees T' and T'', so that $\#V_T = \#V_{T'} + \#V_{T''}$ and $\#E_T = \#E_{T'} + \#E_{T''} + 1$. By the induction hypothesis, $\#V_{T'} = \#E_{T'} + 1$ and $\#V_{T''} = \#V_{T'} + \#E_{T''} + 1$. It follows that $\#V_T = \#E_T + 1$. □

Theorem 1.2.3. *Every nontrivial tree T has at least two vertices of valence* 1.

Proof. By Theorem 1.1.1, for any graph, the sum of the valences is twice the number of edges. By Theorem 1.2.2, $2\#E_T = 2\#V_T - 2$, since T is a tree. By combining these two facts, one obtains the equation

$$\sum_{v \in V_T} \text{valence}(v) = 2\#V_T - 2$$

Since every vertex v of the tree has valence at least 1, there must be at least two vertices of valence exactly 1. □

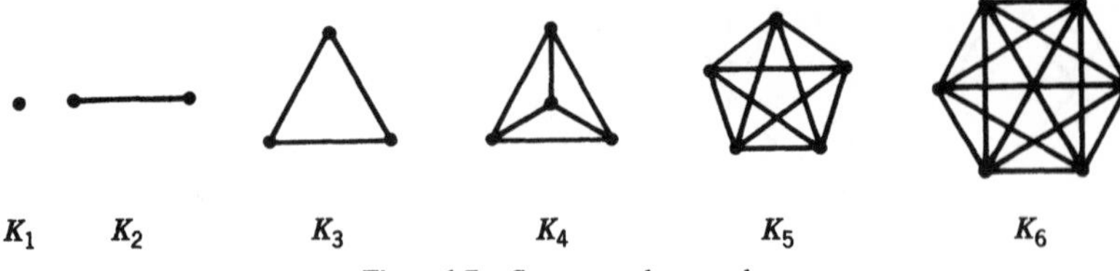

Figure 1.7. Some complete graphs.

1.2.3. Complete Graphs

A simplicial graph is called "complete" if every pair of vertices is adjacent. If two complete graphs have the same number of vertices, then they are isomorphic. The standard model for a complete graph on n vertices is denoted K_n. Figure 1.7 shows the complete graphs K_n, for $n = 1, \ldots, 6$. The graph K_1 is sometimes called the "trivial graph".

1.2.4. Cayley Graphs

Cayley (1878) used graphs to draw pictures of groups, an inspiration that was subsequently refined and used to algebraic advantage by Dehn (1911), Schreier (1927), and others. As preliminary examples, consider the two drawings in Figure 1.8 of the group $\mathscr{Z}_5$ of integers modulo 5.

In both diagrams of Figure 1.8, the elements of the group $\mathscr{Z}_5$ are the vertices. Figure 1.8*a* shows $\mathscr{Z}_5$ as generated by the single residue class 1, so from every vertex v there is a single solid edge to the vertex $v + 1$. Figure 1.8*b* shows $\mathscr{Z}_5$ as generated by the two residue classes 1 and 2, so from each vertex v there is a solid edge to the vertex $v + 1$ and also a dashed edge to the vertex $v + 2$.

Rather than solid lines, dashed lines, and other such devices to distinguish one class of edges from another, Cayley suggested a different color for each generator. From this arose the name "Cayley color graph". A common abuse

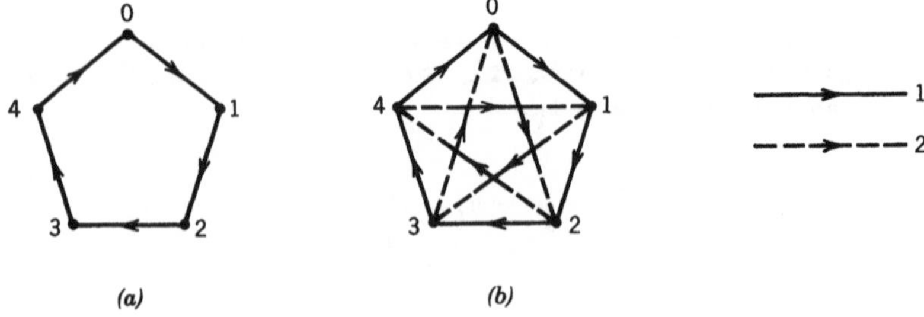

Figure 1.8. Two Cayley color graphs for $\mathscr{Z}_5$: (a) with generating set $\{1\}$; (b) with generating set $\{1, 2\}$.

of terminology, adopted here, is to refer to whatever distinguishing features are used for edges in the drawings as "colors", and also to refer to the generators themselves (when they are regarded as generators) as "colors". Although the construction now defined is more general than Cayley's, his name is retained. As his generating set, Cayley used all the nonidentity elements.

Let $\mathscr{A}$ be a group, and let $X = \{x_1, \ldots, x_n\}$ be a generating set for $\mathscr{A}$. The "(right) Cayley color graph" $C(\mathscr{A}, X)$ has as its vertex set the elements of group $\mathscr{A}$ and as its edge set the cartesian product $X \times \mathscr{A}$. The edge (x, a) has as its endpoints the vertices a and ax, with its plus direction from a to ax. Sometimes it is convenient to write (x, a) in the subscript notation x_a.

The designation of a plus direction and a color for every edge is an intrinsic part of a Cayley color graph $C(\mathscr{A}, X)$. If these designations are suppressed, the result is a graph $C(\mathscr{A}, X)^0$, called the "Cayley graph" for the group $\mathscr{A}$ and the generating set X. Figure 1.9 shows a Cayley color graph, with its edge names given in subscript notation, and the associated Cayley graph. The names of vertices (e.g., a) and edges [e.g., (x, a) or x_a] are the same in a Cayley graph as in the Cayley color graph from which it is derived. Since the name of an edge of a Cayley color graph is readily determined from its color and its initial point, names of edges are hereafter omitted from drawings, thereby decreasing the visual clutter.

A graph map $C(\mathscr{A}, X) \rightarrow C(\mathscr{A}', X')$ between Cayley color graphs is called "color-consistent" if for every pair of edges with the same color in $C(\mathscr{A}, X)$, their images have the same color in $C(\mathscr{A}', X')$. It is called "direction-preserving" if the initial point of every edge in $C(\mathscr{A}, X)$ is mapped onto the initial point of the image edge in $C(\mathscr{A}', X')$. It is called "identity-preserving" if it maps the group-identity vertex of $C(\mathscr{A}, X)$ to the group-identity vertex of

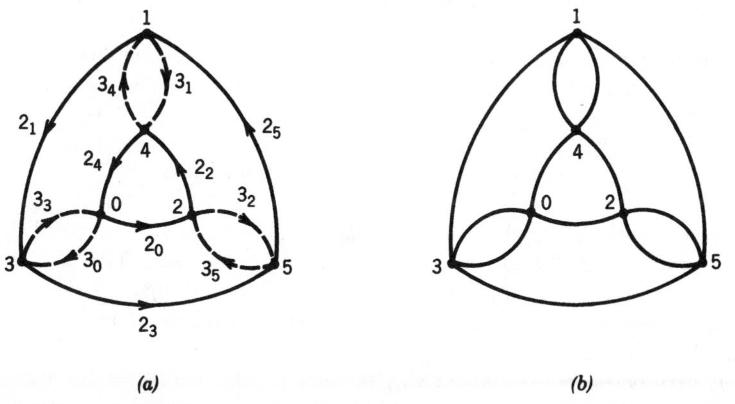

Figure 1.9. (a) The Cayley color graph for the group $\mathscr{Z}_6$ with the generators 2 and 3. (b) The associated Cayley graph $C(\mathscr{Z}_6, \{2, 3\})^0$.

$C(\mathscr{A}', X')$. The following theorem shows the connection between Cayley color graphs and group homomorphisms.

Theorem 1.2.4. *Let $f: C(\mathscr{A}, X) \to C(\mathscr{A}', X')$ be a color-consistent, direction-preserving, identity-preserving graph map. Then its vertex function coincides with a group homomorphism $\mathscr{A} \to \mathscr{A}'$. Conversely, any group homomorphism $h: \mathscr{A} \to \mathscr{A}'$ such that $h(X) \subset X'$ coincides with the vertex function of a color-consistent, direction-preserving, identity-preserving graph map $C(\mathscr{A}, X) \to C(\mathscr{A}', X')$.*

Proof. Assume that the first hypothesis holds. To show that f is a group homomorphism, it suffices to show that $f(ax^{\sigma}) = f(a)f(x)^{\sigma}$ for all $a \in \mathscr{A}$, $x \in X$, and $\sigma = \pm 1$. Since there is an edge in the Cayley graph $C(\mathscr{A}, X)$ from a to ax with the color "x", it follows that in the Cayley graph $C(\mathscr{A}', X')$ there is an edge from $f(a)$ to $f(ax)$. Moreover, this edge has the same color as the edge from $1_{\mathscr{A}'}$ to $f(x)$, since f is color-consistent, direction-preserving, and identity-preserving. The edge from $1_{\mathscr{A}'}$ to $f(x)$ must be colored "$f(x)$", so the edge from $f(a)$ to $f(ax)$ is also colored "$f(x)$". By the definition of Cayley color graph, it follows that $f(ax) = f(a)f(x)$. To show that $f(ax^{-1}) = f(a)f(x)^{-1}$, we now simply observe that $f(a) = f(ax^{-1}x) = f(ax^{-1})f(x)$.

Conversely, let $h: \mathscr{A} \to \mathscr{A}'$ be a group homomorphism such that $h(X) \subset X'$. Define the graph map $f: C(\mathscr{A}, X) \to C(\mathscr{A}', X')$ as follows. If a is a vertex of $C(\mathscr{A}, X)$, let $f(a) = h(a)$. If e is an edge from a to ax (colored x), let $f(e)$ be the edge from $f(a)$ to $f(ax) = f(a)f(x)$ [colored $f(x)$]. It is easily verified that f has the desired properties. □

Remark. *In regard to Theorem* 1.2.4, *it may be observed that the vertex function of a color-consistent, direction-preserving, identity-preserving graph map from one Cayley color graph to another coincides with a group isomorphism if and only if the graph map is a graph isomorphism.*

An alternative definition of a Cayley graph is like the one already given, except that when a generator x has order 2 in the group $\mathscr{A}$, for every element $a \in \mathscr{A}$ the two edges

$$x_a \text{ (from } a \text{ to } ax) \quad \text{and} \quad x_{ax} \text{ (from } ax \text{ to } a)$$

of the Cayley color graph $C(\mathscr{A}, X)$ collapse onto a single edge, denoted either x_a or x_{ax}. The "alternative Cayley graph" is denoted $C(\mathscr{A}, X)^1$. Both $C(\mathscr{A}, X)^0$ and $C(\mathscr{A}, X)^1$ are called "Cayley graphs". Figure 1.10 shows a Cayley color graph for a group with a generator of order 2 and the corresponding alternative Cayley graph.

It may be observed that the Cayley color graphs in Figures 1.9 and 1.10 look quite alike, since their corresponding Cayley graphs are isomorphic, as are their corresponding alternative Cayley graphs, even though the groups $\mathscr{Z}_6$

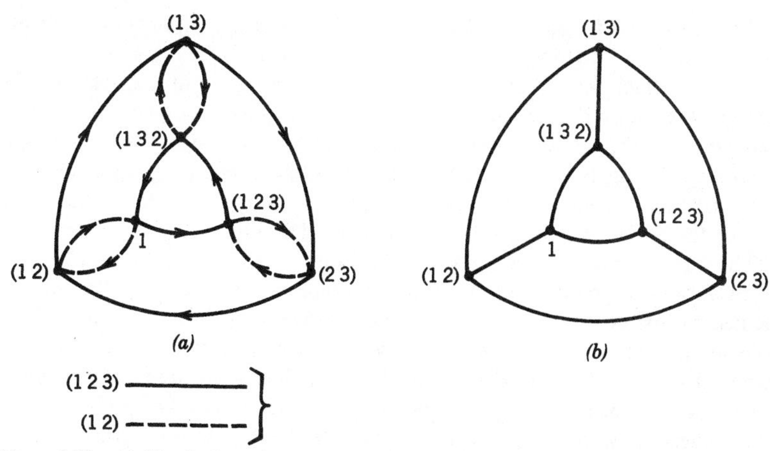

Figure 1.10. (a) The Cayley color graph for the symmetric group $\mathscr{S}_3$ with the generators (1 2 3) and (1 2). (b) The alternative Cayley graph $C(\mathscr{S}_3, \{(123), (12)\})^1$.

and $\mathscr{S}_3$ are not isomorphic. Moreover, it is easy to construct a color-consistent, identity-preserving graph isomorphism between those two Cayley color graphs, but it does not preserve directions, of course, because of Theorem 1.2.4.

A graph is called "n-regular" (or just "regular") if every vertex has valence n. If a group is presented with q generators, then the corresponding Cayley graph is $2q$-regular. The alternative Cayley graph is also regular, but the valence depends on the number of generators of order 2. Regularity of a graph is best understood as a local symmetry condition.

A graph G is called "vertex-transitive" if for every pair of vertices u and v, there is an automorphism of G that carries u to v. Vertex-transitivity is best regarded as a global symmetry condition.

As preparation for proving that a Cayley graph $C(\mathscr{A}, X)^0$ or $C(\mathscr{A}, X)^1$ is vertex-transitive, define for every $a \in \mathscr{A}$ the graph automorphism i_a, according to the rules

$$i_a(u) = au \qquad \text{for every vertex } u \in \mathscr{A}$$

$$i_a((x, u)) = (x, au) \quad \text{for every edge } (x, u) \in X \times \mathscr{A}$$

The graph automorphism i_a is called a "left translation" of the Cayley graph.

Theorem 1.2.5. *Any Cayley graph is vertex-transitive.*

Proof. For any pair of vertices u and v, let $w = vu^{-1}$. Then the left translation i_w takes u to v. □

The converse of Theorem 1.2.5 is not true; for example, the Petersen graph (see Example 1.3.13 and Exercise 2.2.13) is not a Cayley graph, yet it is vertex-transitive. On the other hand, a left translation i_a of a Cayley graph also has the property that it leaves no vertex fixed, unless of course the group element a is the identity. If the automorphism group of a graph G has a subgroup $\mathscr{A}$ that is vertex-transitive, and if every nonidentity element of $\mathscr{A}$ leaves no vertex fixed, then indeed the graph G is a Cayley graph for the group $\mathscr{A}$ (see the corollary to Theorem 2.2.3 and see Theorem 6.2.3).

The Cayley graph associated with a given group and generating set reveals much about the group. Every walk in the graph corresponds to a "word" in the generators; every closed walk corresponds to a "relator", namely a product of generators that is equal to the identity. This relationship has proved to be a powerful tool in combinatorial group theory, which is the study of presentations of groups in terms of generators and relations. It was Dehn (1911) who first focused on this connection. For example, one of Dehn's three famous decision problems for a presentation of a group is to find an algorithm that will determine whether two given words in the generators represent the same element; this is equivalent to finding an algorithm that constructs the Cayley graph corresponding to the given presentation of the group. [Boone (1954) and Novikov (1955) showed this "word problem" was unsolvable for some group presentations.]

Dehn's influence is so great that Cayley graphs are sometimes called "Dehn Gruppenbilder", although Cayley's priority is unquestioned [Burnside's famous *Theory of Groups* (1897, 1911) refers to "Cayley's colour groups" and even has a four-color illustration of a Cayley graph as the frontispiece.] More recent examples of the importance of graphical representations of groups can be found in the work of Stallings (1971) and Serre and Bass (1977) (see also Tretkoff, 1975, 1980). Cannon (1984) carries the word problem one step further by studying presentations for infinite groups that have linear-time algorithms to draw the associated Cayley graph. General monographs for combinatorial group theory have been compiled by Magnus, et al. (1966), by Lyndon and Schupp (1977), and by Coxeter and Moser (1957).

1.2.5. Bipartite Graphs

A graph is called "bipartite" if its vertex set can be partitioned into two subsets U and W such that the vertices in U are mutually nonadjacent and the vertices in W are mutually nonadjacent. If every point of U is adjacent to

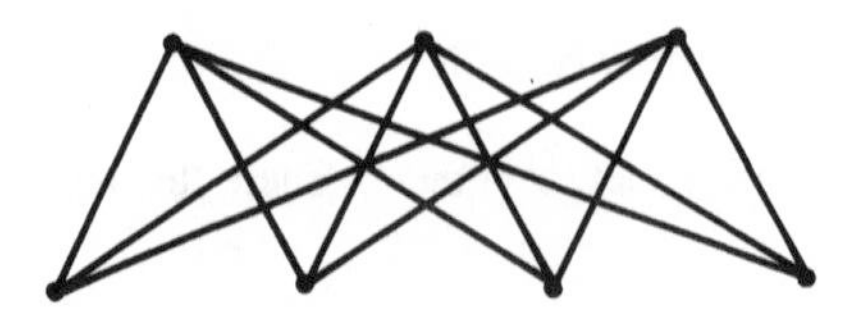

Figure 1.11. The complete bipartite graph $K_{3,4}$.

Figure 1.12. A bouquet of three circles.

every point of W, then the graph is called "complete bipartite" on the sets U and W. The standard model for a complete bipartite graph on sets of m vertices and n vertices is denoted $K_{m,n}$. Figure 1.11 illustrates the complete bipartite graph $K_{3,4}$.

1.2.6. Bouquets of Circles

For topological graph theory, one other class of graphs should be (and is) introduced at the outset. A "bouquet of n circles" is a graph with one vertex and n loops. The standard model is denoted B_n. The bouquet B_3 is illustrated in Figure 1.12.

1.2.7. Exercises

1. In the graph of Figure 1.10, find a path with six vertices and a cycle of length five. Also find a closed walk with four edge-steps that is not a cycle.
2. Prove that if $\mathscr{A}$ and $\mathscr{A}'$ are groups of the same order, they have respective generating sets X and X' such that the Cayley graphs $C(\mathscr{A}, X)^0$ and $C(\mathscr{A}', X')^0$ are isomorphic. (*Hint*: Consider relatively large generating sets.)
3. Show that a graph is connected if and only if for every pair of vertices u and v, there is a walk from u to v.
4. Extend Theorem 1.2.4 to the case where the group $\mathscr{A}$ is infinite.
5. Draw the Cayley color graph for the group $\mathscr{Z}_2 \times \mathscr{Z}_2 \times \mathscr{Z}_2$ with generators $(1,0,0)$, $(0,1,0)$, and $(0,0,1)$.
6. Draw the Cayley graph for the group $\mathscr{Z}_8$ with generating set $\{1,2,3\}$.
7. Draw the alternative Cayley graph for the group $\mathscr{Z}_6$ with generating set $\{1,2,3\}$.
8. Draw the Cayley color graph for the quaternions with generators i and j and relations $i^2 = j^2 = (ij)^2$.
9. Draw the Cayley color graph for the alternating group $\mathscr{A}_4$ with generators $(1\,2\,3)$ and $(2\,3\,4)$.
10. Draw a 4-regular simplicial graph on 11 vertices that is not a Cayley graph. (*Hint*: Use two copies of K_5.)

11. Decide whether the complete bipartite graph $K_{3,3}$ is a Cayley graph. (*Hint*: There are only two groups of order 6.)
12. Prove that every complete graph K_n is a Cayley graph. (*Hint*: Use the group $\mathscr{Z}_n$. When n is even, an alternative Cayley graph construction is required.)
13. Prove that every tree is a bipartite graph.
14. Which n-cycles are bipartite?
15. What is the automorphism group of the complete graph K_n?
16. Verify that a left translation of a Cayley graph is a graph automorphism.
17. Prove that every Cayley graph is connected.

1.3. NEW GRAPHS FROM OLD

There are numerous ways to extract parts, make combinations, or do other operations on graphs so as to obtain new graphs.

1.3.1. Subgraphs

If G and G' are graphs, then G' is called a "subgraph" of G if and only if $V_{G'}$, $E_{G'}$, and $I_{G'}$ are subsets of V_G, E_G, and I_G, respectively. If in addition, $V_{G'} = V_G$, then one says that the subgraph G' "spans" the graph G. Figure 1.13 illustrates a graph at the left and two of its subgraphs.

If a subgraph is a tree and if it spans the graph, it is called a "spanning tree". For instance, the graph in the middle of Figure 1.13 is a spanning tree for the graph at the left. The subgraph at the right does not span, since it does not include the vertex v_4. Nor is it a tree, since it is not connected.

Example 1.3.1. *Every n-vertex simplicial graph is isomorphic to a subgraph of the complete graph K_n. Also, the n-path P_n is isomorphic to a subgraph of the $(n + k)$-path P_{n+k}, for any nonnegative integer k. However, the m-cycle is not isomorphic to a subgraph of the n-cycle unless $m = n$.*

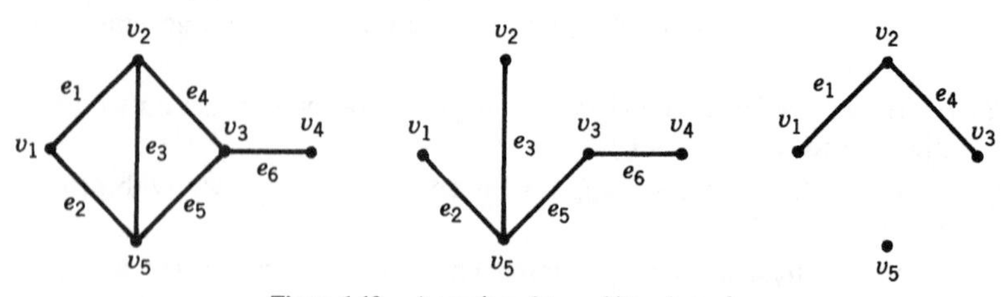

Figure 1.13. A graph and two of its subgraphs.

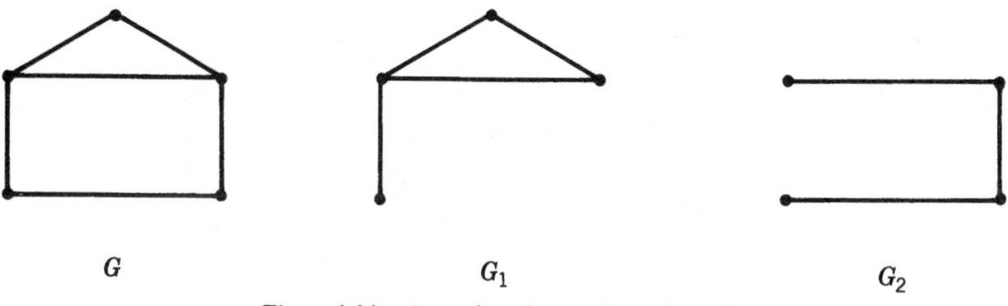

Figure 1.14. A graph and two of its subgraphs.

For any graph G, let $V' \subset V_G$. The "induced subgraph" on the vertex subset V' is the subgraph whose vertex set is V' and whose edge set consists of every edge e of G whose endpoint set $V(e)$ lies in V'. For instance, in Figure 1.14, the subgraph G_1 is an induced subgraph of G, but the subgraph G_2 is not.

There is a big difference between assertions about subgraphs and about induced subgraphs. For example, since the complete graph K_n is, in fact, a Cayley graph (Exercise 1.2.12), it follows that every finite graph is a subgraph of some Cayley graph. The claim, however, that every finite graph is an induced subgraph of some Cayley graph, seems quite surprising. This would imply that although Cayley graphs are globally symmetric, locally they can be arbitrarily complicated. Nevertheless, Babai (1976) has shown this claim to be true. In fact, Godsil and Imrich (1985) prove that given any graph G on n vertices and any group $\mathscr{A}$ of order greater than $(24\sqrt{3})n^3$ there is a Cayley graph for $\mathscr{A}$ containing G as an induced subgraph.

A "component" of a graph is a maximal connected subgraph. Thus, if a graph is connected, it has only one component, itself.

1.3.2. Topological Representations, Subdivisions, Graph Homeomorphisms

Any graph G can be represented by a topological space in the following sense. Each vertex is represented by a distinct point and each edge by a distinct arc, homeomorphic to the closed interval $[0,1]$. Naturally, the boundary points of an arc represent the endpoints of the corresponding edge. (Of course, the interiors of the arcs are mutually disjoint and do not meet the points representing vertices.) Such a space is called a "topological representation" of G.

Subdivision is an operation that does not change the homeomorphism type of a topological representation of a graph. For simplicity, one first considers the effect of the operation on a single edge e of a graph G, with endpoint set $V(e) = \{v_1, v_2\}$. From the informal viewpoint of drawing pictures, one subdivides the edge e into two new edges e' and e'' by putting a new vertex v anywhere in its interior, as shown in Figure 1.15.

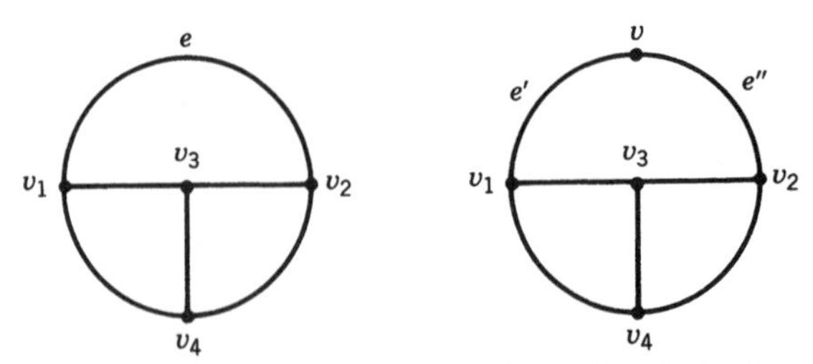

Figure 1.15. The graph at the right is obtained from the graph at the left by subdividing the edge e.

From a formal combinatorial viewpoint, the result of "subdividing the edge" e with endpoints v_1 and v_2 is a new graph whose vertex set is the union of V and $\{v\}$, whose edge set is $(E - \{e\}) \cup \{e', e''\}$, and whose incidence structure is $(I - V(e)) \cup \{V(e'), V(e'')\}$, where $V(e') = \{v_1, v\}$ and $V(e'') = \{v_2, v\}$. In general, a "subdivision of a graph is obtained by a finite sequence of subdivisions of edges.

Among other uses, subdivision can make a graph become simplicial. The first step is to subdivide every loop. The next and last step is to subdivide every edge but one in each multiple adjacency.

The graphs G and H are called "homeomorphic" if they have respective subdivisions G' and H' such that G' and H' are isomorphic graphs. It may be observed that under this combinatorial definition, two graphs are homeomorphic if and only if their topological representations are homeomorphic as topological spaces. Figure 1.16 shows two homeomorphic graphs such that neither is a subdivision of the other. Homeomorphism is actually a point-set topological concept, yet the "Hauptvermutung" [see Papakyriakopoulos (1943), Moise (1952), Bing (1954), and Brown (1969)] enables us to define it combinatorially for low-dimensional complexes. Milnor (1961) proved that Hauptvermutung does not hold in higher dimensions.

Example 1.3.2. *For any two positive integers m and n, the m-cycle C_m and the n-cycle C_n are homeomorphic graphs.*

Example 1.3.3. *For any two integers $m, n \geq 2$, the m-path P_m and the n-path P_n are homeomorphic graphs.*

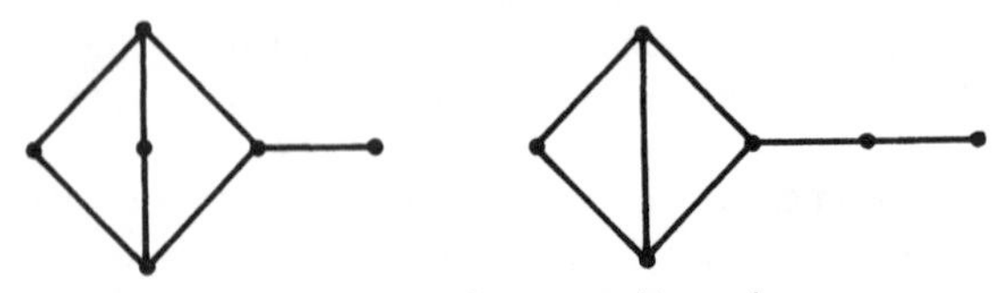

Figure 1.16. Two homeomorphic graphs.

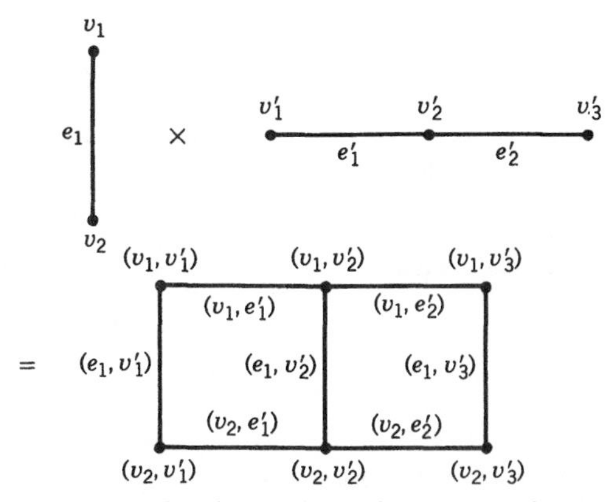

Figure 1.17. The cartesian product of two graphs.

1.3.3. Cartesian Products

The "cartesian product" of the graphs G and G' is denoted $G \times G'$ and defined to be the graph with vertex set $V_G \times V_{G'}$ and edge set $(E_G \times V_{G'}) \cup (V_G \times E_{G'})$. If the edge $(e, v') \in E_G \times V_{G'}$, and if the endpoints of the edge e are v_1 and v_2 then the endpoints of the edge (e, v') are the vertices (v_1, v') and (v_2, v'). If $(v, e') \in V_G \times E_{G'}$, and if the endpoints of the edge e' are v_1' and v_2', then the endpoints of the edge (v, e') are (v, v_1') and (v, v_2'). Figure 1.17 shows the cartesian product of the 2-path and the 3-path.

Example 1.3.4. *The Cayley graph $C(\mathscr{Z}_6, \{2, 3\})^0$ illustrated in Figure* 1.9 *is isomorphic to the cartesian product of the* 2*-cycle C_2 and the* 3*-cycle C_3.*

Example 1.3.5. *The alternative Cayley graph $C(\mathscr{S}_3, \{(123), (12)\})^1$ illustrated in Figure* 1.10 *is isomorphic to a cartesian product of the* 3*-cycle C_3 and the complete graph K_2.*

A graph consisting of the vertices and the edges of a polyhedron, with incidence as on the polyhedron, is called the "1-skeleton" of the polyhedron. It may be observed that the cartesian product of two graphs is the 1-skeleton of their topological product as complexes.

Example 1.3.6. *The "n-cube graph" Q_n is defined to be the* 1*-skeleton of the n-dimensional cube. Equivalently, it may be defined recursively using the operation of cartesian product, with $Q_1 = K_2$, and for $n \geq 1$, $Q_{n+1} = K_2 \times Q_n$.*

1.3.4. Edge-Complements

The "edge-complement" of a simplicial graph G is denoted G^c and is defined to be the graph with the same vertex set as G but such that two vertices are adjacent if and only if they are not adjacent in G.

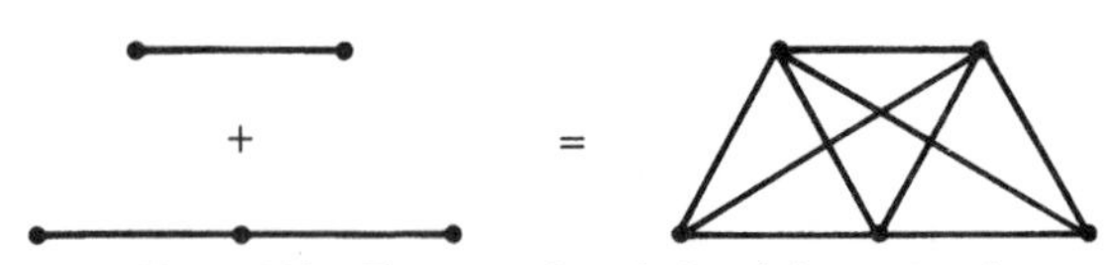

Figure 1.18. The suspension of a 3-path from a 2-path.

Example 1.3.7. *The edge-complement* K_n^{c} *of the complete graph has no edges at all but has n vertices.*

Example 1.3.8. $(K_m \cup K_n)^{\mathrm{c}} = K_{m,n}$.

Example 1.3.9. *The "n-couple cocktail-party graph" is defined to be the edge-complement of the union of n disjoint copies of* K_2. *The explanation of this graph's name, due to Alan Hoffman, is that each of the* $2n$ *persons at the cocktail party talks to each other person except for his or her spouse, and that two vertices are adjacent if the persons they represent talk to each other at the party.*

1.3.5. Suspensions

The "suspension" (elsewhere called the "join") of a graph G from a graph G' is obtained by adjoining every vertex of G to every vertex of G', and is denoted $G + G'$. Thus, its vertex set is $V_G \cup V_{G'}$ and its edge set is $E_G \cup E_{G'} \cup (V_G \times V_{G'})$. The endpoints of an edge of $G + G'$ that arises from E_G or from $E_{G'}$ are exactly as in the graphs G or G', respectively. The endpoints of an edge (v, v') that arises from $V_G \times V_{G'}$ are the vertices v and v'. Figure 1.18 shows the suspension of a 3-path from a 2-path.

Example 1.3.10. *The complete graph* K_{n+1} *is isomorphic to the suspension of the complete graph* K_n *from the graph* K_1.

Example 1.3.11. *The "n-octahedron graph"* O_n *is defined recursively, with* $O_1 = K_2^{\mathrm{c}}$, *and for* $n \geq 1$, $O_{n+1} = O_n + K_2^{\mathrm{c}}$. *It may be observed that the n-octahedron graph is isomorphic to the* 1*-skeleton of the n-dimensional octahedron and is also isomorphic to the n-couple cocktail-party graph.*

1.3.6. Amalgamations

Let G and G' be graphs, and let $f: H \to H'$ be an isomorphism from a subgraph H of G to a subgraph H' of G'. The "amalgamation" $G *_f G'$ is obtained from the union of G and G' by identifying the subgraphs H and H' according to the isomorphism. Figure 1.19 shows the amalgamation of a 3-cycle and a 4-cycle along the thickened edge.

Example 1.3.12. *The bouquet* B_{n+k} *is an amalgamation of the bouquets* B_n *and* B_k *at a vertex.*

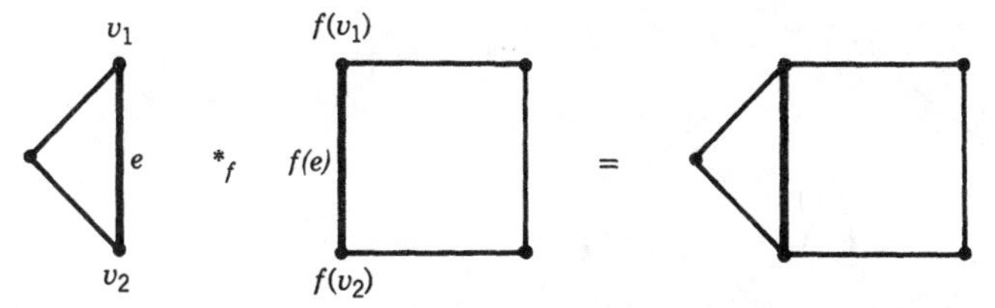

Figure 1.19. The amalgamation of two graphs.

1.3.7. Regular Quotients

Let G be a graph, and let $\mathscr{B}$ be a group such that for each element $b \in \mathscr{B}$, there is a graph automorphism $\phi_b: G \to G$ and such that the following two conditions hold:

i. If 1 is the group identity, then $\phi_1: G \to G$ is the identity automorphism.
ii. For all $b, c \in \mathscr{B}$, $\phi_b \circ \phi_c = \phi_{bc}$.

Then the group $\mathscr{B}$ is said to "act on (the left of) the graph" G. If, moreover, the additional condition

iii. For every group element $b \neq 1$, there is no vertex $v \in V_G$ such that $\phi_b(v) = v$ and no edge $e \in E_G$ such that $\phi_b(e) = e$.

holds, then $\mathscr{B}$ is said to "act without fixed points" (or sometimes "act freely") on G.

For any vertex v of the graph G, the "orbit" $[v]$ (or sometimes $[v]_{\mathscr{B}}$) is defined to be the set $\{\phi_b(v): b \in \mathscr{B}\}$ of all images of v under the group action. Similarly, for any edge e, the orbit $[e]$ (or sometimes $[e]_{\mathscr{B}}$) is defined to be the set $\{\phi_b(e): b \in \mathscr{B}\}$. The sets of vertex orbits and edge orbits are denoted $V/\mathscr{B}$ and $E/\mathscr{B}$, respectively. It may be observed that the orbits partition the vertex set and the edge set of G.

Example 1.3.13. *Figure* 1.20 *contains a drawing of the "Petersen graph" in the plane. For $i \in \mathscr{Z}_5$, let ϕ_i be the rotation of $2\pi i/5$ radians. Under this action of $\mathscr{Z}_5$, there are two vertex orbits, one containing the outer five vertices and the other containing the inner five. There are three edge orbits, one containing the five edges of the outer pentagon, one containing the five edges of the inner star, and one containing the five edges that run between the star and the pentagon.*

Most important, if $\mathscr{A}$ is a group, if $\mathscr{B}$ is a subgroup, and if X is a generating set, then $\mathscr{B}$ acts freely on the Cayley graph $C(\mathscr{A}, X)^0$ according to the rules $\phi_b(a) = ba$ for vertices and $\phi_b(x_a) = x_{ba}$ for edges.

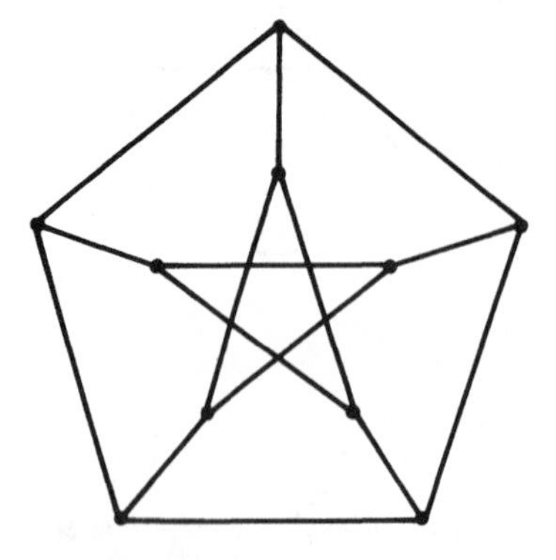

Figure 1.20. The Petersen graph.

Example 1.3.14. *The n-cycle C_n is isomorphic to the Cayley graph $C(\mathscr{Z}_n, \{1\})^0$, so the group $\mathscr{Z}_n$ acts freely on it. There is only one vertex orbit and only one edge orbit. It may be observed that although the dihedral group $\mathscr{D}_n$ is isomorphic to* $\text{Aut}(C_n)$, *it does not act freely, since each reflection fixes either a vertex or an edge.*

The "regular quotient" $G/\mathscr{B}$ is the graph with vertex set $V/\mathscr{B}$ and edge set $E/\mathscr{B}$ such that the vertex orbit $[v]$ is an endpoint of the edge orbit $[e]$ if any vertex in $[v]$ is an endpoint of any edge in $[e]$. It is easy to verify that each edge orbit has one or two vertex orbits as endpoints.

Example 1.3.13 Revisited. *Figure* 1.21 *shows the quotient of the Petersen graph under the prescribed action of $\mathscr{Z}_5$.*

Example 1.3.14 Revisited. *The quotient of the n-cycle C_n under the action of the cyclic group $\mathscr{Z}_n$ is the bouquet B_1 of one circle. In general, when a group acts on a Cayley graph for itself, the quotient is a bouquet of circles, one corresponding to each generator.*

1.3.8. Regular Coverings

Associated with a graph quotient there is a vertex function $v \to [v]$ and an edge function $e \to [e]$ that together are called the "quotient map" $q_{\mathscr{B}}: G \to G/\mathscr{B}$.

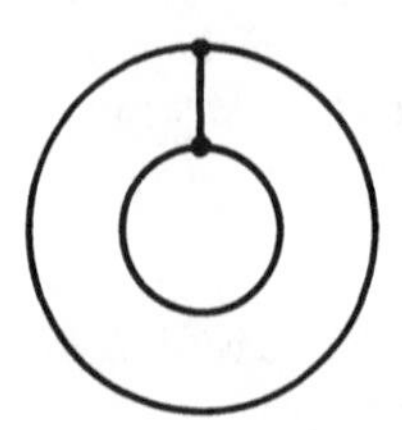

Figure 1.21. A regular quotient of the Petersen graph.

A graph map $p: \tilde{G} \to G$ is called a "regular covering projection" if it is equivalent to a quotient map in the following sense: there exists a group $\mathscr{B}$ that acts freely on the domain $\tilde{G}$, and there exists an isomorphism $i: \tilde{G}/\mathscr{B} \to G$ such that $i \circ q_B = p$, where $q_{\mathscr{B}}: \tilde{G} \to \tilde{G}/\mathscr{B}$. The range graph G is called the "base" (or "base space"), and the domain graph $\tilde{G}$ is called a "regular covering" (or "regular covering space").

Example 1.3.15. *For $n \geq 1$ and $r \geq 1$, there is a regular covering projection $C(\mathscr{Z}_{rn}, \{1\})^0 \to C(\mathscr{Z}_n, \{1\})^0$ defined by the vertex function i modulo $rn \to$ i modulo n and the edge function $(i, j \text{ modulo } rn) \to (i, j \text{ modulo } n)$. The effect of this regular covering projection is to wrap an rn-cycle around an n-cycle r times. The group $\mathscr{Z}_r$ acts freely on the domain Cayley graph.*

For any vertex v in a graph G, the "star" of v, denoted star(v), is the subgraph consisting of the vertex v and all the vertices adjacent to it and of all the edges joining those other vertices to v. If the graph G is simplicial and if the vertex v has valence k, then star(v) is isomorphic to the bipartite graph $K_{1,k}$.

A graph map $f: G' \to G$ is called a "local isomorphism" if for every vertex v in G', the restriction of f to star(v) is an isomorphism onto star($f(v)$). It may be observed that the covering projection of Example 1.3.15 is a local isomorphism for $n \geq 3$ but not for $n = 1$ or 2 and $r \geq 2$. One exercise for this section is to show that if its base space is simplicial, then a covering projection is a local isomorphism. To emphasize that it is more than a local isomorphism, a graph isomorphism is sometimes called a "global isomorphism".

1.3.9. Exercises

1. How many different spanning trees are there in the graph at the left of Figure 1.13? How many different isomorphism types of spanning trees are there? How many isomorphism types of subgraphs are there?
2. Prove that every graph is homeomorphic to a bipartite graph. (*Hint*: Use the operation of subdivision.)
3. How many vertices does the n-cube graph Q_n have? How many edges?
4. How many vertices and edges does the n-octahedron graph O_n have?
5. A suspension $K_1 + C_n$ is called a "wheel". Why?
6. Draw the suspension $K_2{}^c + C_4{}^c$, the suspension $K_2{}^c + C_6$, and the cartesian product $K_2 \times K_{1,3}$.
7. Describe a free action of the cyclic group $\mathscr{Z}_3$ on the complete bipartite graph $K_{3,3}$ and draw the quotient.
8. Describe a free action of the cycle group $\mathscr{Z}_2$ on the 3-cube graph Q_3 whose quotient is isomorphic to the complete graph K_4.
9. Prove that the cyclic group $\mathscr{Z}_7$ cannot have a free action on the Petersen graph.

10. Prove that the cyclic group $\mathscr{Z}_6$ cannot have a free action on the complete bipartite graph $K_{3,3}$.
11. Prove that the even-order cyclic group $\mathscr{Z}_{2n}$ cannot have a free action on the complete graph K_{2n}.
12. Prove that the odd-order cyclic group $\mathscr{Z}_{2n+1}$ has a free action on the complete graph K_{2n+1}.
13. Construct a simplicial graph whose automorphism group acts freely on it and is a nontrivial group.
14. Prove that a local isomorphism from a connected graph onto a tree is a global isomorphism.
15. Prove that every automorphism on a tree has either a fixed vertex or a fixed edge. (*Hint*: Those familiar with the "center" of a tree may find that concept helpful.)
16. Prove that a covering projection $p: \tilde{G} \to G$ with a simplicial base graph G is a local isomorphism.
17. Let G and G' be graphs, and let $v \in V_G$ and $v' \in V_{G'}$. Prove that the induced subgraphs on the vertex sets $V_G \times \{v'\}$ and $\{v\} \times V_{G'}$ are isomorphic to G and G', respectively, in the product graph $G \times G'$.

1.4. SURFACES AND IMBEDDINGS

The central concern of topological graph theory is the placement of graphs on surfaces. Ordinarily, such surfaces are "closed", that is, compact and without boundary. The plane is not closed, of course but since it differs from the sphere by only a single point, it follows that a given graph can be imbedded in the plane if and only if it can be imbedded in the sphere. Accordingly, nothing is lost by considering imbeddings in the sphere, as is the present practice, rather than in the plane.

1.4.1. Orientable Surfaces

There are two kinds of closed surfaces, orientable and nonorientable. The sphere, the torus, the double torus, the triple torus, and so on, as illustrated in Figure 1.22, are orientable. They are commonly denoted $S_0, S_1, S_2, S_3, \ldots$. Indeed, it is proved in Chapter 3 that every closed connected orientable surface is homeomorphic to one of them.

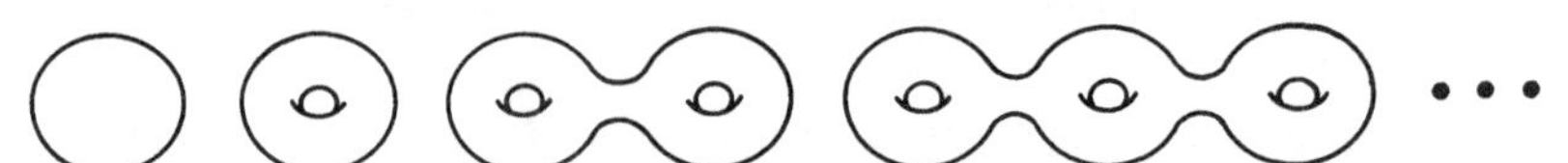

Figure 1.22. The closed orientable surfaces.

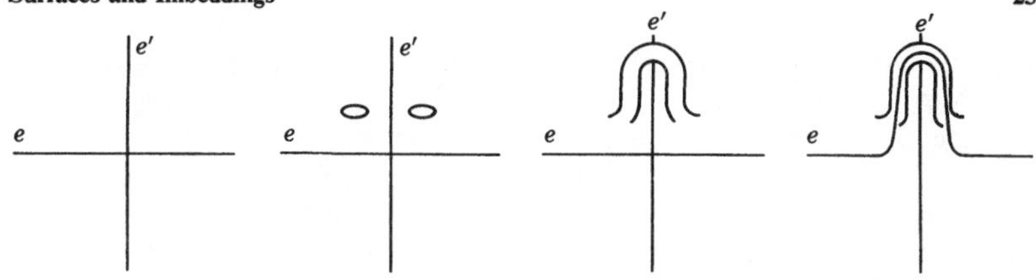

Figure 1.23. Adding a handle to eliminate an edge-crossing.

Another characterization of the closed orientable surfaces is that each one can be obtained by adding some handles to a sphere in 3-space. Adding one handle yields S_1, adding two yields S_2, and so on. From this characterization, it is easily seen that every (finite) graph can be drawn without edge-crossings on some closed surfaces, as follows. First, draw the graph on the sphere, possibly with crossings. Now suppose that the edge e crosses the edge e'. As a surgical operation, cut a hole in the surface on each side of e', near the crossing by e, small enough so that it touches no edge. Next attach a handle from one small hole to the other. Then reroute edge e so that it traverses the handle instead of crossing edge e'. This procedure, illustrated in Figure 1.23, can be repeated until all crossings are eliminated.

1.4.2. Nonorientable Surfaces

The "Möbius band" is a surface that is neither closed nor orientable. To obtain a physical copy of a Möbius band, begin with a rectangular piece of paper. Then, as illustrated in Figure 1.24, give the strip a half-twist, so as to interchange the top and bottom at one side, and finish by pasting the right side to the left. [See Möbius (1865).]

Although a Möbius band is compact, it has a boundary, which is homeomorphic to the circle, so it is not closed. What makes it nonorientable is that if a 2×2 coordinate system specifying a forward direction and a right direction is translated in the forward direction once around the center of the band, then the orientation of the right direction is reversed. It seems from a local viewpoint that a physical copy of a Möbius band is two-sided. However, one can color "both" sides in a single hue without lifting the crayon or crossing the boundary.

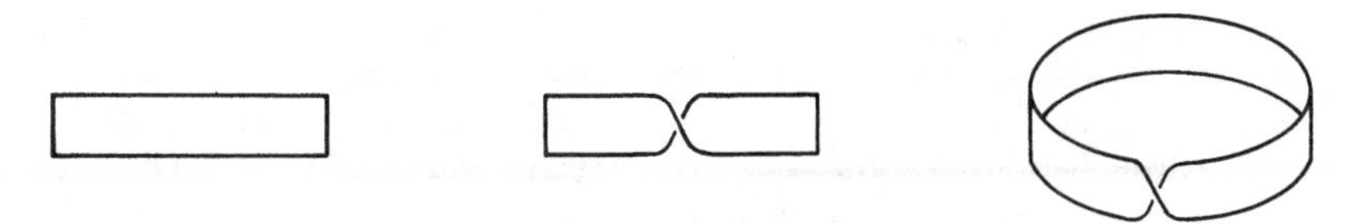

Figure 1.24. Constructing a Möbius band.

The global one-sidedness of the Möbius band in 3-space is a result of the imbedding in 3-space, not an intrinsic property of the Möbius band. For instance, in the imbedding space $Mb \times [0,1]$, where Mb is the Möbius band and $[0,1]$ is the unit interval, the cross-section $Mb \times \{1/2\}$ is two-sided.

If one cuts a hole in a sphere, the resulting boundary is homeomorphic to the circle, just as is the boundary of a Möbius band. The surface obtained by attaching a Möbius band along its boundary to the hole in the sphere, thereby closing off the hole, is called a "projective plane". As the reader may confirm by experimentation with physical models, it is impossible to complete this boundary matching in 3-space.

The "Klein bottle", another nonorientable closed surface, can be obtained by cutting two holes in a sphere and then closing each of them off with a Möbius band. Indeed, as it is proved in Chapter 3, every closed, connected nonorientable surface can be obtained by cutting holes in a sphere and then closing them off with Möbius bands. For $k = 0, 1, \ldots,$ the surface obtained by cutting k holes and closing them off with k Möbius bands is denoted N_k. In this context a Möbius band is often called a "crosscap". Finally, from the discussion here, one may infer that a closed surface is nonorientable if and only if it has a subspace homeomorphic to a Möbius band. See Klein (1911).

1.4.3. Imbeddings

To formalize the notion of a drawing without crossings, we define an "imbedding" of a graph in a surface to be a continuous one-to-one function from a topological representation of the graph into the surface. For most purposes, it is natural to abuse the terminology by referring to the image of the topological representation as "the graph".

If a connected graph is imbedded in a sphere, then the complement of its image is a family of "regions" (or "faces"), each homeomorphic to an open disk. In more complicated surfaces, the regions need not be open disks. If it happens that they are all open disks, then the imbedding is called a "2-cell (or cellular) imbedding". If the boundary circuit of an open disk region has one or more repeated vertices, then the closure of the region is not a closed disk. Nonetheless, whether the imbedding is a 2-cell imbedding depends only on whether all the regions are open disks, not on whether the closures of the regions are closed disks.

A convenient notation for the set of regions of a graph imbedding $i: G \to S$ is F_G, where the letter F reminds one that the regions are something like the faces of a polyhedron. If more than one imbedding of G is under consideration, then the name of the imbedding should appear somewhere in the notation for the set of regions. An even greater convenience, if G is the only graph in the immediate discussion, is to use the notation F instead of F_G.

The "number of sides" (or "size") of a region f is defined to be the number of edge-sides one encounters while traversing a simple circuit just inside the boundary of the region, and is denoted s_f. As illustrated by the exterior region

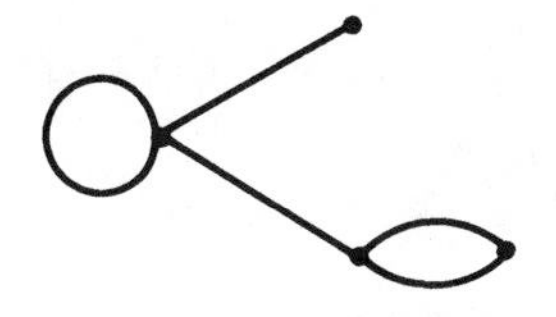

Figure 1.25. A spherical imbedding.

in Figure 1.25, the number of sides of a region need not equal the number of edges in its boundary. Although there are seven sides to the exterior region, there are only five edges in its boundary.

It may be observed that the imbedding in Figure 1.25 has a one-sided region, which is called a "monogon". It also has a two-sided region, which is called a "digon". (Some persons use the words "unigon" and "bigon", which are mixtures of Latin and Greek. Fortunately, these mixtures have not become as firmly entrenched as "hexadecimal".)

If every region of a graph imbedding is three-sided, then one says the imbedding is "triangular". In a "quadrilateral imbedding", every region is four-sided.

1.4.4. Euler's Equation for the Sphere

In a letter to Goldbach, Euler (1750) wrote that he had found a new formula relating the numbers of vertices, edges, and faces of a (spherical) polyhedron:

$$\#V - \#E + \#F = 2.$$

Cauchy (1813) gave basically a graph-theoretic proof of this "Euler equation", but it was Lhuilier (1811) who sorted out the apparent exceptions (faces must be 2-cells and the surface cannot have holes) and generalized the Euler equation to all closed orientable surfaces. Listing (1861) later used the Euler equation to study surfaces in detail; this work is commonly considered the beginning of topology.

The left side of the Euler equation is called the "Euler formula". The right side is called the "Euler characteristic" of the surface. Thus, the number 2 is the Euler characteristic of the sphere. Before proving the Euler equation for the sphere, it may be instructive to evaluate the Euler formula for the two examples depicted in Figure 1.26.

The graph G in Figure 1.26 has five vertices and nine edges. Each of the regions, including the "exterior" region, is three-sided. There are six regions in all, since the exterior region is always counted in computations of the Euler formula. Thus, $\#V - \#E + \#F = 5 - 9 + 6 = 2$. The graph H in Figure 1.26 has four vertices and five edges. The imbedding has three regions. Thus, $\#V - \#E + \#F = 4 - 5 + 3 = 2$.

Theorem 1.4.1. *Let $i: G \to S_0$ be an imbedding of a connected graph in the sphere. Then $\#V - \#E + \#F = 2$.*

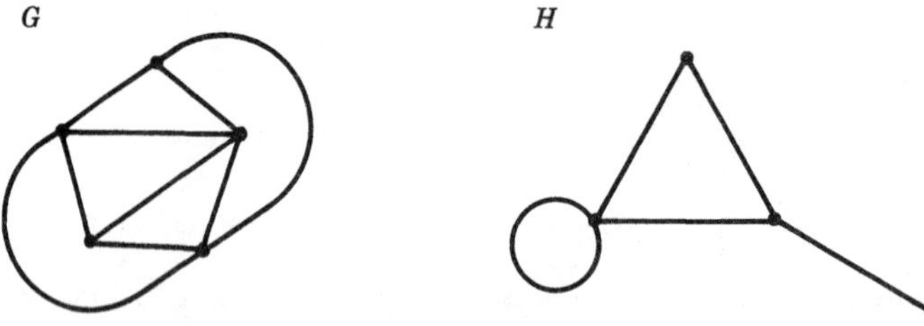

Figure 1.26. Two graph imbeddings in the sphere.

Proof. This proof proceeds by induction on the number $\#F$ of regions. First, observe that if $\#F_G = 1$, then G must be a tree, since the Jordan curve theorem implies that any cycle would separate the sphere. Thus, by Theorem 1.2.2, $\#V_G - \#E_G = 1$, from which it follows that $\#V_G - \#E_G + \#F_G = 2$.

Now suppose that the Euler equation holds when the number of regions is at most n, and suppose that $\#F_G = n + 1$. Then some edge e lies in the boundary circuit of two distinct regions. Since the two regions are distinct, the subgraph G' obtained by removing the edge e is connected. Then $\#F_{G'} = \#F_G - 1 = n$, so by induction, $\#V_{G'} - \#E_{G'} + \#F_{G'} = 2$. Since $\#V_{G'} = \#V_G$, $\#E_{G'} = \#E_G - 1$, and $\#F_{G'} = \#F_G - 1$, it follows that $\#V_G - \#E_G + \#F_G = 2$. □

1.4.5. Kuratowski's Graphs

The Euler equation is often used in conjunction with a relationship between the numbers of edges and regions to prove that certain graphs cannot be imbedded in the sphere. This relationship, called the "edge-region inequality", is established by the following theorem.

Theorem 1.4.2. *Let $i: G \to S$ be an imbedding of a connected, simplicial graph with at least three vertices into any surface. Then $2\#E \geq 3\#F$.*

Proof. The sum $\Sigma_{f \in F} s_f$ of the numbers of sides of the regions counts every edge exactly twice. Thus, $2\#E = \Sigma_{f \in F} s_f$. Since there are no loops or multiple edges in the simplicial graph G, there are no monogons or digons in the imbedding. Thus, for every region f, $s_f \geq 3$. It follows that $2\#E \geq 3\#F$. □

The complete graph K_5 and the complete bipartite graph $K_{3,3}$ are called "Kuratowski's graphs" (or the "Kuratowski graphs") because Kuratowski (1930) proved that they are a complete set of obstructions to imbedding graphs in the sphere, in the following sense:

Kuratowski's Theorem. *The graph G has an imbedding in the sphere if and only if it contains no homeomorph of K_5 or of $K_{3,3}$.*

To see that K_5 is nonspherical, observe that it has five vertices and 10 edges. By the Euler equation $\#V - \#E + \#F = 2$, there would have to be seven regions to any spherical imbedding. However, this would violate the edge-region inequality $2\#E \geq 3\#F$, since $2 \cdot 10 < 3 \cdot 7$.

A proof that $K_{3,3}$ is nonspherical begins in the same way. Since $K_{3,3}$ has six vertices and nine edges, the Euler equation implies that a spherical imbedding would have five regions. However, the edge-region inequality is satisfied, because $2 \cdot 9 \geq 3 \cdot 5$. This proof is continued after a strengthening of the edge-region inequality, which involves the notion of girth.

The "girth" of a graph is the number of edges in its shortest cycle. If a graph has a loop, then its girth is 1. Otherwise, it is at least 2. If it also has no multiple edges, then its girth is at least 3. The girth of a tree is infinite.

The key fact in establishing the edge-region inequality for simplicial graphs is that the girth of a simplicial graph is at least 3. Since a bipartite graph has no odd cycles, its girth cannot be 3. Thus, the girth of a bipartite simplicial graph is at least 4. In particular, the girth of $K_{3,3}$ is 4, from which it follows that $2\#E \geq 4\#F$. However, since a spherical imbedding of $K_{3,3}$ would have five regions, this stronger inequality would be violated, because $2 \cdot 9 < 4 \cdot 5$. For future use, it is convenient to isolate the "strong edge-region inequality", as follows.

Theorem 1.4.3. *Let G be a connected graph that is not a tree, and let $i: G \to S$ be an imbedding. Then* $2\#E \geq \text{girth}(G) \cdot \#F$.

One direction of Kuratowski's theorem is now easily proved. Since K_5 and $K_{3,3}$ are nonspherical, a graph G with an imbedding in the sphere cannot possibly contain a homeomorph of either of them. A proof of the converse appears in Section 1.6.

1.4.6. Genus of Surfaces and Graphs

If a surface is orientable, its "genus" is defined to be the number of handles one must add to the sphere to obtain its homeomorphism type. That is, the genus of the orientable surface S_g is g. If a surface is nonorientable, then its "crosscap number" is defined to be the number of Möbius bands one must attach to the sphere to obtain its homeomorphism type. Thus, the crosscap number of the nonorientable surface N_k is k for $k \geq 1$. The notation N_0 means the 2-sphere, even though the 2-sphere is orientable.

The "(orientable) genus" of a graph G, denoted $\gamma(G)$ or simply γ if G is the only graph in context, is defined to be the smallest number g such that the graph G imbeds in the orientable surface S_g. The "crosscap number" of the graph G, denoted $\bar{\gamma}(G)$ or $\bar{\gamma}$, is the smallest number k such that G imbeds in the surface N_k. Sometimes one calls $\bar{\gamma}(G)$ the "nonorientable genus" of G.

Since neither K_5 nor $K_{3,3}$ imbeds in the sphere, they both have positive genus and positive crosscap number. Figure 1.27 establishes that both K_5 and

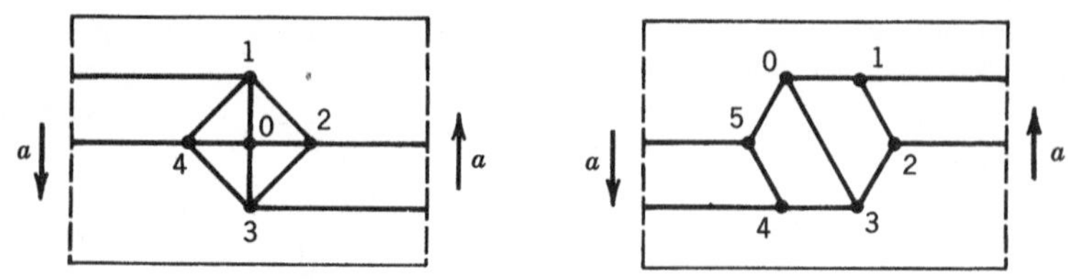

Figure 1.27. Imbeddings of K_5 and $K_{3,3}$ in the Möbius band.

$K_{3,3}$ have imbeddings in the Möbius band, and consequently in the projective plane N_1. Thus they both have crosscap number 1. A calculation of their orientable genera appears later in this section.

The imbedding of K_5 in Figure 1.27 shows the vertices and all but two of the edges in the "center" of the Möbius band. The edge from vertex 2 to vertex 4 runs through the middle of the pasting edge, and the edge from vertex 1 to vertex 3 also passes through the pasting edge. It is precisely the half-twist before pasting that makes the imbedding possible. The imbedding of $K_{3,3}$ depends on the same trick.

The method described earlier in this section for using handles to eliminate crossings does not give a very good upper bound on the orientable genus, because it is usually highly inefficient to send only one edge across a handle. For instance, it is impossible to draw K_6 in the sphere with fewer than three crossings, but only one handle is needed to eliminate them all (see Exercise 1.4.3).

During the 1960s and 1970s there was an intensity of interest in finding minimum imbeddings for graphs, largely due to the outstanding work of Gerhard Ringel, who calculated the genus of the n-cube graphs (see Ringel, 1955), the genus of the complete bipartite graphs (see Ringel, 1965), and the nonorientable genus of the complete graphs (see Ringel, 1959). After calculating the orientable genus of one-third of the complete graphs, he teamed with J. W. T. Youngs to finish them off, thereby solving the Heawood map-coloring problem. An exposition of this solution, noting the contributions of other persons, is given in Chapter 5.

1.4.7. The Torus

Just as the Möbius band can be cut open and flattened into a rectangle, so can the torus, thereby permitting one to draw imbeddings on a flat piece of paper. To retrieve the torus from a rectangle, first paste the top of the rectangle to the bottom, in order to obtain a tube. Then paste the right end of the tube to the left, which yields the torus, as illustrated in Figure 1.28.

In order to show an imbedding of a graph on a torus, one may draw some of the edges of the graph "through" the sides of a rectangle, taking care in each instance to continue at the corresponding position on the opposite side. For instance, Figure 1.29 shows imbeddings of K_5 and $K_{3,3}$ on the torus.

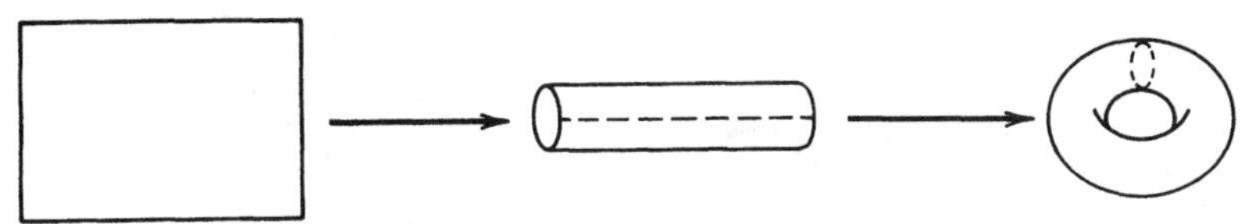

Figure 1.28. Constructing a torus from a rectangle.

It has already been proved that neither K_5 nor $K_{3,3}$ imbeds in the sphere. From Figure 1.29 it now follows that both have orientable genus 1. Although K_5 has the same nonorientable genus as orientable genus, as does $K_{3,3}$, for most graphs the orientable genus and the nonorientable genus differ.

1.4.8. Duality

Given a connected graph G, a closed surface S, and a 2-cell imbedding $i: G \to S$, there is an idea due to Poincaré for constructing what are called a "dual graph" and a "dual imbedding". First, for each region f of the imbedding $i: G \to S$, place a vertex f^* in its interior. Then, for each edge e of the graph G, draw an edge e^* between the vertices just placed in the interiors of the regions containing e. (If both sides of the edge e lie in the same region f, then the dual edge e^* is a loop based at the dual vertex f^*.) The resulting graph with vertices f^* and edges e^* is called the "dual graph" for the imbedding $i: G \to S$ and is denoted G^{*i}, or simply G^*, if i is the only imbedding under consideration. The resulting imbedding of the graph G^* in the surface S is called the "dual imbedding". In this context, the original graph G and the original imbedding i are called the "primal graph" and the "primal imbedding". Figure 1.30 shows how two different imbeddings of the same graph in the same surface can yield different dual graphs.

In Figure 1.30 the primal graph is drawn with solid dots for vertices and solid lines for edges, whereas the dual graphs are drawn with open dots for vertices and dashed lines for edges. Since the dual graph at the left has a vertex of valence 7, and the maximum valence in the dual graph at the right is

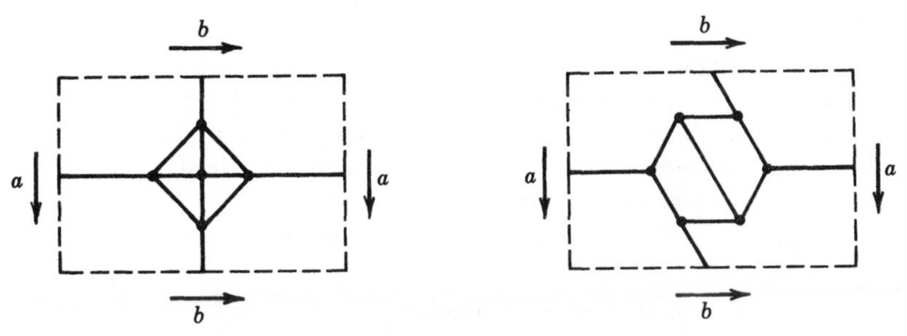

Figure 1.29. Imbeddings of K_5 and $K_{3,3}$ on the torus.

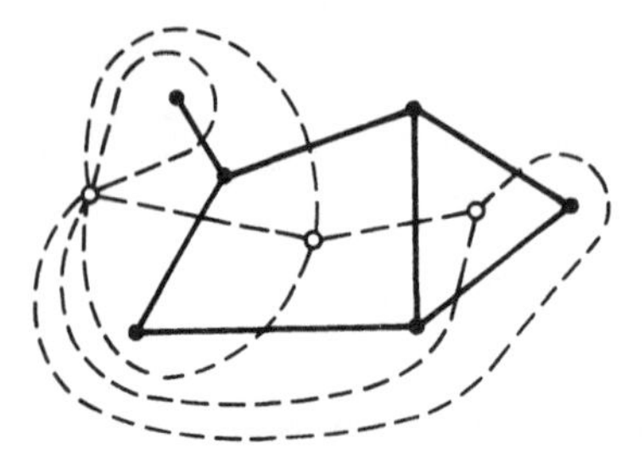

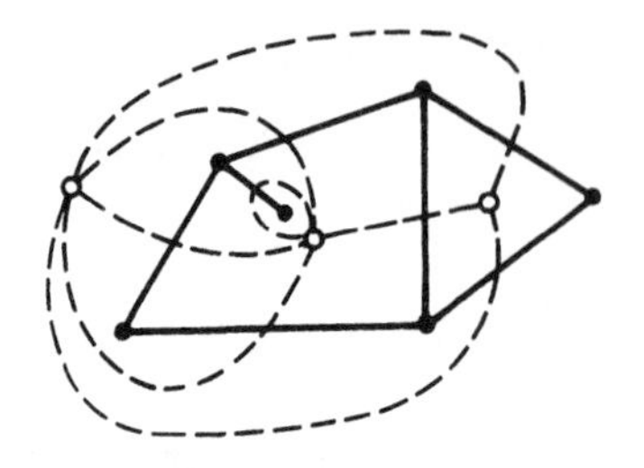

Figure 1.30. Two different duals of the same graph in the sphere.

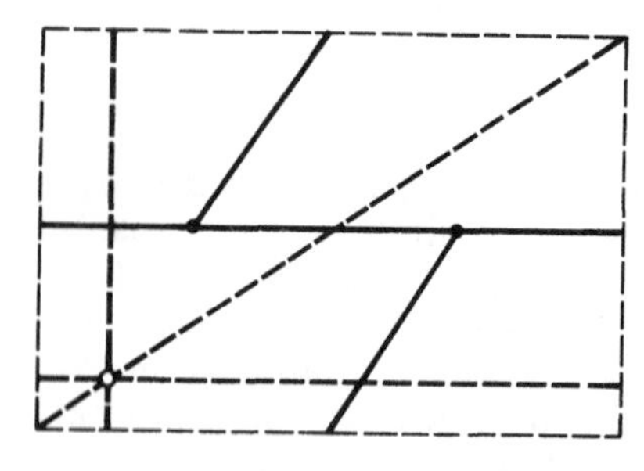

Figure 1.31. A graph in the torus and its dual.

6, the two duals are not isomorphic. It may be observed that for any region f of the primal imbedding, the valence of the dual vertex f^* equals the number of sides of f.

Figure 1.31 shows the duality construction on a torus. The primal graph has two vertices and three edges between them, and there is only one primal region. The dual graph has only one vertex, at which three loops are based.

It should be noted that if one were to construct the dual of the dual imbedding, then the primal imbedding of the primal graph would be restored. This fact is what justifies the use of the word "dual".

1.4.9. Exercises

1. Cut out a rectangular strip of paper about 3 cm wide and about 30 cm long. On both sides, draw a lengthwise center line. Next, paste the strip into a Möbius band. Then cut the strip along the center line. How many pieces are obtained? Next, cut each of the bands obtained in the previous step along their center lines. What happens?
2. Draw K_5 in the sphere so that there is only one edge-crossing. Do the same for $K_{3,3}$.
3. Draw an imbedding of K_6 on the torus.
4. Draw K_6 on the sphere with only three edge-crossings.

5. Prove that no simplicial graph with seven vertices and 16 edges can be imbedded in the sphere.
6. Suppose that a simplicial graph has p vertices. What is the maximum number of edges it can have and still have an imbedding in the sphere?
7. Suppose that a simplicial bipartite graph has p vertices. What is the maximum number of edges it can have and still have an imbedding in the sphere?
8. Prove that there are exactly five different simplicial graphs with six vertices and 12 edges. Decide which ones can be imbedded in the sphere. (Give proofs, of course.)
9. Draw three cellular imbeddings of different graphs in the torus, and then calculate the Euler formula in each case. What number is its value? This number is the Euler characteristic of the torus.
10. Find a subgraph of the Petersen graph (recall Figure 1.20) that is a subdivision of $K_{3,3}$, thereby proving that the Petersen graph is nonspherical.
11. Using the girth of the Petersen graph, prove it is nonspherical.
12. Draw the dual for the imbedding of Figure 1.25.
13. Draw 2-cell imbeddings for $K_5 - e$ (i.e., the result of removing one edge from K_5) in the sphere and in the torus. Then draw their duals.
14. Show that the 3-cube graph Q_3 is isomorphic to a dual of the 3-octahedron graph O_3 in the sphere.
15. Prove that there are exactly four isomorphism types of connected graphs with three vertices and three edges, and that exactly two of them have triangular imbeddings in the sphere.
16. Let G be a graph. The "line graph" $L(G)$ has vertex set E_G, the edges of G. Two members of E_G are adjacent in L_G if and only if they have a common endpoint in G. Give an example of a graph G such that

$$\gamma(L(G)) > \gamma(G)$$

Also, prove that if the maximum valence of a graph G is less than or equal to 3, then

$$\gamma(L(G)) \leq \gamma(G)$$

17. Prove that a nonconnected graph cannot have a 2-cell imbedding in a connected surface.

1.5. MORE GRAPH-THEORETIC BACKGROUND

This section gives some explicit attention to a few of the standard topics in graph theory that arise most frequently in topological developments.

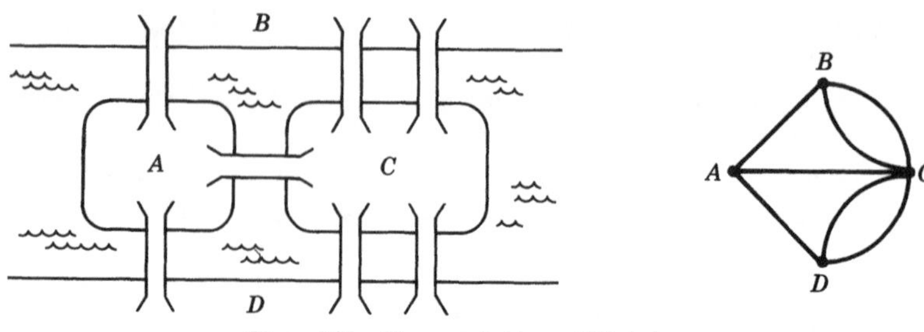

Figure 1.32. The seven bridges of Königsberg.

1.5.1. Traversability

In the earliest known paper on graph theory, Euler (1736) proved that it was impossible to cross each of the seven bridges of Königsberg once and only once on a walk through the town. Figure 1.32 shows the original Königsberg problem, with two land areas on the opposite sides of the Pregel River and two islands in the river, and also its graph-theoretic abstraction, in which the four land areas are represented by vertices and the seven bridges by edges.

From the Königsberg bridge problem arises the definition that a walk in a connected graph is "eulerian" if it traverses each edge exactly once. A closed eulerian walk is called an "eulerian circuit". A graph containing an eulerian circuit is called an "eulerian graph".

Theorem 1.5.1. *A connected graph G is eulerian if and only if every vertex has even valence.*

Proof. First suppose that the closed walk W is an eulerian circuit in G that starts and stops at the vertex v. Then each occurrence in W of a vertex of G, other than v, contributes an addend of 2 to the valence of that vertex. It follows that the valence of any vertex of G except v equals twice the number of times it occurs on W. The valence of v is 2 for the beginning and end of W plus twice the number of times that v occurs elsewhere in W.

Conversely, suppose that every vertex of the graph G has even valence. By way of induction, assume that G is the smallest such graph that has no eulerian circuit. The graph G is not K_1, which is eulerian. Moreover, G has no vertices of valence 1. Thus, from Theorem 1.2.3, the graph G is not a tree; so it must contain a cycle C. Since every vertex of the graph $G - C$ obtained by removing the edges of the cycle C from the graph G has even valence, and since each component of $G - C$ is smaller than G, it follows that each component of $G - C$ has an eulerian circuit.

Let $G_1, \ldots, G_n$ be the components of $G - C$, indexed in the order in which they are encountered on a traversal of the cycle C. An eulerian circuit for the graph G is constructed as follows. Start traversing the cycle C. As soon as the

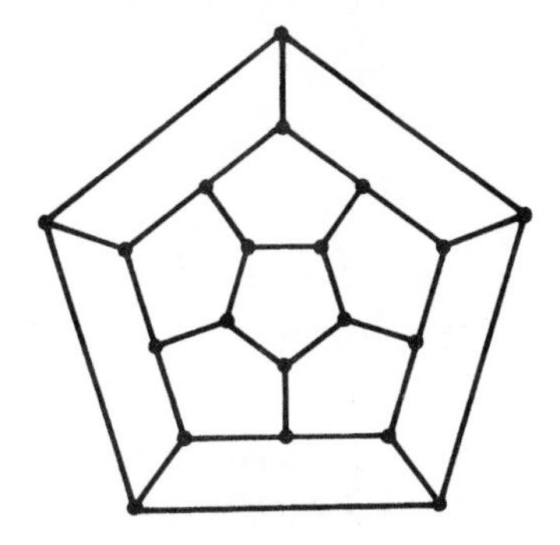

Figure 1.33. The dodecahedron graph.

component G_1 is encountered, interrupt the traversal of C to do a complete traversal of the eulerian circuit for G_1. Then resume the traversal of the cycle C until G_2 is encountered. Once again interrupt the traversal of C, this time to traverse completely the eulerian circuit for G_2, and so on. □

Theorem 1.5.1 does not quite solve the Königsberg bridge problem, as originally stated. Making the necessary modifications to solve that problem is left as an exercise at the end of this section.

A more difficult problem than finding an eulerian circuit has grown out of a puzzle invented by Sir William Hamilton (see Biggs et al., 1976). The "dodecahedron" is the unique 12-sided regular three-dimensional polyhedron. Each of its sides is a pentagon. The 1-skeleton of this polyhedron, illustrated in Figure 1.33, is called the "dodecahedron graph". From our perspective, Hamilton's puzzle was to find a cycle containing every vertex of the dodecahedron graph.

In general, a cycle that contains every vertex of a graph is called a "hamiltonian cycle". A graph that has a hamiltonian cycle is called a "hamiltonian graph". Whereas there are very fast ways to construct an eulerian circuit in a given graph or to determine that none exists, it is usually much more difficult to construct a hamiltonian cycle or to determine that none exists (see Karp, 1972).

1.5.2. Factors

The edge set of a spanning subgraph for a graph G is called an "n-factor" for G if that spanning subgraph is regular of valence n. If the edges of a graph G can be partitioned into n-factors for G, then G is called "n-factorable". For instance, Figure 1.34 shows a 1-factorization of the complete graph K_4 and a 2-factorization of the complete graph K_5.

In general, the edges of a 2-factor form a family of mutually disjoint cycles that includes every vertex. Petersen (1891) observed that every 4-regular graph is 2-factorable, as demonstrated by the following argument. Since every vertex

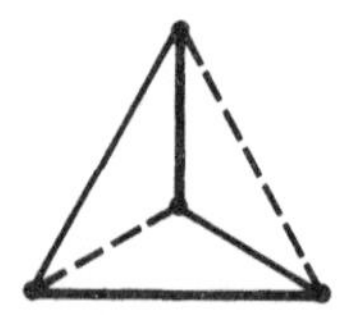
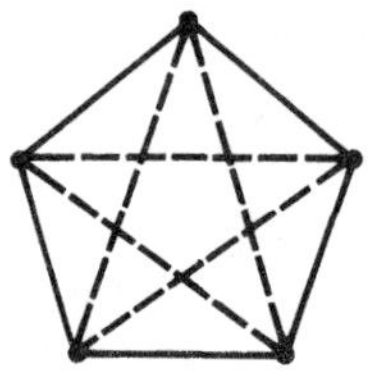

Figure 1.34. A 1-factorization of K_4 and a 2-factorization of K_5.

has even valence, a 4-regular graph G has an eulerian circuit W. As one traverses the circuit W, one may color the edges along it alternately red and blue. Except for the starting point of the circuit, it is obvious that each vertex will thus obtain red valence 2 and blue valence 2. By Euler's theorem on valence sum (Theorem 1.1.1), the number of edges of G is even, which implies that the final color assigned along the eulerian circuit is different from the starting color, so that the starting vertex also has red valence 2 and blue valence 2. Thus the red edges form a 2-factor, as do the blue edges, thereby yielding a 2-factorization. By reiteration of this edge-coloring argument, one may prove that every 2^n-regular graph is 2-factorable. Exercises 18 and 19 outline Petersen's method to extend this result to all regular graphs of even valence.

Theorem 1.5.2 (Petersen, 1891). *Every regular graph of even valence is 2-factorable.*

1.5.3. Distance, Neighborhoods

The "length" of a walk is defined to be the number of edges it contains (when considered as a sequence). Thus, the length of the path P_n is $n - 1$, and the length of the n-cycle C_n is n.

The "distance" between two vertices u and v of a graph G is defined to be the length of the shortest path between them. If u and v are in separate components, then the distance is infinite.

A "neighbor" of a vertex v is any vertex at distance 1 from v. The "neighborhood" of v is the induced subgraph on the neighbors of v. For instance, in the complete graph K_n, every vertex has a neighborhood isomorphic to K_{n-1}. In the n-cycle C_n, if $n \geq 4$, then every vertex has a neighborhood isomorphic to K_2^c. In a regular graph, every vertex has the same number of neighbors, but unlike the case in a vertex-transitive graph, the neighborhoods in a regular graph need not be isomorphic.

Example 1.5.1. *Let G be the complement in K_7 of a 2-factor consisting of a 3-cycle and a 4-cycle. The neighborhood of a vertex on the missing 3-cycle is isomorphic to the disjoint union of two copies of K_2, and the neighborhood of a vertex on the missing 4-cycle is isomorphic to $K_{1,3}$.*

1.5.4. Graph Colorings and Map Colorings

An "n-coloring" for a graph G is a map from the vertices of G to a set of n distinct objects called "colors", such that no two adjacent vertices are assigned the same color. (Graphs with loops are noncolorable.) The "chromatic number" $\text{chr}(G)$ is defined to be the smallest number n such that G has an n-coloring. A graph G with chromatic number n is called "n-chromatic".

Obviously, a graph is 1-colorable if and only if it has no edges. Also, a graph is 2-colorable if and only if it is bipartite. The simplest example of a graph that requires at least three colors is a cycle of odd length. As indicated by the following theorem, this example is the fundamental obstruction to 2-colorings.

Theorem 1.5.3. *A graph is 2-colorable if and only if it has no odd-length cycles.*

Proof. In any partition of the vertices of an odd-length cycle into two classes, at least one class must contain two adjacent vertices. Thus, an odd-length cycle requires more than two colors, and consequently, no graph containing an odd-length cycle is 2-colorable.

Conversely, let G be a graph with no odd-length cycles. Since the components of a graph may be colored independently of each other, it suffices to assume that G is connected. Select any vertex v_1 of G, and let V_1 be the subset of V_G consisting of all vertices at even distance from v_1. Let $V_2 = V_G - V_1$. If two vertices v and v' in V_1 are adjacent, then let P and P' be respective minimum length paths to them from v_1. Let u be the last vertex of G in which P and P' meet. Then the segment of P from u to v plus the segment of P' from u to v' plus the edge between v and v' forms an odd-length cycle, a contradiction. Similarly, no two vertices of V_2 are adjacent. Thus, the graph G is 2-colorable, with color classes V_1 and V_2. □

Example 1.5.2. *A suspension of a cycle from K_1 is called a "wheel". For any graph G, the chromatic number* $\text{chr}(K_1 + G)$ *of the suspension of G from K_1 is* $\text{chr}(G) + 1$. *Thus, the chromatic number of a wheel formed from an odd-length cycle is four.*

Let $i: G \to S$ be a graph imbedding. A "map coloring" for the imbedding i is an assignment of colors to the regions such that no two regions with a common edge have the same color. Thus, a map coloring for the imbedding i is equivalent to a graph coloring for the dual graph G^{*i}.

The "chromatic number $\text{chr}(S)$ of the surface S" is defined to be the minimum number of colors needed to color any map corresponding to a 2-cell imbedding of a graph into S. Equivalently, it is defined to be the maximum of the set of chromatic numbers of graphs that can be imbedded in S. (This equivalence is realized through the duality construction.) Since multiple edges between two vertices can be collapsed onto a single edge without changing the chromatic number of a graph, one need consider only the chromatic numbers of simplicial graphs imbeddable in S. By the Ringel–Youngs (1968) solution to the Heawood map-coloring problem,

$$\mathrm{chr}(S_g) = \left\lfloor \frac{7 + \sqrt{1 + 48g}}{2} \right\rfloor \quad \text{for every genus } g \geq 1$$

One may observe that for $g = 0$, the value[3] of the right-hand side of this equation is 4. The equation $\mathrm{chr}(S_0) = 4$ was verified separately by Appel and Haken (1976), who thereby solved the four-color problem. Map-coloring problems are discussed at length in Chapter 5.

1.5.5. Edge Operations

If H is either a subgraph of G or a subset of edges of G, then the notation $G - H$ means the graph obtained by deleting from G all the edges of H. The graph $G - H$ is called the "complement of H in the graph G". One may recall that if G is a complete graph, one also uses the notation H^c.

Two subgraphs H_1 and H_2 of a graph G are called "equivalent" if there is an automorphism of G that carries H_1 onto H_2. If the vertices and edges of a graph H do not actually lie in G, but if all isomorphic copies of H in G are equivalent, then the notation $G - H$ is used to denote the complement in G of any one of them.

Example 1.5.3. *The result of deleting an edge from K_5 is often denoted $K_5 - K_2$.*

Example 1.5.4. *The graph of Example* 1.5.1 *might be denoted $K_7 - (C_3 \cup C_4)$.*

To any graph G, one may associate a maximal simplicial spanning subgraph G^{simp}, obtained by deleting all the loops of G and by removing all but one edge of each multiple adjacency. It is easy enough to extend any imbedding $G^{\mathrm{simp}} \to S$ to an imbedding $G \to S$, by drawing each loop close to the vertex at which it is based, in any region incident on that vertex, and by drawing the removed edges from each multiedge of G close alongside the one that remained in G^{simp}. Figure 1.35 illustrates such an extension of imbeddings.

It follows from this principle for extending imbeddings of G^{simp} to imbeddings of G that $\gamma(G^{\mathrm{simp}}) = \gamma(G)$. However, as explained in Chapter 4, sometimes one uses nonminimum imbeddings of a nonsimplicial graph G to obtain a minimum (or other) imbedding of a (possibly simplicial) covering

[3] The self-explanatory Iverson notations $\lfloor x \rfloor$ ("floor" of x) and $\lceil x \rceil$ ("ceiling" of x) are used in preference to the arbitrary notations $[x]$ and $\{x\}$, respectively, to indicate the greatest integer less than or equal to the number x and the least integer greater than or equal to x.

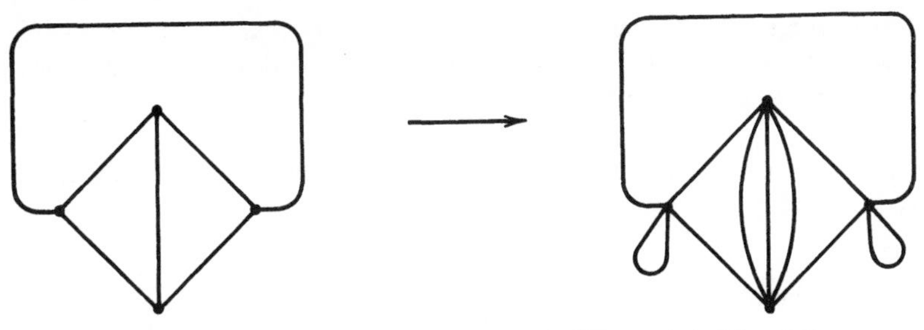

Figure 1.35. The extension of an imbedding of G^{simp} to an imbedding of G.

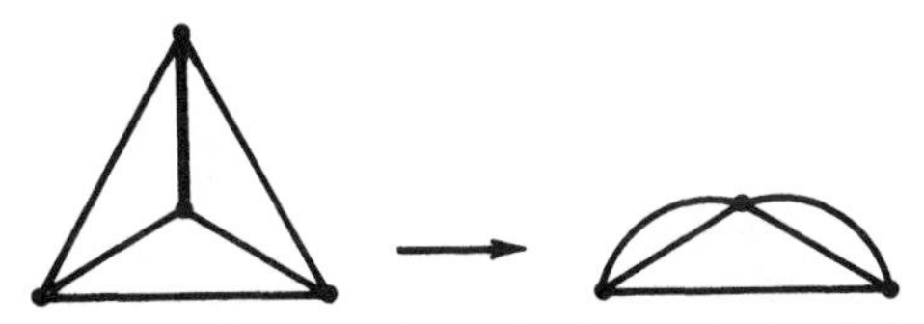

Figure 1.36. The contraction of K_4 along the darkened edge.

space of G. Thus, this imbedding extension principle does not eliminate nonsimplicial graphs from imbedding theorems.

A "contraction" of a graph G along the edge e is the result of deleting the edge e from G and then identifying its endpoints. We image that the edge e is topologically shrunk ("contracted") to a point, and we denote the result by G/e. Figure 1.36 illustrates the contraction of K_4 along an edge. The word "contraction" refers to the operation as well as to the resulting graph. Contraction of an edge can introduce multiple adjacencies. In the context of simplicial graphs, extra adjacencies are often discarded; that is, $(G/e)^{\mathrm{simp}}$ is sometimes considered, in an abuse of terminology, to be the "contracted graph".

If a graph G is imbedded in a surface, then the contraction of G along any edge may be realized in the surface. Accordingly, the genus of a graph is at least as large as the genus of any graph obtained by a sequence of contractions applied to it.

Example 1.5.5. *Since Petersen's graph can be contracted to K_5, by a sequence of contractions along the edges of the 1-factor that runs between the inner star and the outer pentagon, its genus must be at least 1, that is, at least $\gamma(K_5)$.*

Warning 1.5.1. *One of the standard mistakes in topological graph theory is to proceed as if the inverse of a contraction does not increase genus. It is easy to draw examples in which the obvious inverse contractions do not increase genus, and to overlook temporarily the fact that every connected graph contracts to a bouquet of circles, or, in the case of simplicial contraction, to K_1.*

Hadwiger (1943) formulated a conjecture to help characterize n-chromatic graphs. It is known to be true for $n \le 5$ and for nearly all graphs. (See Bollobas and Catlin, 1980.)

Hadwiger Conjecture. *Every connected n-chromatic graph G contracts to K_n, or to a copy of K_n with some multiple adjacencies.*

1.5.6. Algorithms

There are numerous known algorithms to decide whether a graph is spherical, and there are several to calculate the genus. A naive sphericity algorithm is obtained from Kuratowski's theorem, with the aid of a definition that permits modular algorithm structure.

Let u be a vertex of valence 2 in a graph G, and let e_1 and e_2 be the edges incident on u, with $V(e_1) = \{u, v\}$ and $V(e_2) = \{u, w\}$. The graph obtained by "smoothing at u" has for its vertex set $V_G - \{u\}$ and for its edge set $E_G - \{e_1, e_2\}$ plus a new edge between v and w. Subdividing this new edge would invert the operation of smoothing at u.

Naive Sphericity Algorithm. *Let G be a graph. For each subset of E_G whose removal leaves only one nontrivial component H, construct the graph H^{smooth} by successively smoothing over every vertex of valence 2 in H, except possibly for a last vertex if H is a cycle. If some H^{smooth} is isomorphic to K_5 or to $K_{3,3}$ then the graph is nonspherical, but otherwise, by Kuratowski's theorem, G is spherical.*

Since there are $2^{\#E}$ subsets of edges to consider, an execution of this naive algorithm could be quite lengthy. In Section 1.6, we derive a polynomial-time algorithm for sphericity. A general (exponential-time) algorithm for genus is given in Chapter 3.

1.5.7. Connectivity

A "cutpoint" of a graph G is a vertex whose removal would increase the number of components, where it is understood that removing a vertex from G means also removing every edge incident on that vertex, so that the result of the operation is a graph. A maximal subgraph for G with no cutpoints is called a "block".

The "connectivity" (G) of a graph G is defined to be the minimum number of points whose removal results in a disconnected graph, a trivial graph, or a bouquet of circles. It may be observed that G has the same connectivity as G^{simp}.

A graph whose connectivity is k or greater is said to be "k-connected". For instance, by a 3-connected graph, we mean graphs of connectivity 3, 4, 5, and so on.

1.5.8. Exercises

1. Prove that a connected graph G has an open eulerian walk if and only if exactly two of the vertices have valence 1.
2. Use Theorem 1.5.1 and Exercise 1 to solve the Königsberg bridge problem.
3. Draw a hamiltonian circuit in the dodecahedron graph.
4. How many vertices and edges does the dodecahedron graph have? Restriction on method: Do not count the vertices or the edges. Using the fact that a dodecahedron is a 12-sided polyhedron, each side a pentagon, it is possible to calculate $\#V$ and $\#E$.
5. The "icosahedron" is the unique 20-sided regular three-dimensional polyhedron. Each side is a triangle. The "icosahedron graph" is its 1-skeleton. It is known that the icosahedron is dual to the dodecahedron. Draw the icosahedron graph as the dual to a spherical imbedding of the dodecahedron graph. How many vertices and edges does the icosahedron graph have? Restriction on method: Do not count the vertices or the edges of the icosahedron graph.
6. Prove that the Petersen graph is not 1-factorable.
7. Find a 2-factorization of the complete graph K_9.
8. Let G and H be vertex-transitive graphs such that the neighborhood of any vertex of G is isomorphic to the neighborhood of any vertex in H. Are G and H necessarily isomorphic graphs? Explain.
9. Prove that any imbedding of a wheel in the sphere has for its dual graph an isomorphic wheel.
10. What is the chromatic number of Petersen's graph?
11. Draw a map on the Klein bottle that requires six colors.
12. Draw a map on the torus that requires seven colors.
13. The Hajós conjecture is that every n-chromatic graph contains a subdivision of K_n. Prove the Hajós conjecture for $n = 1$, 2, and 3. (The Hajós conjecture is false for nearly all graphs. See Catlin (1979) and Erdös and Fajtlowicz (1981).
14. Prove the Hajós conjecture for $n = 4$.
15. Prove that a contraction of a tree is a tree.
16. Prove the Hadwiger conjecture for $n = 2$.
17. Give an example to show that a contraction can increase the chromatic number of a graph.
18. Let e_1 and e_2 be proper edges with a common endpoint w in a graph G, let u be the other endpoint of e_1, and let v be the other endpoint of e_2. Let G' denote the graph obtained from G by deleting edges e_1 and e_2 and then adding a loop at the vertex w and a new edge between the

vertices u and v (i.e., another new loop if $u = v$). Use Petersen's theorem for 4-regular graphs to prove that if G' is 2-factorable, then so is G.

19. Using Exercise 18 and an induction on the number of proper edges, prove Theorem 1.5.2.

1.6. PLANARITY

The study of planar imbeddings has a long and rich history that intertwines with chromatic graph theory, algorithmic analysis, enumeration, and much else. However, unlike higher-genus surfaces, the plane and sphere are homologically trivial, which is to say, overly simple in a topological sense. Accordingly, the methods used to study planar imbeddings are less topological than the ones of primary interest to us here. Nonetheless, Kuratowski's theorem might well be the most famous theorem in all of graph theory, and it might be remiss to offer no proof of it.

Of the many published proofs, we have selected one whose organization is due to Thomassen (1980). It is not the shortest, and it is not the easiest. However, it is illuminating, in that it also yields other results about planar imbeddings. One is Fary's theorem that the edges of a planar imbedding of a simplicial graph can be chosen to be straight lines. Another is Tutte's theorem that the regions of a planar, 3-connected graph imbedding can be chosen to be convex. Still another is Whitney's theorem that a 3-connected simplicial planar graph has only one planar imbedding. Moreover, Thomassen's proof is readily converted into a planarity algorithm that is polynomial in the number of vertices, a great improvement over the exponential Naive Algorithm given in Section 1.5.

1.6.1. A Nearly Complete Sketch of the Proof

Thomassen's proof of Kuratowski's theorem depends on the use of simplicial edge contractions. As noted in Warning 1.5.1, edge contraction can be a dangerous tool, since a single edge contraction can lower the genus of a graph dramatically. On the other hand, for planar graphs, where the genus is already as low as it can go, edge contraction is safe and natural, because an edge contraction of a planar graph yields a planar graph.

We have already proved in Section 1.4 that a graph containing a homeomorph of either the complete graph K_5 or the complete bipartite graph $K_{3,3}$ cannot be planar. Therefore, Kuratowski's theorem is proved if we establish for every graph G that the following statement is true:

(KT) If the graph G contains no homeomorph of K_5 or of $K_{3,3}$, then G is planar.

The proof is by induction on the number of vertices in the graph G.

Basis step. *The statement (KT) is vacuously true of all graphs with four or fewer vertices.*

Induction Hypothesis. *We assume that the statement (KT) is true of all graphs with fewer than n vertices, where n is any number greater than or equal to* 5.

Induction Step. *Let G be a graph with n vertices, and let G contain no homeomorph of K_5 or $K_{3,3}$. We shall prove that G is planar.*

Let e be any edge in the graph G, and suppose that its endpoints are the vertices u and v. Let G' be the graph obtained from G by a simplicial contraction of edge e, and let us give the name v' to the vertex to which e is contracted. We momentarily interrupt the proof of Kuratowski's theorem to establish a lemma.

Lemma. *Let G be a graph that contains no homeomorph of K_5 or $K_{3,3}$, and let e be any edge in G. Then the result of simplicially contracting the graph G on edge e is a graph G′ that contains no homeomorph of K_5 or of $K_{3,3}$.*

Proof of Lemma. If we already knew that Kuratowski's theorem is true, we would imbed G in the plane, then simplicially contract e within the imbedding, and thereby obtain an imbedding of the contracted graph G'. However, our task is to prove this lemma without using Kuratowski's theorem.

Suppose that the contracted graph G' contained a homeomorph of K_5 or of $K_{3,3}$. Then G would contain a subgraph H such that the contraction of a single edge e of H produces a homeomorph of K_5 or of $K_{3,3}$. Using reversal of contraction, we see that there are only three essentially different possibilities for H, up to homeomorphism; they are illustrated in Figure 1.37. In each case, the subgraph H itself contains a homeomorph of $K_{3,3}$ or of K_5, contradicting the hypothesis for the original graph G. On the left, the subgraph H is already homeomorphic to $K_{3,3}$, and in the center H is homeomorphic to K_5. On the right, simply delete the two horizontal edges to obtain a homeomorph of $K_{3,3}$, in which the vertices v, x, y form one half of the bipartition. □

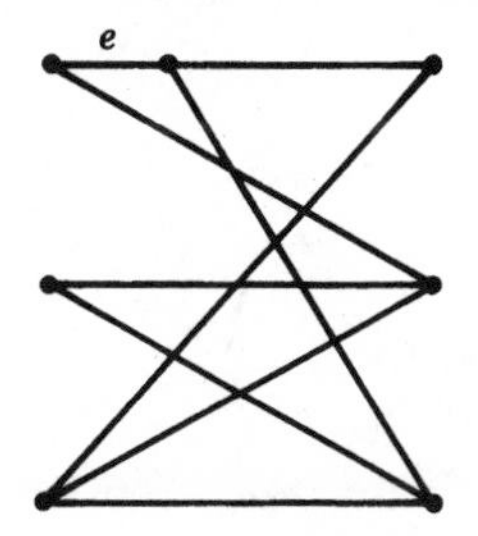

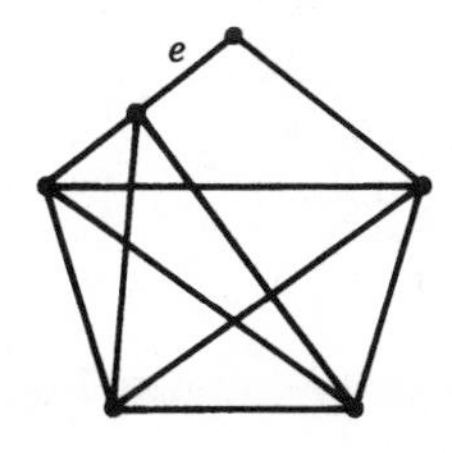

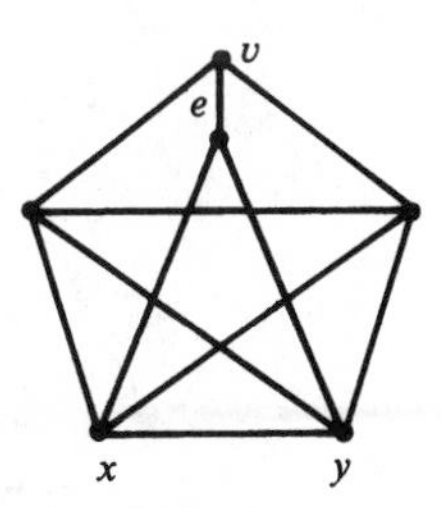

Figure 1.37. The subgraph H obtained by reversing the contraction of edge e.

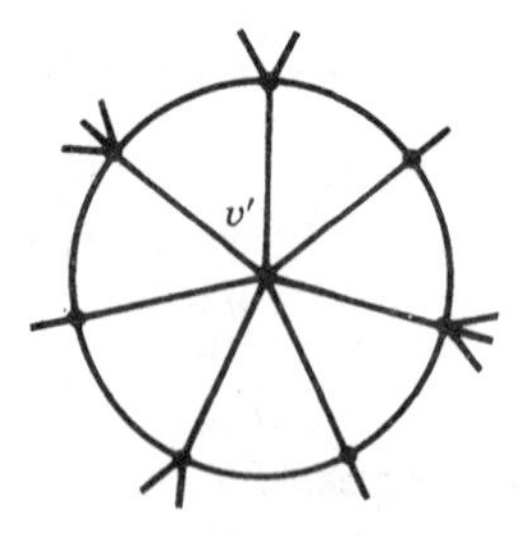

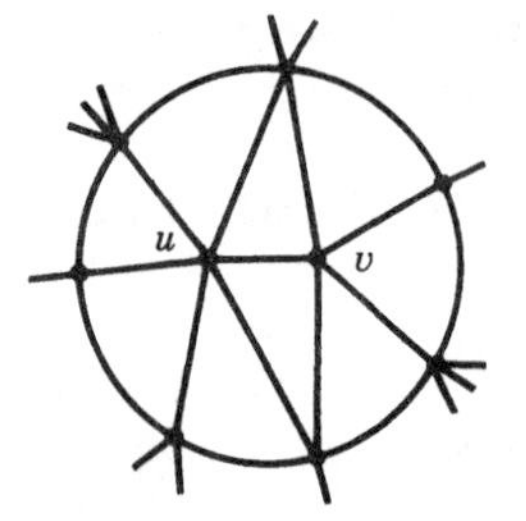

Figure 1.38. Splitting vertex v' back into vertices u and v.

Having established the lemma, we resume the proof of Kuratowski's theorem, now certain that the graph G' contains no homeomorph of K_5 or of $K_{3,3}$. By the induction hypothesis, we know that there is a planar imbedding of G'.

A key step in Thomassen's approach is assertion and proof that there is a planar imbedding for the contracted graph G' that, near the vertex v', looks like a cycle C with "spokes" to the vertex v'. (See the wheel on the left of Fig. 1.38.) We should like to split the vertex v' back into the original vertices u and v and attach them to the appropriate vertices on the cycle C, all the while maintaining a planar imbedding. (See the right of Fig. 1.38.)

We postpone the proof of this key assertion until our completion of the sketch. For the time being, assume that the key assertion has been proved.

There are three types of vertices on the cycle C: those adjacent to vertex u in the original graph G, those adjacent to vertex v, and those adjacent to both u and v. As Figure 1.38 indicates, we should like there to be at most two vertices adjacent to both u and v. If there were three such vertices, however, then G would contain the homeomorph of K_5 shown on the left of Figure 1.39 (note the five vertices of valence 4). We should also like all the vertices adjacent to u to occur contiguously around the cycle C, and the same for all the vertices adjacent to v. If strings of u-neighbors and v-neighbors alternated, however, then the graph G would contain the homeomorph of $K_{3,3}$ shown on the right of Figure 1.39, in which vertices s, t, u form one half of the bipartition and vertices v, x, y form the other half. Since by hypothesis, the

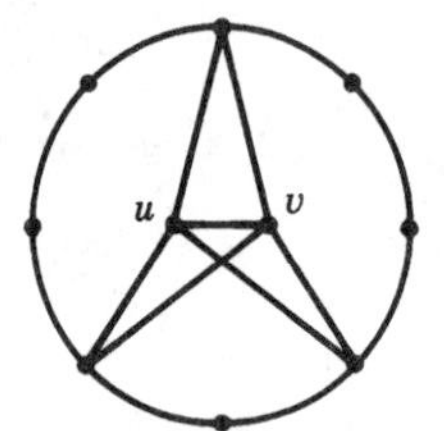

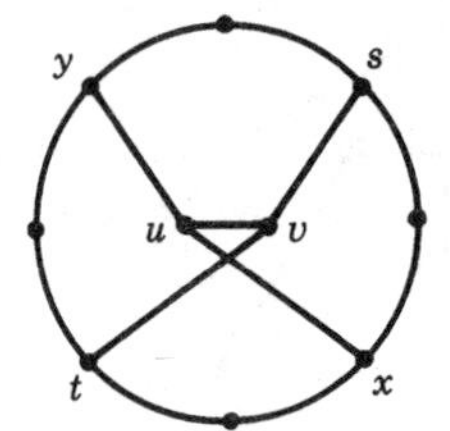

Figure 1.39. Obstacles at the vertex v'.

graph G contains no homeomorph of K_5 or of $K_{3,3}$, the desired planar imbedding for G can be constructed as in Figure 1.38. This concludes our sketch of the proof of Kuratowski's theorem. The proof will be entirely complete once we have verified the key assertion.

This proof is particularly insightful because the roles played by K_5 and $K_{3,3}$ are so clear. The two Kuratowski graphs are exactly the two obstructions to inverting an edge contraction.

The key assertion is a sticky point. Just how do we know that the imbedding of the contracted graph G' near v' looks like Figure 1.38? In order for this to be true, it would have to be the case that when the vertex v' and its incident edges are deleted from the imbedding of G', the resulting region has a cycle in G' as its boundary. Unfortunately, if the graph obtained from G' by deleting the vertex v' is not 2-connected, then the resulting region might very well not have a simple cycle as its boundary, in which case the whole proof collapses. Thus, to preserve the sketch, we shall need G' and hence G to be at least 3-connected. As a result, the 1-connected and 2-connected cases must be dealt with separately, and, what is more important, a delicate question about 3-connectivity and edge contraction must be considered. In what follows, we first investigate the relationship between connectivity, region boundaries, and edge contractions. Next we prove Kuratowski's theorem for 3-connected graphs. The 1-connected and 2-connected cases are treated last.

1.6.2. Connectivity and Region Boundaries

When picturing the regions of a planar imbedding, one tends to imagine "round" (i.e., convex) polygons whose boundaries are cycles of the imbedded graph, without any repeated vertices. Of course, the unbounded region is not convex, but one still pictures the boundary of this region as a simple cycle. This picture of region boundaries as simple cycles is not wholly accurate. For planar imbeddings, the problem is easily pinpointed by the following theorem.

Theorem 1.6.1. *Every region of a planar imbedding of a graph G has a simple cycle for its boundary if and only if G is 2-connected.*

Proof. Suppose that the boundary of some region r is not a cycle, but some other kind of closed walk, so that the vertex v occurs twice, as on the left of Figure 1.40. Then there is a simple closed path in the plane that leaves from vertex v between two edges of the boundary of r, stays the whole time within the region r, and later comes back to vertex v between a different pair of edges, as shown on the right-hand side of Figure 1.40. This closed path separates the plane into two pieces, both of which contain parts of the graph G. Since the path intersects the graph G only at vertex v, it follows that v is a cutpoint of the graph G. Therefore, G is not 2-connected.

Conversely, suppose that G has a cutpoint. Then G may be viewed as the amalgamation of two graphs H and K at the vertex v. In any imbedding of G,

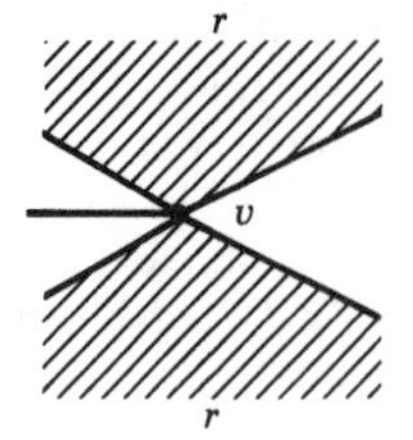

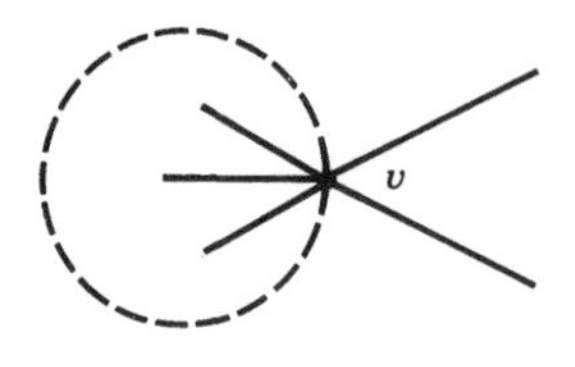

Figure 1.40. The region r (shaded) and vertex v.

on a traversal of a small circle centered around the vertex v, one must encounter somewhere an edge d of H (with endpoint v) followed immediately by an edge e of K (with endpoint v). Consider the region r with the corner at v between edges d and e. Since the edge d lies in subgraph H and since v is a cutpoint, no edges of subgraph K are encountered until the boundary returns to v. Since the edge e from K must occur somewhere on that boundary traversal, and since that closed walk must return to vertex v, we conclude that the boundary of region r is not a simple cycle. □

The situation for higher-genus surfaces is more complicated. Half of Theorem 1.6.1 generalizes; that is, if G has a cutpoint, then every imbedding of G in any surface has a region whose boundary is not a cycle, since the proof given above applies verbatim. On the other hand, it is easy to give an imbedding of a 2-connected graph in the torus having a region whose boundary is not a cycle (Exercise 1.6.1).

Surprisingly enough, it is not even known whether every 2-connected graph has an imbedding in some surface such that every face boundary is a cycle. The "double cover" conjecture that such an imbedding always exists may have had its origins in a paper by Tutte (1949) related to 1-factorizations of 3-regular graphs. It is mentioned at the end of a paper of Haggard (1980). Archdeacon (1984) proves the conjecture true for 4-connected graphs. One might guess that a minimal genus imbedding of a 2-connected graph should have every face boundary a cycle, but it is not hard to give a counterexample.

1.6.3. Edge Contraction and Connectivity

Suppose that the graph G' is obtained from the graph G by contracting the edge e with endpoints u and v to the single vertex v', as in the sketch of the proof of Kuratowski's theorem. If G is k-connected, then G' is at least $(k-1)$-connected, since any set of m vertices disconnecting G' can be turned into a set of at most $m+1$ vertices disconnecting G by splitting the vertex v' back into the vertices u and v. On the other hand, it is possible that G' is not k-connected. In fact, if the identified vertices u and v are in a set of k vertices that disconnects G, then the same set of vertices, with u and v replaced by v', forms a set of $k-1$ vertices that disconnects G'. Figure 1.41 illustrates an

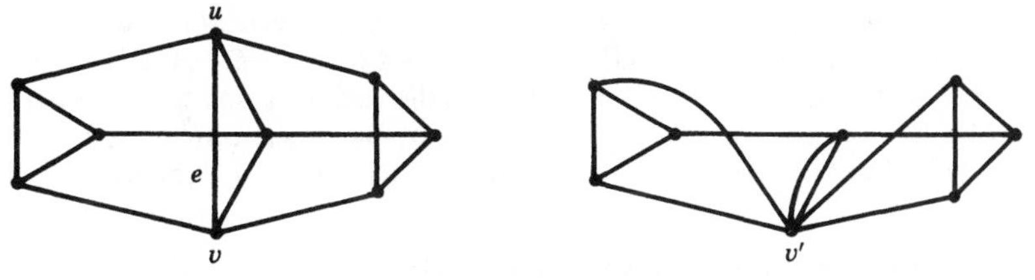

Figure 1.41. An edge contraction of a 3-connected graph to a 2-connected graph.

example in which the original graph G is 3-connected and the contracted graph is 2-connected.

In order for Thomassen's proof to work for 3-connected graphs, it is sufficient that the graph G' obtained by contracting an edge e of the original graph G be still 3-connected. The previous discussion shows that this might not be the case if the edge e were chosen at random. Thus the following theorem is an essential ingredient in Thomassen's proof.

Theorem 1.6.2. *Let G be a 3-connected graph with five or more vertices. Then there is some edge e of G such that the graph G/e obtained by contracting e is also 3-connected.*

Proof (Thomassen). Suppose, by way of contradiction that, for every edge e, the contracted graph G/e has a set of two vertices that disconnects it. One of those two vertices must be the vertex obtained by identifying the two endpoints of the edge e; otherwise, the same set of two vertices would also disconnect G, thereby contradicting the 3-connectivity of G. Thus for every edge e, the endpoints u and v of e together with some third vertex w disconnect G. Accordingly, let us choose an edge e and a vertex w such that the largest component H of the graph $G - \{u, v, w\}$ is the largest possible, for any disconnecting set consisting of three vertices, two of which are adjacent. Let x be a vertex adjacent to w such that x lies in a component of $G - \{u, v, w\}$ other than the maximum component H. Since vertices w and x are the endpoints of an edge of G, it follows that G has a disconnecting set of the form $\{w, x, y\}$.

We claim that some component of $G - \{w, x, y\}$ is larger than H, a contradiction. To see this, let H' be the subgraph of G induced by the vertices of H together with u and v. Since both u and v are adjacent to vertices of H (otherwise G would not be 3-connected), the subgraph H' is connected. On one hand, perhaps the vertex y is not in H'. Since w and x are not in H' either, it follows that H' is contained in a component of $G - \{w, x, y\}$, contradicting the maximality of H. On the other hand, perhaps y is in H'. If $H' - y$ is connected, then there is again a contradiction of the maximality of H, since $H' - y$ has one more vertex than H. If $H' - y$ were not connected, then one component of $H' - y$ would contain both the vertices u and v, since u is adjacent to v; and, therefore, all the other components of $H' - y$ are connected to the rest of the graph G through the vertices y and w. This would imply that $\{y, w\}$ disconnects G, contradicting the 3-connectivity of G. We conclude that for some edge e, the contracted graph G/e is 3-connected. □

1.6.4. Planarity Theorems for 3-Connected Graphs

We are now ready to prove Kuratowski's theorem for 3-connected graphs.

Theorem 1.6.3. *If G is a 3-connected graph containing no homeomorph of K_5 or $K_{3,3}$, then G is planar.*

Proof (Thomassen). The proof is by induction on the number of vertices of G just as in the sketch given at the beginning of this section. By Theorem 1.6.2, the contracted edge can be chosen so that the new graph G' is still 3-connected. This implies that the 3-connected part of the induction hypothesis holds for G'. It also means the graph obtained from G' by deleting the vertex v' is 2-connected. It follows from Theorem 1.6.1 that the resulting face containing vertex v' is bounded by a cycle, and hence the rest of the original proof does apply. The other part of the induction hypothesis, that G' contains no homeomorph of K_5 or $K_{3,3}$, holds because of the lemma given early in the sketch. Thus, Kuratowski's theorem holds for all 3-connected graphs. □

The proof of Kuratowski's theorem in the 3-connected case can be refined so as to retain geometric properties of imbeddings. Suppose that the bounded regions of the planar imbedding for the contracted graph G' are all convex and that the edges are all straight line segments. Then when vertex v' is split back into vertices u and v, if u and v are chosen close enough to v', the resulting imbedding for G can also be made to have convex regions and straight-line edges. Thus, with this minor adjustment, Thomassen's proof of Kuratowski's theorem also yields inductive proofs of the following theorems, in which the basis step for the induction is provided by the obvious planar imbedding of K_4.

Theorem 1.6.4 (Tutte, 1960). *Any planar 3-connected graph has a planar imbedding such that every bounded region is convex.*

Theorem 1.6.5 (Fary, 1948). *Any planar 3-connected graph has a planar imbedding such that every edge is a straight line segment.*

It is obvious from Theorem 1.6.1 that Tutte's theorem does not hold for 1-connected planar graphs. Moreover, it does not hold even for 2-connected planar graphs (Exercise 1.6.2). On the other hand, Fary's theorem holds for all planar graphs, as we shall show later in this section.

There is another observation worth making about the proof for Kuratowski's theorem in the 3-connected case. When the vertex v' is split back into vertices u and v, there really is no choice of how the resulting imbedding looks at u and v. Thus, there is essentially only one way to imbed G in the plane. By "essentially one way", we mean up to a homeomorphism of pairs. Further discussion of this concept is deferred to Chapter 3. Anyone familiar both with Thomassen's proof and that concept of uniqueness will recognize that Thomassen's proof of Kuratowski's theorem can be adapted to prove the following result of Whitney, by once again using an induction based on K_4.

Theorem 1.6.6 (Whitney, 1933). *There is only one way to imbed a 3-connected planar graph in the plane.*

Proof. Omitted. □

1.6.5. Graphs That Are Not 3-Connected

The following theorem is a basic tool for deriving results about planar graphs that are not 3-connected. Although it might appear intuitively obvious, the proof does have details that need to be checked.

Theorem 1.6.7. *Let H and K be planar graphs. Then the graph obtained by amalgamating H and K either at a single vertex v or along a single edge e is planar.*

Proof. If the graphs H and K are imbedded disjointly in the plane as in Figure 1.42, so that both copies of the vertex v or of the edge e lie on the "exterior" region, it would be easy to construct a planar imbedding of the amalgamation. For a vertex amalgamation, one simply pulls each copy of the vertex v out away from the rest of the graph H or K, along a path not touching the rest of its respective graph, while "dragging" along the edges incident on v. For an edge amalgamation, there is a preliminary step of topologically shrinking each copy of e until it is scarcely larger than a vertex, after which one does the same pulling away as with a vertex.

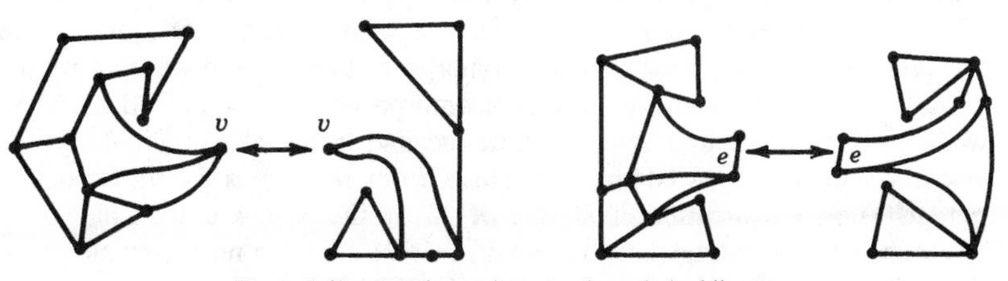

Figure 1.42. Amalgamating two planar imbeddings.

To get the respective copies of v or of e to the exterior region, imagine that H and K are first imbedded in separate copies of the 2-sphere, and select in each copy a region that contains v or e on its boundary. Next, delete the interior of a closed disk from both selected regions. Then paste the two resulting surfaces together by identifying the boundaries of the removed disks, thereby forming a new 2-sphere, in which the graphs H and K are imbedded so that both copies of v or e lie on the boundary of the same region. (A fully rigorous topological proof is readily formulated along this plan, with the aid of the Jordan curve theorem.) □

Theorem 1.6.7 now leads directly to a proof of Kuratowski's theorem for all graphs, 3-connected or not, starting from Theorem 1.6.3, the version for 3-connected graphs. However, we should like to prove Fary's theorem as well, and the proof of Theorem 1.6.7 wreaks havoc with straight line segments. Accordingly, we take a different tack, as follows. If a graph contains no homeomorph of K_5 or $K_{3,3}$, and it is not 3-connected, then we add as many edges to the graph as possible without creating K_5 or $K_{3,3}$. If the extra edges make the resulting graph 3-connected, then Kuratowski's theorem and Fary's theorem will hold for the original graph, because it is a subgraph of a 3-connected graph. The following theorem is what is needed.

Theorem 1.6.8. *Let G be a graph containing no homeomorph of K_5 or $K_{3,3}$ and such that the addition of any edge to G creates such a homeomorph. Then G is 3-connected.*

Proof. The proof is by induction on the number n of vertices of G. As a basic step, we observe that the theorem is true for $n = 5$, since $K_5 - e$ is 3-connected. Now assume that the theorem is true for all graphs with fewer than n vertices, where $n > 5$, and further assume that the graph G has n vertices.

Suppose, by way of contradiction, that G is only 2-connected, so that it is the amalgamation of graphs H and K at two vertices u and v. We assert that the vertices u and v are adjacent in the graph G. Suppose not. Then the graph H' obtained from H by adding an edge e from u to v still would not contain a homeomorph of K_5 or $K_{3,3}$, since any such homeomorph would be contained in G (with a path through K playing the role of edge e). Also, the addition of any edge to H' would create a homeograph of K_5 or $K_{3,3}$ in the original graph G; since that homeomorph cannot be disconnected by the removal of vertices u and v, that homeomorph must be contained in H' as well. Thus by induction, H' is 3-connected and hence planar. Similarly, the graph K' obtained by adding an edge e from u to v is 3-connected and planar. The amalgamation of H' and K' along the edge e is then planar by Theorem 1.6.7. In particular, that amalgamation contains no homeomorph of K_5 or $K_{3,3}$, by the arguments of Section 1.4. This contradicts the maximality of the original graph G, since the amalgamation of H' and K' along the edge e is simply G with the edge e added.

We therefore assume that G is the amalgamation of graphs H and K along the edge e and that H and K share the same maximality property as G with respect to homeomorphs of K_5 and $K_{3,3}$. By the induction hypothesis, the graphs H and K are 3-connected, and hence planar. By Theorem 1.6.7, the graph G is planar. It follows that in any planar imbedding of G, H, or K, there can be no region with four or more sides. Otherwise a "diagonal" edge across the region could be added, creating a planar graph with one more edge, contradicting the maximality of the given graph (there is a slight problem if the vertices we want to join by a diagonal are already joined by an edge elsewhere in the imbedding; see Exercises 3 and 4). Since the graphs in question have no loops or multiple edges, every region must have three sides. By Euler's equation it follows that $\#E = 3\#V - 6$ for the edge and vertex sets of any of the graphs G, H, and K (see Exercise 5). But since G is the amalgamation of H and K along a single edge, $\#V(G) = \#V(H) + \#V(K) - 2$ and $\#E(G) = \#E(H) + \#E(K) - 1$. Therefore

$$\begin{aligned}\#E(G) &= \#E(H) + \#E(K) - 1\\ &= 3\#V(H) + 3\#V(K) - 13\\ &= 3(\#V(G) + 2) - 13\\ &\neq 3\#V(G) - 6\end{aligned}$$

a contradiction. We conclude that G cannot be the amalgamation of two graphs H and K at two vertices. A similar but easier argument applies if G is 1-connected but not 2-connected. □

Corollary. *Kuratowski's theorem and Fary's theorem hold for graphs that are not 3-connected.*

1.6.6. Algorithms

The number of steps required by the Naive Sphericity (or Planarity) Algorithm suggested in Section 1.5 is an exponential function of the number n of vertices of a graph G. We shall see that Thomassen's design for a proof of the Kuratowski theorem leads to an algorithm whose execution time is a polynomial function of n, thereby a great improvement.

We first assume that the graph G to be tested is 3-connected. In that case, according to Theorem 1.6.2, there exists an edge e such that contraction of G on e yields a 3-connected graph. This step is iterated on that new 3-connected graph, and reiterated until a 4-vertex graph is obtained.

Execution time analysis: It takes $n - 4$ edge contractions to convert an n-vertex graph into a 4-vertex graph. Let us suppose that prior to each edge contraction, we find it necessary to make an exhaustive search over all edges before we encounter one such that contraction will preserve 3-connectedness. Since the original graph G has n vertices, its number of edges cannot exceed $(n^2 - n)/2$. Indeed if G has more than $3n - 6$ edges, then G is not planar (see Exercise 5).

Perhaps, for each edge e in a graph created by the contraction sequence, we shall need to do a preliminary contraction to see if the subsequent resulting graph would be 3-connected. The number of steps needed to construct the graph that would result from the contraction is, at worst, of the same order of magnitude as the size of the graph itself. Naively, one might think of removing every possible 2-vertex subset and checking whether the result is connected. However, the only 2-vertex subsets that must be checked are the ones that contain the new vertex to which the edge e was contracted. The time needed to check whether the result of removing such a 2-vertex subset is connected is at most linear in n, if you use a depth-first search to check connectedness.

Thus, the total number of steps needed to contract iteratively the original graph G down to a 4-vertex graph is, at worst, of the order of magnitude of the polynomial

$$(n - 4) * (3n - 6) * n * n$$

In other words, the number of steps needed for the edge-contracting part of the algorithm is at most quartic in the number of vertices.

Of course, since the complete graph K_4 is the only 3-connected 4-vertex graph, it must be the graph we ultimately obtain from the sequence of contractions. At this point, we imbed K_4 in the plane, and iteratively reverse each of the contractions, either until some reversal yields a copy of a Kuratowski graph, or until we have a planar imbedding of the original graph G in the plane.

At each reverse contraction, when a vertex v' is split back into two vertices u and v, the cycle C of vertices adjacent to v' is checked for the following patterns:

1. three vertices in cycle C adjacent to both u and v
2. four vertices in cycle C arranged so that when C is traversed, the ones adjacent to u alternate with the ones adjacent to v

Patterns 1 and 2 lead to homeomorphs of K_5 and $K_{3,3}$, respectively. If neither pattern is encountered, then splitting v back into u and v leads at once to a planar imbedding of the resulting graph.

Execution-time analysis: Even the most naive algorithm to check for patterns 1 and 2 requires at most quartic time in the number of vertices on cycle C.

Far more efficient algorithms for planarity exist. For instance, Demoucron et al. (1964) gave one of order n^2. Later, Hopcroft and Tarjan (1974), in one of the most famous early papers on computational complexity, reduced this to order n. Filotti et al. (1979) have found polynomial-time algorithms to determine if a given graph can be imbedded in a surface of given genus g. However, the degree of the polynomial is an increasing function of g, so it is still not known at present whether there is a polynomial-time algorithm for computing the genus of a given graph.

1.6.7. Kuratowski Graphs for Higher Genus

It is natural to ask whether there is an analog to Kuratowski's theorem for surfaces other than the sphere. Given a closed surface S, let $K(S)$ be the set of all graphs G such that G has no 2-valent vertices and such that G is not imbeddable in S but $G - e$ is, for any edge e. Thus $K(S)$ is the minimal set of obstructions to imbedding a graph in the surface S. Kuratowski's theorem asserts that $K(S_0) = \{K_5, K_{3,3}\}$. Must the obstruction set $K(S)$ be finite for an arbitrary, closed surface S?

Glover et al. (1979) and Archdeacon (1981) were the first to prove $K(S)$ finite for a surface other than the sphere, when they showed that $K(N_1)$ has 103 elements for the projective plane N_1. Unpublished work of Haggard, Decker, Glover, and Huneke recorded more than 300 members of $K(S_1)$. Archdeacon and Huneke (1985) showed that $K(N)$ contains only finitely many 3-regular graphs for any nonorientable surface N. Finally, Robertson and Seymour, in a series of papers deriving highly general results about families of graphs, proved that $K(S)$ is indeed finite for all surfaces. Although the size of $K(S)$ is in general enormous, there seems to be some structure. Glover has conjectured that every element of $K(S_g)$ is the union of at most $2g + 1$ copies of the original Kuratowski graphs K_5 and $K_{3,3}$.

1.6.8. Other Planarity Criteria

There are many other planarity criteria besides Kuratowski's. We give two more here that are due to mathematicians most famous for their work in topology and algebra, Whitney and MacLane. Although these criteria are not especially useful, they do involve elegant applications of duality and the cycle structure of a graph. Tutte's proofs of the theorems of Whitney and MacLane are given in Exercises 6–11.

Consider the dual graph G^* for a planar imbedding of the graph G. Every cycle C of the graph G separates the regions of the imbedding into two collections: those inside the cycle and those outside the cycle. In terms of the dual graph, the collection C^* of edges dual to those in cycle C, when deleted from the graph G^*, separate G^* into two disjoint parts such that every edge in C^* goes from one part to the other. In general, such a collection of edges is called a "bond". Neither of the two parts is required to be connected. It is a little harder to see, but still true that, conversely, every bond for G^* corresponds to a cycle in G. What is most surprising is that this relationship between a graph and one of its planar duals characterizes planarity.

Theorem 1.6.9 (Whitney, 1932). *A 2-connected graph G is planar if and only if there is a graph G^* whose edges are in one-to-one correspondence with those of G such that cycles in G correspond to bonds in G^*.*

Proof. See Exercises 11 and 12. □

MacLane's planarity condition depends on the cycle structure of a graph as well. Given a finite set X of size n, let $S(X)$ be the set of all formal sums $a_1x_1 + \cdots + a_nx_n$, where each a_i is 0 or 1. Add two formal sums by adding corresponding coefficients of two elements. It is easiest to think of each formal sum in $S(X)$ as a subset of X, namely, those elements of X with coefficient 1. If s corresponds to subset A and t to subset B, then $s + t$ corresponds to the symmetric difference $A \cup B - (A \cap B)$.

Given a graph G with edge set E, the "cycle space" of G, $C(G)$, is the subspace of $S(E)$ spanned by the cycles of G. Each element of $C(G)$ is either a cycle or a union of edge-disjoint cycles. If G is a 2-connected graph imbedded in the plane, then every region boundary is a cycle of G. Moreover, the outside (unbounded) region boundary cycle is the sum of all the inside region boundary cycles (all interior edges occur in two cycles and hence cancel them out in the sum). In fact, any cycle in the graph G is the sum of the region boundary cycles contained in the interior of the cycle. Thus the set of inside region boundary cycles form a basis for the cycle space $C(G)$. Moreover, this basis has the property that every edge of G is contained in at most two elements of the basis. In general, call a collection B of cycles in a graph H a "2-basis" if B is a basis for the cycle space of H and if every edge of H is contained in at most two elements of B. The discussion so far shows that every 2-connected planar graph has a 2-basis. MacLane proved that the converse is true as well:

Theorem 1.6.10 (MacLane, 1937). *A 2-connected graph is planar if and only if it has a 2-basis.*

Proof. See Exercises 6–10. □

1.6.9. Exercises

1. Give an imbedding of a 2-connected graph in the torus such that some region has a boundary that is not a cycle.
2. Find a planar 2-connected graph that has no planar imbedding with every bounded region convex.
3. Prove that in any planar imbedding of the complete graph K_4, every region is 3-sided. Conclude that if a simple graph G has a planar imbedding with four or more vertices all lying on the same region boundary, then an edge can be added to G (without creating multiple edges) and still maintain planarity.

4. Use Exercise 3 to prove that if G is planar and the addition of any edge to G creates a nonplanar graph, then every region in any planar imbedding of G has size 3. (*Caution*: The number of distinct vertices in a region boundary can be less than the number of sides of the region.)
5. Prove, using Euler's equation, that any planar imbedding for a connected simplicial graph with edge set E such that $\#E \geq 2$ and vertex set V satisfies $\#E \leq 3\#V - 6$, with equality if and only if every region is 3-sided.
6. Given a graph G with vertex set $V = \{v_1, \dots, v_n\}$ and edge set E, define the linear transformation $d: S(E) \to S(V)$ by $d(e) = u + v$, where u and v are the endpoints of the edge e. Show that the cycle space $C(E)$ is the kernel (nullspace) of d, and that if P is a path of edges from vertex u to vertex v, then $d(P) = u + v$. Show that if G is connected, then $v_1 + v_2, v_1 + v_3, \dots, v_1 + v_n$ are all in the range of d, but v_1 is not. Conclude that if G is connected, the range of d has dimension $n - 1$ and therefore the dimension of the cycle space $C(E)$ is $\#E - \#V + 1$.
7. Prove that $K_{3,3}$ has no 2-basis. (*Hint*: The dimension of $C(K_{3,3})$ is 4, by Exercise 6. Show that at least seven edges of $K_{3,3}$ must be each contained in two elements of a 2-basis for $K_{3,3}$ and then consider the sum of all elements of the 2-basis.)
8. Prove that K_5 has no 2-basis (see Exercise 7).
9. Prove that if the graph G has a 2-basis, then so does $G - e$ where e is any edge of G.
10. Use Exercises 7–9 and Kuratowski's theorem to prove MacLane's theorem.
11. Given a 2-connected graph G with edge set E, let the bond space $B(E)$ be the subspace of $S(E)$ spanned by the bonds of G. Let star(v) be the subgraph induced by collection of all edges incident to the vertex v. Prove that $\{E(\text{star}(V)) \mid v \in V\}$ is a spanning set for the bond space $B(E)$. [If $A \subset E$ and $G - A$ has two components H and K, consider $\Sigma\, E(\text{star}(V))$ where the sum is taken over all vertices of H.]
12. Use Exercise 11 and MacLane's theorem to prove Whitney's theorem.
13. Find an algorithm that determines in at most kn steps, k a constant, whether a given n-vertex graph G with at most $3n$ edges is connected. (Assume you are given for each vertex a list of the adjacent vertices.)

2

Voltage Graphs and Covering Spaces

There is an obvious problem in trying to represent an imbedding of an arbitrarily large graph in a surface, namely, to avoid inundating the reader in details. If all graphs deserved equal interest, then the prospects for obtaining a satisfactory representation would be dim. Fortunately, as a matter of nature and esthetics, the graphs of greatest interest have symmetries that can be exploited in developing an economical description.

One mathematical structure that permits an economical description of graphs and their imbeddings is a covering space, which is a topological generalization of a Riemann surface. As we demonstrate in this chapter, complete graphs, Cayley graphs, n-cube graphs, and many other kinds of graphs are derivable as covering spaces of graphs with only one vertex—that is, of bouquets of circles—or are derivable by deleting a 1-factor from a covering of such a bouquet.

The details for constructing a covering space of a graph are efficiently encoded in what is called a "voltage graph". In an electrical network, there is a voltage difference (perhaps zero) experienced in traversing any line. Although the mathematical voltages here are elements of a finite group, and not real numbers, the assignment of a voltage to every directed edge of a graph provides a clear analogy to electrical networks, with an important exception: two directed paths in parallel, that is, with the same initial and terminal vertices, need not be assigned the same mathematical voltage.

As we have stated previously, the central concern of topological graph theory is the placement of graphs on surfaces. Given the problem of placing a large graph on a surface, the value of the voltage graph construction is that it may enable one to reduce the problem to placing a quotient of the given graph in some kind of quotient of the given surface and assigning appropriate voltages. Since the quotient graph is a fraction of the size of the given graph, designing a suitable placement of the quotient might be easier. Thus, one may well wish to know which graphs have quotients (other than themselves) and how to reconstruct a graph from one of its quotients.

2.1. ORDINARY VOLTAGES

Let G be a graph whose edges have all been given plus and minus directions, and let $\mathscr{A}$ be a group, assumed to be finite unless explicitly stated to be infinite. A set function α from the plus-directed edges of G into the group $\mathscr{A}$ is called an "ordinary voltage assignment" on G, and the pair $\langle G, \alpha \rangle$ is called an "ordinary voltage graph". The values of α are called "voltages", and $\mathscr{A}$ is called the "voltage group". The ordinary voltage-graph construction was first suggested by Gross (1974) and immediately improved by Gross and Tucker (1974). Its advantage over various formalistic "covering graph" constructions, all essentially equivalent, is largely its visual suggestiveness.

2.1.1. Drawings of Voltage Graphs

Voltage graphs are usually given by pictures, rather than by combinatorial descriptions. For instance, suppose the graph G has vertex set $V = \{u, v\}$ and edge set $E = \{d, e\}$, with d a u-based loop and and e a proper edge from u to v. Suppose that voltages $\alpha(d) = 1$ and $\alpha(e) = 0$ are assigned in the cyclic group $\mathscr{Z}_3$. Then the voltage graph $\langle G, \alpha \rangle$ is presented as shown at the left of Figure 2.1.

The purpose of assigning voltages to the graph G, called the "base graph" or simply the "base", is to obtain an object called the "(right) derived graph", denoted G^α, which is illustrated at the right of Figure 2.1. The vertex set of G^α is the cartesian product $V_G \times \mathscr{A}$, and the edge set of G^α is the cartesian product $E_G \times \mathscr{A}$. Although the pair notations (v, a) and (e, a) are sometimes used for the vertices and edges of G, more frequently the subscripted notations v_a and e_a are used instead. If the directed edge e^+ of the base graph G runs from the vertex u to the vertex v, and if v is the voltage assigned to e^+, then

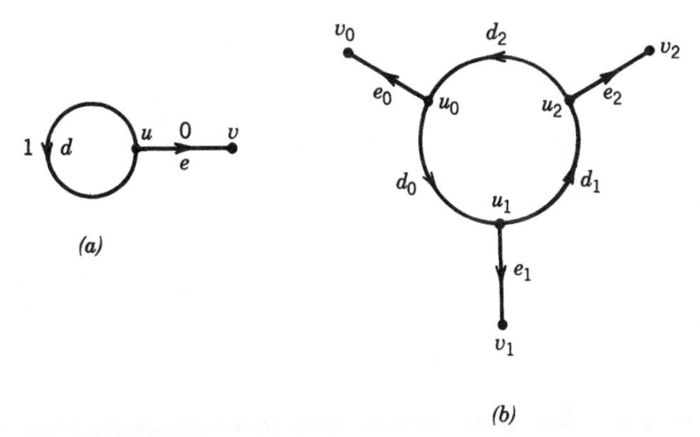

Figure 2.1. (a) A graph with voltages assigned in the cyclic group $\mathscr{Z}_3$ ($\langle G, \alpha \rangle$) and (b) the associated derived graph (G^α).

the directed edge e_a^+ of the derived graph G^α runs from the vertex u_a to the vertex v_{ab}.

Remark. *In the definition of the derived graph, note that whereas e_a^+ runs from u_a to v_{ab}, the reverse edge e_a^- runs from v_{ab} to u_a. In particular, the subscript on the minus-directed edge e_a^- is the product of the subscript on its initial vertex v_{ab} and the group element b^{-1} that is inverse of the voltage on e^+. Thus, whereas the subscript of a plus-directed edge of the derived graph always agrees with the subscript on its initial vertex, the subscript on a minus-directed edge always agrees with the subscript on its terminal vertex.*

Example 2.1.1. *Since the base graph in Figure 2.1 has two vertices, u and v, and two edges, d and e, and since the voltage group $\mathscr{Z}_3$ has three elements, 0, 1, and 2, it follows that the derived graph has six vertices, u_0, u_1, u_2, v_0, v_1, and v_2, and six edges, d_0, d_1, d_2, e_0, e_1, and e_2. Since the u-based loop d^+ of the base graph G is assigned voltage 1, it follows that for $i = 0, 1, 2$, the edge d_i^+ of the derived graph G runs from the vertex u_i to the vertex u_{i+1}. Since the edge e^+ is assigned the voltage 0, it follows that for $i = 0, 1, 2$, the edge e_i^+ of the derived graph runs from the vertex u_i to the vertex v_i.*

Example 2.1.2. *Figure 2.2 shows how the Petersen graph might be derived by an assignment of voltages in the cyclic group $\mathscr{Z}_5$ to a particular base graph with two vertices and three edges, which is sometimes called a "dumbbell" graph. As indicated, the directions on the edges of the derived graph are induced by the directions of the corresponding edges in the base graph.*

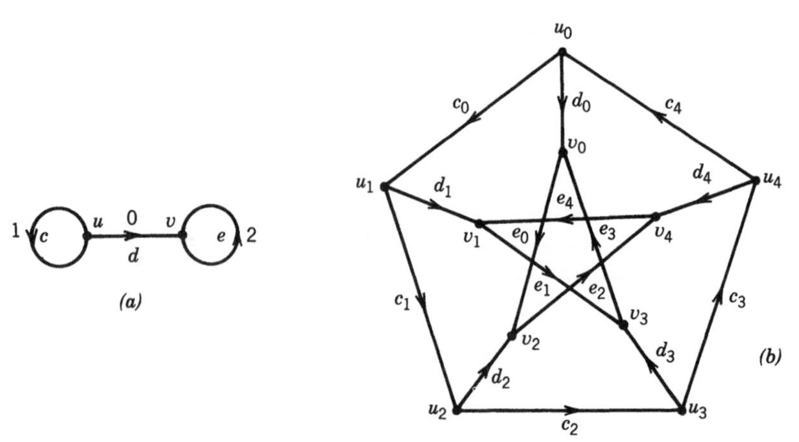

Figure 2.2. A derivation (b) of the Petersen graph from a voltage assignment in $\mathscr{Z}_5$ to a dumbbell graph (a).

Since the dumbbell graph in Figure 2.2 has two vertices, u and v, and since the voltage group is $\mathscr{Z}_5$, the vertices of the derived graph are u_0, u_1, u_2, u_3, v_4, v_0, v_1, v_2, v_3, and v_4. Since the dumbbell graph has edges c, d, and e, the derived graph has edges c_0, c_1, c_2, c_3, c_4, d_0, d_1, d_2, d_3, d_4, e_0, e_1, e_2, e_3, and e_4. The u-based loop c^+ has voltage 1 in the base graph. Thus, for $i \in \mathscr{Z}_5$, the edge c_i^+ runs from the vertex u_i to the vertex u_{i+1}, as shown on the outer pentagon of the derived Petersen graph. The v-based loop e^+ in the base graph has voltage 2. Thus, for $i \in \mathscr{Z}_5$, the edge e_i^+ runs from the vertex v_i to the vertex v_{i+2}, as shown on the inner star. The proper edge d^+ has initial vertex u and terminal vertex v in the dumbbell and is assigned voltage 0. Thus, for $i \in \mathscr{Z}_5$, the edge d_i^+ in the derived graph runs from the vertex u_i to the vertex v_i.

If one uses a different cyclic group $\mathscr{Z}_n$ as the voltage group and assigns voltages 1 and k to the loops c and e, then one obtains the generalized Petersen graph $G(n, k)$ (Frucht et al., 1971).

Example 2.1.3. *Figure 2.3 shows a 3-runged "Möbius ladder" as a base graph, with voltages in $\mathscr{Z}_2$. Since this base has six vertices and nine edges, and since the voltage group has order 2, the derived "circular ladder" $C_6 \times P_2$ has 12 vertices and 18 edges.*

To avoid cluttering the diagram, the names of the edges have been omitted from Figure 3.3. This does not affect the construction of the derived graph.

Quite often, the genus of the derived graph is higher than the genus of the base graph, as illustrated by Example 2.1.2. However, the genus of the derived graph need not be larger. Indeed, as Example 2.1.3 proves, it may be smaller than the genus of the base graph, for we recognize that the 3-runged Möbius ladder is isomorphic to the Kuratowski graph $K_{3,3}$, whereas the 6-runged circular ladder is visibly planar.

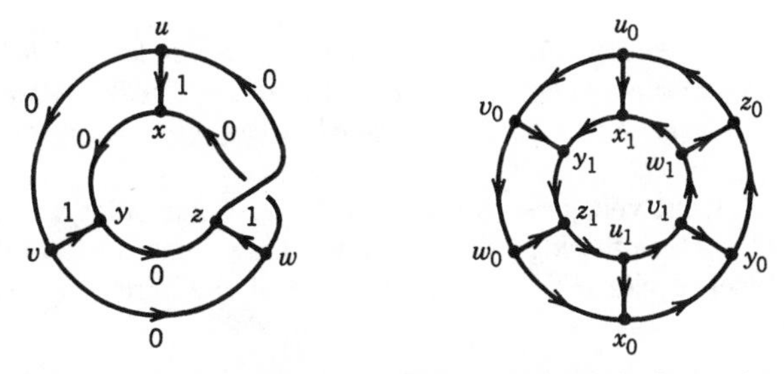

Figure 2.3. A circular ladder graph derived by a voltage assignment in $\mathscr{Z}_2$ to a 3-runged Möbius ladder.

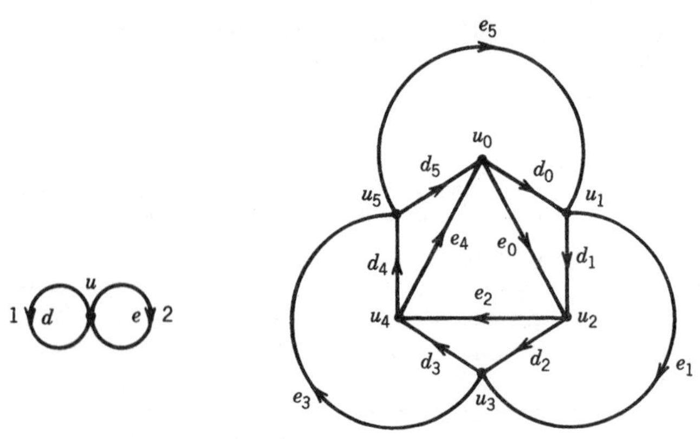

Figure 2.4. The 3-octahedron graph O_3 derived by assigning voltages in $\mathscr{Z}_6$ to the bouquet B_2.

2.1.2. Fibers and the Natural Projection

If voltages are assigned in a group $\mathscr{A}$ to a base graph G, then for every vertex v of G, the set of vertices v_a in the derived graph is called the "fiber" over v. Also, for every edge e of G, the set of edges e_a in the derived graph is called the "fiber" over e.

If e is a proper edge running from vertex u to vertex v in the graph G and if e^+ is assigned voltage b, then each edge e_a in the fiber over e runs from the vertex u_a in the fiber over the initial point u to the vertex v_{ab} in the fiber over the terminal point v. It follows from elementary group theory that the edge fiber over e matches the vertices in the vertex fiber over u one-to-one onto the vertices of the vertex fiber over v. Thus, the fiber over a proper edge is isomorphic to the disjoint union of $\#\mathscr{A}$ copies of K_2. For instance, in Example 2.1.2, the fiber over the proper edge d is the 1-factor between the outer pentagon and the inner star of the graph. However, as illustrated by Figure 2.4, the fiber over a loop always forms a set of cycles if the voltage group is finite.

Example 2.1.4. *The $\mathscr{Z}_6$-voltage* 1 *on the loop d of the base bouquet in Figure* 2.4 *has order* 6, *and the edge fiber over d forms one* 6-*cycle. The $\mathscr{Z}_6$-voltage* 2 *on the loop e of the bouquet has order* 3, *and the edge fiber over e forms two* 3-*cycles.*

In general, if the voltage b on a v-based loop e^+ has order n in the group $\mathscr{A}$, then each cycle in the edge fiber over e must have length n, and there must be $\#\mathscr{A}/n$ such cycles. For instance, starting at the vertex v_a there is the n-cycle

$$v_a, e_a^+, v_{ab}, e_{ab}^+, v_{ab^2}, e_{ab^2}^+, \ldots, v_{ab^{n-1}}, e_{ab^{n-1}}^+, v_{ab^n} = v_a$$

Since only the edge e_c^+ of the fiber over the loop e^+ starts at the vertex v_c of the fiber over the vertex v, only one of these n-cycles passes through v_c, which implies that these n-cycles are mutually disjoint.

The "natural projection" from the derived graph to the base graph is the graph map $p: G^\alpha \to G$ that maps every vertex in the fiber over v to the vertex v, for all $v \in V_G$, and every edge in the fiber over e to the edge e, for all $e \in E_G$. In terms of the notation, the natural projection wipes out the subscripts.

2.1.3. The Net Voltage on a Walk

The "voltage on a minus-directed edge" e^- is understood to be the group inverse of the voltage on its reverse edge e^+, that is, $\alpha(e^-) = (\alpha(e^+))^{-1}$. This enables us to define the "net voltage" on a walk

$$W = e_1^{\sigma_1}, \ldots, e_n^{\sigma_n} \quad (\text{each } \sigma_i = + \text{ or } -)$$

to be the product $\alpha(e_1^{\sigma_1}) \cdots \alpha(e_n^{\sigma_n})$ of the voltages on the edges of W in the order and direction of that walk. For instance, in Example 2.1.2, the net voltage on the walk e^+, d^-, c^-, d^+ is

$$\alpha(e^+) + \alpha(d^-) + \alpha(c^-) + \alpha(d^+) = 2 + 0 + (-1) + 0 = 1$$

A "lift" of a walk $W = e_1^{\sigma_1}, \ldots, e_n^{\sigma_n}$ in the base graph G is a walk

$$\tilde{W} = \tilde{e}_1^{\sigma_1}, \ldots, \tilde{e}_n^{\sigma_n}$$

in the derived graph G^α such that for $i = 1, \ldots, n$ the edge $\tilde{e}_i$ is in the fiber over the edge e_i.

Example 2.1.5. *In Figure 2.5, the voltage group is the cyclic group $\mathscr{Z}_3$. There are three lifts of the walk c^-, e^+ from the vertex u to the vertex v of the base graph, namely, the following walks of the derived graph:*

$$\begin{array}{ll} c_1^-, e_1^+ & \text{from } u_0 \text{ to } v_2 \\ c_2^-, e_2^+ & \text{from } u_1 \text{ to } v_0 \\ c_0^-, e_0^+ & \text{from } u_2 \text{ to } v_1 \end{array}$$

One important observation about Example 2.1.5 is that although the walk c^-, e^- of the base graph includes the same edges as the walk c^-, e^+ and although it has the same endpoints, its lifts are different. The lifts of the walk c^-, e^- are

$$c_1^-, e_0^- \quad c_2^-, e_1^- \quad \text{and} \quad c_0^-, e_2^-.$$

Continuity is the reason that the lift of the walk c^-, e^+ starting at the vertex

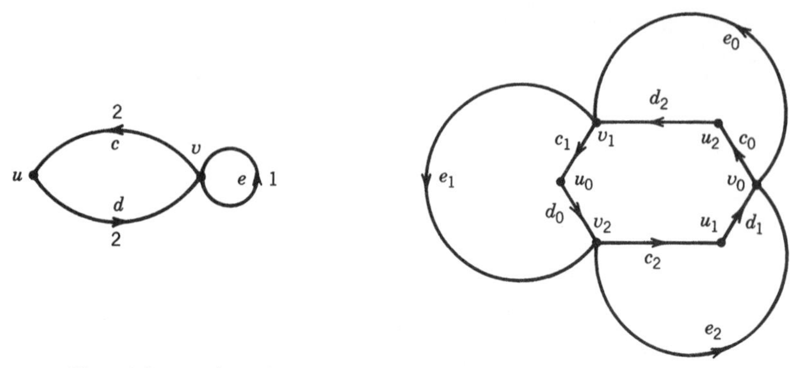

Figure 2.5. Another voltage graph and its derived graph. The voltage group is $\mathscr{Z}_3$.

u_i in the derived graph is the walk $c_{i+1}{}^-, e_{i+1}{}^+$, whereas the lift of c^-, e^- starting at u_i is $c_{i+1}{}^-, e_i{}^-$. Since the voltage on the directed edge c^- is 1, the next vertex on either lifted walk is v_{i+1}. When e^+ is the next edge of the base walk, the next lifted edge must be the unique edge in the fiber over e whose plus direction originates at v_{i+1} (i.e., e_{i+1}). When e^- is the next edge of the base walk, the next lifted edge must be the unique edge in the fiber over e whose minus direction originates at v_{i+1} (i.e., e_i). This is one reason that in topological graph theory, the direction in which one traverses a loop cannot be ignored.

2.1.4. Unique Walk Lifting

Another observation about Example 2.1.5 is that starting at each of the three vertices in the fiber over the initial vertex u of the base walk c^-, e^+, there is a unique lifted walk. That is, one lift starts at u_0, one at u_1, and the other at u_2. The following theorem indicates the generality of this observation.

Theorem 2.1.1. *Let W be a walk in an ordinary voltage graph $\langle G, \alpha \rangle$ such that the initial vertex of W is u. Then for each vertex u_a in the fiber over u, there is a unique lift of W that starts at u_a.*

Proof. Consider the first directed edge of the walk W, say e^+ or e^-. If it is e^+, then, since only one plus-directed edge of the fiber over e starts at the vertex u_a (i.e., the edge $e_a{}^+$), that edge must be the first edge in the lift of W starting at u_a. If it is e^-, then, since only one minus-directed edge of the fiber over e starts at u_a (i.e., the edge $e_{a\alpha(e^-)}{}^-$), it follows that the edge must be the first edge in the lift of W starting at u_a. Similarly, there is only one possible choice for a second edge of the lift of W, since the initial point of that second edge must be the terminal point of the first edge, and since that second edge of the lift must lie in the fiber over the second edge of the base walk W. This uniqueness holds, of course, for all the remaining edges as well. □

Notation. *According to Theorem* 2.1.1 *it makes sense to designate the lift of a walk W starting at the vertex u_a by W_a.*

Theorem 2.1.2. *Let W be a walk from u to v in an ordinary voltage graph $\langle G, \alpha \rangle$, and let b be the net voltage on W. Then the lift W_a starting at u_a terminates at the vertex v_{ab}.*

Proof. Let $b_1, \ldots, b_n$ be the successive voltages encountered on a traversal of the walk W. Then the subscripts of the successive vertices of the lift W_a are $a, ab_1, ab_1b_2, \ldots, ab_1 \ldots b_n = ab$. Since W_a terminates in the fiber over v, its final vertex is v_{ab}. □

Example 2.1.5 Revisited. *The net voltage on the walk $W = c^-, e^+$ from u to v is $(-2) + 1 = 1 + 1 = 2$. The walk W_0 $(= c_1{}^-, e_1{}^+)$ ends at $v_{0+2} = v_2$, the walk W_1 $(= c_2{}^-, e_2{}^+)$ ends at $v_{1+2} = v_0$, and the walk W_2 $(= c_0{}^-, e_0{}^-)$ ends at $v_{2+2} = v_1$.*

2.1.5. Preimages of Cycles

The "net voltage on a directed cycle C", based at a vertex u, is defined to be the product of the voltages encountered on a traversal of C, starting with the edge that originates at u and then proceeding in the preferred direction. If the group is abelian, then the net voltage on a directed cycle is independent of the choice of a base point, so we usually do not bother to designate one. If the group is nonabelian, then the different net voltages on a directed cycle corresponding to different choices of a base point are all conjugate group elements, so they all have the same order in the group. Moreover, the net voltage on a directed cycle has the same order as the net voltage on the reverse cycle, because those net voltages are group inverses of each other. Since the main concern is nearly always the order and not the net voltage itself, we are usually relieved of the burden of specifying either a basepoint or a direction on a cycle.

Theorem 2.1.3. *Let C be a k-cycle in the base space of an ordinary voltage graph $\langle G, \alpha \rangle$ such that the net voltage on C has order m in the voltage group $\mathscr{A}$. Then each component of the preimage $p^{-1}(C)$ is a km-cycle, and there are $\#\mathscr{A}/m$ such components.*

Proof. Let the cycle C be represented by a closed u-based walk W, let b be the net voltage on W, and let u_a be a vertex in the fiber over u. Then the component of $p^{-1}(C)$ containing u_a is formed by the edges in the walks $W_a, W_{ab}, \ldots, W_{ab^{m-1}}$, which attach end-to-end to form a km-cycle. For each of

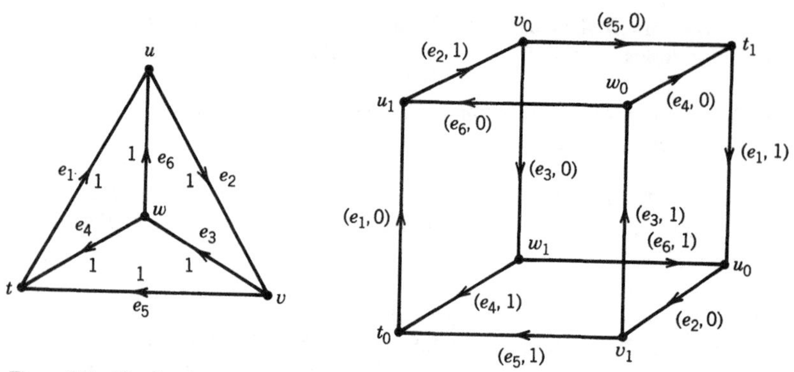

Figure 2.6. The 3-cube graph Q_3 derived by assigning $\mathscr{Z}_2$-voltages to the complete graph K_4.

the $\#\mathscr{A}/m$ left cosets of the cyclic group generated by the net voltage b, there is a unique component of $p^{-1}(C)$. □

Example 2.1.6. *Figure 2.6 shows how the 3-cube graph Q_3 may be derived using $\mathscr{Z}_2$-voltages on the complete graph K_4. One observes that the sum (i.e., the product operation for the group $\mathscr{Z}_2$) of the voltages on the 4-cycle e_1, e_2, e_3, e_4 is 0 modulo 2, which has order 1, and that the preimage of that 4-cycle is the union of the two 4-cycles $(e_1, 0), (e_2, 1), (e_3, 0), (e_4, 1)$ and $(e_1, 1), (e_2, 0), (e_3, 1), (e_4, 0)$. On the other hand, one notes that the 3-cycle e_1, e_2, e_5, in the base graph has voltage sum 1 modulo 2, which has order 2, and thus the preimage in the derived graph of that 3-cycle is the 6-cycle $(e_1, 0), (e_2, 1), (e_5, 0), (e_1, 1), (e_2, 0), (e_5, 1)$.*

Example 2.1.5 Revisited Again. *The net voltage on the 2-cycle c^+, d^+ in Figure 2.5 is 1 modulo 3. According to Theorem 2.1.3, the preimage of that 2-cycle should be one 6-cycle. In fact, it is the 6-cycle $c_0{}^+, d_2{}^+, c_1{}^+, d_0{}^+, c_2{}^+, d_1{}^+$. Also, the net voltage on the 1-cycle e^- is 2 modulo 3, and its preimage should be one 3-cycle. In fact, it is the 3-cycle $e_2{}^-, e_1{}^-, e_0{}^-$.*

2.1.6. Exercises

Figure 2.7 shows four voltage graphs, whose derived graphs are to be constructed in Exercises 1–4.

1. Draw the derived graph corresponding to the voltage graph in Figure 2.7*a*.
2. Draw the derived graph corresponding to the voltage graph in Figure 2.7*b*.
3. Draw the derived graph corresponding to the voltage graph in Figure 2.7*c*.

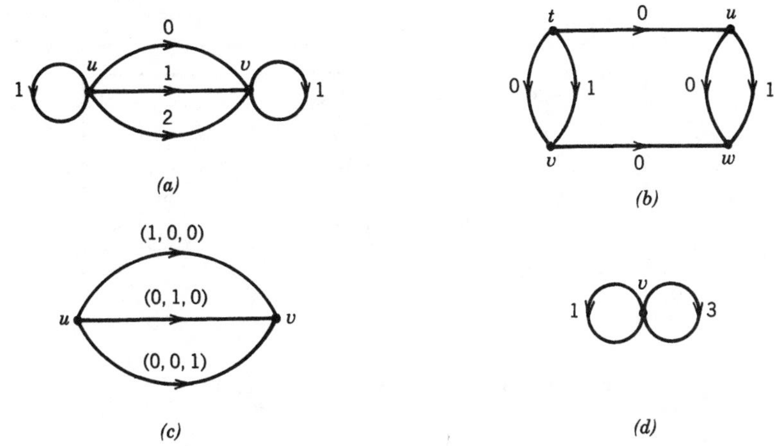

Figure 2.7. Four voltage graphs. (a) In $\mathscr{Z}_3$, (b) in $\mathscr{Z}_2$, (c) in $\mathscr{Z}_2 \times \mathscr{Z}_2 \times \mathscr{Z}_2$, (d) in $\mathscr{Z}_5$.

4. Draw the derived graph corresponding to the voltage graph in Figure 2.7*d*.
5. Draw copies of all the quotient graphs of the 3-cube graph Q_3.
6. Assign voltages to each of the graphs obtained in Exercise 5 so that the derived graph in each case is Q_3.
7. Draw copies of all the quotient graphs of the complete graph K_6.
8. Assign voltages to each of the graphs obtained in Exercise 7 so that the derived graph in each case is K_6.
9. Draw a representative of each of the isomorphism types of graph that can be derived by assigning voltages $\mathscr{Z}_3$ to the dumbbell graph of Figure 2.2.
10. What is the net voltage on the walk e_1^+, e_6^-, e_3^- in Figure 2.6?

Exercises 11–15 all pertain to the voltage graphs of Figure 2.8.

11. What is the net voltage on the walk e_2^+, e_3^-, e_6^- in Figure 2.8*a*? Construct the unique lift that begins at the vertex u_4.
12. How many components does the preimage of the cycle e_4^+, e_5^+, e_6^- in Figure 2.8*a* have? What is the length of each?
13. What is the net voltage on the walk $e_1^+, e_5^+, e_3^-, e_2^+$ in Figure 2.8*b*? Construct the unique lift that begins at the vertex $t_{(12)}$.
14. The preimage of the cycle e_3^+, e_4^+, e_2^- in Figure 2.8*b* has how many components? What is the length of each?
15. The preimage of the cycle e_1^+, e_5^+, e_3^- in Figure 2.8*b* has how many components? What is the length of each? Construct the one that contains the vertex $w_{(234)}$.

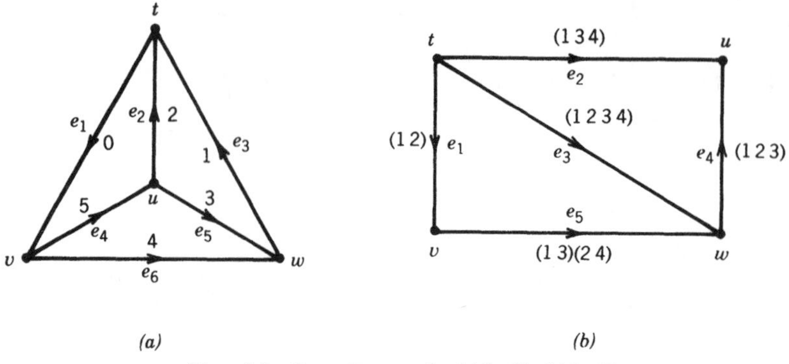

Figure 2.8. Two voltage graphs. (a) In $\mathscr{Z}_6$, (b) in $\mathscr{S}_4$.

2.2. WHICH GRAPHS ARE DERIVABLE WITH ORDINARY VOLTAGES?

We have seen in Section 2.1 that the Petersen graph, the octahedral graph O_3, the 3-cube graph Q_3, and several other graphs can all be obtained as the derived graph of an ordinary voltage-graph construction. The unifying characteristic of these derivable graphs is a certain kind of global symmetry, precisely described as the existence of a group of automorphisms that acts freely. Immediately below, we prove that a given graph is derivable with ordinary voltages from a proposed base graph if and only if the proposed base graph is a regular quotient of the given graph. This facilitates proof that some graphs, such as trees, are derivable only from themselves, whereas other graphs, such as Cayley graphs, are derivable from small base graphs.

2.2.1. The Natural Action of the Voltage Group

Let $\langle G, \alpha \rangle$ be a voltage graph. For every element c of the voltage group $\mathscr{A}$, let ϕ_c: $G^\alpha \to G^\alpha$ denote the graph automorphism defined by the rules $\phi_c(v_a) = v_{ca}$ on vertices and $\phi_c(e_a) = e_{ca}$ on edges. Obviously, ϕ_1: $G^\alpha \to G^\alpha$ is the identity automorphism, and no other ϕ_c fixes any vertex or edge. Moreover, it is a straightforward matter to verify that the composition $\phi_c \circ \phi_d$ is the same graph map as the automorphism ϕ_{cd}. Thus, the voltage group $\mathscr{A}$ acts freely on the left of the derived graph G^α, and this is called the "natural (left) action" of the voltage group.

Theorem 2.2.1. *The vertex orbits of the natural action of the voltage group on an ordinary derived graph are the vertex fibers, and the edge orbits are the edge fibers.*

Proof. For any vertex v_a of the derived graph G^α, the orbit of v_a is defined to be the set $\{\phi_c(v_a) \mid c \in A\}$, which is the same as the set $\{v_{ca} \mid c \in A\}$ =

$\{v_c \mid c \in A\}$. That is, the orbit of v_a is the fiber over v. Similarly, the orbit of the edge e_a of the derived graph is the fiber over the edge e. □

Corollary. *The natural projection $p: G^\alpha \to G$ of the derived graph onto the base graph is a covering projection.*

Proof. The composition of the quotient map $q_A: G^\alpha \to G^\alpha/A$ with the graph isomorphism $G^\alpha/A \to G$, which takes the fiber over a vertex or edge to that vertex or edge, is equal to the natural projection $p: G^\alpha \to G$. □

2.2.2. Fixed-Point Free Automorphisms

An automorphism of a graph G is called "fixed-point free" if it maps no vertex or edge of G to itself. Thus, a free group action on a graph may be identified with a subgroup of Aut(G) in which every automorphism except the identity is fixed-point free. In particular, every regular quotient arises from such a subgroup, and if the only such subgroup is trivial, then the only regular quotient is the given graph itself.

Example 2.2.1. *The only nontrivial automorphism of the n-path P_n fixes the central vertex if n is odd, or the central edge if n is even. Thus the n-path has no regular quotient except itself.*

Example 2.2.2. *The automorphism group of the complete graph K_4 is isomorphic to the symmetric group $\mathcal{S}_4$. The only fixed-point free automorphisms of K_4 are 4-cycles on both the vertices and the edges. For instance, the vertex map $(a\,b\,c\,d)$ corresponds to a fixed-point free automorphism. However, any subgroup containing such an automorphism would also contain its square, which fixes two edges. For instance, the square of $(a\,b\,c\,d)$ is $(a\,c)(b\,d)$, whose corresponding edge map would fix the edge between the vertices a and c and the edge between the vertices b and d. Thus, no nontrivial subgroup of* Aut(K_4) *acts freely on K_4, and accordingly, the complete graph K_4 has no regular quotients except itself.*

On the other hand, the subgroup of automorphisms on the Petersen graph in Figure 2.2 generated by the rotation of $2\pi/5$ radians acts freely, and the corresponding quotient is isomorphic to the dumbbell graph. One observes that the voltage group $\mathscr{Z}_5$ used to derive the Petersen graph from the dumbbell graph is isomorphic to this subgroup of automorphisms. Moreover, the free action of the voltage group on the derived graph is consistent with the free action of this subgroup. Theorem 2.2.2 is a generalization of these observations.

Theorem 2.2.2 (Gross and Tucker, 1974). *Let $\mathscr{A}$ be a group acting freely on a graph G, and let G be the resulting quotient graph. Then there is an assignment α*

of ordinary voltages in $\mathscr{A}$ to the quotient graph G and a labeling of the vertices of $\tilde{G}$ by the elements of $V_G \times \mathscr{A}$ such that $\tilde{G} = G^\alpha$ and that the given action of $\mathscr{A}$ on $\tilde{G}$ is the natural left action of $\mathscr{A}$ on G^α.

Proof. First, choose positive directions for the edges of the graphs G and $\tilde{G}$ so that the quotient map $q_{\mathscr{A}}: \tilde{G} \to G$ is direction-preserving and that the action of $\mathscr{A}$ on G preserves directions. Next, for each vertex v of G, label one vertex of the orbit $q_{\mathscr{A}}^{-1}(v)$ in $\tilde{G}$ as v_1, and for every group element $a \neq 1$, label the vertex $\phi_a(v_1)$ as v_a. Now one may label edges. If the edge e of G runs from v to w, one assigns the label e_a to the edge of the oribt $a_{\mathscr{A}}^{-1}(e)$ that originates at the vertex v_a. Since the group $\mathscr{A}$ acts freely on G, there are $\#\mathscr{A}$ edges in the orbit $q_{\mathscr{A}}^{-1}(e)$, one originating at each of the $\#\mathscr{A}$ vertices in the vertex orbit $q_{\mathscr{A}}^{-1}(v)$. Thus, the choice of an edge to be labeled e_a is unique. Finally, if the terminal vertex of the edge e_1 is w_b, one assigns voltage b to the edge e of the quotient graph G; that is, $\alpha(e^+) = b$. To show that this labeling of edges in the orbit $q_{\mathscr{A}}^{-1}(e)$ and the choice of a voltage b for the edge e really yields an isomorphism $\tilde{G} \to G^\alpha$, one must show for all $a \in \mathscr{A}$ that the edge e_a terminates at the vertex w_{ab}. However, since $e_a = \phi_a(e_1)$, the terminal vertex of the edge e_a must be the terminal vertex of the edge $\phi_a(e_1)$, which is

$$\phi_a(w_b) = \phi_a\phi_b(w_1) = \phi_{ab}(w_1) = w_{ab}$$

Under this labeling process, the isomorphism $\tilde{G} \to G^\alpha$ identifies orbits in $\tilde{G}$ with fibers of G^α. Moreover, it is defined precisely so that the action of an automorphism in A on $\tilde{G}$ is consistent with its natural action on the derived graph G^α. □

As a result of Theorem 2.2.2, one knows that a given graph can be derived from a smaller graph using ordinary voltages if and only if there is a nontrivial free action on the given graph. To decide whether such an action exists, one observes that the vertex orbits and the edge orbits under such a free action would have to be of the same size as the group. Equivalently, in a derived graph, all the vertex fibers and edge fibers have the same cardinality as the voltage group. Thus, the order of a possible voltage group to derive a given graph G must be a common divisor of $\#V_G$ and $\#E_G$. However, this common-divisor criterion is not always sufficient.

Example 2.2.3. *A tree cannot be derived from voltages on some smaller graph, because the number of vertices and the number of edges of a tree are relatively prime. A more difficult proof of the fact that trees have no nontrivial regular quotients might be obtained using Exercise* 15 *of Section* 1.3.

Example 2.2.4. *Since the 3-cube graph Q_3 has eight vertices and* 12 *edges, it might have one or more regular quotients with four vertices and six edges or with* *two vertices and three edges, but there are no other possibilities. See Exercise* 5 *of Section* 2.1.

Example 2.2.2 Revisited. *The complete graph* K_4 *has four vertices and six edges. Although the number* 2 *is a common divisor of* 4 *and* 6, *there is no quotient of* K_4 *with two vertices and three edges, because* K_4 *has no regular quotients itself, as proved in the original discussion of Example* 2.2.2.

Remark. *One additional fact that may be used to find the quotients of a graph is that all vertices in a fiber have the same valence, because the voltage group acts transitively on the set of neighborhoods of all the vertices in the fiber over any vertex of the base graph.*

2.2.3. Cayley Graphs Revisited

To counterbalance the possibly disappointing news that not all interesting graphs are derivable with ordinary voltages, we now consider the spectacularly encouraging example of Cayley graphs. Theorem 2.2.3 identifies Cayley graphs with regular coverings of bouquets B_n. This suggests that, in developing a strategy of solving a graph-placement problem by first seeking placements of its quotients, one might find Cayley graphs of special interest.

Theorem 2.2.3. *Let* $\mathcal{A}$ *be a group, and let* $X = \{x_1, \ldots, x_n\}$ *be a generating set for* $\mathcal{A}$. *Then the Cayley graph* $C(\mathcal{A}, X)^0$ *is a regular covering of the bouquet* B_n. *Conversely, if the graph* G *is a regular covering of the bouquet* B_n *and if* $\mathcal{A}$ *is the associated group that acts freely on* G, *then there is an assignment* α *of voltages in* $\mathcal{A}$ *to the bouquet* B_n *such that the voltages assigned generate* $\mathcal{A}$ *and that the derived graph* $B_n{}^\alpha$ *is isomorphic to* G.

Proof. Let v denote the vertex of the bouquet B_n and $e_1, \ldots, e_n$ the edges. Given the group $\mathcal{A}$ and the generating set X, assign the generator x_i as the voltage on the loop $e_i{}^+$, for $i = 1, \ldots, n$. Then the derived graph $B_n{}^\alpha$ is isomorphic to the Cayley color graph $C(\mathcal{A}, X)$ under the vertex map $v_a \to a$ and the edge map $(e_i, a) \to (x_i, a)$. The converse is a special case of Theorem 2.2.2. □

Corollary (Sabidussi, 1958). *A connected graph* G *is isomorphic to a Cayley graph* $C(\mathcal{A}, X)^0$ *if and only if some subgroup of* Aut(G) *acts freely on* G *and transitively on* V_G.

Proof. Let G be isomorphic to a Cayley graph $C(\mathcal{A}, X)^0$. Then the group of left translations on $C(\mathcal{A}, X)^0$ composes with the isomorphism $G \to C(\mathcal{A}, X)^0$ to act freely on G and transitively on V_G. Conversely, if some group of automorphisms acts freely on G and transitively on V_G, then V_G is the only vertex orbit and the resulting quotient is a bouquet of circles, from which it follows that G is isomorphic to a Cayley graph. □

Example 2.2.5. *The complete graph K_{2n+1} is isomorphic to the Cayley graph $C(\mathscr{Z}_{2n+1}, \{1,\ldots,n\})^0$. Equivalently, if one assigns the voltages $1,\ldots,n$ modulo $2n+1$ to the loops of the bouquet B_n, then the derived graph is isomorphic to K_{2n+1}.*

Example 2.2.6. *The octahedron graph O_n is isomorphic to the Cayley graph $C(\mathscr{Z}_{2n}, \{1,\ldots,n-1\})^0$. Equivalently, if one assigns the voltages $1,\ldots,$ $n-1$ modulo $2n$ to the loops of the bouquet B_{n-1}, then the derived graph is isomorphic to O_n.*

One might recall that the alternative Cayley graph $C(\mathscr{A}, X)^1$ is obtained from the Cayley graph $C(\mathscr{A}, X)^0$ by deleting a 1-factor for each generator of order 2.

Example 2.2.7. *The complete graph K_{2n} is isomorphic to the alternative Cayley graph $C(\mathscr{Z}_{2n}, \{1,\ldots,n\})^1$.*

Example 2.2.8. *The n-cube graph Q_n is isomorphic to the alternative Cayley graph $C(\mathscr{Z}_2{}^n, X)^1$, where X is the canonical set of generators $\{(1,0,\ldots,0),$ $(0,1,0,\ldots,0),\ldots,(0,\ldots,0,1)\}$.*

2.2.4. Automorphism Groups of Graphs

In general, a Cayley graph $C(\mathscr{A}, X)^0$ has additional automorphisms besides the group of left translations $\{\phi_a: a \in A\}$. For instance, the Cayley graph $C(\mathscr{Z}_3, \{1\})^0$ is a 3-cycle, and although the group of left translations includes all the rotations of that 3-cycle, the full automorphism group is the dihedral group $\mathscr{D}_3$, which also includes the reflections. Thus, it is not obvious which finite groups can serve as the automorphism groups of graphs.

A classical theorem of Frucht (1938) is that every finite group is isomorphic to the automorphism group of some graph. The method in Frucht's proof is to make a modification on each edge of a Cayley graph for a given group $\mathscr{A}$ so that the automorphisms of the modified graph are forced to respect the directions and the colors on the edges of the Cayley color graph $C(\mathscr{A}, X)$. For instance, Figure 2.9 shows how to do this for the group $\mathscr{Z}_3$.

All the edges of the Cayley color graph $C(\mathscr{Z}_3, \{1\})$ have the same color. An "arrow" is placed on each, in effect, by attaching a 2-path near the terminal vertex and a 3-path near the initial vertex. If there were a second color for some edges, those edges would be marked with a 4-path and a 5-path, so they would be distinguished from edges of the first color, and so forth.

Frucht's Theorem (1938). *Let $\mathscr{A}$ be a finite group. Then there is a group G such that* Aut(G) *is isomorphic to $\mathscr{A}$.*

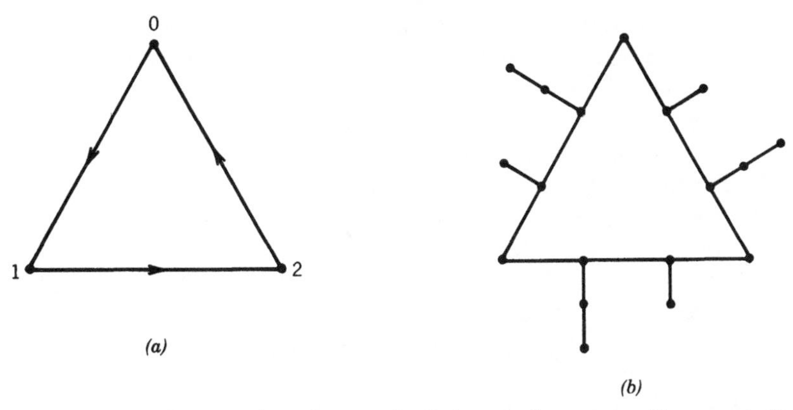

Figure 2.9. (a) A Cayley graph for the group $\mathscr{Z}_3$. (b) A graph whose automorphism group is $\mathscr{Z}_3$.

Proof. Let $X = \{x_1, \ldots, x_n\}$ be a generating set for the group $\mathscr{A}$. It is immediately clear that every left translation ϕ_a on the Cayley color graph $C(\mathscr{A}, X)$ is color-consistent and direction-preserving. It is first to be proved that every color-consistent, direction-preserving automorphism on $C(\mathscr{A}, X)$ is a left translation. For this purpose, suppose that the automorphism f is color-consistent and direction-preserving, and suppose that the image $f(1_{\mathscr{A}})$ of the identity vertex is the vertex b. It is to be proved that $f = \phi_b$.

Since the graph automorphism $\phi_b \circ f^{-1}$ on $C(\mathscr{A}, X)$ is color-consistent, direction-preserving, and identity-preserving, Theorem 1.2.4 implies that its vertex function coincides with a group homomorphism $g\colon A \to A$. Since $\phi_b \circ f^{-1}$ is color-preserving on edges, it follows that $g(x_i) = x_i$ for every generator $x_i \in X$. Therefore, g is the identity group automorphism, for which it follows that $f = \phi_b$.

The next step is to modify the graph $C(\mathscr{A}, X)^0$ into a graph G so that the features of color and direction from $C(\mathscr{A}, X)$ are somehow present in G. To do this, first insert two new vertices on each edge of $C(\mathscr{A}, X)^0$. Then for $i = 1, \ldots, n$ and for each edge in $C(\mathscr{A}, X)^0$ corresponding to color x_i, attach a $2i$-path at the new vertex near the terminal end and a $(2i + 1)$-path at the new vertex near the initial end. The automorphism group of the resulting graph G is isomorphic to the group of left translations on $C(\mathscr{A}, X)$, which is, of course, isomorphic to the group $\mathscr{A}$. □

2.2.5. Exercises

1. Prove that the "wheel" $K_1 + C_n$ has no nontrivial quotient.
2. Let G and H be graphs without cutpoints. Prove that the result of amalgamating G and H at a single vertex has no nontrivial regular quotients.

3. Prove that K_8 has no nontrivial quotients.
4. Prove that the complement of a tree with three or more vertices has no nontrivial quotients.
5. Prove that $K_{3,4}$ has no nontrivial quotients.
6. Prove that the valence of a vertex in the base graph equals the valence of any vertex in the fiber over it.
7. Assign voltages in $\mathscr{Z}_2$ to the Kuratowski graph K_5 so that the resulting derived graph is planar.
8. Assign voltages in $\mathscr{Z}_2$ to the Kuratowski graph $K_{3,3}$ so that the resulting derived graph is planar.
9. Prove that for every positive integer n, the regular bipartite graph $K_{n,n}$ is a Cayley graph.
10. Prove that if the graph G and H are both Cayley graphs, then their product $G \times H$ is also a Cayley graph.
11. Prove or disprove: The edge-complement of a Cayley graph is a Cayley graph.
12. Prove or disprove: A regular covering space of a Cayley graph is a Cayley graph.
13. Prove that the Petersen graph is not a Cayley graph. (*Hint*: There are only two groups of order 10.)
14. Construct a graph whose automorphism group is isomorphic to $\mathscr{Z}_2 \times \mathscr{Z}_2$.
15. Construct a graph whose automorphism group is isomorphic to $\mathscr{Z}_3 \times \mathscr{Z}_3$.
16. One observes that the graph G constructed in the proof of Frucht's theorem has many more vertices than the order of its automorphism group $\mathscr{A}$. Moreover, the group $\mathscr{A}$ does not act transitively on the vertices of the graph G. A graph H is called a "graphical regular representation" for the group $\mathscr{A}$ if $\#V_H = \#\mathscr{A}$ and if $\mathscr{A}$ acts transitively on V_H. For instance, the complete graph K_2 is a graphical regular representation for the cyclic group $\mathscr{Z}_2$. Prove that the dihedral group $\mathscr{D}_3$ has no graphical regular representation.

2.3. IRREGULAR COVERING GRAPHS

In topology, a continuous function $f: X \to Y$ is called a "covering projection" if every point of the topological space Y has a neighborhood N_y that is evenly covered, that is, such that every component of $f^{-1}(N_y)$ is mapped homeomorphically by f onto N_y. For compact spaces, such as the topological representations of finite graphs, it is equivalent to stipulate that the function f be a local homeomorphism onto the range graph. The covering projections one obtains this way that are not regular are called "irregular", of course. By a straightforward generalization of ordinary voltage graphs, we now obtain the complete combinatorial counterpart to topological covering projections, including all the irregular ones. A graph-theoretic motivation for making such a generalization lies in Schreier graphs and, more generally, in "relative" voltage-graph constructions.

2.3.1. Schreier Graphs

Let $\mathcal{B}$ be a subgroup of a group $\mathcal{A}$, and let $X = \{x_1, \ldots, x_n\}$ be a generating set for $\mathcal{A}$. The "(right) Schreier color graph" $S(\mathcal{A}: \mathcal{B}, X)$ has for vertices the right cosets of $\mathcal{B}$ in $\mathcal{A}$, the set of which is denoted, $\mathcal{A}: \mathcal{B}$. The edge $(x, \mathcal{B}a)$ has as its endpoints the vertices $\mathcal{B}a$ and $\mathcal{B}ax$, with its plus direction from $\mathcal{B}a$ to $\mathcal{B}ax$. The subscript $x_{\mathcal{B}a}$ is sometimes used for the edge $(x, \mathcal{B}a)$.

If the choice of plus directions and the edge colors are ignored in $S(\mathcal{A}: \mathcal{B}, X)$, the result is the "Schreier graph" $S(\mathcal{A}: \mathcal{B}, X)^0$, also called a "Schreier coset graph". Schreier graphs are an obvious generalization of Cayley graphs, in the sense that if $\mathcal{B} = \{1_{\mathcal{A}}\}$, then the Schreier color graph $S(\mathcal{A}: \mathcal{B}, X)$ is isomorphic to the Cayley color graph $C(\mathcal{A}, X)$. Moreover, it is easily verified (see Exercise 13) that if $\mathcal{B}$ is any normal subgroup of $\mathcal{A}$, then the Schreier graph $S(\mathcal{A}: \mathcal{B}, X)$ is isomorphic to the Cayley graph $C(\mathcal{A}/\mathcal{B}, X/\mathcal{B})$, where $\mathcal{A}/\mathcal{B}$ denotes the quotient group and $X/\mathcal{B}$ denotes the set of images of generators under the quotient homomorphism.

Example 2.3.1. *The main interest in Schreier graphs is for the case in which the subgroup is not normal. Consider the subgroup $\mathcal{B} = \{1, (2\,3)\}$ of the dihedral group $\mathcal{D}_3$, with generating set $X = \{(1\,2\,3), (1\,2)\}$. Then $S(\mathcal{D}_3: \mathcal{B}, X)$ has as vertices the right cosets*

$$\mathcal{B} = \{1, (2\,3)\}$$

$$\mathcal{B}(1\,2) = \{(1\,2), (1\,2\,3)\}$$

$$\mathcal{B}(1\,3) = \{(1\,3), (1\,3\,2)\}$$

and edges as illustrated in Figure 2.10.

It is obvious that if $\#X = q$, then the Schreier graph $S(\mathcal{A}: \mathcal{B}, X)^0$ is $2q$-regular. However, it is also obvious from Figure 2.10 that a Schreier graph need not be vertex-transitive. In particular, no automorphism can map the vertex $\mathcal{B}(1\,3)$ onto any other vertex, because $\mathcal{B}(1\,3)$ is the only vertex at which a loop is based.

The "alternative Schreier graph" $S(\mathcal{A}: \mathcal{B}, X)^1$ is the obvious generalization of the alternative Cayley graph. That is, if the generator x has order 2, then for every right coset $\mathcal{B}_a$, the two edges $x_{\mathcal{B}a}$ and $x_{\mathcal{B}ax}$ collapse into a single edge.

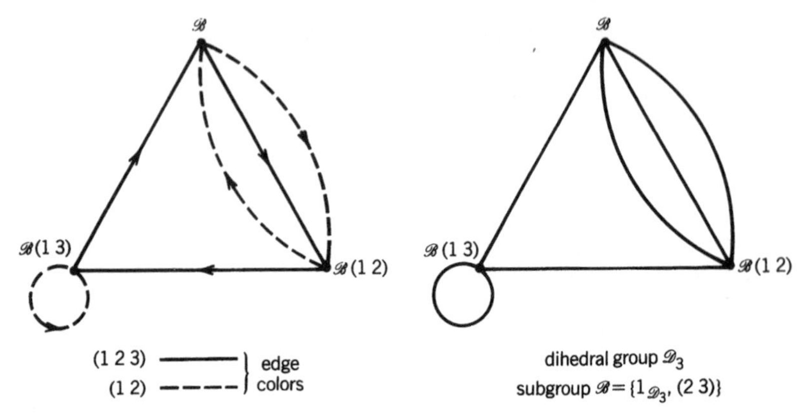

Figure 2.10. (left) The Schreier color graph for Example 2.3.1. (right) The associated Schreier graph.

2.3.2. Relative Voltages

Since the definition of a Schreier graph is so similar to that of a Cayley graph, one might think that a Schreier graph also covers a bouquet of circles. In the topological sense mentioned at the beginning of this section, it does; but since a Schreier graph need not be vertex-transitive, it follows that such a cover need not be regular. Thus, to construct all Schreier graphs from bouquets, we need something more general than ordinary voltage assignments. The more general object is called a "relative voltage assignment", and it was introduced by Gross and Tucker (1977).

Let G be a graph whose edges have all been given plus and minus directions, let $\mathscr{A}$ be a group, and let $\mathscr{B}$ be a subgroup. A "voltage assignment" for G in $\mathscr{A}$ "relative" to $\mathscr{B}$ is a set function α from the set of plus-directed edges of G to the group $\mathscr{A}$. The pair $\langle G, \alpha/\mathscr{B}\rangle$ is called a "relative voltage graph".

The "(right) derived graph" $G^{\alpha/\mathscr{B}}$ has as its vertex set the cartesian product $V_G \times (\mathscr{A}: \mathscr{B})$ and for its edge set the cartesian product $E_G \times (\mathscr{A}: \mathscr{B})$. If the directed edge e^+ of the base graph G runs from vertex u to vertex v, and if $\alpha(e^+) = c$ is the voltage assigned to e^+, then the directed $e_{\mathscr{B}a}{}^+$ runs from the vertex $u_{\mathscr{B}a}$ to the vertex $v_{\mathscr{B}ac}$.

Example 2.3.2. *Let $\mathscr{B}$ be the subgroup $\{1_{\mathscr{D}_3}, (2\,3)\}$ of the dihedral group $\mathscr{D}_3$. Figure 2.11 shows a voltage assignment on a dumbbell graph in the group $\mathscr{D}_3$ relative to the subgroup $\mathscr{B}$, and the resulting derived graph.*

Example 2.3.3. *Let $\mathscr{B}$ be a subgroup of a group $\mathscr{A}$, and let $X = \{x_1, \ldots, x_n\}$ be a generating set for $\mathscr{A}$. Assign directions to the edges of the bouquet B_n and*

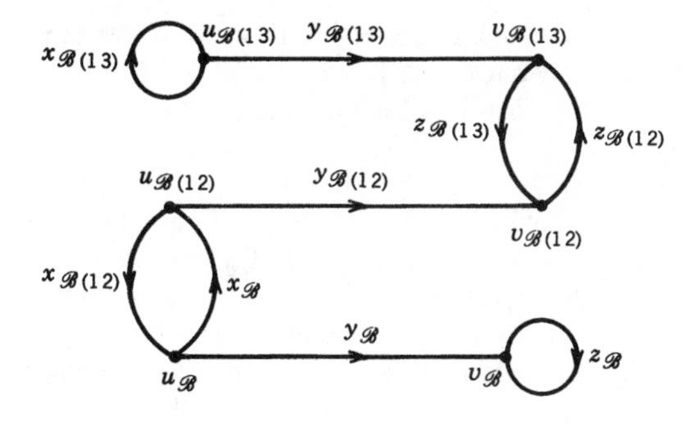

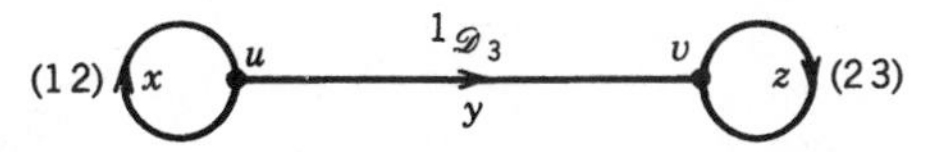

Figure 2.11. A relative voltage assignment on a dumbbell graph and the resulting derived graph.

assign relative voltages $x_1, \ldots, x_n$ *to the respective plus-directed edges. Then the resulting derived graph is isomorphic to the Schreier graph* $S(\mathscr{A}\colon \mathscr{B}, X)^0$.

Given a relative voltage graph $\langle G, \alpha/\mathscr{B} \rangle$ with voltages in the group $\mathscr{A}$, fibers in the relative derived graph are defined as for the ordinary voltage-graph construction. That is, for any vertex v of the base graph G, the "fiber" over v is the vertex set $\{v\} \times (\mathscr{A}\colon \mathscr{B})$ in the derived graph. Similarly, for any edge e of G, the "fiber" over e is the edge set $\{e\} \times (\mathscr{A}\colon \mathscr{B})$. The "natural projection" of the relative derived graph onto the base graph is the graph map that takes the fiber over every vertex onto that vertex and the fiber over every edge onto that edge. If the subgroup $\mathscr{B}$ is not normal, then the relative derived graph $G^{\alpha/\mathscr{B}}$ does not admit a natural fiber-consistent free group action, so that the natural projection is not a regular covering. However, it may be called an irregular covering, in a sense we can now make precise.

2.3.3. Combinatorial Coverings

Let $f\colon G \to H$ be a direction-preserving graph map, with H connected. Then f is called a "(combinatorial) covering projection" if for every vertex v of V_G the following conditions hold: f maps the set of edges originating at v one-to-one onto the set of edges originating at $f(v)$, and f maps the set of edges terminating at v one-to-one onto the set of edges terminating at $f(v)$. In the event that plus and minus edge-directions have not been chosen for the graphs G and H, we call f a "covering projection" if there exists a choice of edge directions for G and H such that f is direction-preserving and that the condition given above holds. It is a straightforward matter to verify that if that condition holds for any one choice of edge-directions for G and H such that f is direction-preserving, then it holds for all such choices of domain and range edge-directions.

A "topological representation" of a graph map $f: G \to H$ is a continuous function from a topological representation of G to a topological representation of H whose restriction to the interior of any edge of G is a homeomorphism and is consistent with the graph map f. Persons familiar with topology will recognize that a graph map f is a combinatorial covering projection if and only if a topological representation of f is a topological covering projection.

Example 2.3.4. *Let $p: G_\alpha \to G$ be the natural projection for an ordinary voltage graph $\langle G, \alpha \rangle$. If $e_1, \ldots, e_m$ are the edges originating at the vertex u of the base graph G, then for any element $\mathscr{A}$ of the voltage group, the edges originating at the vertex u_a of the fiber over u are $(e_1, a), \ldots, (e_m, a)$. Thus, the natural projection maps the set of edges originating at u_a bijectively to the set of edges originating at u. Moreover, if $d_1, \ldots, d_n$ are the edges terminating at the vertex v, and if $b_1, \ldots, b_n$ are the respective voltages on those edges, then $(d_1, ab_1{}^{-1}), \ldots, (d_n, ab_n{}^{-1})$ are the edges terminating at v_a in the derived graph. The natural projection $p: G^\alpha \to G$ also maps the edges terminating at v_a one-to-one onto the edges terminating at v. Thus, the natural projection p is a covering projection. Since, according to Theorem 2.2.2, every regular covering projection is equivalent to the natural projection of an ordinary derived graph onto its base graph, it follows that the topological representation of every regular covering projection of graphs is a covering projection, as one might expect from the terminology.*

Warning 2.3.1. *Naive attempts to simplify the definition of a covering projection lead to different constructions that fail to have some important properties of covering projections. For instance, as Example 2.3.5 illustrates, it is not sufficient to specify that the graph map f preserves valence, or even (see Exercise 15) that the inverse images of every vertex and edge of the range have the same cardinality.*

Example 2.3.5. *The graph map f illustrated in Figure 2.12 that is defined by the dropping of subscripts preserves valence, since all the vertices have valence 3. However, the two edges d_1 and d_2 originating at u_1 are both mapped onto the same edge d, originating at $f(u_1) = u$. Thus, f is not a covering projection, because its restriction to edges originating at u_1 is not one-to-one. Moreover, changing edge-directions on the graphs will not help at all.*

Example 2.3.6. *Let $p: G^{\alpha/\mathscr{B}} \to G$ be the natural projection for a relative derived graph. Then p is a covering projection, by the same reasoning as in Example 2.3.4.*

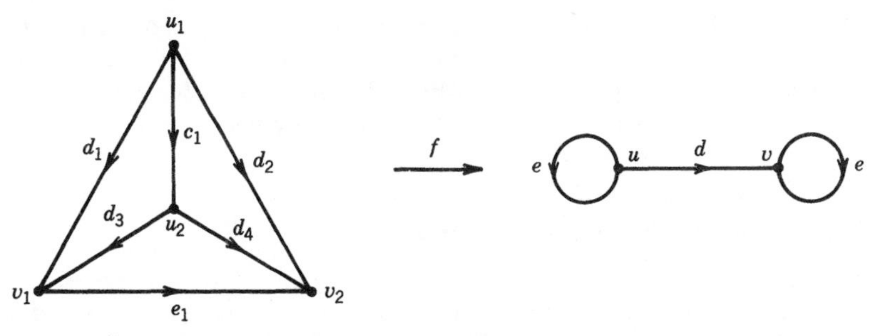

Figure 2.12. A valence-preserving graph map that is not a covering projection.

Example 2.3.3 Revisited. *As a special case of Example 2.3.6, observe that a Schreier graph $S(\mathscr{A}: \mathscr{B}, X)^0$ is a covering space of the bouquet B_n, where n is the cardinality of the generating set X.*

Example 2.3.7. *Let $\tilde{G}$ denote the edge-complement of the disjoint union of a 3-cycle and a 4-cycle. Then although $\tilde{G}$ is 4-regular, it is not vertex-transitive. Thus, the graph $\tilde{G}$ is not a Cayley graph, and it cannot be derived by assigning ordinary voltages to the bouquet of circles B_2. However, the graph map illustrated in Figure 2.13, which takes solid edges to solid edges and dashed edges to dashed edges and preserves directions, is a covering projection.*

The domain of a covering projection $f: G \to H$ of graphs is called the "covering graph" (or "covering space") and the range is called the "base graph" (or "base space"). Any covering projection that is not regular may be called "irregular", and its domain may be called an "irregular covering space" (or "irregular covering graph") of the base space (or base graph). For instance, Example 2.3.7 shows that the graph $(C_3 \cup C_4)^c$ is an irregular covering space

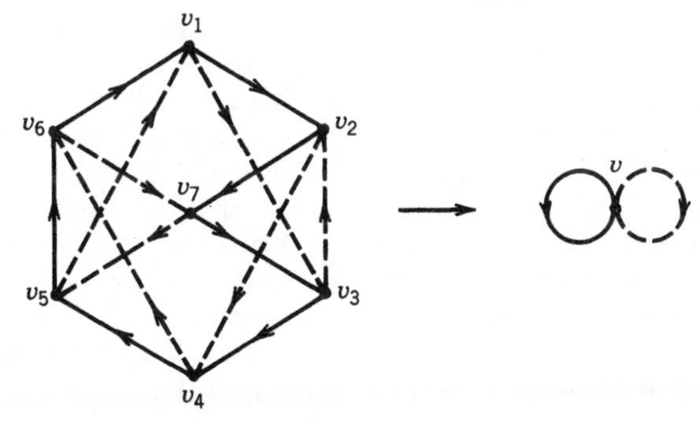

Figure 2.13. An irregular covering projection of the 4-regular graph $(C_3 \cup C_4)^c$ onto the bouquet B_2.

of the bouquet B_2. It is possible for the same graph to be both a regular covering space and an irregular covering space of a given base graph (by two different covering projections; see Exercise 2.4.8). However, the graph $(C_3 \cup C_4)^c$ cannot, under any graph map, be a regular covering of the bouquet B_2, because it is not vertex-transitive.

2.3.4. Most Regular Graphs Are Schreier Graphs

In examining Example 2.3.7 and the accompanying Figure 2.13, one might notice that the solid edges in the domain graph form a hamiltonian cycle and the dashed edges also form a hamiltonian cycle. With the solid-edge cycle we associate the permutation $\pi_1 = (1\,2\,7\,3\,4\,5\,6)$, corresponding to the order of the indices of vertices encountered as one traverses the hamiltonian cycle in its order of orientation. With the dashed-edge cycle we may in like fashion associate the permutation $\pi_2 = (1\,3\,2\,4\,6\,7\,5)$.

Now let $\mathscr{A}$ be the group of permutations generated by π_1 and π_2, and let $\mathscr{B}$ be the subgroup of $\mathscr{A}$ that fixes the symbol 1. For $i = 1, \ldots, 7$, let f_i be a permutation in $\mathscr{A}$ such that $f_i(1) = i$. For instance, it is possible to obtain f_i as a power either of π_1 or of π_2. Then, for $i = 1, \ldots, 7$, the right coset $\mathscr{B}f_i$ contains all the elements of $\mathscr{A}$ that permute 1 onto i. Rename the vertices of the graph in Figure 2.13 so that, for $i = 1, \ldots, 7$, the right coset $\mathscr{B}f_i$ replaces the label v_i. Also, every edge of the solid cycle is labeled π_1, and every edge of the dashed cycle is labeled π_2. Then the resulting graph is a Schreier graph $S(\mathscr{A}: \mathscr{B}, X)^0$, where $x = \{\pi_1, \pi_2\}$. This observation leads to the following result.

Theorem 2.3.1 (Gross, 1977). *Let G be a connected regular graph of even valence. Then G is isomorphic to a Schreier coset graph.*

Proof. Suppose that G is $2n$-regular and that its vertices are $v_1, \ldots, v_p$. By Theorem 1.5.2 (Petersen's theorem) the edges of G can be partitioned into n 2-factors, $L_1, \ldots, L_n$. Assign an arbitrary orientation to each cycle of each 2-factor (thereby inducing plus directions on all the edges of G). If a 2-factor L_i is a hamiltonian cycle, then we could associate a cyclic permutation with it exactly as in the preceding discussion of Figure 2.10. If the 2-factor L_i has several component cycles, $C_1, \ldots, C_m$, then with each of these edge cycles we may associate a cyclic permutation corresponding to the order of the indices encountered as one makes a traversal in the order of orientation, and the permutation π_i to be associated with the 2-factor L_i is the product of the disjoint cycle permutations corresponding to its component cycles.

Let $\mathscr{A}$ be the group generated by the permutations $\pi_1, \ldots, \pi_n$ corresponding to the respective 2-factors $L_1, \ldots, L_n$, and let $\mathscr{B}$ be the subgroup of $\mathscr{A}$ consisting of all elements of $\mathscr{A}$ that fix the symbol 1. Then the graph G is isomorphic to the Schreier graph $S(\mathscr{A}: \mathscr{B}, X)^0$, where $X = \{\pi_1, \ldots, \pi_n\}$. To construct the isomorphism, first assign each edge in the 2-factor L_i the color

π_i, for $i = 1, \ldots, n$. Then replace each vertex label v_j by the right coset of $\mathscr{B}$ that permutes 1 onto j, for $j = 1, \ldots, p$. The subgroup $\mathscr{B}$ has such cosets because G is connected. In particular, the product of the "signed colors" one traverses on an edge path from v_1 to v_j is a permutation f_j such that $f_i(1) = j$, so that $\mathscr{B}f_j$ is the right coset that permutes 1 onto j. □

Corollary. *Let B be a connected regular graph of odd valence such that G has a 1-factor. Then G is isomorphic to an alternative Schreier graph.*

Proof. The edge-complement in G of the 1-factor is a regular graph of even valence. Thus, by Theorem 1.5.2, the edge set of G can be partitioned into a 1-factor plus some 2-factors. By an argument similar to the proof of Theorem 2.3.1, it is possible to construct an isomorphism of G to an alternative Schreier graph, where the 1-factor is colored by a permutation of order 2 and the respective 2-factors are each colored by other permutations. □

Unfortunately, not all regular graphs of odd valence have a 1-factor (see Exercise 5). Biggs et al. (1976) give a historical account of 1-factors, including Tait's "theorem" that 2-connected, 3-valent graphs are 1-factorable and Petersen's counterexample, the graph that now bears his name. Akiyama and Kano (1985) survey more recent results on graph factorization. Tutte (1947) gives a necessary and sufficient condition for the existence of a 1-factor, from which he obtains the following theorem.

Theorem 2.3.2 (Tutte, 1947). *Let G be an $(n-1)$-connected regular graph of valence n with an even number of vertices. Then G has a 1-factor.*

Proof. Omitted. □

Corollary. *Let G be a $2r$-connected regular graph of valence $2r+1$. Then G is isomorphic to an alternative Schrier graph.*

Proof. From Theorem 1.1.1 (Euler's theorem on valence sum), it follows that G has an even number of vertices. From Theorem 2.3.2, we infer that G has a 1-factor. The conclusion then follows from the corollary to Theorem 2.3.1. □

2.3.5. Exercises

1. Let $\mathscr{B}$ be the subgroup generated by the element $(1\,2\,4)$ of the alternating group $\mathscr{A}_4$, and let X be the generating set $\{(1\,2\,3), (2\,3\,4)\}$. Draw the Schreier graph $S(\mathscr{A}_4 : \mathscr{B}, X)^0$.

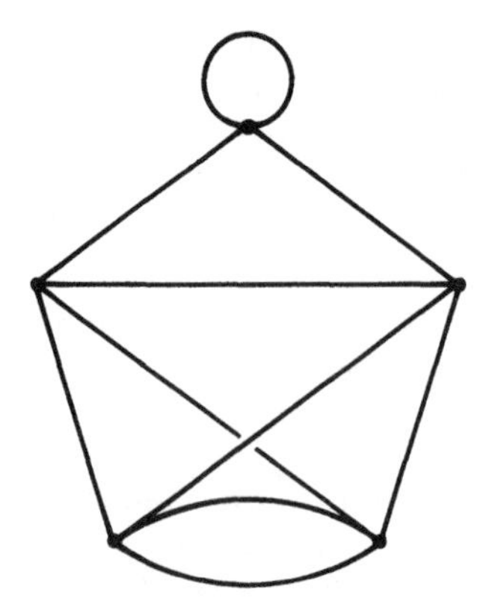

Figure 2.14. A 4-regular graph.

2. Let $\mathscr{B}$ be the cyclic subgroup generated by the element (1 2 3 4) of the symmetric group $\mathscr{S}_4$, and let X be the generating set $\{(1\,2\,4\,3), (1\,2)\}$. Draw the alternative Schreier graph $S(\mathscr{S}_4: \mathscr{B}, X)^1$.
3. Let $\mathscr{B}$ be the subgroup generated by the element $(1, 1, 1)$ of the group $\mathscr{Z}_2{}^3$, and let X be the canonical generating set $\{(1, 0, 0), (0, 1, 0), (0, 0, 1)\}$. Draw the alternative Schreier graph $S(\mathscr{Z}_2{}^3: \mathscr{B}, \mathrm{X})^1$.
4. Label the graph in Figure 2.14 as a Schreier coset graph.
5. Construct a 3-regular graph that has no 1-factor.
6. Prove that every regular bipartite graph of positive valence has a 1-factor (König, 1916). (*Hint*: Consider the result of doubling every edge.)
7. Use Exercise 6 to prove that every regular bipartite graph is 1-factorable.
8. Prove that every graph has a bipartite double covering space.
9. Use Exercises 6 and 8 to prove that every connected regular graph has a double covering that is a Schreier graph (or an alternative Schreier graph).
10. Prove that any two Schreier graphs of the same valence have a common covering space. (*Hint*: Consider the product group.)
11. Prove the theorem of Angluin and Gardiner (1980) that any two connected n-regular graphs have a common covering. (*Hint*: Use Exercises 9 and 10.)
12. Does every graph of maximum valence 3 or less have a planar covering space? From Example 2.1.3, we recall that the Kuratowski graph $K_{3,3}$ has a planar double covering.
13. Let $\mathscr{B}$ be a normal subgroup of a group $\mathscr{A}$ and X a generating set for $\mathscr{A}$. Prove that the Schreier graph $S(\mathscr{A}: \mathscr{B}, X)$ is isomorphic to the Cayley graph $C(\mathscr{A}/\mathscr{B}, X/\mathscr{B})$, where $X/\mathscr{B}$ denotes the set of images of generators under the quotient homomorphism.

14. Is every covering projection a local isomorphism? Is every local isomorphism a covering projection?
15. Give an example of a graph map $f: G \to H$, with G and H connected, such that the inverse image of every vertex or edge of H has the same number of elements but that f is not a covering projection.

2.4. PERMUTATION VOLTAGE GRAPHS

The relative-voltage-graph construction is a natural extension of the ordinary voltage-graph construction, but the subgroup makes it somewhat awkward to use. In this section we introduce another computational device, the permutation voltage graph, which is equivalent to a relative voltage graph but is more convenient to apply. The permutation voltage-graph construction is also due to Gross and Tucker (1977).

2.4.1. Constructing Covering Spaces with Permutations

Let G be a graph whose edges have all been assigned plus and minus directions. A "permutation voltage assignment" for G is a function α from the plus-directed edges of G into a symmetric group $\mathscr{S}_n$. The elements of the image G^{α} are called "permutation voltages", and the subscripted pair $\langle G, \alpha \rangle_n$ is called a "permutation voltage graph". The graph G is called the "base graph" or "base space".

To each such permutation voltage graph there is associated a "permutation derived graph" G^{α}, whose vertex set is the cartesian product $V_G \times \{1, \dots, n\}$ and whose edge set is the cartesian product $E_G \times \{1, \dots, n\}$. As for ordinary derived graphs, one uses either the pair notations, (v, i) and (e, i), or the subscripted notations, v_i and e_i, whichever alternative is more convenient in the context at hand. If the edge e of the base graph G runs from the vertex u to the vertex v, and if the voltage on e^+ is the permutation π, then for $i = 1, \dots, n$, the edge e_i of the derived graph G^{α} runs from the vertex u_i to the vertex $v_{\pi(i)}$.

The natural projection $p: G^{\alpha} \to G$ for permutation voltage graph $\langle G, \alpha \rangle_n$ is the graph map that takes any vertex v_i or edge e_i of the derived graph to the vertex v or edge e of the base graph. The set of vertices $\{v_i \mid i = 1, \dots, n\}$ is called the "fiber" over v and the set of edges $\{e_i \mid i = 1, \dots, n\}$ is called the "fiber" over e.

Warning 2.4.1. *When α assigns permutation voltages to a graph in a symmetric group $\mathscr{S}_n$, it is absolutely necessary to distinguish $\langle G, \alpha \rangle_n$ from the ordinary voltage graph $\langle G, \alpha \rangle$, because the respective derived graphs are quite different. Whereas the cardinality of each fiber in the ordinary derived graph would be $n!$, the order of the group $\mathscr{S}_n$, the cardinality of each fiber in the permutation derived graph would be n, the number of permuted objects.*

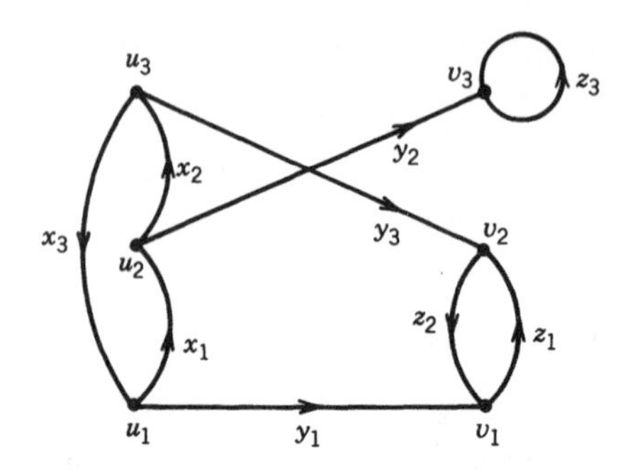

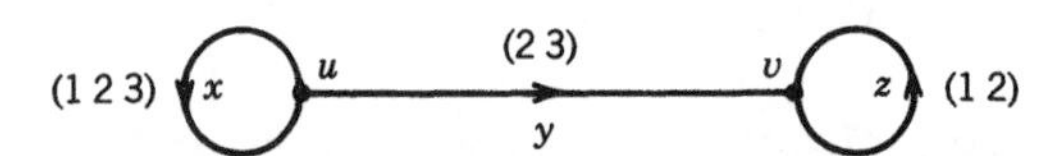

Figure 2.15. A permutation voltage assignment and the derived graph.

Example 2.4.1. *Figure* 2.15 *shows a permutation voltage assignment to the dumbbell graph in the symmetric group* $\mathcal{S}_3$ *and the associated permutation derived graph.*

2.4.2. Preimages of Walks and Cycles

The permutation voltage on a minus-directed edge e^- is understood to be the inverse permutation of the voltage on e^+. This enables us to define the "net permutation voltage on a walk" as the product of the voltages encountered in a traversal of that walk, exactly as for ordinary voltages. For instance, the net permutation voltage on the walk y^-, x^-, y^+ of Figure 2.15 is $(2\,3)(1\,3\,2)(2\,3) = (1\,2\,3)$. A lift of a walk W in the base graph G is a walk $\tilde{W}$ in the permutation derived graph such that the natural projection p maps the edges of $\tilde{W}$ onto the edges of W precisely in the order of traversal. It is implicit that direction is preserved on every edge of a lift, which is an important requirement if loops occur.

Theorem 2.4.1. *Let W be a walk in the base space of a permutation voltage graph $\langle G, \alpha\rangle_n$ such that the initial vertex of W is u. Then for each vertex u_i in the fiber over u, there is a unique lift of W that starts at u_i.*

Proof. This theorem is the analog of Theorem 2.1.1, and its proof is exactly analogous to the proof of that theorem. □

Notation. *According to Theorem 2.4.1, it is appropriate to designate the lift of a walk W starting at the vertex u_i as $\tilde{W}_i$.*

Example 2.4.1 Revisited. *The walk $W = y^-, x^-, y^+$ in the base graph of Figure 2.15 has three lifts, namely, the walks $\tilde{W}_1 = y_1{}^-, x_3{}^-, y_3{}^+$; $\tilde{W}_2 = y_3{}^-, x_2{}^-, y_2{}^+$; and $\tilde{W}_3 = y_2{}^-, x_1{}^-, y_1{}^+$.*

Theorem 2.4.2. *Let W be a walk in the base space of a permutation voltage graph $\langle G, \alpha \rangle_n$, and let π be the net voltage on W. Then the lift $\tilde{W}_i$ starting at the vertex u_i terminates at the vertex $v_{\pi(i)}$.*

Proof. Let $\pi_1, \ldots, \pi_r$ be the successive permutation voltages encountered on a traversal of the walk W. Then the subscripts of the successive vertices of the lift $\tilde{W}_i$ are

$$i, \quad \pi_1(i), \quad \pi_2(\pi_1(i)), \quad \ldots, \quad \pi_r(\ldots \pi_2(\pi_1(i))) = \pi(i)$$

Since $\tilde{W}_i$ terminates in the fiber over v, its final vertex is $v_{\pi(i)}$. □

The "net voltage on a permutation directed cycle" C, based at a vertex u, is defined to be the product of the voltages encountered on a traversal of C, starting with the edge that originates at u and then proceeding in the given direction. We recall from Theorem 2.1.3 that in an ordinary voltage graph, a k-cycle with net voltage of order m in the voltage group $\mathscr{A}$ lifts to $\#\mathscr{A}/m$ disjoint km-cycles. In a permutation voltage graph, although the preimage of a cycle in the base graph is still a union of disjoint cycles in the derived graph, the cycles in the preimage are not generally all of the same length. This is one of the most important differences between permutation voltage graphs and ordinary voltage graphs. In a permutation voltage graph, the cycle structure of the net permutation voltage on the base cycle is what determines the number and lengths of the component cycles of the preimage in the derived graph.

Let π be a permutation in the symmetric group $\mathscr{S}_n$. One recalls that any permutation has an essentially unique decomposition as a product of disjoint cyclic permutations. For $j = 1, \ldots, n$, let c_j denote the number of j-cycles in the decomposition of π. Then the n-tuple $(c_1, \ldots, c_n)$ is called the "cycle structure" of π.

Theorem 2.4.3. *Let C be a k-cycle in the base space of a permutation voltage graph $\langle G, \alpha \rangle_n$ with net voltage π, and let $(c_1, \ldots, c_n)$ be the cycle structure of π. Then the preimage of c in the derived graph G has $c_1 + \cdots + c_n$ components, including for $j = 1, \ldots, n$ exactly c_j kj-cycles.*

Proof. Choose any vertex u of the cycle C and regard C as a walk from u to itself. From Theorem 2.4.2, it follows that for $i = 1, \ldots, n$ the lift of C starting at u_i terminates at $u_{\pi(i)}$. Thus, the component of $p^{-1}(C)$ that contains u_i also contains $u_{\pi(i)}, u_{\pi^2(i)}, \ldots, u_{\pi^{j-1}(i)}$, where j is the length of the cyclic permutation containing the object i in the decomposition of π. Therefore, that component is a kj-cycle. The conclusion now follows readily. □

Example 2.4.2. *Suppose that C is a u-based k-cycle with net permutation voltage $(1\,4\,5)(2)(3\,9)(6\,7\,8)$ in $\mathcal{S}_9$. Then the preimage $p^{-1}(C)$ has four components, namely, a k-cycle containing the vertex u_2, a $2k$-cycle containing the vertices u_3 and u_9, and two $3k$-cycles, one containing the vertices u_1, u_4, and u_5, the other containing the vertices u_6, u_7, and u_8.*

2.4.3. Which Graphs Are Derivable by Permutation Voltages?

The next two theorems indicate that the set of connected graphs that can be obtained from a permutation voltage assignment of a given graph G is precisely the set of covering graphs of G.

Theorem 2.4.4. *The natural projection $p: G^\alpha \to G$ associated with any permutation voltage graph $\langle G, \alpha \rangle_n$ is a covering projection on each component of its domain.*

Proof. This is a straightforward exercise in applying the definitions of permutation derived graph and covering projection, so details are omitted. □

Theorem 2.4.5 (Gross and Tucker, 1977). *Let the graph map $q: \tilde{G} \to G$ be a covering projection. Then there is an assignment α of permutation voltages to the base graph G such that the derived graph G^α is isomorphic to $\tilde{G}$.*

Proof. Choose a spanning tree T in the base graph G, pick a root vertex u, and assign plus directions to the edges of T so that it is possible to travel from the root u to any other vertex of G entirely along plus-directed edges of T. Assign plus directions to the other edges of G arbitrarily. Then assign directions to the edges of G so that the map q is direction-preserving.

Suppose that the preimage $q^{-1}(u)$ has n vertices. Label them $u_1, \ldots, u_n$ arbitrarily. Then choose an edge e of T originating at u and terminating, say, at v. By the definition of covering projection, the preimage $q^{-1}(e)$ consists of n edges, one originating at each of the vertices $u_1, \ldots, u_n$. Moreover, exactly one of these terminates at each vertex of $q^{-1}(v)$, so that there are also n vertices in $q^{-1}(v)$. Label the edge of $q^{-1}(e)$ that originates at u_i as e_i, for $i = 1, \ldots, n$. Also, label the n vertices of $q^{-1}(v)$ as $v_1, \ldots, v_n$ in any way. If we match $q^{-1}(u)$ to $q^{-1}(v)$, the edges of $q^{-1}(e)$ define a permutation π in the symmetric group $\mathcal{S}_n$. That is, if u_i is matched to v_j then $\pi(i) = j$. The permutation voltage π is assigned to the edge e of the base graph G. (If the vertices $v_1, \ldots, v_n$ are labeled so that for $i = 1, \ldots, n$, the edge e_i terminates at v_i, then the voltage assigned to the edge e is the identity permutation.)

Continue this procedure until all edges of T have been assigned permutation voltages, always selecting as the next edge of T one whose initial point lies on a path from the root vertex u whose every edge already has a permutation voltage assigned. (By the parenthetic remark at the end of the preceding paragraph, it is possible to assign the identity permutation to every edge of the maximal tree T.) The last step is to assign permutation voltages to the edges of G not in T. Suppose that d is such an edge, running from a vertex s to a vertex t. Then $q^{-1}(d)$ matches $s_1, \ldots, s_n$ to $t_1, \ldots, t_n$, so we may assign names $d_1, \ldots, d_n$ to the edges in $q^{-1}(d)$ in agreement with the subscripts on their initial points, and we may assign a permutation voltage to the edge d according to the matching. □

Corollary 1. *Let the graph map $p: \tilde{G} \to G$ be a covering projection. Let u and v be any two vertices of the base graph G and e any edge of G, not necessarily incident either on u or on v. Then $\#p^{-1}(u) = \#p^{-1}(v) = \#p^{-1}(e)$. That is, there is a number n such that all vertex fibers and all edge fibers have cardinality n.* □

Corollary 2. *Let the graph map $p: \tilde{G} \to G$ be a covering projection on every component of its domain, and let the inverse images of all vertices of G have the same cardinality n. Then there is a permutation voltage assignment α in $\mathscr{S}_n$ such that the derived graph G^α is isomorphic to $\tilde{G}$.* □

2.4.4. Identifying Relative Voltages with Permutation Voltages

Since every component of a relative derived voltage graph covers the base graph by the natural projection, it follows from Theorem 2.4.5 that it could also be derived by permutation voltages. An explicit way to convert a relative voltage graph $\langle G, \alpha/\mathscr{B}\rangle$ to a permutation voltage graph begins with the observation that the voltage group $\mathscr{A}$ acts as a permutation group on the right cosets of $\mathscr{B}$. Thus, if $n = \#(\mathscr{A}: \mathscr{B})$, then the permutation action of $\mathscr{A}$ on $\mathscr{A}: \mathscr{B}$ imbeds in the symmetric group action on $\{1, \ldots, n\}$. As a special case, we observe that an ordinary voltage graph converts to a permutation voltage graph because the action of a group on its own elements imbeds in a symmetric group action.

Conversely, if $\langle G, \alpha\rangle_n$ is a permutation voltage graph, then let $\mathscr{A} = \mathscr{S}_n$, and let $\mathscr{B}$ be the subgroup that fixes the symbol 1. The right cosets

$$\mathscr{B}, \quad \mathscr{B}(1\,2), \quad \mathscr{B}(1\,3), \quad \ldots, \quad \mathscr{B}(1\,n)$$

behave under right multiplication by an element of π of $\mathscr{A}$ exactly as the symbols $1, 2, 3, \ldots, n$ behave under the permutation action of π. Thus, the relative voltage graph $\langle G, \alpha/\mathscr{B}\rangle$ is equivalent to the permutation voltage graph $\langle G, \alpha\rangle$.

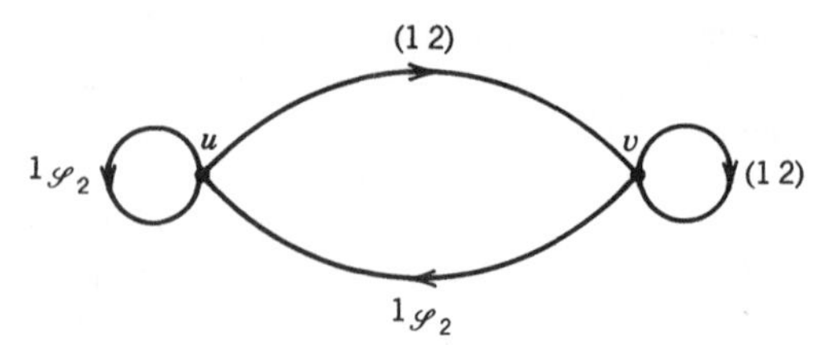

Figure 2.16. A graph with permutation voltages in $\mathscr{S}_2$.

2.4.5. Exercises

1. Construct the derived graph corresponding to the permutation voltage graph in Figure 2.16.
2. Prove that the composition of two covering projections is a covering projection.
3. Prove that the composition of two regular covering projections can sometimes be an irregular covering projection. (*Hint*: Use Exercise 1.)
4. Let G be a graph obtained from the Petersen graph by doubling the 1-factor that extends from the star to the pentagon. (That is, for each edge e in the 1-factor, attach another edge e' with the same endpoints.) Assign permutation voltages in $\mathscr{S}_{10}$ to the bouquet B_2 so that the derived graph is isomorphic to G.
5. Use Exercise 4 to prove that the Petersen graph is an alternative Schreier graph, by explicitly exhibiting the group, the subgroup, and the generating set.
6. Prove that any two graphs that cover the same graph have a common covering space.
7. Let C be a hamiltonian cycle in the base space of a permutation voltage graph. Under what circumstance is the preimage of C (under the natural projection) a hamiltonian cycle in the derived graph?
8. Show that there exists both a regular covering projection and an irregular covering projection of the complete graph K_7 onto the bouquet B_3.
9. Let F be an n-factor in the base space of a covering projection $p: \tilde{G} \to G$. Prove that the preimage $p^{-1}(F)$ is an n-factor in the covering space.
10. Prove or disprove: If the base graph of a (finite) covering projection is n-connected, then so is each component of the covering graph.
11. Prove that any 2-sheeted covering projection is a regular covering projection.

2.5. SUBGROUPS OF THE VOLTAGE GROUP

To decide whether two vertices in an ordinary derived graph are in the same component or to count the number of components, one considers a form of isotropy group called a local voltage group, which is a subgroup of the voltage group itself. The considerations involved here also lead to the proof of Babai's theorem, that given a subgroup $\mathscr{B}$ of a group $\mathscr{A}$ and a Cayley graph G for $\mathscr{A}$, there is a Cayley graph H for $\mathscr{B}$ such that G contracts onto H.

2.5.1. The Fundamental Semigroup of Closed Walks

Let W and W' be walks in a graph G such that the initial vertex of W' is the terminal vertex of W. Then the product walk WW' is constructed by extending the sequence of edges in W by the sequence of edges in W'. If W and W' are both u-based closed walks, then so is WW'. Under this product operation, the set of all u-based closed walks forms a semigroup, called the "fundamental semigroup of u-based closed walks".

If u is any vertex of the base graph of either an ordinary voltage graph or a permutation voltage graph, then the set of net voltages occurring on u-based closed walks forms a subgroup of the voltage group, as originally observed (in a current group formulation—see Chapter 4) by Alpert and Gross (1976, p. 297). Following Stahl and White (1976), who first reinterpreted it for voltage assignments, we adopt the name "local (voltage) group" at u, and we denote it $\mathscr{A}(u)$.

Example 2.5.1. *Since every u-based closed walk in the base graph of Figure 2.17 traverses the edge between u and v an even number of times, it follows that the local group at u is the subgroup* $\{0, 2, 4\}$ *of the voltage group* $\mathscr{Z}_6$. *Similarly, the local group at v is also* $\{0, 2, 4\}$.

Theorem 2.5.1 (Alpert and Gross, 1976). *Let u be a vertex in the base space of an ordinary voltage graph* $\langle G, \alpha \rangle$ *with voltages in the group* $\mathscr{A}$. *Then the vertices* u_a *and* u_b *are in the same component of the derived graph* G^α *if and only if the voltage group element* $a^{-1}b$ *lies in the local group* $\mathscr{A}(u)$.

Proof. If the vertices u_a and u_b are in the same component of the derived graph G^α, then there is a path from u_a to u_b. The image of that path under the

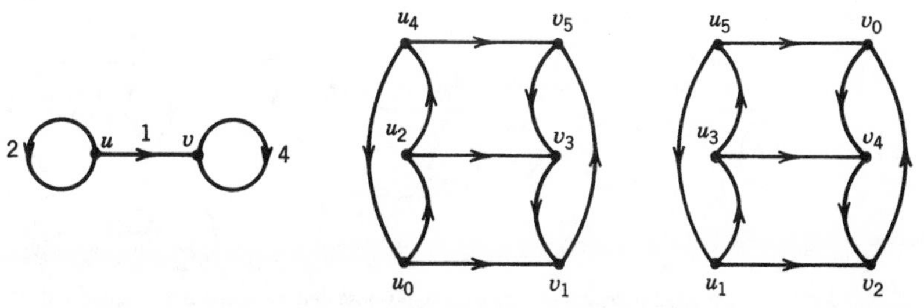

Figure 2.17. A connected ordinary voltage graph whose derived graph is not connected. The voltages are in $\mathscr{Z}_6$.

natural projection is a u-based closed walk in the base graph G, whose net voltage lies in the local group $\mathscr{A}(u)$. The value of that net voltage must be $a^{-1}b$, by Theorem 2.1.2. Reversing these steps yields the converse conclusion. □

Corollary 1. *Let u and v be the vertices in the base graph G and W a walk from u to v with net ordinary voltage c. Then the vertices u_a and v_b of the derived graph lie in the same component if and only if the voltage group element $a^{-1}b$ lies in the right coset $\mathscr{A}(u)c$ of the local group at u.* □

Corollary 2. *The number of components of an ordinary derived graph G^α is $\#(\mathscr{A}: \mathscr{A}(u))$, where u is any vertex of a connected base graph G.* □

Since the voltage group $\mathscr{A}$ acts transitively on an ordinary derived graph G^α, it follows that the components of G^α are mutually isomorphic. Example 2.5.2 illustrates that the components of a permutation derived graph need not be isomorphic. Example 2.5.2 also illustrates another fact. If c is the ordinary or permutation net voltage on a walk from a vertex u to a vertex v in the base graph, then the local group at u is conjugate to the local group at v. In particular, $\mathscr{A}(u) = c^{-1}\mathscr{A}(u)c$.

Example 2.5.2. *In the permutation voltage graph of Figure 2.18, the local group at u is the direct product of the subgroup of all permutations on $\{1,2\}$ with the subgroup of all permutations on $\{3,4,5\}$. The local group at v is the direct product of the subgroup of all permutations on $\{1,3\}$ with the subgroup of all permutations on $\{2,4,5\}$.*

For the most part, there are permutation voltage analogs to the theorems about ordinary voltage graphs. For instance, Theorem 2.5.2 and its corollaries are analogs to Theorem 2.5.1 and its corollaries.

Theorem 2.5.2 (Ezell, 1979). *Let u be a vertex in the base space of the permutation voltage graph $\langle G, \alpha\rangle_n$. Then the vertices u_i and u_j are in the same*

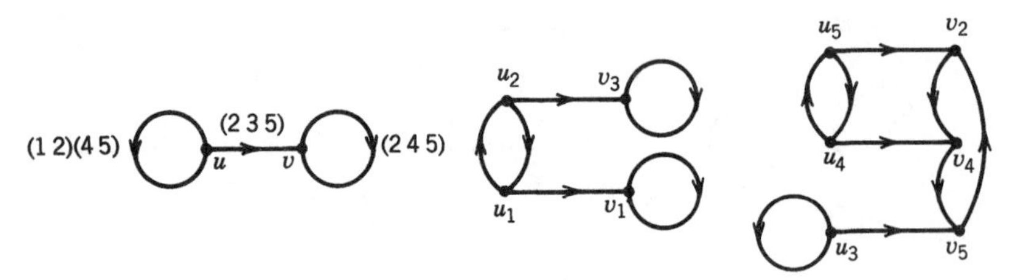

Figure 2.18. A connected permutation voltage graph whose derived graph is not connected. The voltages are in $\mathscr{S}_5$.

component of the derived graph G if and only if the symbols i and j are in the same orbit of the action of the local group at u on $\{1, \ldots, n\}$.

Proof. The vertices u_i and u_j are in the same component of the derived graph if and only if there is a path from u_i to u_j. The image of such a path under the natural projection is a u-based closed walk in the base graph, whose net voltage lies in the local group at u. The value of that net voltage must be a permutation taking i onto j, by Theorem 2.4.2. □

Corollary 1. *Let u and v be vertices in the base graph G and W a walk from u to v with net permutation voltage* π *in* $\mathcal{S}_n$. *Then the vertices* u_i *and* v_j *of the derived graph lie in the same component if the right coset* $\mathcal{S}_n(u)\pi$ *contains a permutation that maps i onto j.* □

Corollary 2. *Let* $\langle G, \alpha \rangle_n$ *be a connected permutation voltage graph, and let u be a vertex of the base graph. The number of components of the derived graph equals the number of orbits induced by the local group at u on the set* $\{1, \ldots, n\}$. □

2.5.2. Counting Components of Ordinary Derived Graphs

If the base graph G of an ordinary voltage graph is a bouquet of circles, then the local group at the vertex is simply the subgroup generated by the voltages on all the loops. If the base graph is not a bouquet of circles, then a preliminary procedure is required to isolate the generators of the local group before one has any hope of counting the components. For instance, even though the proper edge in the base graph of Example 2.5.1 has voltage 1 modulo 6, the local group does not include 1 modulo 6. Thus, the local group is not necessarily generated by the image of the voltage assignment.

The preliminary procedure is, in effect, to convert the base graph G into a bouquet of circles. In the discussion that follows, we suppose that the base graph is connected, for otherwise we would simply iterate the procedure for every component of G. The first step is to select an arbitrary spanning tree T and to regard any vertex for which one wants the local group as the root. Although the case illustrated is for ordinary voltages, the same procedure also applies to permutation voltages.

Example 2.5.3. *Figure* 2.19*a shows a voltage graph* $\langle G, \alpha \rangle$ *with voltage group* $\mathcal{Z}_{12}$. *Figure* 2.19*b shows the choice of a root u and a spanning tree T.*

For every vertex v in the base graph G, there is a unique path in the tree T from the root u to v. Define the T-potential $\alpha(v, T)$ of the vertex v to be the net voltage along that path. The second step is to compute the T-potential of every vertex of G. Figure 2.20 shows the result of this computation for the voltage graph of Example 2.5.3.

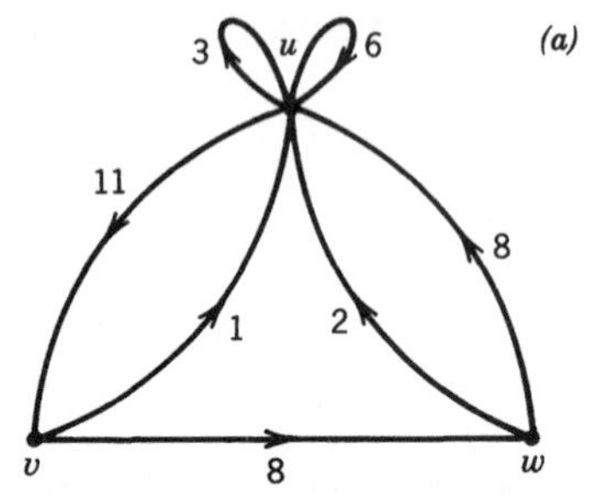

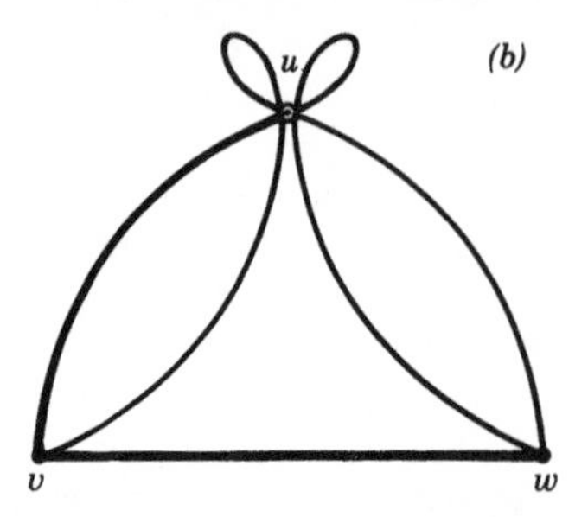

Figure 2.19. (a) A graph with voltages in the cyclic group $\mathscr{Z}_{12}$. (b) The choice of a root and a spanning tree.

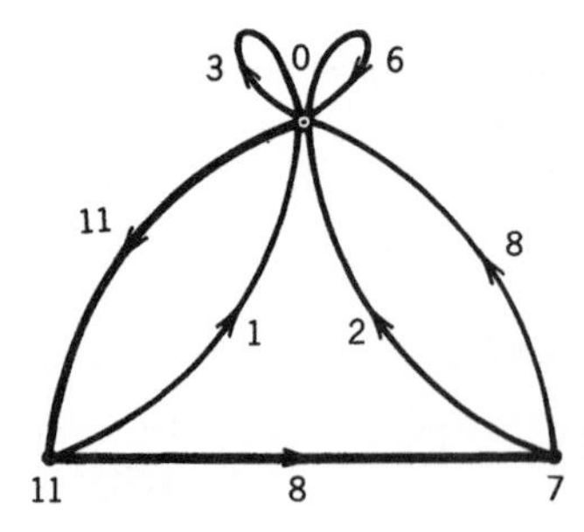

Figure 2.20. T-potentials for the vertices of G.

If the plus direction of the edge e in the base graph G has initial vertex v and terminal vertex w, then the "T-voltage" $\alpha_T(e)$ is defined to be the product

$$\alpha(v, T)\alpha(e)\alpha(w, T)^{-1}$$

The third step is to compute the T-voltage of every edge in G. It should be noted that the T-voltage of every edge in the spanning tree T is the group identity. Figure 2.21 shows the T-voltages for the edges of the voltage graph of Example 2.5.3.

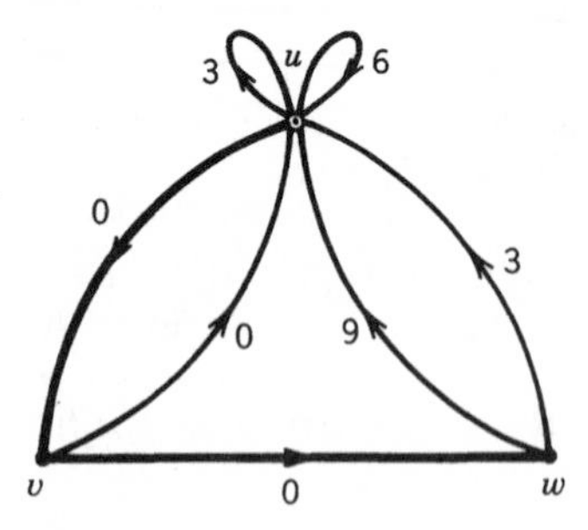

Figure 2.21. T-voltages for the edges of G.

If there is an edge between two vertices whose voltage is the identity, then those two vertices have the same local group, by Corollary 1 to Theorem 2.5.1 (or by Corollary 1 to Theorem 2.5.2, for permutation voltages). If both of those vertices are identified with the new vertex obtained as a result of contracting the base graph along that edge, then it follows that the local group at every vertex is preserved in the resulting voltage graph. Thus, the local group at the root vertex u with respect to the T-voltages could be obtained from the bouquet that results from contracting the base graph G along the entire spanning tree T. That is, the voltages on that bouquet generate the local group at u with respect to the T-voltage assignment. However, these generators are simply the set of T-voltages on those edges of the base graph G that are not in the spanning tree T, so one need not actually carry out the contraction.

For Example 2.5.3, the local group at u with respect to the T-voltages is now calculated to be the subgroup $\{0, 3, 6, 9\}$. The purpose of this calculation is to show that the local group at u with respect to the original voltage assignment α is the same subgroup, as proved by the following theorem.

Theorem 2.5.3 (Gross and Tucker, 1979a). *Let G be a graph, let α be either an ordinary or a permutation voltage assignment for G, and let u be any vertex of G. Then the local group at u with respect to the T-voltages, for any choice of a spanning tree T, is identical to the local group at u with respect to α.*

Proof. The net voltage on any u-based closed walk is the same with respect to T-voltages as with respect to α. □

Theorem 2.5.4. *Let G be a graph, let α be either an ordinary or a permutation voltage assignment for G, and let T be a spanning tree for G with root vertex u. Then the derived graph G^{α_T} corresponding to the T-voltages is isomorphic to the derived graph G^{α} corresponding to the original assignment.*

Proof. A relabeling procedure is implicit in the parenthetic comments in the proof of Theorem 2.4.5. We now make it explicit. For each vertex v of G, relabel the vertices in the fiber over v according to the rule $i \to ic^{-1}$, where c is the net ordinary voltage on the unique path from the root u to the vertex v, or according to the rule $i \to \pi^{-1}(i)$, if that path has net permutation voltage π. If e is an edge originating at v, then also change the subscripts of edges e_i in the fiber over e so that they agree with the new subscripts of their initial points. This relabeling defines an isomorphism $G^{\alpha} \to G^{\alpha_T}$. □

Corollary. *The number of components of the derived graph G^{α} equals the index in the voltage group of the subgroup generated by the T-voltages.* □

2.5.3. The Fundamental Group of a Graph

Let W be a u-based closed walk in a connected graph G. If no directed edge in W is followed by its reverse, we say that W is a "reduced walk".

Two u-based closed walks W and W' are called "equivalent" if there is a sequence

$$W = W^{(0)}, W^{(1)}, \dots, W^{(r)} = W'$$

such that for $k = 1, \dots, r$, the walk $W^{(k)}$ differs from its predecessor $W^{(k-1)}$ by either the insertion or the removal of a directed edge followed by its reverse. By induction, one readily verifies that every u-based closed walk is equivalent to a unique reduced walk.

The equivalence classes formed in the fundamental semigroup of u-based closed walks form a group, in which the inverse of the equivalence class of a u-based closed walk is the equivalence class of the reverse walk. This group is called the "fundamental group of the graph G based at u", and is denoted $\pi_1(G, u)$. For a connected graph G, the isomorphism type of its fundamental group is independent of the choice of the base vertex u.

There is an important correspondence between voltage assignments for a graph G and representations of its fundamental group. Since the net voltage on any two equivalent u-based closed walks is the same, any voltage assignment for G in a group $\mathscr{A}$ induces a homomorphism $\pi_1(G, u) \to \mathscr{A}$. That is, every element of $\pi_1(G, u)$ is mapped onto the net voltage assigned to any u-based closed walk representing that element.

Conversely, every homomorphism $\pi_1(G, u) \to \mathscr{A}$ is induced by some voltage assignment in $\mathscr{A}$ on the graph G. Given such a homomorphism, first choose a spanning tree T and assign to every edge of T the identity voltage. For any edge of G not in T, suppose that its initial point is v and that its terminal vertex is w. Let W be the u-based closed walk that begins with the unique path in T from u to v, next traverses the edge e, and then concludes with the unique path in T that runs from w to u. Assign as a voltage to the edge e the image in the group $\mathscr{A}$ of the equivalence class in $\pi_1(G, u)$ of the walk W.

From the correspondence just described, one may obtain for graphs all of the standard topological theorems on the relationship between fundamental groups and covering spaces.

2.5.4. Contracting Derived Graphs onto Cayley Graphs

The technique of selecting a spanning tree in a voltage graph and using it to obtain an equivalent voltage assignment is useful for other purposes, in addition to counting components of derived graph. The following theorem is equivalent to Lemma 3 of Babai (1977).

Theorem 2.5.5. *Let G be a graph with a voltage assignment α in a group $\mathscr{A}$ such that the derived graph G^{α} is connected. Then G^{α} contracts onto a Cayley graph for $\mathscr{A}$.*

Proof. Choose a spanning tree T for the base graph G. By Theorem 2.5.4, the derived graph G^{α_T} corresponding to the T-voltages is isomorphic to G^{α}, so it is sufficient to prove that G^{α_T} contracts onto a Cayley graph for $\mathscr{A}$. Since $\alpha_T(T) = 0$, by Theorem 2.1.3, the preimage $p^{-1}(T)$ of T under the natural projection has $\#\mathscr{A}$ components, each mapped isomorphically by p onto the spanning tree T. In particular, for each element $\mathscr{A}$ of the voltage group $\mathscr{A}$, one component of $p^{-1}(T)$ contains all of the vertices of G^{α_T} with the subscript a. If each component of $p^{-1}(T)$ is contracted onto a single vertex, which is identified with the common subscript of all the vertices in that component, then the resulting graph is a Cayley graph for $\mathscr{A}$. One might observe that one can also contract the base graph G along the spanning tree T, thereby obtaining a bouquet of circles. The voltages on these circles generate $\mathscr{A}$, because G^{α_T} is connected. Then the contracted derived graph is precisely the derived graph for the contracted voltage graph. □

Corollary 1 (Babai, 1977). *Let $\mathscr{B}$ be a subgroup of a group $\mathscr{A}$, and let G be a Cayley graph for $\mathscr{A}$. Then G contracts onto some Cayley graph for $\mathscr{B}$.*

Proof. Since the group $\mathscr{A}$ acts freely on the Cayley graph G, so does its subgroup $\mathscr{B}$. It follows that G may be derived by assigning voltages in $\mathscr{B}$ to the quotient of G under the action of $\mathscr{B}$. By the theorem, it follows that G contracts onto a Cayley graph for $\mathscr{B}$. □

The "genus of a group" is defined to be the least genus occurring among all its Cayley graphs. White (1973, p. 80) asked whether a finite group $\mathscr{A}$ can have a subgroup $\mathscr{B}$ whose genus exceeds the genus of $\mathscr{A}$. It is an immediate consequence of Corollary 1 that it cannot.

Corollary 2 (Babai, 1977). *The genus of a finite group is greater than or equal to the genus of each of its subgroups.* □

2.5.5. Exercises

Exercises 1–4 concern the voltage graphs in Figure 2.22.

1. Calculate the local group at the vertex u of the voltage graph in Figure 2.22a. How many components are there in the derived graph?
2. Calculate the local group at the vertex u of the voltage graph in Figure 2.22b. How many components are there in the derived graph?
3. Calculate the local groups at vertices u and v of the voltage graph in Figure 2.22c. How many components are there in the derived graph?

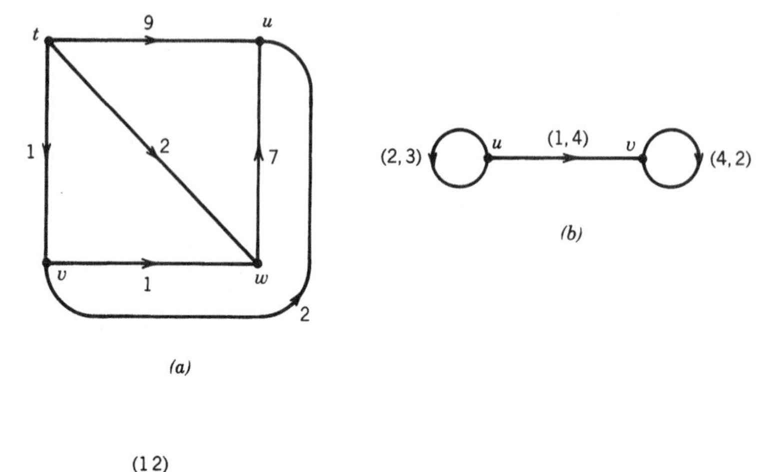

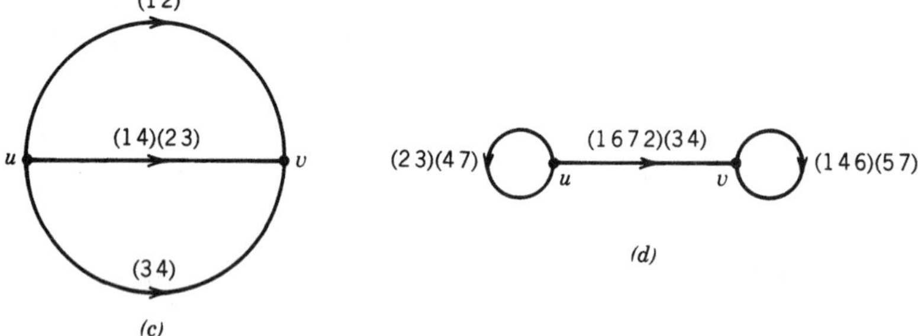

Figure 2.22. Four voltage graphs. (a) Ordinary voltages in $\mathscr{Z}_{18}$, (b) ordinary voltages in $\mathscr{Z}_6 \times \mathscr{Z}_6$, (c) ordinary voltages in $\mathscr{S}_4$, (d) permutation voltages in $\mathscr{S}_7$.

4. Calculate a generating set for the local group at the vertex u of the voltage graph in Figure 2.22d. How many components does the derived graph have?
5. Prove that any graph map $f: G \to H$ induces a homomorphism $f_*: \pi_1(G, u) \to (H, f(u))$, defined as follows: If $[w]$ is the equivalence class of any closed u-based walk in G, then $f_*([w]) = [f(w)]$.
6. Prove that if $p: G \to H$ is a covering projection, then $p_*: \pi_1(G, u) \to \pi_1(H, p(u))$ is one-to-one.
7. Let $\mathscr{B}$ be a subgroup of finite index in $\pi_1(G, u)$. Prove that there is a covering $p: \tilde{G} \to G$ and a vertex $\tilde{u} \in p^{-1}(u)$ such that $p_*(\pi_1(\tilde{G}, \tilde{u})) = \mathscr{B}$. Prove that G is unique up to graph isomorphism. (Thus, $\tilde{G}$ can be called the covering space of G corresponding to the subgroup $\mathscr{B}$.)

3

Surfaces and Graph Imbeddings

It is a classical theorem of Rado (1925) that any compact surface can be dissected into a finite number of triangular regions. The incidence structure of these regions determines a combinatorial object called a "complex", which is a higher-dimensional analog of a graph. Whereas a graph has only zero-dimensional "cells" called vertices and one-dimensional "cells" called edges, a complex may include pieces of any dimension. In particular, a 2-complex also contains some two-dimensional cells. Often these cells correspond to triangular regions, but more generally they may be arbitrary polygons, with any number of sides. Knowing that any closed surface can be represented by such a finite combinatorial object enables us to classify the closed surfaces, that is, to construct a countable list that includes them all and contains no repetitions.

The combinatorial viewpoint also enables us to decide which graphs can be imbedded in which surfaces. Even more significantly, it enables us to construct all of the essentially different possibilities for imbedding an arbitrary graph in an arbitrary surface.

3.1. SURFACES AND SIMPLICIAL COMPLEXES

A topological space M is called an "n-manifold" if M is Hausdorff and is covered by countably many open sets, each one of which is homeomorphic either to the n-dimensional open ball

$$\{(x_1, \ldots, x_n) | x_1^2 + \cdots + x_n^2 < 1\}$$

or the n-dimensional half-ball

$$\{(x_1, \ldots, x_n) | x_1^2 + \cdots + x_n^2 < 1 \text{ and } x_n \geq 0\}.$$

Thus every point in an n-manifold appears, at least locally, to be sitting in the middle of n-space or at the edge of half of n-space. [The need for the Hausdorff condition is explained in Munkres (1975).] The "boundary" of an n-manifold M is the collection of all points in M that do not have neighborhoods homeomorphic to the n-dimensional open ball. For example, the annulus $\{(x, y)|1 \leq x^2 + y^2 \leq 4\}$ is a 2-manifold whose boundary is the pair of circles of radius 1 and 2 centered at the origin. A manifold is "closed" if it is compact and its boundary is empty. By a "surface" we usually mean a closed, connected 2-manifold, such as the sphere, the torus, or the Klein bottle. When we are considering a surface with boundary, such as the Möbius band, or a noncompact surface such as the plane, it is clear from the context or explicitly stated.

3.1.1. Geometric simplicial complexes

A "(geometric) k-simplex" is the convex hull of $k + 1$ affinely independent points or "vertices" in Euclidean n-space $\mathbb{R}^n$. For example, the triangular region determined by three noncollinear points in $\mathbb{R}^2$ or $\mathbb{R}^3$ is a 2-simplex and the line segment determined by two distinct points in $\mathbb{R}^n$ is a 1-simplex. Let $\langle v_0, \ldots, v_k \rangle$ denote the k-simplex determined by the vertices $v_0, \ldots, v_k$. Then the simplex determined by any subset of $\{v_0, \ldots, v_k\}$ is called a "face" of the simplex $\langle v_0, \ldots, v_k \rangle$. Thus the faces of the 1-simplex $\langle v_0, v_1 \rangle$ are the empty set, the endpoints $\langle v_0 \rangle$ and $\langle v_1 \rangle$, and the 1-simplex $\langle v_0, v_1 \rangle$ itself.

A (geometric) "simplicial complex" K is a finite collection of simplexes in $\mathbb{R}^n$ satisfying these two conditions:

1. Every face of every simplex in K is a simplex in K.
2. The intersection of any two simplexes in K is a simplex in K.

The point set

$$|K| = \bigcup_{S \in K} S$$

is called the "carrier" of the simplicial complex K. It is possible that the same point set in $\mathbb{R}^n$ can be the carrier of more than one simplicial complex, as illustrated in Example 3.1.1. If m is the largest integer such that K contains an m-simplex, then K is an "m-complex". The collection of all k-simplexes of K for $k \leq r$ is called the "r-skeleton" of K and is denoted $K^{(r)}$.

Example 3.1.1. *Figure 3.1 shows three simplicial 2-complexes in* $\mathbb{R}^3$. *The complex on the left has one 2-simplex:* $\langle v, w, x \rangle$; *nine 1-simplexes:* $\langle u, v \rangle$, $\langle u, w \rangle$, $\langle u, x \rangle$, $\langle v, w \rangle$, $\langle v, x \rangle$, $\langle v, y \rangle$, $\langle w, x \rangle$, $\langle w, y \rangle$, $\langle x, y \rangle$; *and five 0-simplexes:* $\langle u \rangle$, $\langle v \rangle$, $\langle w \rangle$, $\langle x \rangle$, $\langle y \rangle$. *The center 2-complex has two 2-simplexes:* $\langle u, v, w \rangle$, $\langle u, w, x \rangle$; *nine 1-simplexes:* $\langle u, v \rangle$, $\langle u, w \rangle$, $\langle u, x \rangle$, $\langle u, y \rangle$, $\langle v, w \rangle$, $\langle v, z \rangle$, $\langle w, x \rangle$, $\langle w, y \rangle$, $\langle x, z \rangle$; *and six 0-simplexes:* $\langle u \rangle$, $\langle v \rangle$, $\langle w \rangle$, $\langle x \rangle$, $\langle y \rangle$, $\langle z \rangle$. *Note that the right-hand 2-complex has the same carrier as the one on the left, but this time the 2-simplex* $\langle v, w, x \rangle$ *is replaced by the three 2-simplexes* $\langle v, w, z \rangle$, $\langle w, x, z \rangle$, $\langle x, v, z \rangle$ *and their faces. In each instance, the 1-skeleton is the graph that remains when the 2-simplexes are discarded. (One discards only the 2-simplexes, not their various faces.)*

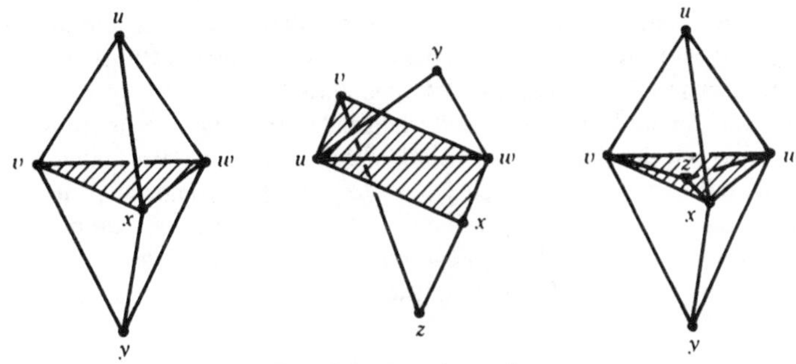

Figure 3.1. Some 2-complexes.

3.1.2. Abstract Simplicial Complexes

An obvious combinatorial structure can be refined from a simplicial complex. Define an "abstract simplicial complex" (V, C) to be a finite set V together with a collection C of subsets of V whose union is V and such that every subset of a member of C is itself a member of C. Then with any (geometric) simplicial complex K, one associates the abstract simplicial complex (K^0, C) where a subset of the 0-skeleton K^0 is in the collection C if and only if the subset forms the vertex set of a simplex in K. In fact, every abstract simplicial complex (V, C) can be associated in this way with some (geometric) simplicial complex K: if the set V has n elements, simply let K^0 be n affinely independent points in $\mathbb{R}^{n-1}$. One may observe that an abstract simplicial complex is a special kind of "hypergraph".

It should not be surprising that the topological type of the carrier of a simplicial complex is determined by the associated abstract simplicial complex. To be precise, define an isomorphism of abstract simplicial complexes (V, C) and (V', C') to be a bijection $f: V \to V'$ such that $f(s) \in C'$ if and only if $s \in C$. Then any isomorphism f of the abstract complexes associated with the geometric complexes K and K' extends to a homeomorphism of their respective carriers $|K|$ and $|K'|$. (The homeomorphism is defined one simplex at a time, starting with the 0-skeleton, then extending to the 1-skeleton, next the 2-skeleton, and so on. The crucial observation is that any homeomorphism from the boundary of one n-simplex to the boundary of another extends, say radially, to a homeomorphism of the interiors.)

Is the converse true? Namely, if the carriers $|K|$ and $|K'|$ are homeomorphic as topological spaces, must the associated abstract simplicial complexes be isomorphic? The answer is clearly no, since one can always subdivide a simplex into smaller simplices as in Example 3.1.1. But suppose one is allowed to subdivide K and K': if $|K|$ and $|K'|$ are homeomorphic, are there subdivisions L and L' of K and K', respectively, such that L and L' are isomorphic? This fundamental question is called the "Hauptvermutung". It says, in effect, that if a topological space has the structure of a simplicial complex, then that structure is unique up to subdivision. The subdivision in question can be restricted to be barycentric (see the next section) and can be interpreted in purely combinatorial terms; thus the topology of a simplicial complex can be faithfully combinatorialized, if the Hauptvermutung holds. Unfortunately, it does not hold for all complexes: Milnor (1961) gives pairs of n-dimensional complexes for each $n \geq 6$, that are homeomorphic but combinatorially distinct. On the other hand, for low-dimensional complexes, the Hauptvermutung does hold. Papakyriakopoulos (1943) proves it for 2-complexes, and Brown (1969) for 3-complexes.

3.1.3. Triangulations

A "triangulation" of a topological space X is a homeomorphism h from the carrier of some simplicial complex K to the space X. The image of a simplex of K under h is called a simplex of the triangulation. The aforementioned theorem of Rado states that every compact surface has a triangulation.

Example 3.1.2. *On the left of Figure* 3.2 *is a triangulation of a torus, where the sides of the rectangle are to be identified in pairs, as in Chapter* 1, *so that vertices match up. Even though all its faces are three-sided, the right side of Figure* 3.2 *is not a triangulation of a torus since no two different simplexes of a triangulation have the exact same vertices. Also an n-simplex of a triangulation must have* $n + 1$ *distinct vertices, and this "triangulation" has only one vertex.*

The vertices and edges of a graph in $\mathbb{R}^3$ are not the 0-simplexes and 1-simplexes of a triangulation if the graph has any multiple edges or loops. This can always be rectified, of course, by a subdivision of the graph inserting one vertex on each multiple edge and two vertices on each loop.

The process of subdividing self-loops and multiple edges to make a nonsimplicial graph simplicial extends to higher dimensions. We might think of a loop as an "inadequately subdivided" edge, and we might also think of a nontriangular polygon (or a polygon with repeated vertices or edges) as

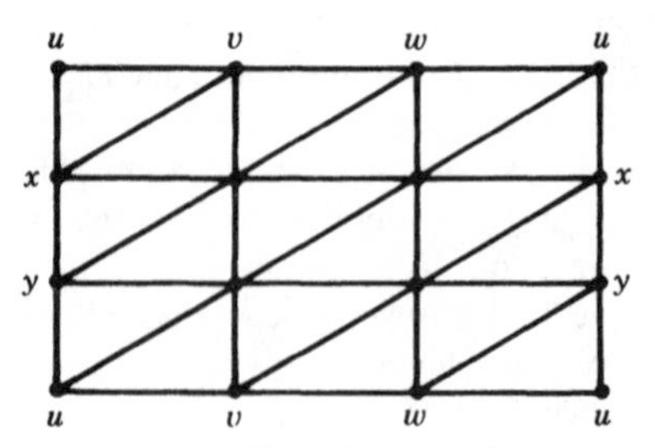

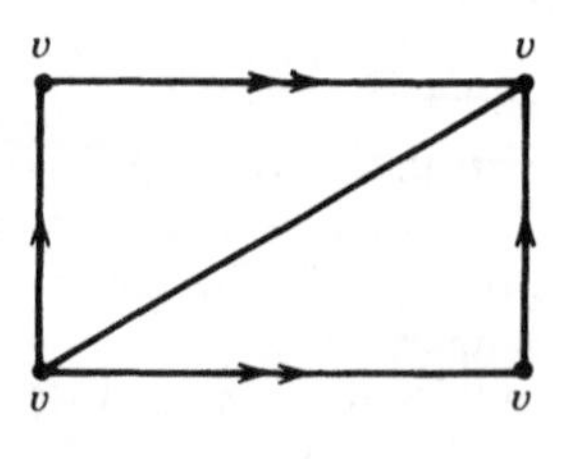

Figure 3.2. A triangulation and a non-triangulation of a torus.

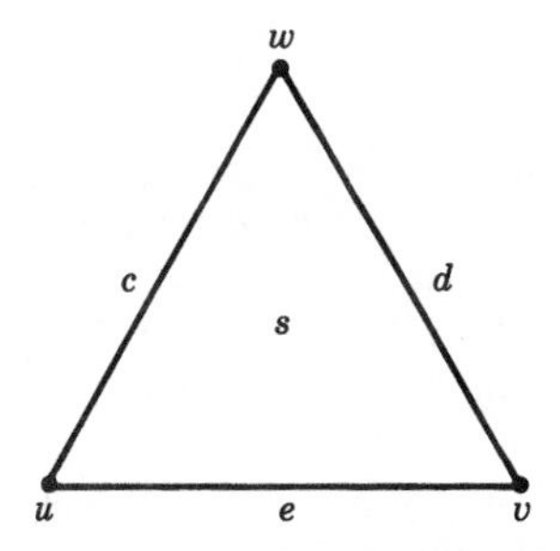

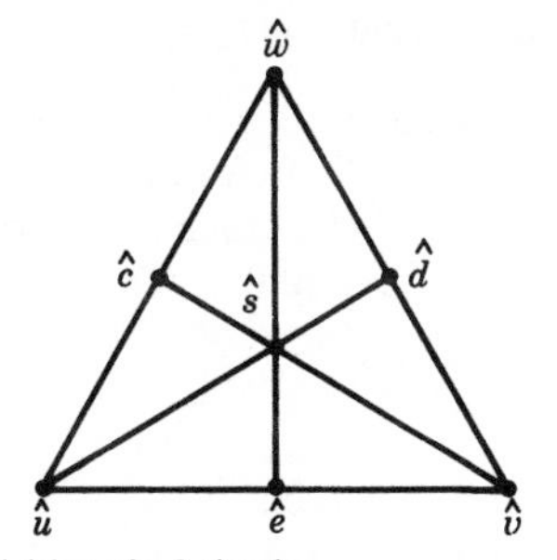

Figure 3.3. The first barycentric subdivision of a 2-simplex.

inadequately subdivided. The higher-dimensional analog of a nonsimplicial complex is called a "cell complex", and just as the operation of subdivision can make a nonsimplicial graph simplicial, it can also do the same for cell complexes of arbitrarily high dimension. We shall first define a particular kind of subdivision operation as it applies to simplicial complexes and then extend the concept.

Let $s = \langle v_0, \ldots, v_k \rangle$ be a k-simplex in $\mathbb{R}^n$. The point $s = (v_0 + v_1 + \cdots + v_k)/k$ is the "barycenter" of s (addition here is usual vector addition in $\mathbb{R}^n$). The "first barycentric subdivision" of a simplicial complex K is the collection of all simplexes of the form $\langle s_0, s_1, \ldots, s_m \rangle$ where each s_i is a simplex of K and s_i is a face of s_{i+1}. Figure 3.3 shows the first barycentric subdivision of the 2-complex consisting of a 2-simplex and its faces.

An s-sided polygonal region, where $s \geq 4$, is an example of a nonsimplicial two-dimensional cell. If the region is parametrized by a convex s-gon, then there is a "center of gravity". We can make a barycentric subdivision by first subdividing every edge on the boundary of the s-gon and then drawing an edge from the center of gravity to each vertex, old and new, on the boundary. Even a one-sided or two-sided region can be convex, if its sides are curved, so there is little problem in extending the concept of barycentric subdivision to all 2-complexes. The same process is also applicable to polyhedra of arbitrarily high dimension.

The operation of barycentric subdivision may be iterated. We state, without proof, the standard fact of combinatorial topology that the second barycentric subdivsion of any cell complex is a simplicial complex. One subdivision might not be enough, even in dimension 1, since the first barycentric subdivision of a self-loop yields two edges between the same two vertices.

The definitions given here can be generalized to allow infinitely many vertices in a simplicial complex. However, the topological considerations are more subtle than the reader might expect, even if one demands that every vertex lie on finitely many simplexes (see Exercise 6). As our interest is now in finite graphs, we shall not pursue the matter further.

3.1.4. Cellular imbeddings

Although a graph is defined as a purely combinatorial object in Chapter 1, it is now helpful to augment the definition by including a fixed topological representation of the graph. For a simplicial graph, this might be the carrier of a geometric 1-complex. This permits a more fluent discussion. Moreover, we often abuse the terminology altogether by referring to a topological representative as the "graph", particularly when we discuss an imbedding of the graph.

An "imbedding" of a graph G in the surface S is a continuous one-to-one function $i: G \to S$. In nearly all cases, the graph G is assumed to be a subset of the surface S, and the function $i: G \to S$ is the inclusion map. The imbedding is then denoted simply $G \to S$. It can be shown as a generalization of the Schoenfliess theorem that for any imbedding $G \to S$, the graph G is contained in the 1-skeleton of a triangulation of the surface S. Thus graph imbeddings in surfaces are "tame" and can be analyzed by combinatorial methods. [The Schoenfliess theorem, a strengthening of the Jordan curve theorem, states that any simple closed curve in the plane can be carried onto the unit circle by some homeomorphism of the plane; for a proof, see Newman (1954). The Alexander horned sphere (1924) shows that the Schoenfliess theorem does not hold for spheres in three-dimensional space.]

Given an imbedding $G \to S$, the components of the complement $S - G$ are called "regions". If each region is homeomorphic to an open disk, the imbedding is said to be a "2-cell (or cellular) imbedding" and the regions are also called "faces" of the imbedding. The closure in the surface S of a region in the 2-cell imbedding $G \to S$ need *not* be homeomorphic to the closed disk.

Example 3.1.3. *The imbedding of a bouquet of two circles in a torus on the left of Figure* 3.4 *is not cellular, since one region is homeomorphic to an open annulus. The imbedding on the right is cellular. Observe that the closure in the surface of the one and only face is not a closed disk, but instead, the whole torus.*

Unless we explicitly say otherwise, any imbedding in this book is assumed to be a 2-cell imbedding. In particular, we shall not usually imbed a disconnected graph in a connected surface or imbed a tree in any surface other than the sphere. Also, an imbedding of a nonempty finite graph in the plane cannot

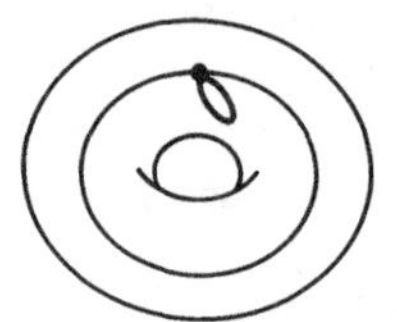

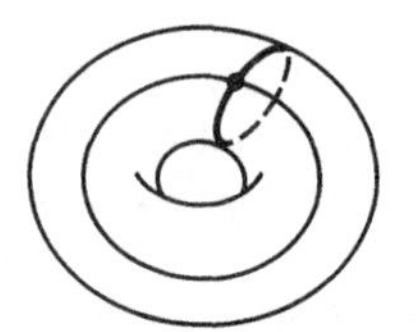

Figure 3.4. Two imbeddings of the bouquet B_2 on a torus.

be cellular. Thus, when an imbedding in the sphere is drawn in the plane, the reader is expected to remember the necessary point at infinity.

Given a 2-cell imbedding $G \to S$, the collection of all edges and vertices lying in the closure of a region can be organized into a closed walk in the graph G by traversing a simple closed curve just inside the region. This closed walk is unique up to the choice of initial vertex and direction, and is called the "boundary walk" of the region. A face of a 2-cell imbedding can usually be specified unambiguously simply by giving its boundary walk (see Exercise 17). For this reason, we shall frequently abuse the language with phrases such as "the face $abcd$" where $abcd$ is in fact the boundary walk of the face. An occurrence of an edge in the boundary walk of a region is called a "side" of the region. The "size" of a region is the length of its boundary walk. Since individual edges may occur twice in a boundary walk, the size of a region can be greater than the number of edges that are sides of the region. An n-sided face in a 2-cell imbedding is called an "n-gon", although 3-gons and 4-gons are also called triangles and quadrilaterals, respectively.

Example 3.1.4. *The imbedding in the sphere given in Figure* 3.5 *has three regions. There is a quadrilateral* $abcd$, *a triangle* eef, *and a* 9-*gon* $abggcdhfh$. *The* 9-*gon has only seven different edges as its sides.*

Two imbeddings of a given graph $i: G \to S$ and $j: G \to T$ are called "equivalent" if there is an orientation-preserving homeomorphism $h: S \to T$ such that $hi = j$. It is important to observe that by this definition, it does not suffice simply to find a homeomorphism from S to T that takes the image $i(G)$ to the image $j(G)$; for the imbeddings to be equivalent, the homeomorphism must also respect a labeling of the vertices and edges of G.

Example 3.1.5. *Four different imbeddings of the same tree in the sphere* (*plane*) *are given in Figure* 3.6. *A rotation by* 90° *gives an equivalence between the first and second imbeddings. A reflection in a horizontal line through the vertex* y *yields an equivalence of the second and third imbeddings. However, the fourth imbedding is not equivalent to any of the previous three.*

Two imbeddings $i: G \to S$ and $j: G \to T$ are called "weakly equivalent" if there is a homeomorphism $h: S \to T$ such that $h(i(G)) = j(G)$. For instance,

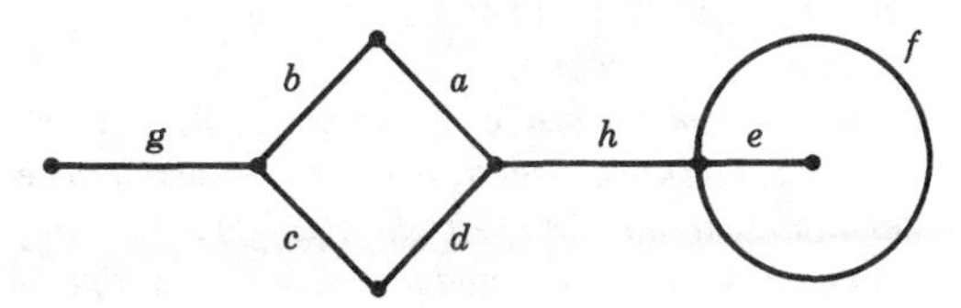

Figure 3.5. An imbedding in the sphere.

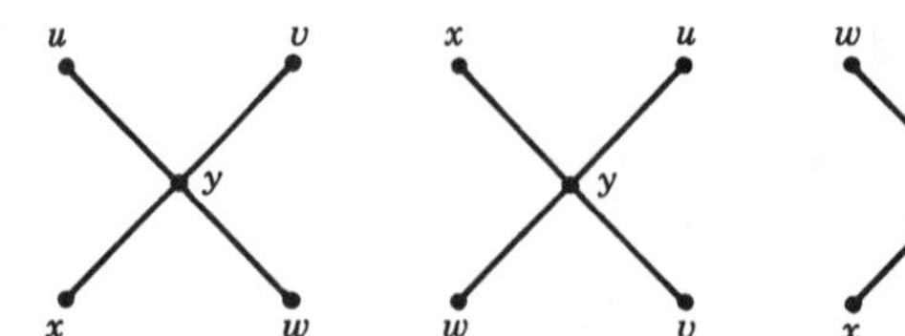

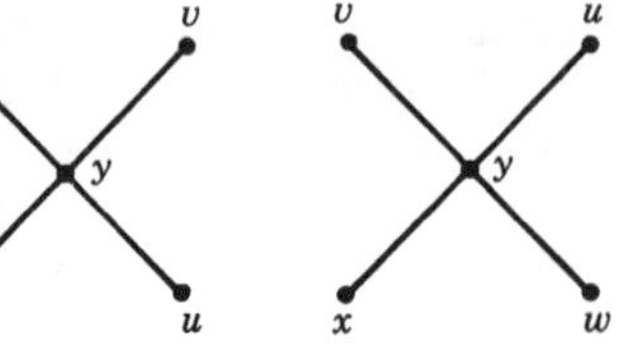

Figure 3.6. Four imbeddings of the same tree.

the fourth imbedding in Figure 3.6 is weakly equivalent to the other three. It will become apparent in Section 3.2 that equivalence is easier to analyze than weak equivalence, although it may seem at first less natural. In any case, it must be remembered that any discussion of equivalence here presumes that the vertices and edges of the imbedded graph have been labeled. "Congruent" is a possible synonym for "weakly equivalent".

Given a 2-cell imbedding $G \to S$, the "dual imbedding" $G^* \to S$ is defined as in Chapter 1: the vertices of G^* correspond to regions of the imbedding $G \to S$ and vice versa. The dual graph gives an efficient description of the incidence relationship among the regions of an imbedding, for which purpose it is used extensively in Section 3.3.

3.1.5. Representing Surfaces by Polygons

One way to present a surface with a given triangulation is simply to list (or draw) every triangle with each of its sides directed and labeled. The surface is then assembled by identifying the two sides that have the same label so that the directions agree. The homeomorphism doing the identifying for a particular edge does not matter, since any two triangulated surfaces obtained from the same labels and directions have the same underlying abstract simplicial complex and hence are homeomorphic. If desired, one may "preassemble" some of the triangles into larger polygons whose labeled sides are to be identified later to obtain the given surface. In fact, such preassembly permits one to represent any compact connected surface by a single polygon with labeled and directed sides (similarly to what was done in Chapter 1 for the torus). A label that appears only once corresponds to an edge in the boundary of the surface.

Example 3.1.6. *In Chapter* 1 *the projective plane was defined to be the surface obtained by identifying the boundary of a disk to the boundary of a Möbius band. In Figure* 3.7, *polygon* 2 *becomes a Möbius band when the b-edges are identified. Thus the hexagon formed by polygons* 1, 2, *and* 3 *represents the projective plane. If the consecutive sides a, b, c are considered to be just one edge, we get the simpler representation on the right of Figure* 3.7. *In particular, the projective plane can be viewed as the closed unit disk with antipodal points on the unit circle identified in pairs. Thus, the definition given here for a projective plane*

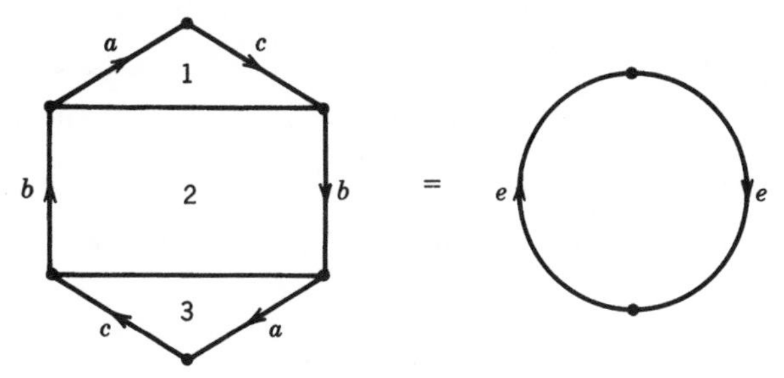

Figure 3.7. Two representations of the projective plane.

corresponds to that given in geometry as the space of lines through the origin in R^3 *(represented by points on the northern hemisphere of the unit sphere with antipodal points on the equator identified).*

Example 3.1.7. *The Klein bottle was defined in Chapter* 1 *as the surface obtained by identifying the boundaries of two Möbius bands to the boundary of a sphere with two holes. Since a sphere with two holes is homeomorphic to an annulus and identifying one boundary component of an annulus to a Möbius band still leaves a Möbius band (with a "collar" on its boundary), the Klein bottle can also be viewed as the surface obtained by identifying the boundaries of two Möbius bands alone. In Figure* 3.8, *polygon* 2 *becomes a Möbius band when the b-edges are identified. Polygons* 1 *and* 3 *also form a single Möbius band when their edges are identified (first identify the d-edges to get a rectangle like polygon* 2). *Thus the polygons* 1, 2, *and* 3 *together represent the Klein bottle. On the right in Figure* 3.8 *is a simpler representation.*

If the direction of one copy of edge e in Figure 3.7 is reversed, the resulting surface is the sphere. Similarly, if the direction of one copy of edge e in Figure 3.8 is reversed, the resulting surface is the torus.

Example 3.1.8. *If all the identifications are performed on the two polygons in Figure* 3.9 *except on the sides labeled e, one obtains two tori each with a puncture. When the e-sides are then identified, the result is a surface of genus* 2.

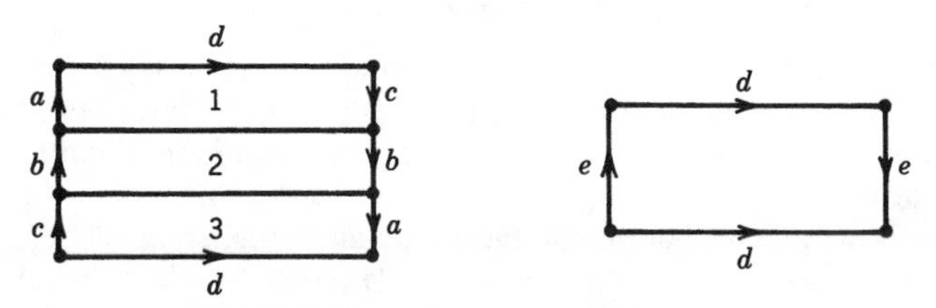

Figure 3.8. Two representations of the Klein bottle.

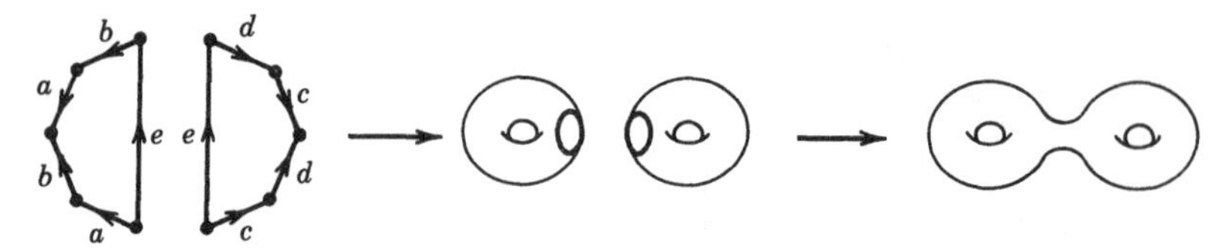

Figure 3.9. How to obtain a genus two surface from polygons.

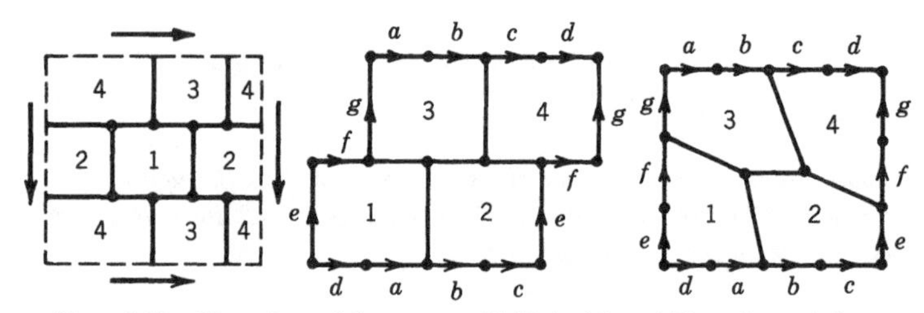

Figure 3.10. Three views of the same toroidal imbedding of $K_{4,4}$ minus a 1-factor.

If the e-sides were identified first, a representation of a genus-2 surface by a single eight-sided polygon would be obtained.

A picture of a graph imbedding can be drawn on a polygon with identified sides, as in Chapter 1. When a graph edge hits a side it continues at the other identified side. The sides of the so-called outer polygon are sometimes allowed to be edges of the graph also.

Example 3.1.9. *An imbedding in a torus of $K_{4,4}$ with a 1-factor removed is given on the left in Figure 3.10. The sides of the "outer" rectangle are not edges of the graph. The four six-sided regions of this imbedding (labeled 1, 2, 3, 4) can be reassembled to give the middle imbedding, in which the outside edges are to be identified in pairs as usual. This time the "outer" polygon's sides are edges of the graph. On the right is the same imbedding with the path efg straightened.*

3.1.6. Pseudosurfaces and Block Designs

A "pseudosurface" is the carrier of a 2-complex in which every 1-simplex is a face of exactly two 2-simplexes. Every point in a pseudosurface except for the underlying points of some vertices has a neighborhood homeomorphic to the interior of the unit disk. A neighborhood of the underlying point of a vertex, however, is in general homeomorphic to a finite number m of disks all of whose centers have been identified to the single vertex. If $m > 1$ then the point is called a "singular point of the pseudosurface". Thus a pseudosurface can be

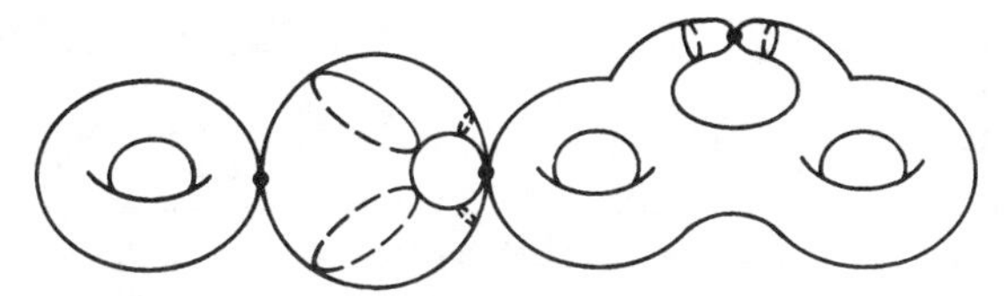

Figure 3.11. A pseudosurface with its singular points darkened.

viewed as a surface in which finite sets of points are identified. Figure 3.11 shows a typical pseudosurface.

Alpert (1975) showed that covering space techniques could be useful in constructing triangulated pseudosurfaces that represent "twofold triple systems", thereby reviving interest in the connection between block design and topological graph theory explored earlier by Heffter (1897) and Emch (1929). Subsequent advances in exploiting this connection were achieved by White (1978) and Garman (1979). The basic ideas are easy enough to explain.

A "block design" consists of a set V of objects and a set B of subsets of V, called "blocks", such that (i) each object appears in the same number of blocks, called the "replication number", and (ii) each block has the same number of objects, called the "blocksize". If every pair of objects appears in the same number of blocks, called the "balancing index", then the design is called "balanced". A trivial kind of balanced block design has every object in every block, and is called a "complete" block design. If the blocksize is less than $\#V$ then the design is called "incomplete".

Balanced incomplete block designs, abbreviated BIBDs, are important to statisticians because they facilitate the design of experiments on samples of limited size. They are also of considerable interest to algebraists and combinatorial analysts.

A twofold triple system is a BIBD with blocksize equal to 3 and balancing index equal to 2. Alpert proved that there is a bijective correspondence between the twofold triple systems and the (equivalence classes of) triangulation of closed pseudosurfaces by complete graphs. This topological insight facilitates an easy proof that some classical twofold triple systems of Bhattacharya (1943) and Hanani (1961) are distinct and that they are also distinct from new twofold triple systems that Alpert developed from the Ringel–Youngs constructions in cases 1, 6, 9, and 10 of the Heawood map-coloring problem.

White and Garman both explored the extension to "partially" balanced incomplete block designs with two association classes of objects. These association classes partition the objects so that pairs of objects from one class have one balancing index while pairs from the other class have another balancing index. To obtain his results, Garman formally extended the imbedded voltage graph construction described in Chapter 4 to voltage graphs imbedded in pseudosurfaces.

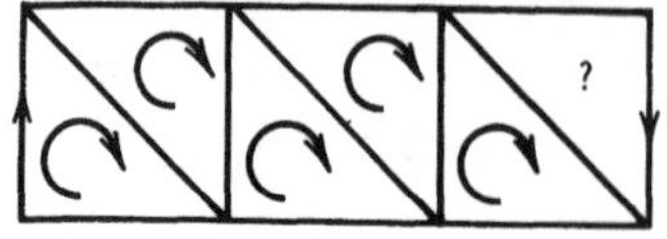

Figure 3.12. An orientation for an annulus and an attempt at an orientation for a Möbius band.

3.1.7. Orientations

Each face of an imbedding $G \to S$ has two possible directions for its boundary walk. A face is assigned an "orientation" by choosing one of these two directions. An "orientation for the imbedding" $G \to S$ is an assignment of orientations to all the faces so that adjacent regions induce opposite directions on every common edge. If the graph G is the 1-skeleton of a triangulation of the surface S, then an orientation for the imbedding $G \to S$ is called an "orientation of the triangulation". A surface is "orientable" if it has an imbedding with an orientation.

It is a simple matter to test whether a given connected imbedding $G \to S$ has an orientation. First, assign an arbitrary orientation to a particular face. This forces an orientation of each face that shares an edge with the original face. Since the surface is connected, the process can be continued until the orientation of each face has been forced. Either the result is an orientation for the imbedding or else the given imbedding has no orientation.

Example 3.1.10. *An orientation for a triangulation of an annulus is given on the left in Figure* 3.12. *On the right is an attempt at orienting a Möbius band. Since the attempt fails, this triangulation has no orientation. Indeed, it is clear that no sequence of adjacent 2-simplexes along the central horizontal line of a Möbius band can be oriented to achieve an orientation for the surface. Thus a Möbius band is not orientable.*

Example 3.1.10 is, in fact, quite general. If a surface contains a subset homeomorphic to a Möbius band, it cannot be orientable. Conversely, suppose a triangulation of a surface has no orientation. Then starting from an orientation of a fixed 2-simplex, a chain of adjacent 2-simplexes can be formed that is homeomorphic to a Möbius band.

3.1.8. Stars, Links, and Local Properties

Two important types of subcomplexes enable us to define "local" properties of a graph. Let us suppose that K is a simplicial complex of arbitrary dimension.

The "star" of a vertex v is defined to be minimal subcomplex of K that contains every simplex of K that is incident on v. In particular, if K is a graph, then for any vertex v, star(v) is the subgraph consisting of all the edges incident on v and the endpoints of those edges. Such a graph does actually look something like a heavenly star, which is the inspiration for the terminology. A topological space X is a closed surface if and only if there is a triangulation of X in which the carrier of the star of each vertex is homeomorphic to a closed disk.

The "link" of a vertex v is defined to be the maximal subcomplex of star(v) that does not contain v itself. If K is a graph, then link(v) is the totally disconnected graph comprising the neighbors of v and no edges. If the carrier of K is a surface, then the link of every vertex is a cycle through its neighbors. If the carrier of K is a pseudosurface, then the link of a singular vertex is the disjoint union of two or more cycles.

The standard definition of a "local property" of a complex K is to say it is a property that is true of the star or the link of every vertex. In the case of graphs, we sometimes associate it with the subgraph induced by the link of a vertex.

For instance, we might say that a graph is "locally hamiltonian" if for every vertex v, the subgraph induced by link(v) is hamiltonian. The principal of this concept is that a graph cannot triangulate a surface unless it is locally hamiltonian.

3.1.9. Exercises

1. Which points in the given set X fail to have a neighborhood homeomorphic to the unit disk?
   (a) $X = \{(x, y, z) \mid xyz = 0\}$.
   (b) $X = \{(x, y, z) \mid x^2 + y^2 = z^2\}$.
2. Prove that the r-skeleton of a simplicial complex is a simplicial complex.
3. Figure 3.3 shows the first barycentric subdivision of the 2-complex consisting of a 2-simplex and its faces. Prove that the first barycentric subdivision is itself a simplicial complex.
4. Draw imbeddings in the plane of the 1-skeleton of the first two 2-complexes in Figure 3.1. Show that the 1-skeleton of the first barycentric subdivisions of these 2-complexes do not have planar imbeddings. (*Hint:* Show that the first contains $K_{3,3}$ and that the second contains K_5.)
5. Find a 2-complex that cannot be imbedded in the sphere (or for that matter in any surface) but whose first barycentric subdivision has a 1-skeleton that is imbeddable in the plane.
6. Let $v_n = (\cos 2\pi/n, \sin 2\pi/n) \in \mathbb{R}^2$ for $n = 1, 2, \ldots$. Consider the collection K of all simplexes of the form $\langle v_n, v_{n+1}\rangle$ or $\langle v_n\rangle$ for $n = 1, 2, \ldots$. If the restrictions of finiteness were deleted from the definition of simplicial complex, would K be a simplicial complex? Could K triangulate the unit circle?

7. Draw a triangulation of the projective plane, trying to use as few triangles as possible.

8. The directions and edge labels for a polygonal representation of a surface can be coded simply by giving a closed walk around the boundary of each polygon. Thus the "outer" polygon on the left of Figure 3.8 could be coded as $dcbad^{-1}cba$. The projective plane polygon on the right of Figure 3.7 would be ee. Thus to any list of "words" using edge labels and their inverses, such that each edge label is used exactly twice, we can associate a surface. What surface does the list $\{abac^{-1}, bc\}$ represent? What surface does $\{aba^{-1}c^{-1}, bc\}$ represent?

9. It should be clear that cyclically permuting or inverting any word coding a polygon (see Exercise 8) corresponds only to choosing a different initial edge or direction for the closed walk around the polygon and hence does not change the surface being represented. Explain why the following operations do not change the surface either.
   (a) Replace the two words $Ae, e^{-1}B$ by the single word AB (a picture will suffice).
   (b) Replace the word $ABA^{\epsilon}C$ by the word $xBx^{\epsilon}C$ where $\epsilon = \pm 1$.
   (c) Replace $Aee^{-1}B$ by AB.

10. Use the operations described in Exercise 9 to find out what surface the list $\{aba^{-1}cd, c^{-1}ef^{-1}, b^{-1}e^{-1}fd^{-1}\}$ represents.

11. How does one recognize when a surface represented by one word (Exercise 8) is nonorientable? Suggest an algorithm for determining whether a connected surface represented by a list of more than one word is nonorientable.

12. Prove that equivalence and free equivalence of imbeddings are equivalence relations.

13. Find an imbedding of the tree in Example 3.1.5 that is not equivalent to any of the imbeddings shown in Figure 3.6.

14. Prove that any two imbeddings of a tree having largest valence 3 are weakly equivalent. Give an example of a tree having two imbeddings that are not freely equivalent.

15. Draw an imbedding of the complete graph K_5 on the torus as a polygon whose sides are edges of K_5 (see Example 3.1.9). Do the same for K_6.

16. Draw a 2-cell imbedding of K_5 in a polygonal representation of a surface of genus 2 (see Example 3.1.8). Be sure the imbedding is cellular. How many faces does it have?

17. Prove that if two regions of an imbedding $G \to S$ have the same boundary walk, then the graph G is a cycle.

18. Give an imbedding of K_5 in the pseudosurface obtained by identifying two points on the sphere. Prove that K_6 cannot be imbedded in this pseudosurface.

3.2. BAND DECOMPOSITIONS AND GRAPH IMBEDDINGS

When a surface is formed as in Section 3.1 by pasting polygons to each other, the vertices and edges of the polygons fit together to form a graph imbedded in that surface. Alternatively, one might begin with a graph and attach polygons to it so that a surface is formed. This can be achieved, as we see in this section, without explicit mention of the polygons. Instead, we begin with a graph and add some combinatorial structure that implicitly describes the polygons, the surface to be obtained, and the imbedding.

3.2.1. Band Decomposition for Surfaces

Let $G \to S$ be a 2-cell imbedding of a graph in a surface. One can surround each vertex of the graph G by a small disk in the surface S and each edge of G by a thin band so that the union of all disks and bands is a neighborhood of G in S whose shape preserves that of the graph itself. The complement in S of this neighborhood consists of a family of disks, one just inside each face of the imbedding. This decomposition of the surface into bands and (two kinds of) disks is now described in detail.

Define a "1-band" to be a topological space b together with a homeomorphism $h: I \times I \to b$, where I denotes the unit interval $[0,1]$. The arcs $h(I \times \{j\})$ for $j = 0,1$ are called the "ends" of the band b, and the arcs $h(\{j\} \times I)$, for $j = 0,1$, are called the "sides" of b. A "0-band" is simply a homeomorph of the unit disk, as is a "2-band".

A "band decomposition" of the surface S is a collection B of 0-bands, 1-bands, and 2-bands satisfying these conditions:

1. Different bands intersect only along arcs in their boundaries.
2. The union of all the bands is S.
3. Each end of each 1-band is contained in a 0-band.
4. Each side of each 1-band is contained in a 2-band.
5. The 0-bands are pairwise disjoint and the 2-bands are pairwise disjoint.

The corresponding "reduced band decomposition" B omits the 2-bands.

Example 3.2.1. *Figure* 3.13 *shows a graph imbedding and the associated reduced band decomposition. The* 0*-bands assume the names of their respective vertices and the* 1*-bands assume the names of their respective edges.*

A band decomposition of a surface is the two-dimensional version of the topological construction known as a "handle decomposition" of an n-manifold. The manipulation of bands as used in this chapter to classify the surfaces has higher-dimensional analogs that were instrumental in Smale's proof (1961) of the Poincaré conjecture in higher dimensions [see also Milnor (1965a)]. Use

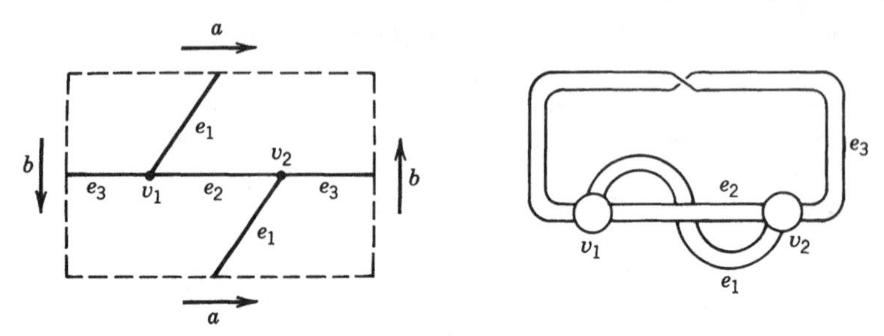

Figure 3.13. (left) An imbedding of a graph with two vertices and three edges in the Klein bottle. (right) The corresponding reduced band decomposition.

of band decompositions in the study of surfaces can be found implicitly in Griffiths (1976). For the more traditional approach of sewing together labeled polygons, see Blackett (1967).

3.2.2. Orientability

A band decomposition is called "locally oriented" if each 0-band is assigned an orientation. Then a 1-band is called "orientation-preserving" if the directions induced on its ends by adjoining 0-bands are the same as those induced by one of the two possible orientations of the 1-band; otherwise the 1-band is called "orientation-reversing". These two possibilities are illustrated in Figure 3.14.

An edge e in the graph imbedding associated with a locally oriented band decomposition is said to have "(orientation) type 0" if its corresponding 1-band is orientation preserving, and "(orientation) type 1" otherwise. A walk in the associated graph has type 1 if it has an odd number of type-1 edges and has type 0 otherwise.

Remark. *The designation of orientation reversing for a* 1*-band depends only on the orientations chosen for its adjoining* 0*-bands and does not necessarily imply nonorientability of the associated surface. Rather, the surface is nonorientable if*

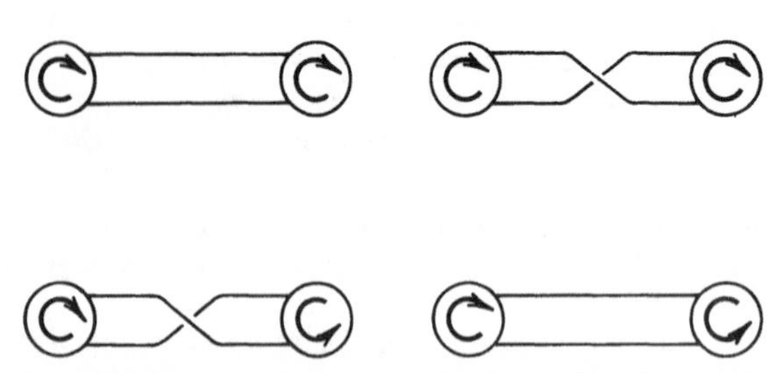

Figure 3.14. Two orientation preserving bands (left) and two orientation reversing bands (right).

*and only if there exists a type-*1 *cycle in the associated graph, which implies immediately that the surface contains a Möbius band. For instance, the cycle* e_2, e_3 *in Figure* 3.13 *has type* 1, *no matter what the choice of local orientations.*

In searching for type-1 cycles, it is often advantageous first to reverse the orientation of some 0-bands. This obviously does not change the surface. However, when the orientation of a 0-band is reversed, this changes the orientation type of every proper edge incident on the corresponding vertex (the orientation type of a loop does not change).

Example 3.2.2. *Consider the reduced band decomposition shown on the left in Figure* 3.15. *First, reverse the orientation of the* 0*-band labeled* u. *Then reverse the orientation of the* 0*-band* v. *The resulting decomposition has no orientation reversing* 1*-bands, so the underlying surface is orientable, despite all the twisted* 1*-bands in the leftmost drawing.*

Orientability Algorithm. *To determine whether the surface* S *corresponding to a given band decomposition is orientable or not, first choose a spanning tree* T *for the associated graph* G. *Next choose a root vertex* u *for* T *and an orientation for the* 0*-band of* u. *Then for each vertex* v *adjacent to* u *in the tree* T, *choose the orientation for the* 0*-band of* v *so that the edge of* T *from* u *to* v *has type* 0. *Continue this process on the tree* T *until every* 0*-band has an orientation. That is, if* v *and* w *are adjacent in* T *and if the orientation at* v *has already been determined, then choose whichever orientation at* w *makes the edge from* v *to* w *have type* 0. *No conflicts in choice of orientation arise since* T *is a tree. The result is a band decomposition such that every edge in the tree* T *has type* 0. *Any type-*1 *edge in* $G - T$ *forms a closed walk with a path in* T *joining its endpoints. Since every walk in* T *has type* 0, *such a closed walk would have type* 1. *Thus the surface* S *is nonorientable if and only if some edge in the complement* $G - T$ *has type* 1.

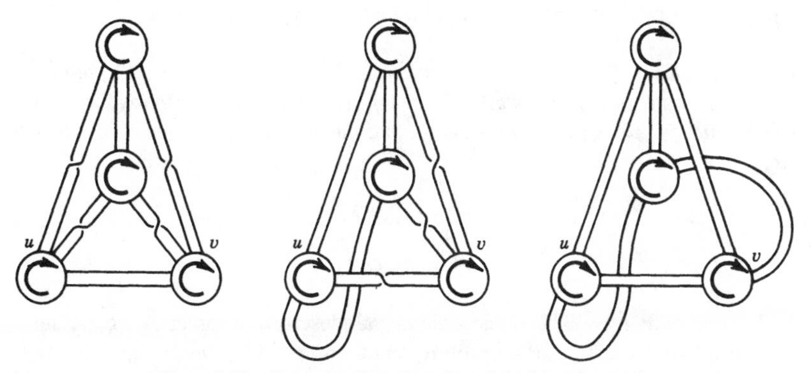

Figure 3.15. Reversing the orientation of the 0-bands u and v.

3.2.3. Rotation Systems

We wish to develop a convenient method of describing a band decomposition or, equivalently, a graph imbedding $G \to S$. One first observes that the 2-bands are not really needed to define the surface S, since the union of the 0-bands and 1-bands is a surface with boundary, and since there is essentially only one way to fill in the faces to complete the closed surface. The second part of this explanation follows from Lemma 3.2.1, whose truth may seem intuitively obvious, but whose proof is actually rather delicate and involves more topology than we care to consider here.

Lemma 3.2.1. *Let S and T be surfaces with the same number of boundary components. Let S' and T' be closed surfaces obtained by identifying the boundaries of disks to the boundary components of S and T, respectively. Then the surfaces S and T are homeomorphic if and only if the surfaces S' and T' are homeomorphic.* □

Observe that Lemma 3.2.1 validates the omission of 2-bands as in Figures 3.13 and 3.15. Since the surface defined by the 0-bands and 1-bands can be drawn in 3-space, while the whole surface sometimes cannot, the boon to visual intuition is considerable.

Thus, to describe a graph imbedding $G \to S$ or its band decomposition, one need specify only how the ends of the 1-bands are to be attached to the 0-bands. This is done by labeling a disjoint collection of arcs in the boundary of each 0-band by the names of edges incident on the vertex surrounded by that 0-band. Each arc of 1-band attachment is directed by the assigned orientation of the given 0-band. The ends of each 1-band are then directed and identified to the arcs labeled by the corresponding edge. The lengths of the arcs do not matter. As complicated as this sounds, it means that one need only provide for each vertex an ordered list, unique up to a cyclic permutation, of the edges incident on that vertex and a designation of orientation preserving or reversing to each 1-band (to tell which way the ends are pasted), or equivalently, whether the corresponding edge is type 0 or type 1.

Example 3.2.3. *For instance, if we orient the 0-bands of Figure 3.13 both to be clockwise on the page, we have the following combinatorial description, where the superscript "1" designates a type-1 edge and the absence of a superscript a type-0 edge:*

$$v_1. \quad e_1\, e_2\, e_3^{\,1}$$

$$v_2. \quad e_3^{\,1}\, e_1\, e_2$$

To make this method of combinatorial description precise, we define a "rotation" at a vertex v of a graph to be an ordered list, unique up to a cyclic permutation, of the edges incident on that vertex. Naturally, every v-based loop is mentioned twice in the rotation at v. Let a "rotation system" on a graph be an assignment of a rotation to each vertex and a designation of orientation type for each edge. Then the preceding discussion can be summarized by the following theorem, used extensively by Ringel in the 1950s. The first formal proof was published by Stahl (1978).

Theorem 3.2.2. *Every rotation system on a graph G defines (up to equivalence of imbeddings) a unique locally oriented graph imbedding $G \to S$. Conversely, every locally oriented graph imbedding $G \to S$ defines a rotation system for G.* □

3.2.4. Pure Rotation Systems and Orientable Surfaces

Let the graph G have an imbedding in an oriented surface S. If the 0-bands of the associated band decomposition are given local orientations consistent with the orientation of the surface S, then every 1-band will be orientation-preserving. Conversely, any rotation system for the graph G such that every edge has type 0 induces an imbedding in an orientable surface (obtained by supplying the 2-bands) and a specific orientation of that surface. Theorem 3.2.3 summarizes this discussion. For the sake of brevity we define a "pure rotation system" for a graph to be one in which every edge has type 0.

Theorem 3.2.3. *Every pure rotation system for a graph G induces (up to orientation-preserving equivalence of imbeddings) a unique imbedding of G into an oriented surface. Conversely, every imbedding of a graph G into an oriented surface induces a unique pure rotation system for G.* □

Like Ringel (1974), we think it appropriate to ascribe credit for Theorem 3.2.3 jointly to Heffter (1891) and Edmonds (1960). See the Historical Note at the end of the section.

3.2.5. Drawings of Rotation Systems

There is a particularly simple way to incorporate into a drawing of a graph a rotation system for the graph: just be sure the clockwise order of the edges incident on a vertex in the drawing agrees with the assigned rotation at the vertex. The easiest way to do this is to draw first a dot for each vertex with spokes radiating from the dot labeled in clockwise order according to the rotation at the vertex. Then curves are drawn joining spokes with the same label. Finally, all type-1 edges are marked with a cross. The resulting drawing is called a "projection" of the given rotation. For instance, Figure 3.16 shows a projection of the pure rotation system defined by the toroidal imbedding of $K_{4,4}$ minus a 1-factor given in Figure 3.10. For this projection a judicious choice of the "under" edge at each crossing helps make the surface "visible".

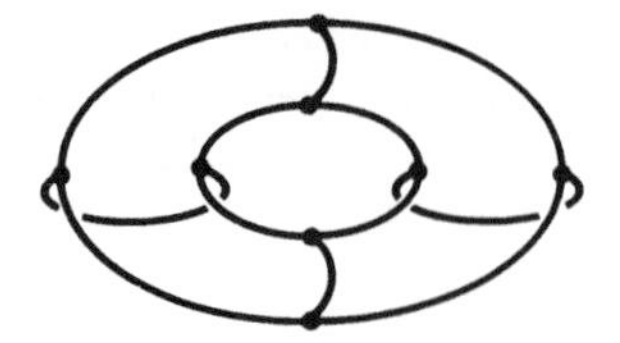

Figure 3.16. A rotation projection with no type-1 edges.

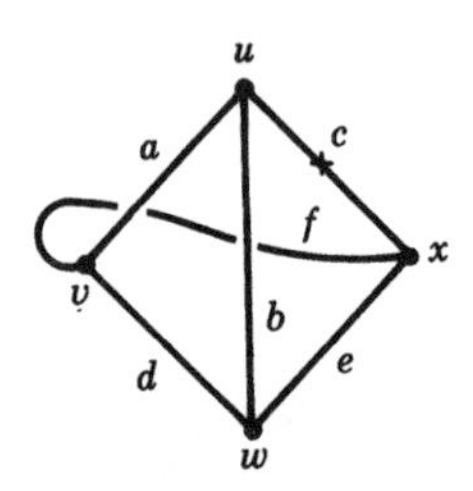

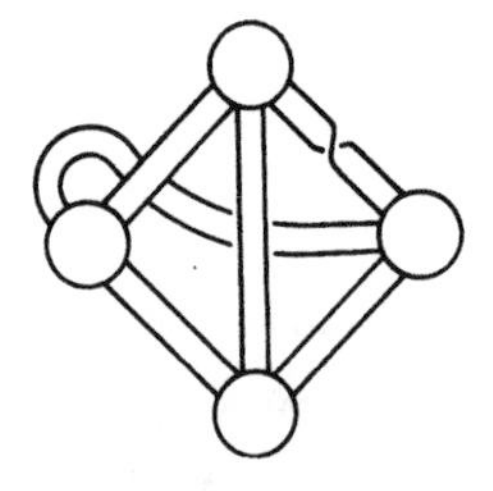

Figure 3.17. A rotation projection for K_4 and the corresponding reduced band decomposition.

If a given graph is simplicial, then the list format of a rotation system may give adjacent vertices instead of edges. This is called the "vertex form of a rotation system". In this context, it is also natural to list only the sequence of vertices in a boundary walk.

Example 3.2.4. *The rotation system for K_4 whose projection is illustrated on the left in Figure 3.17 can be given a list format in either the edge form or the vertex form.*

$$
\begin{array}{llll}
u. & c^1\,b\,a & u. & x^1\,w\,v \\
v. & f\,a\,d & v. & x\,u\,w \\
w. & d\,b\,e & w. & v\,u\,x \\
x. & e\,f\,c^1 & x. & w\,v\,u^1
\end{array}
$$

The corresponding reduced band decomposition is given at the right. By tracing along the boundary of this reduced band decomposition surface, one easily verifies that the imbedding has two faces $uvxwuxwvx$ and uvw in vertex form, or $afebcedfc$ and adb in edge form.

3.2.6. Tracing Faces

Given a rotation system for a graph, one frequently needs to obtain a listing or enumeration of the boundary walks of the reduced faces. If the rotation system is given by a projection, then the procedure is well illustrated in Example 3.2.4. First, thicken each edge into a 1-band, and give the band a twist if the edge is type 1. Then simply trace out, say with a pencil, the boundary components of the resulting surface. On the other hand, if the rotation system is given in list format, for example, as computer input, then this geometric method is impossible. Some thought about the geometric method should convince the reader that the following Face Tracing Algorithm is correct. We first introduce some helpful terminology. If the rotation at vertex v is $\ldots de \ldots$ (up to a cyclic permutation), then we say that d is "the edge before e at v", that e is "the edge after d at v", and that the edge pair (d, e) is a "corner at v with second edge e".

Face Tracing Algorithm. *Assume that the given graph G has no 2-valent vertices. Choose an initial vertex v_0 of G and a first edge e_1 incident on v_0. Let v_1 be the other endpoint of e_1. The second edge e_2 in the boundary walk is the edge after (resp., before) e_1 at v_1 if e_1 is type 0 (resp., type 1). If the edge e_1 is a loop, then e_2 is the edge after (resp., before) the other occurrence of e_1 at v_1. In general, if the walk traced so far ends with edge e_i at vertex v_i, then the next edge e_{i+1} is the edge after (resp., before) e_i at v_i if the walk so far is type 0 (resp., type 1). The boundary walk is finished at edge e_n if the next two edges in the walk would be e_1 and e_2 again. To start a different boundary walk, begin at the second edge of any corner that does not appear in any previously traced faces. If there are no unused corners, then all faces have been traced.*

Observe that the walk does not necessarily stop when the first edge e_1 is encountered a second time; we might not be on the same side of e_1 as at the beginning. The followup by the edge e_2 is what confirms that we are on the original side of e_1, assuming of course that the vertex v_1 does not have valence 2 (see Exercise 14).

Example 3.2.5. *Consider the rotation system*

$$
\begin{array}{ll}
u. & a^1\, f\, b\, d^1\, a^1\, e^1\, b\, c \\
v. & c\, f\, g \\
w. & e^1\, d^1\, g
\end{array}
$$

Begin the first face at vertex u with the edge a at the corner (c, a). The next edge in the boundary walk is d, the edge before (since the walk a is type 1) the other occurrence of the edge a at the vertex u. The next edge is g, the edge after d (since the walk $a\,d$ is type 0) at the vertex w. The next edge is c. Since the following two edges would be a and d again, the walk terminates with c, yielding the face $a\,d\,g\,c$. Since the corner (a, f) has not yet appeared, we can begin a second boundary walk with edge f at vertex u. This time the face $f\,g\,e\,a$ is obtained. The third and fourth faces $b\,c\,f$ and $d\,e\,b$ are obtained in a similar fashion by starting with edge b from corner (f, b) and edge d from corner (b, d). Figure 3.18 shows these four faces, first separately, then assembled together.

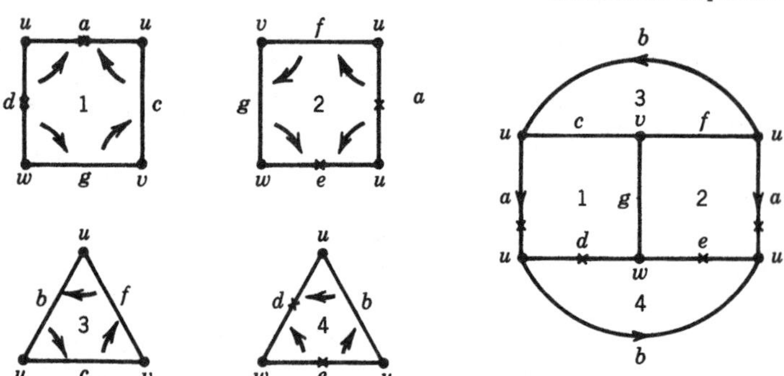

Figure 3.18. The faces from Example 3.2.5 assembled into a Klein bottle, so that the local orientations at vertices u, v, and w are correctly completed.

Example 3.2.6. *The rotation system*

$$\begin{array}{ll} t. & d^1\, b\, e^1 \\ u. & f\, d^1\, a^1 \\ v. & c^1\, b\, a^1 \\ w. & c^1\, e^1\, f \end{array}$$

has as its faces the polygons dfe, bad, ecb, and fca. When these faces are assembled, they yield a 2-sphere. Thus, as we have mentioned, the existence of orientation reversing edges need not mean that the surface is nonorientable, provided that no cycle contains an odd number of orientation reversing edges.

3.2.7. Duality

Let B be a band decomposition of the surface S. Suppose each 2-band has been given a specific orientation. The dual band decomposition of B, denoted B^*, is defined by letting the i-bands of B be the $(2-i)$-bands of B^*, for $i = 0, 1, 2$; the ends of each 1-band of B become the sides of the corresponding 1-band of B^* and vice versa. If $G \to S$ is the graph imbedding associated with the decomposition B, then the imbedding associated with B^* is naturally the dual imbedding $G^* \to S^*$.

Let e be an edge of the imbedding $G \to S$ associated with the decomposition B. The orientation type of the dual edge e^* depends on the choice of orientations for the 2-bands of the primal imbedding or, equivalently, on the direction given to the closed walk around the boundary of each primal face. The edge e appears twice in the course of listing all directed face boundaries. If the two appearances have opposite directions, then the dual edge e^* is type 0; otherwise it is type 1. Thus, in Example 3.2.4, if the directed boundaries are $uvxwuxwvx$ and uvw, then $(ux)^*$, $(vw)^*$, and $(wx)^*$ are type 0, whereas

$(uv)^*$, $(uw)^*$, and $(vx)^*$ are type 1. If the direction uwv replaces direction uvw for the second face, then $(uv)^*$ and $(uw)^*$ become type 0; the dual edge $(vw)^*$ becomes type 1, and the loops $(ux)^*$, $(wx)^*$, and $(uv)^*$ remain the same.

If the surface S is orientable and all face orientations are chosen to agree with one orientation for S, then all the dual edges are type 0.

3.2.8. Which 2-Complexes Are Planar?

One possible way to view the 2-cells in a 2-complex K (simplicial or not) is as restrictions on the possible rotation systems for its 1-skeleton. That is, if two edges are consecutive on the boundary of any 2-cell of K, then they must be consecutive in every rotation at their common vertex, or else the rotation system would not extend the 2-complex K to a surface.

Using rotation systems, it is not difficult to prove that if the star of every vertex of the 2-complex K is planar—in which case we say that K is "locally planar"—then K has an imbedding in some closed surface. Harary and Rosen (1976) give a complete proof of the folk theorem that characterizes the absence of local planarity in terms of a single obstruction. A "thumbtack" is a topological space homeomorphic to the following subset of $\mathbb{R}^3$:

$$\{(x, y, 0) | x^2 + y^2 \leq 1\} \cup \{(0, 0, z) | 0 \leq z \leq 1\}$$

The characterization is that a 2-complex is not locally planar if and only if it contains a thumbtack.

One might guess naively that a locally planar 2-complex is planar if and only if its 1-skeleton is planar. However, Figure 3.1 contains a counterexample to that assertion. In particular, the 2-complex on the left has a planar 1-skeleton, isomorphic to $K_5 - e$. However, the 2-complex on the right has the same topological space as its carrier, and its 1-skeleton contains $K_{3,3}$.

Gross and Rosen (1981) prove that a simplicial 2-complex is imbeddable in the sphere if and only if it is locally planar and the 1-skeleton of its first barycentric subdivision is planar. If the 2-complex in question is not simplicial, it can first be subdivided into a simplicial complex before applying this test. In a sequel, Gross and Rosen (1979) prove that reducing the planarity question for an arbitrary 2-complex to testing the planarity of a graph can be accomplished in linear time in the number of vertices of the 2-complex.

Historical Note. *The earliest form of Theorem* 3.2.3 *was implicitly given in a dualized vertex form by Heffter* (1891) *and used extensively by Ringel in the* 1950*s. Evidently unaware of the existence of this work, Edmonds* (1960) *invented the primalized vertex form, and its details were subsequently made accessible by Youngs* (1963). *The full generalization to all graphs, simplicial or not, requires the switch from the vertex form to the directed-edge form, and it was developed by Gross and Alpert* (1974).

3.2.9. Exercises

1. Consider the pure rotation system

$$\begin{array}{llll} u. & cbd & x. & abg \\ v. & ajh & y. & idf \\ w. & ecfe & z. & hijg \end{array}$$

Draw a projection of this rotation system and list the faces.

2. Consider the pure rotation projection given in Figure 3.19. Write out the rotation system and list the faces.
3. Suppose that the edges a, d, e, and h in Exercise 1 were type 1. Draw the 0-bands and 1-bands of the corresponding band decomposition. Then list the faces.
4. Suppose that the edges a, d, h, and i in Exercise 2 were type 1. List the faces of the resulting imbedding.
5. Draw a rotation projection for an imbedding of K_6 on a torus.
6. Suggest a way of describing a rotation system using an incidence matrix.
7. Suggest a way of describing a rotation system for a simplicial graph using an adjacency matrix.
8. How many inequivalent vertex-labeled imbeddings in orientable surfaces does K_7 have?
9. Show that there are at most $5!(3!)^6$ labeled imbeddings of K_7 in orientable surfaces having six three-sided faces meeting at vertex 1. Conclude that the fraction of labeled orientable imbeddings of K_7 that have all faces triangular is less than 1.6×10^{-8}.

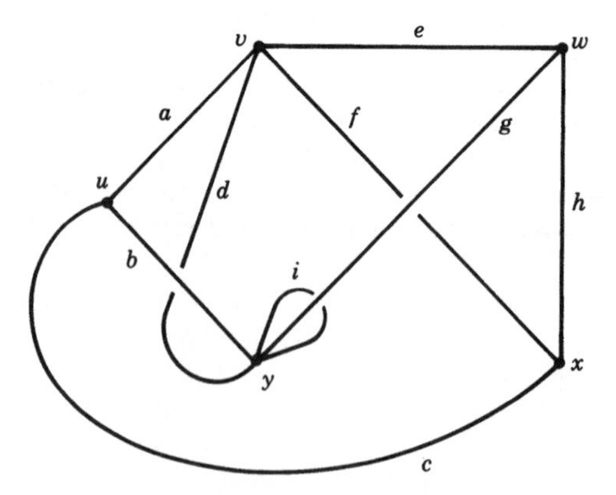

Figure 3.19. A rotation projection.

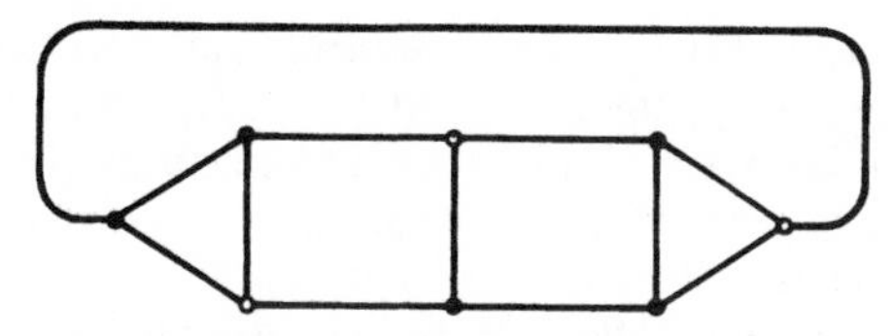

Figure 3.20. A rotation system for a cubic graph. Hollow dots have counterclockwise rotation.

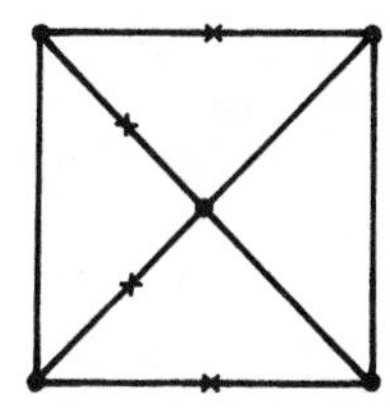

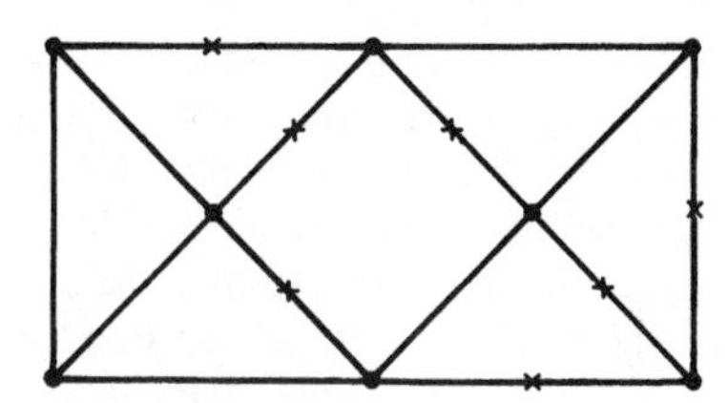

Figure 3.21. Two rotation projections.

10. Let $G \to S$ be an imbedding. Suppose that f_1 and f_2 are two distinct faces of the imbedding whose boundaries share the edge e. Let $G' \to S'$ be the imbedding obtained from the rotation system for $G \to S$ by "giving the edge e an extra twist" (that is, by changing the orientation type of e). Show that the imbedding $G \to S'$ has one less face than the imbedding $G \to S$.

11. Use Exercise 10 to show that every connected graph has an imbedding with only one face (Edmonds, 1965).

12. Let G be a regular graph of valence 3. Each vertex has exactly two possible rotations. Thus a rotation system for G can be given from a drawing of G in the plane by, say, a hollow dot for a vertex whose rotation is counterclockwise in the drawing and a solid dot if clockwise. How many faces are there in the rotation system given by this method in Figure 3.20?

13. Two rotation projections are given in Figure 3.21. Choose a spanning tree in each graph and reverse some vertex orientations so that each edge in the spanning tree is type 0. Then determine which of the two surfaces is orientable.

14. Show by an example that the procedure given in the Face Tracing Algorithm for terminating a boundary walk may stop the walk too soon if the vertex v_1 is 2-valent. Suggest a way of handling 2-valent vertices.

3.3. THE CLASSIFICATION OF SURFACES

It is a remarkable fact of low-dimensional topology that the surfaces $S_0, S_1, S_2, \ldots$ form a complete set of representatives of the homeomorphism types of closed, connected, orientable surfaces, and moreover, that they are distinguishable by a single integer-valued invariant, the Euler characteristic. Similarly, the surfaces $N_1, N_2, N_3, \ldots$ form a complete set of representatives of the homeomorphism types of closed, connected, nonorientable surfaces, and they too are distinguishable by Euler characteristic.

The proof of the classification theorem is facilitated by a designation of convenient models of the surfaces. As a model of S_g we take the surface of an unknotted g-holed solid doughnut in 3-space.

A "meridian" in this model is a topological loop on S_g that bounds a disk in the g-holed solid that does not separate the solid. The topological closure of the complement of the g-holed solid is also a g-holed solid doughnut. A "longitude" on S_g is a topological loop that bounds a disk in the complementary solid that does not separate the complementary solid. Figure 3.22 shows a three-holed solid doughnut and standard sets of meridians and longitudes.

As a model of N_k we take the surface obtained from a sphere by first deleting the interiors of k disjoint disks and then closing each resulting boundary component with a Möbius band. No closed nonorientable surface is imbeddable in 3-space.

The classifications are derived in three steps. The first step is to calculate an Euler characteristic—relative to a particular cellular imbedding of a particular graph—for each of the standard surfaces. Second, we show that the Euler characteristic is an invariant of every standard surface, that is, it is independent of the choice of a graph and of the choice of an imbedding (provided that the imbedding is cellular).

Since the standard orientable surfaces have different Euler characteristics, it follows from the second step that no two of them are homeomorphic. Similarly, the nonorientable standard surfaces are pairwise distinct. In the third step, we introduce the concept of "surgery" to prove that every closed surface is homeomorphic to one of the standard surfaces.

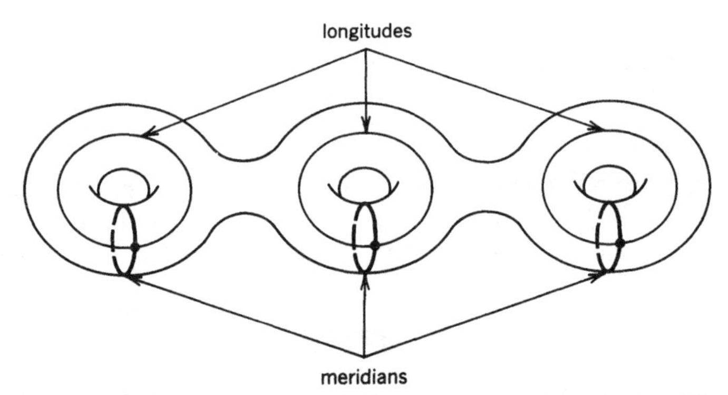

Figure 3.22. An unknotted 3-holed solid doughnut in 3-space, with standard meridians and longitudes.

3.3.1. Euler Characteristic Relative to an Imbedded Graph

The "(relative) Euler characteristic of a cellular imbedding" $G \to S$ of a connected graph into a closed, connected surface is the value of the Euler formula $\#V - \#E + \#F$, and it is denoted $\chi(G \to S)$. In the course of proving the classification theorems we will prove that this number depends only on the homeomorphism type of the surface S and not on the choice of a graph G or its imbedding.

Theorem 3.3.1. *For each orientable surface S_g ($g = 0, 1, 2, \ldots$) there exists a connected graph G and a cellular imbedding $G \to S_g$ whose Euler characteristic satisfies the equation $\chi(G \to S_g) = 2 - 2g$.*

Proof. Theorem 1.4.1 implies this result for $g = 0$. For $g > 0$, let the vertices of the graph G be the g intersection points of a standard set of g meridian–longitude pairs on the surface S_g. The edges of G include the g meridians, the g longitudes, and a set of $g - 1$ "bridges" that run directly between consecutive intersection points. Thus, the graph G has g vertices and $3g - 1$ edges. One verifies by an induction on g that there is only one face and that it is a 2-cell. It follows that $\chi(G \to S_g) = g - (3g - 1) + 1 = 2 - 2g$. □

Theorem 3.3.2. *For each nonorientable surface N_k ($k = 0, 1, 2, \ldots$) there is a graph G and a cellular imbedding of G into the surface N_k such that $\chi(G \to N_k) = 2 - k$.*

Proof. For $k = 0$, this is true by Theorem 1.4.1. For $k > 0$, choose a point on the center loop of each of the k crosscaps as a vertex of G. The edges of G are the k center loops plus $k - 1$ bridges to link the vertices to each other. Thus, there are k vertices and $2k - 1$ edges. An induction argument on the number k proves that there is only one face and that it is a 2-cell. Therefore,

$$\chi(G \to N_k) = k - (2k - 1) + 1 = 2 - k. \quad \square$$

3.3.2. Invariance of Euler Characteristic

The choice in Theorem 3.3.1 and in Theorem 3.3.2 of particular cellular imbeddings in the standard surfaces was designed to simplify the calculation of their Euler characteristics. We shall now see that for every one of the standard surfaces, the value of the Euler formula $\#V - \#E + \#F$ is independent of the selections of a graph and of a cellular imbedding.

It follows immediately from the invariance of the Euler characteristic of the standard surfaces that they are mutually distinct, because if two orientable surfaces differ in genus or if two nonorientable surfaces differ in crosscap number, then they differ in Euler characteristic. Establishing the invariance of the Euler characteristic, and thereby, the distinctness of the standard surfaces, is the second step in the classification of surfaces.

Theorem 3.3.3 (The Invariance of Euler Characteristic of Orientable Surfaces). *Let $G \to S_g$ be a cellular imbedding, for any $g = 0, 1, 2, \ldots$. Then $\chi(G \to S_g) = 2 - 2g$.*

Proof. Theorem 1.4.1 implies that the conclusion holds for the sphere S_0. As an inductive hypothesis, assume that it holds for the surface S_g, and suppose that the graph G is cellularly imbedded in S_{g+1}.

First, draw a meridian on S_{g+1} so that it meets the graph G in finitely many points, each in the interior of some edge of G and each a proper intersection, not a tangential intersection. If necessary, we subdivide edges of G so that no edge of G crosses the meridian more than once, which does not change the relative Euler characteristic of the imbedding into S_{g+1}. Next thicken the meridian to an annular region R, as illustrated in Figure 3.23.

A homeomorphic copy of S_g can be obtained by excising from S_{g+1} the interior of region R and by then capping off the two holes with disks. We now construct a cellular imbedding $G' \to S_g$. The vertex set of the graph G' is the union of the vertex set V_G and the intersection set of G with the boundary of R. If the graph G intersects the meridian in p places and if the region R is selected to be adequately narrow, then $\#V_{G'} = \#V_G + 2p$.

The edge set $E_{G'}$ contains all the $\#E_G - p$ edges of G that do not cross the meridian. Furthermore, it contains the arc segments on the boundary components of the region R that connect adjacent points of intersection of boundary(R) and G. Since G intersects the meridian in p places, there are p such arcs on each boundary component, for an additional subtotal of $2p$. The edge set $E_{G'}$ also contains the segments of edges of G that run from vertices of G to intersections of G and boundary(R). There are $2p$ such segments in all. Thus,

$$\#E_{G'} = (\#E_G - p) + 2p + 2p = \#E_G + 3p$$

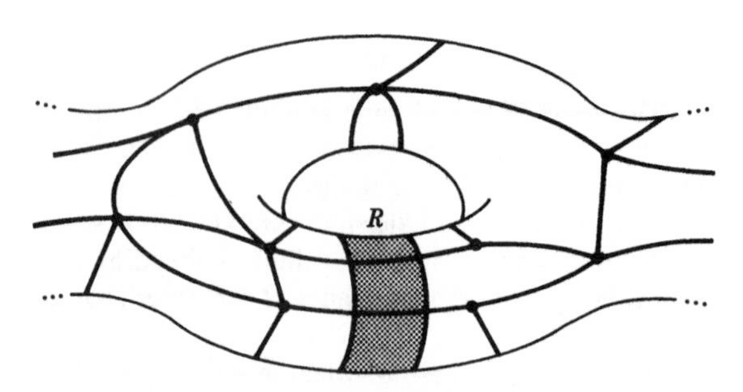

Figure 3.23. Thickening the meridian yields the shaded region R.

Each face of the imbedding $G \to S_{g+1}$ is cellular. It follows from the Jordan curve theorem that each of the p subarcs of the meridian subtended by G, considered in succession, cuts a region into two parts, thereby yielding p additional faces in the imbedding $G' \to S_g$, it follows that

$$F_{G'} = \#F_G + p + 2$$

Therefore, we may compute that

$$\begin{aligned}\chi(G \to S_g) &= \#V_{G'} - \#E_{G'} + \#F_{G'}\\ &= (\#V_G + 2p) - (\#E_G + 3p) + (\#F_G + p + 2)\\ &= \#V_G - \#E_G + \#F_G + 2\\ &= \chi(G \to S_{g+1}) + 2\end{aligned}$$

According to the induction hypothesis,

$$\chi(G' \to S_g) = 2 - 2g$$

We immediately infer that

$$\chi(G \to S_{g+1}) = -2g = 2 - 2(g+1) \quad \square$$

Corollary. *Let i and j be distinct nonnegative integers. Then the surfaces S_i and S_j are not homeomorphic.*

Proof. A homeomorphism $f: S_i \to S_j$ would carry a cellular imbedding $G \to S_i$ of relative Euler characteristic $\chi(G \to S_i) = 2 - 2i$ to a cellular imbedding $G \to S_j$ of relative Euler characteristic $\chi(G \to S_j) = 2 - 2i$, in violation of Theorem 3.3.3, which implies that $\chi(G \to S_j) = 2 - 2j$. $\square$

Theorem 3.3.4 (The Invariance of Euler Characteristic of Nonorientable Surfaces). *Let $G \to N_k$ be a cellular imbedding, for any $k = 0, 1, 2, \ldots$. Then $\chi(G \to N_k) = 2 - k$.*

Proof. The conclusion holds for the 2-sphere N_0 by Theorem 1.4.1. By way of an induction, assume that it holds for the surface N_k, and suppose that the graph G is cellularly imbedded in N_{k+1}.

The induction step here is analogous to the one for the orientable case. This time draw a center loop on one of the crosscaps of N_{k+1} so that it meets the graph G in finitely many points, and subdivide some edges if needed, so that no edge of G crosses that center loop more than once. Such subdivision does not alter the relative Euler characteristic of the imbedding. When the center loop is thickened, the result is a Möbius band B, rather than an annulus.

A homeomorphic copy of N_k can be obtained by excising from N_{k+1} the interior of the Möbius band B and then capping off the resulting hole with a disk. Continuing the analogy, we construct an imbedding $G' \to N_k$. If we assume that the graph G intersects the center loop in p points, then the vertex set $V_{G'}$ contains $\#V_G$ vertices from G plus $2p$ points of intersection of G with boundary(R).

The edge set $E_{G'}$ contains $\#E_G + 3p$ edges, as before. However, since only one hole results from excising the interior of the Möbius band B, the face set $F_{G'}$ contains $\#F_G + p + 1$ faces. It follows that

$$\begin{aligned}\chi(G \to N_k) &= \#V_{G'} - \#E_{G'} + \#F_{G'} \\ &= (\#V_G + 2p) - (\#E_G + 3p) + (\#F_G + p + 1) \\ &= \#V_G - \#E_G + \#F_G + 1 \\ &= \chi(G \to N_{k+1}) + 1\end{aligned}$$

The induction hypothesis implies that $\chi(G \to N_k) = 2 - k$, which enables us to conclude that

$$\chi(G \to N_{k+1}) = 1 - k = 2 - (k + 1) \quad \square$$

Corollary. *Let i and j be distinct nonnegative integers. Then the surfaces N_i and N_j are not homeomorphic.* □

3.3.3. Edge-Deletion Surgery and Edge Sliding

What remains to complete the classification of closed surfaces is the third step, which is to prove that every closed surface is homeomorphic to one of the standard surfaces. To accomplish this, we introduce some topological techniques called "surgery". This is the most difficult step in the classification.

Let $G \to S$ be a cellular imbedding of a graph in a surface, and let e be an edge of G. To perform "edge-deletion surgery" at e, one first deletes from a band decomposition (not reduced) for $G \to S$ the 2-band or 2-bands that meet the e-band; one next deletes the e-band itself; and finally one closes the hole or holes with one or two new 2-bands, as needed.

Remark 3.3.1. *Suppose that the imbedding $G \to S$ induces the rotation system R and that the imbedding $G' \to S'$ results from edge-deletion surgery at edge e. Then the rotation system R' induced by $G' \to S'$ can be obtained simply by deleting both occurrences of edge e from R. For a rotation projection, this means erasing the edge e. For a reduced band decomposition, it means deleting the e-band.*

There are three different possible occurrences when the edge e is deleted, which are called cases i, ii, and iii. The effect of the surgery depends on the case.

i. The two sides of the edge e lie in different faces, f_1 and f_2. Then deleting the f_1-band, the f_2-band, and the e-band leaves one hole, which is closed off with one new 2-band.

ii. Face f is pasted to itself along edge e without a twist, so that deleting the f-band and the e-band leaves two distinct holes (and possibly disconnects the surface). These holes are closed off with two new 2-bands.

iii. Face f is pasted to itself with a twist along edge e, so that the union of the f-band and the e-band is a Möbius band. Then deleting the f-band and the e-band leaves only one hole, which is closed off with one new 2-band.

The following theorem follows immediately from the preceding description.

Theorem 3.3.5. *Let $G \to S$ be a cellular graph imbedding, and let e be an edge of the graph. Let F be the set of faces for $G \to S$, and let F' be the set of faces of the imbedding obtained by edge-deletion surgery at e. Then in case* i, $\#F' = \#F - 1$, *and the resulting surface is homeomorphic to S; in case* ii, $\#F' = \#F + 1$; *and in case* iii, $\#F' = \#F$. □

Another variety of imbedding modification is simpler, since it changes only the imbedded graph, not the surface. Let f be a face of an imbedding $G \to S$, let d and e be consecutive edges in the boundary walk of f with common vertex v, and let u and w be the other endpoints of d and e, respectively. To "slide d along e", one adds a new edge d' from u to w and places its image across face f so that it is adjacent to d in the rotation at u and adjacent to e in the rotation at w, and one then removes edge d from the graph. This is illustrated in Figure 3.24.

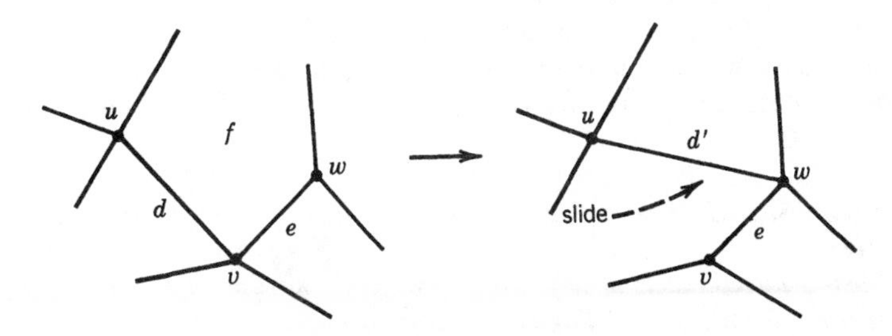

Figure 3.24. Sliding edge d along edge e.

Remark 3.3.2. *Suppose that the edge-sliding operation is performed on a locally oriented surface. If edge d is type* 0 *and edge e is type* 1, *then the new edge d′ that results from sliding d along e is type* 1.

3.3.4. Completeness of the Set of Orientable Models

The following lemma is a crucial ingredient of the proof that the surfaces $S_0, S_1, \ldots$ represent all the orientable types. We hope that its proof is intuitively obvious.

Lemma 3.3.6. *Let T be an orientable surface with two boundary components. Let T_1 be an orientable surface obtained from T by identifying one end of a* 1*-band $I \times I$ to an arc in one boundary component and identifying the other end to an arc in the other boundary component. Suppose that T_2 is another orientable surface obtained from T in a similar way. Then T_1 and T_2 are homeomorphic.*

Proof. Suppose that a_i and b_i are the arcs in boundary(T) to which the band for T_i is attached, for $i = 1, 2$. Then there is a homeomorphism of T to itself taking a_1 to a_2 and b_1 to b_2. This homeomorphism can be extended to the attached bands by the product structure of $I \times I$, since neither band is "twisted" (the surfaces T, T_1, and T_2 are all orientable by assumption). The result is a homeomorphism from T_1 to T_2. □

Corollary. *Let $G \to S$ and $H \to T$ be two different one-face imbeddings in orientable surfaces, and let d and e be arbitrary edges of the graphs G and H, respectively. If edge-deletion surgery at d and edge-deletion surgery at e yield homeomorphic surfaces S' and T', respectively, then the original surfaces S and T are homeomorphic.*

Proof. Surgery on an edge of a one-face imbedding in an orientable surface must be of type ii. Thus, it follows from theorem 3.3.5 that the imbeddings $(G - d) \to S'$ and $(H - e) \to T'$ both have two faces, so that their reduced band decompositions have two boundary components. By Remark 3.3.1 on edge-deletion surgery, one can obtain reduced band decompositions for the imbeddings $G \to S$ and $H \to T$, respectively, simply by restoring to each a band from one boundary component to the other. By Lemma 3.3.6, the reduced band decompositions for $G \to S$ and $H \to T$ are homeomorphic surfaces. It follows from Lemma 3.2.1 that S and T are homeomorphic surfaces. □

Example 3.3.1. *It is not visually obvious that the two reduced band decomposition surfaces in Figure* 3.25 *are homeomorphic. However, one can easily verify by tracing along the boundary walks that both surfaces have one boundary component. Moreover, deleting band c from both yields two surfaces that have two boundary components each and are obviously homeomorphic. It follows from*

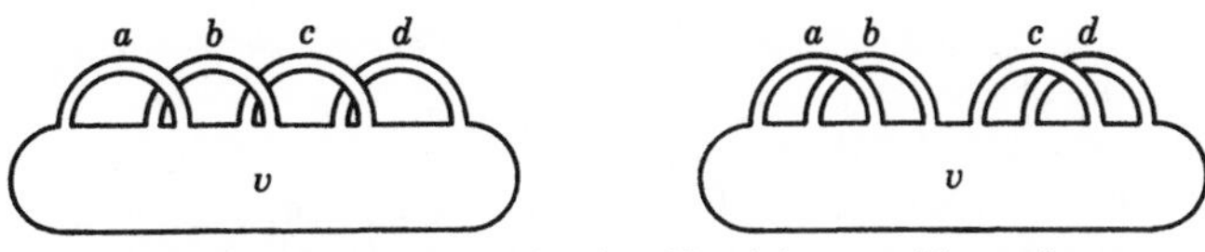

Figure 3.25. Homeomorphic reduced band decomposition surfaces.

Lemma 3.3.6 *that the two surfaces illustrated are also homeomorphic. The corollary implies that the imbedding surfaces induced by the rotation systems*

$$v.\ \ a\,b\,a\,c\,b\,d\,c\,d \quad \text{and} \quad v.\ \ a\,b\,a\,b\,c\,d\,c\,d$$

are homeomorphic.

Theorem 3.3.7. *Let $G \to S$ and $H \to T$ be cellular graph imbeddings into closed, connected, orientable surfaces, such that $\chi(G \to S) = \chi(H \to T)$. Then the surfaces S and T are homeomorphic. Moreover, $\chi(G \to S)$ and $\chi(H \to T)$ are even numbers less than or equal to* 2.

Proof. As the basis for an induction, suppose that $\#E_G + \#E_H = 0$. Then the graphs G and H have only one vertex each, and the imbeddings $G \to S$ and $H \to T$ have one face each. Hence, the surfaces S and T are both 2-spheres, and $\chi(G \to S) = \chi(H \to T) = 2$.

Assume that the inclusion holds when $\#E_G + \#E_H = n$, and suppose that $G \to S$ and $H \to T$ are imbeddings such that $\#E_G + \#E_H = n + 1$. If the imbedding $G \to S$ had more than one face, then the graph G would have an edge e common to two faces. By case i of Theorem 3.3.5, the imbedding $(G - e) \to S$ obtained by surgery at e would have Euler characteristic equal to $\chi(G \to S)$. Thus, the conclusion would follow by induction. Therefore, one may assume that the imbedding $G \to S$ has only one face. Likewise, one may assume that the imbedding $H \to T$ also has only one face. Furthermore, the same argument applied to the dual imbeddings permits one to assume that the graphs G and H have but one vertex each.

Now let d and e be arbitrary edges of the graphs G and H, respectively, and let $G' \to S'$ and $H' \to T'$ be the imbeddings that result from surgery at d and e, respectively. Case ii of Theorem 3.3.5 implies that $\chi(G' \to S') = \chi(G \to S) + 2$ and that $\chi(H' \to T') = \chi(H \to T) + 2$. Thus, $\chi(G' \to S') = \chi(H' \to T')$. Since G and H are both one-vertex graphs, it follows that G', H', S', and T' are each connected. Therefore, by induction, $\chi(G' \to S')$ and $\chi(H' \to T')$ are even numbers less than 2, from which it follows that $\chi(G \to S)$ and $\chi(H \to T)$ are even numbers less than 2. By induction, it follows that the surfaces S' and T' are homeomorphic. By the corollary to Lemma 3.3.6, we infer that the surfaces S and T are homeomorphic. □

Corollary. *Every closed, connected, orientable surface is homeomorphic to one of the standard surfaces* $S_0, S_1, \ldots$.

Proof. Let S be a closed, connected, orientable surface. By Theorem 3.3.7, the surface S has a relative Euler characteristic $\chi(G \to S)$ that is even and less than or equal to 2. (Indeed, Theorem 3.3.7 states that all the relative Euler characteristics for S are even and less than or equal to 2.) Let $g = 1 - 1/2 \cdot \chi(G \to S)$. By Theorem 3.3.1, the surface S_g has a relative Euler characteristic $\chi(H \to S_g) = 2 - 2g = \chi(G \to S)$. It follows from Theorem 3.3.7 that S is homeomorphic to S_g. □

3.3.5. Completeness of the Set of Nonorientable Models

The proof of the completeness of the set $N_1, N_2, \ldots$ is parallel to the orientable case. It begins with an analogy to Lemma 3.3.6.

Lemma 3.3.8. *Let T be a surface with one boundary component, and let T_1 be a surface also with one boundary component, obtained from T by identifying the ends of a* 1*-band to disjoint arcs in the boundary. Let T_2 be another surface with one boundary component obtained from T in the same way. Then T_1 and T_2 are homeomorphic.*

Proof. The proof is essentially the same as that for Lemma 3.3.6, but this time one may observe that both attached bands must have a "twist" so that the surfaces T_1 and T_2 still have only one boundary component. □

Corollary. *Let $G \to S$ and $H \to T$ be two different one-face imbeddings, and let d and e be arbitrary edges of the graphs G and H, respectively. If edge-deletion surgery at d and edge-deletion surgery at e yield one-face imbeddings into homeomorphic surfaces S' and T', respectively, then the original surfaces S and T are homeomorphic.*

Proof. Since the surfaces S' and T' are homeomorphic, and since the imbeddings $(G - d) \to S'$ and $(H - e) \to T'$ are both one-faced, the reduced band decomposition surfaces for $(G - d) \to S'$ and for $(H - e) \to T'$ are homeomorphic surfaces with one boundary component. By Remark 3.3.1 and Lemma 3.3.8, it follows that the reduced band decomposition surfaces for $G \to S$ and $H \to T$ are homeomorphic surfaces with one boundary component each. Lemma 3.2.1 now implies that S and T are homeomorphic. □

Example 3.3.2. *Both the surfaces in Figure* 3.26 *have one boundary component. Moreover, when the b-band is deleted from both, the resulting surfaces are obviously homeomorphic. It follows from Lemma* 3.3.8 *that the two surfaces illustrated are homeomorphic, even though this might not be immediately obvious.*

Figure 3.26. Homeomorphic nonorientable reduced band decomposition surfaces.

The Corollary implies that the imbedding surfaces induced by the rotation systems

$$v. \quad a^1\, b\, a^1\, b \quad \text{and} \quad v. \quad a^1\, a^1\, b^1\, b^1$$

are homeomorphic.

Theorem 3.3.9. *Let $G \to S$ and $H \to T$ be cellular, locally oriented graph imbeddings into closed, connected, nonorientable surfaces, such that $\chi(G \to S) = \chi(H \to T)$. Then the surfaces S and T are homeomorphic. Moreover, $\chi(G \to S)$ and $\chi(H \to T)$ are less than or equal to* 1.

Proof. As for orientable surfaces, the proof is by induction on the sum $\#E_G + \#E_H$. The induction begins with $\#E_G + \#E_H = 2$, since by hypothesis, the imbeddings $G \to S$ and $H \to T$ must have at least one type-1 edge apiece. If the sum is 2, then in both graphs G and H, the single edge must be an orientation-reversing loop. Thus, the reduced band decomposition surfaces for the imbeddings $G \to S$ and $H \to T$ are both Möbius strips, from which it follows that S and T are both homeomorphic to the projective plane N_1. It also follows that $\chi(G \to S) = \chi(H \to T) = 1 - 1 + 1 = 1$.

Now suppose that the conclusion holds whenever $\#E_G + \#E_H \leq n$ for some $n \geq 2$. Furthermore, suppose that $G \to S$ and $H \to T$ are imbeddings satisfying the hypotheses of the theorem and the condition $\#E_G + \#E_H = n + 1$. As in Theorem 3.3.7 a type-i edge-deletion surgery argument enables us to assume that both G and H have one vertex and that both imbeddings $G \to S$ and $H \to T$ have one face.

Since $\#E_G + \#E_H > 2$ and since $\chi(G \to S) = \chi(H \to T)$, both the bouquets G and H have more than one loop. Since S and T are nonorientable, both imbeddings have at least one orientation-reversing loop. By Remark 3.3.2 on edge sliding, it may be assumed that all the edges of G and H are orientation-reversing loops.

Now let d and e be edges of the graphs G and H, respectively, whose duals d^* and e^* are orientation-reversing loops in the respective dual imbeddings. (Since the surfaces S and T are nonorientable, there must exist at least one orientation-reversing loop in both of the respective dual imbeddings.) Next let $G' \to S'$ and $H' \to T'$ be the imbeddings that result from edge-deletion surgery at d and e, respectively. Since the dual loops d^* and e^* are both orientation-reversing, it follows that both surgeries are type iii, which implies, by Theorem 3.3.5, that the resulting imbeddings $G' \to S'$ and $H' \to T'$ have one face each.

The surfaces S' and T' are both nonorientable, since the original imbeddings $G \to S$ and $H \to T$ had at least one orientation-reversing loop each besides the deleted loops d and e, respectively. Since $\chi(G' \to S') = \chi(G \to S) + 1$ and $\chi(H' \to T') = \chi(G \to T) + 1$, it follows that $\chi(G' \to S') = \chi(H' \to T')$. Moreover,

$$\#E_{G'} + \#E_{H'} = (\#E_G - 1) + (\#E_H - 1) = \#E_G + \#E_H - 2 = n - 1 < n$$

By the induction hypothesis, the surfaces S' and T' are homeomorphic, and

$$\chi(G' \to S') = \chi(H' \to T') \leq 1$$

The corollary to Lemma 3.3.8 now implies that the surfaces S and T are homeomorphic. Moreover,

$$\chi(G \to S) = \chi(H \to T) = \chi(G' \to S') - 1 \leq 0 \leq 1. \quad \square$$

Corollary. *Every closed, connected, nonorientable surface is homeomorphic to one of the standard surfaces $N_1, N_2, \ldots$.*

Proof. Let S be a closed, connected, nonorientable surface. By Theorem 3.3.2 it has a relative Euler characteristic $\chi(G \to S)$ less than or equal to 1. Let $k = 2 - \chi(G \to S)$. By Theorem 3.3.2, the surface N_k has a relative Euler characteristic $\chi(H \to N_k) = 2 - k = \chi(G \to S)$. Theorem 3.3.7 implies that S is homeomorphic to N_k. $\square$

Theorems 3.3.3 and 3.3.4 and the corollaries to Theorems 3.3.7 and 3.3.9 justify the definition of the "Euler characteristic" $\chi(S)$ of a surface as the value of the formula $\#V - \#E + \#F$ for any cellular imbedding. In particular, $\chi(S_g) = 2 - 2g$ and $\chi(N_k) = 2 - k$.

3.3.6. Exercises

1. For each of the three surfaces in Figure 3.27, decide whether it is orientable, calculate its Euler characteristic, and state to which of the standard surfaces $S_0, S_1, \ldots$ or $N_1, N_2, \ldots$ it is homeomorphic.
2. List the faces of the imbedding obtained by edge-deletion surgery on the edge uv in the imbedding of Figure 3.17. List the faces obtained when the surgery is on edge vw instead.
3. Let e be an edge of the imbedding $G \to S$ whose dual edge e^* is not a loop. Let the imbedding $H \to T$ be the result of surgery on e. Show that the dual imbedding $H^* \to T^*$ can be obtained from the imbedding $G^* \to S^*$ by contracting the edge e^*.

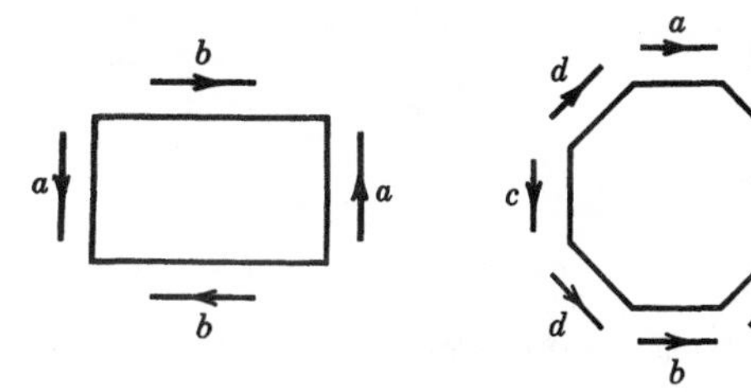

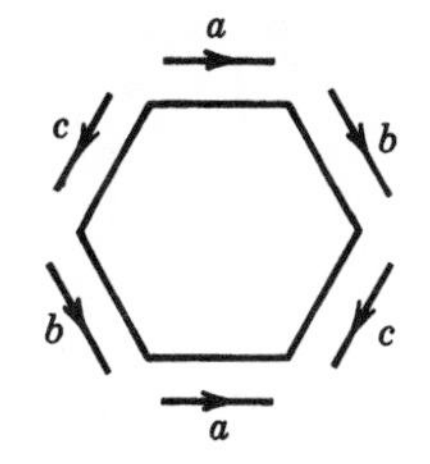

Figure 3.27. Three surfaces.

4. Let e be an edge of a connected graph G such that the graph $G - e$ is disconnected. Show that for every imbedding of G into a surface, the dual edge e^* is a loop. (*Hint:* Use Exercise 3 and connectivity in primal and dual imbeddings.)
5. Give an example of a one-face imbedding $G \to S$ with an edge e such that surgery on e disconnects G and hence S. Why is it important in the proof of Theorem 3.3.7 that we may assume the graphs G and H have only one vertex each?
6. Give an example of a one-face imbedding $G \to N_3$ with an edge e such that surgery on e leaves an orientable surface. Give an example of a one-face imbedding $H \to N_3$ such that every surface obtainable by edge-deletion surgery is nonorientable.
7. Give an example of a one-face, one-vertex imbedding $G \to S$ all of whose edges are type 1 such that surgery on some edge of G results in a two-face imbedding. (*Hint:* It is easiest to start with the two-face imbedding and then to attach a 1-band.)
8. Find cellular imbeddings of the bouquet of three circles in the surfaces N_1, N_2, and N_3. Can imbeddings be found in each case that have all edges type 1?
9. Consider the partial rotation system

$$\begin{array}{ll} u. & \ldots de \ldots \\ v. & \ldots e \ldots \end{array}$$

Give the rotation at vertices u and v after sliding the edge d along the edge e, first when e is type 0, then when e is type 1.
10. Use your answer to Exercise 9 to give a sequence of edge slides that turns the one-vertex imbedding with rotation $ababc^1c^1$ into the one-vertex imbedding with rotation $a^1a^1b^1b^1c^1c^1$. (For example, sliding b along c gives $ab^1ac^1b^1c^1$ and then sliding c along a gives $ab^1ac^1c^1b^1$.)
11. The "disk sum" of two connected, closed surfaces S and T, denoted $S \circ T$, is the surface obtained by removing the interior of a disk from

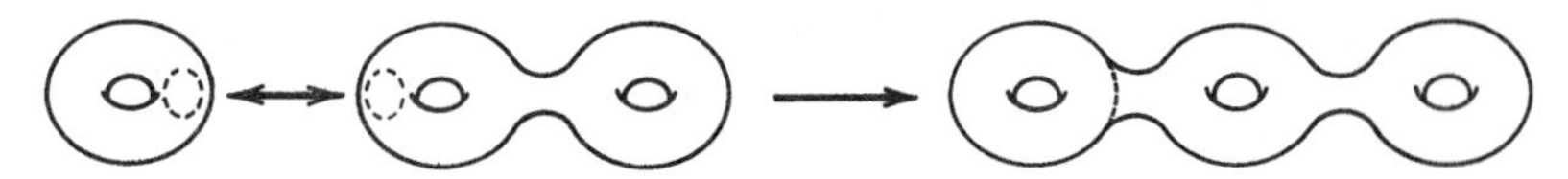

Figure 3.28. The disk sum of two surfaces.

both S and T and then identifying the two resulting boundary components (see Figure 3.28). Prove that $\chi(S \circ T) = \chi(S) + \chi(T) - 2$.

12. Prove that the disk sum defined in Exercise 11 is an associative operation on closed, connected surfaces. What is its unit element? Is the operation commutative?
13. Define the connected surface S to be prime if S is not the sphere and if $S = R \circ T$ implies that either R or T is a sphere. What are the prime surfaces? Give two different factorizations of N_3 into primes. Do orientable surfaces have unique factorizations?
14. Prove that if the surface S contains k disjoint Möbius bands, then $\chi(S) \leq 2 - k$. (*Hint:* Use the fact that if the surface S contains a Möbius band, it can be factored into $S' \circ N_1$.) Conclude that for any nonorientable surface S, $\chi(S) = 2 - k$ if and only if S contains k disjoint Möbius bands but no more.
15. Explain how to imbed a graph in the surface N_{2k+1} so that there exists an edge e on which surgery yields the orientable surface S_k. (*Hint:* See Exercise 13.) What is the minimum number of edge-deletion surgery steps required to make the surface N_{2k} orientable, for $k \geq 1$?

3.4. THE IMBEDDING DISTRIBUTION OF A GRAPH

The "genus range" of a graph G, denoted GR(G), is defined to be the set of numbers g such that the graph G can be cellularly imbedded in the surface S_g. Of course, the minimum number in GR(G) is the genus $\gamma(G)$ of the graph. Sometimes we call $\gamma(G)$ the "*minimum genus*", to create direct contrast with the maximum number in GR(G), which is called the "maximum genus" of G, and is denoted $\gamma_M(G)$. Whereas the study of minimum genus dates back into the 19th century, interest in maximum genus began with Nordhaus, Ringeisen, Stewart, and White (1971, 1972).

Suppose that a certain graph G can be cellularly imbedded in both the surfaces S_m and S_n, and that r is an integer such that $m < r < n$. The "interpolation theorem"—to be proved here as Theorem 3.4.1—for orientable surfaces states that then there exists a cellular imbedding $G \to S_r$.

For any two integers m and n such that $m \leq n$, we define the "interval of integers" $[m, n]$ to be the set of all integers r such that $m \leq r \leq n$. Thus, according to the interpolation theorem, the genus range $GR(G)$ is the interval $[\gamma(G), \gamma_M(G)]$.

Analogously, the "crosscap range" $CR(G)$ is defined to be the set of numbers k such that the graph G can be cellularly imbedded in the surface N_k. The crosscap number $\bar{\gamma}(G)$ is the minimum value in this range. The maximum number in $CR(G)$ is called the "maximum crosscap number" of G, and is denoted $\bar{\gamma}_M(G)$. There is also an interpolation theorem for nonorientable surfaces, from which we infer that $CR(G) = [\bar{\gamma}(G), \bar{\gamma}_M(G)]$.

Because of the interpolation theorems, computations of the limiting values of the imbedding ranges have received great emphasis. In general, the genus and the crosscap number of a graph are difficult to compute, although they have been determined for various special classes of graphs. By way of contrast, the maximum genus is thoroughly tractable, mainly because of its characterization by Xuong (1979). The time needed for computation of the maximum genus is bounded by a polynomial function of the number of vertices, as proved by Furst, Gross, and McGeoch (1985a).

In this section we are concerned primarily with establishing the interpolation theorems, with some relationships between genus and crosscap number, and with computing the maximum crosscap number and maximum genus. Some general statistical questions about the set of all imbeddings will also be raised.

3.4.1. The Absence of Gaps in the Genus Range

One calls the graph imbeddings $G \to S$ and $G \to T$ "adjacent" if there is an edge e in G such that the two imbeddings $(G - e) \to S'$ and $(G - e) \to T'$ obtained by edge-deletion surgery on e are equivalent. Under such circumstances, one also calls the rotation systems for $G \to S$ and $G \to T$ adjacent. In accordance with Remark 3.3.1, this means that two rotation systems are adjacent if and only if the result of erasing some edge symbol e in both of them yields identical rotation systems for the graph $G - e$.

Theorem 3.3.5 implies that the numbers of faces in adjacent imbeddings $G \to S$ and $G \to T$ both differ by at most one from the common number of faces in the equivalent imbeddings $(G - e) \to S'$ and $(G - e) \to T'$. Thus, the numbers of faces in $G \to S$ and $G \to T$ differ by at most 2. It follows that the Euler characteristics $\chi(S)$ and $\chi(T)$ differ by at most 2. Thus, if the imbedding surfaces S and T are orientable, their genera $\gamma(S)$ and $\gamma(T)$ differ by at most 1.

Example 3.4.1. *The three pure rotation projections illustrated in Figure* 3.29 *correspond to mutually adjacent imbeddings, since edge-deletion surgery on the edges in the "nine o'clock" position would make them all equivalent. The left and middle imbedding surfaces visibly have genus* 1. *The third surface has genus* 2.

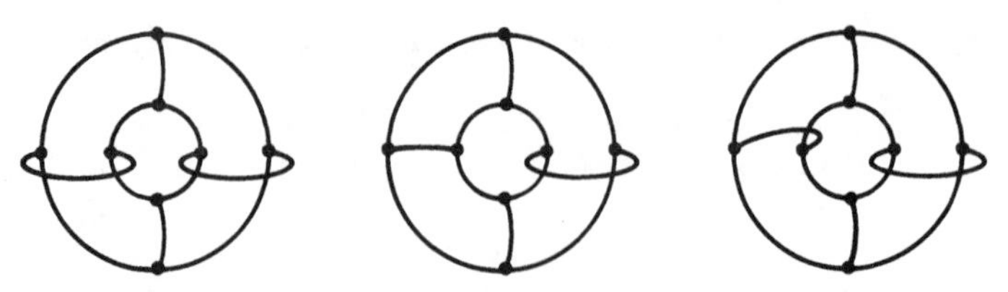

Figure 3.29. Projections of three mutually adjacent rotation systems.

Theorem 3.4.1 [Orientable Interpolation Theorem (Duke, 1966)]. *Let G be a connected graph. Then the genus range* GR(G) *is an unbroken interval of integers.*

Proof. Let L_1 and L_2 be list formats for any two rotation systems of G. Of course, L_2 may be obtained from L_1 by permuting the entries in the various rows. Any such permutation may be accomplished by a sequence of steps that moves one edge symbol at a time. The consecutive list formats in such a sequence represent adjacent imbeddings, by Remark 3.3.1. Thus, there exists a sequence of adjacent cellular imbeddings of the graph G starting with one in the surface of genus $\gamma(G)$ and ending with one in the surface of maximum genus $\gamma_M(G)$. Since adjacent imbedding surfaces differ by at most one in genus, the conclusion follows. □

3.4.2. The Absence of Gaps in the Crosscap Range

In order to obtain the nonorientable interpolation theorem, it is helpful to introduce another form of surgery on an imbedding $G \to S$. To perform "edge-twisting surgery" at the edge e, one first deletes from a band decomposition for $G \to S$ the 2-band or 2-bands that meet the e-band and gives the e-band an extra twist. One then closes the resulting surface with one or two 2-bands, as required.

Remark 3.4.1. *Suppose that edge-twisting surgery at edge e of the imbedding $G \to S$ yields the imbedding $G \to S'$ and that the corresponding rotation systems are R and R', respectively. Then rotation system R' can be obtained from rotation system R simply by changing the orientation type of edge e, that is, either from type 0 to type 1 or from type 1 to type 0.*

Edge-twisting surgery on e falls into only two cases, illustrated in Figure 3.30.

i. The two sides of e lie in separate faces f_1 and f_2. Then deleting the f_1-band and the f_2-band leaves two holes, which are merged into a single hole by twisting the e-band. Since the union of the e-band and

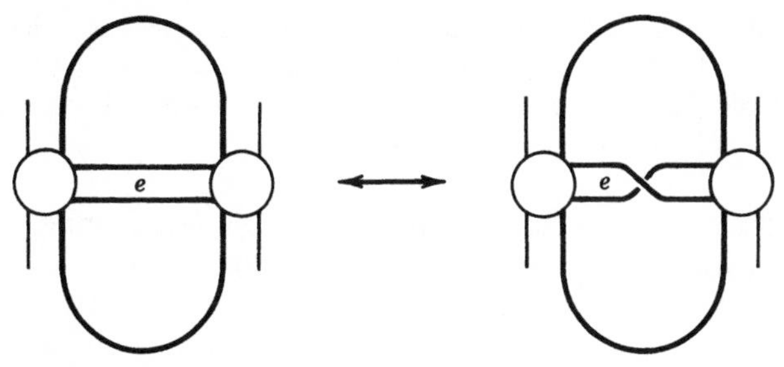

Figure 3.30. Case i of edge-twisting surgery goes left to right, whereas case ii goes right to left. The 2-band boundaries are thickened.

the new 2-band used to close the hole is a Möbius band, the resulting surface is nonorientable, regardless of whether the starting surface is orientable.

ii. Both sides of e lie on the same face f. Then deleting the f-band leaves one hole, which is changed to two holes by twisting the e-band.

Theorem 3.4.2 follows at once from this description.

Theorem 3.4.2. *Let $G \to S$ be a cellular graph imbedding, and let e be an edge of the graph. Let F be the set of faces of $G \to S$, and let F' be the set of faces of the imbedding $G \to S'$ obtained by edge-twisting surgery at e. Then in case* i, $\#F' = \#F - 1$, *and* $\chi(S') = \chi(S) - 1$; *in case* ii, $\#F' = \#F + 1$, *and* $\chi(S') = \chi(S) + 1$. □

Edge-twisting surgery yields not only a proof that there are no gaps in the crosscap range but also an easy formula for the maximum crosscap number. The "Betti number" (also called the "cycle rank") of a connected graph G, denoted $\beta(G)$, is defined by the equation $\beta(G) = 1 - \#V + \#E$, and it equals the number of edges in the complement of a spanning tree for G.

Theorem 3.4.3 (Edmonds, 1965). *Let G be a connected graph. Then* $\bar{\gamma}_M(G) = \beta(G)$.

Proof. If the graph G is a tree, then $\bar{\gamma}_M(G) = 0$, because its maximum crosscap number is achieved by an imbedding in the sphere. By Theorem 1.2.2, the Betti number of any tree is 0. If G is not a tree, then G has an imbedding $G \to S$ with two or more faces. One might construct such an imbedding, for instance, by fitting a polygon boundary to a cycle and then extending the induced partial rotation system in an arbitrary manner to a complete rotation system. For such an imbedding, one may choose an edge e that lies on two different faces and perform edge-twisting surgery, so that the result is a nonorientable manifold, and so that the number of faces is reduced by 1. Iteration of edge-twisting surgery yields a one-face cellular imbedding in a nonorientable surface N, so that $\chi(N) = \#V - \#E + 1$. Consequently, $\bar{\gamma}(N) = 2 - \chi(N) = 1 - \#V + \#E = \beta(G)$. □

Theorem 3.4.4 [Nonorientable Interpolation Theorem (Stahl 1978)]. *Let G be a connected graph. Then the crosscap range* CR(G) *is an unbroken interval of integers.*

Proof. If G is a tree, then CR(G) is the degenerate interval [0, 0]. Otherwise, starting with an imbedding $G \to N$ into a surface that realizes the crosscap number, one performs a succession of edge-twisting surgeries. At each stage, one selects an edge that lies in two faces of the imbedding for that stage. Thus, in $\bar{\gamma}_M(G) - \bar{\gamma}(G)$ steps, one has obtained a cellular imbedding into every surface N_r such that r lies in the interval $[\bar{\gamma}(G), \bar{\gamma}_M(G)]$. □

The Betti number of a graph also gives an upper bound for the maximum genus of a graph, namely, $\gamma_M(G) \leq \lfloor \beta(G)/2 \rfloor$ with equality occurring only if the graph G has an orientable imbedding having one or two faces. However, unlike the maximum crosscap number, this upper bound for the maximum genus is not always achieved. One might hope that the Betti number also provides a useful upper bound on the minimum genus. Indeed, Duke (1971) once conjectured that $\gamma(G) \leq \lfloor \beta(G)/4 \rfloor$. Many counterexamples were found to Duke's conjecture. Milgram and Ungar (1977) found perhaps the most discouraging result: given any $\epsilon > 0$, there are infinitely many graphs G such that $\gamma(G)/\beta(G) > \frac{1}{2} - \epsilon$. In other words, we cannot obtain an upper bound for $\gamma(G)$ of the form $c\beta(G)$ that is any better than $\beta(G)/2$.

3.4.3. A Genus-Related Upper Bound on the Crosscap Number

Edge-twisting surgery now enables us to establish an upper bound on the crosscap number of a graph, based on its genus. It follows that the crosscap range is always nearly twice as large as the genus range of a graph or larger.

Theorem 3.4.5. *Let G be a connected graph. Then $\bar{\gamma}(G) \leq 2\gamma(G) + 1$.*

Proof. If G is a tree, then $\gamma(G) = \bar{\gamma}(G) = 0$. Otherwise, let e be an edge that lies on two faces of an imbedding $G \to S$ into an orientable surface of genus $\gamma(G)$. Then edge-twisting surgery on e yields an imbedding of G into a nonorientable surface of crosscap number $2\gamma(G) + 1$, by Theorem 3.4.2. □

Corollary. *Let G be a connected graph. Then*

$$\#\mathrm{CR}(G) \geq 2\#\mathrm{GR}(G) - 2$$

Proof. If the maximum genus of G is realized by an imbedding in the surface S, then

$$2 - 2\gamma_M(G) = \chi(S) = \#V - \#E + \#F \geq \#V - \#E + 1$$

from which it follows that $\bar{\gamma}_M(G) = \beta(G) \geq 2\gamma_M(G)$. Thus,

$$\begin{aligned}\#\mathrm{CR}(G) &= \bar{\gamma}_M(G) - \bar{\gamma}(G) + 1\\ &\geq 2\gamma_M(G) - (2\gamma(G) + 1) + 1\\ &= 2(\gamma_M(G) - \gamma(G) + 1) - 2\\ &= 2\#\mathrm{GR}(G) - 2 \quad \square\end{aligned}$$

3.4.4. The Genus and Crosscap Number of the Complete Graph K_7

The upper bound on $\gamma(G)$ given by Theorem 3.4.5 is essentially unimprovable. What this means is that sometimes a graph can be imbedded in an orientable surface S_g, of Euler characteristic $2 - 2g$, but cannot be imbedded in the nonorientable surface N_{2g} of the same characteristic. In particular, Heffter (1891) proved that the complete graph K_7 can be imbedded in the torus S_1, while Franklin (1934) proved that K_7 cannot be imbedded in the Klein bottle N_2. We combine these two results in the following theorem.

Theorem 3.4.6. *The genus of the complete graph* K_7 *is* 1, *and the crosscap number of* K_7 *is* 3.

Proof. By Theorem 1.4.2, no imbedding of K_7 has more than $(2/3)\#E = (2/3)(21) = 14$ faces. Thus, the Euler characteristic of an imbedding surface for K_7 is at most $7 - 21 + 14 = 0$. Hence, $\gamma(K_7) \geq 1$ and $\bar{\gamma}(K_7) \geq 2$. Figure 3.31 shows an imbedding $K_7 \to S_1$, which completes the proof that $\gamma(K_7) = 1$. Figure 5.2 contains a more elegant depiction of a weakly equivalent imbedding.

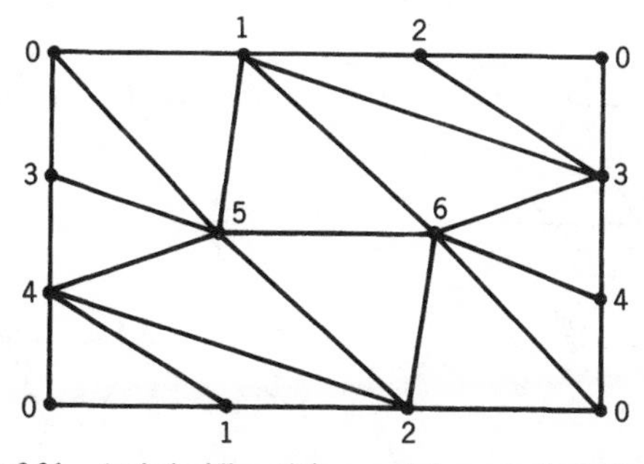

Figure 3.31. An imbedding of the complete graph K_7 in the torus.

It follows from Theorem 3.4.5 that $\bar{\gamma}(K_7) \leq 3$. Thus, to complete the proof that $\bar{\gamma}(K_7) = 3$, it is sufficient to show that K_7 has no imbedding in N_2. By the invariance of Euler characteristic, if K_7 did have an imbedding in N_2, there would have to be 14 faces. Since K_7 has no 1-cycles or 2-cycles, each of the 14 faces would have to be three-sided. Therefore, suppose that $K_7 \to S$ is an imbedding such that every face is three-sided. It is now to be proved that the surface S is orientable, thereby establishing that K_7 is not imbeddable in the Klein bottle N_2.

Regarding K_7 as a Cayley graph for the cyclic group $\mathscr{Z}_7$, we label its vertices $0, 1, \ldots, 6$. It may be assumed, without loss of generality, that the rotation at vertex 0 is

0. 123456

The edges incident on vertex 0 form a spanning tree, and we locally orient the vertices so that all six of these edges are type 0. The left side of Figure 3.32 shows a clockwise partial rotation projection "centered" at vertex 0.

Since each of the six faces incident on vertex 0 is three-sided, we may infer from the given rotation at vertex 0 and the Face Tracing Algorithm the following partial rotation system (in vertex format):

0. 123456
1. 206 ...
2. 301 ...
3. 402 ...
4. 503 ...
5. 604 ...
6. 105 ...

If the edge 12 had type 1, then the sequence 10, 02, 21 in an execution of the Face Tracing Algorithm would be followed by some edge $1x$, $x \neq 0$, thereby creating a face that is not three-sided. Thus, the edge 12 must have type 0.

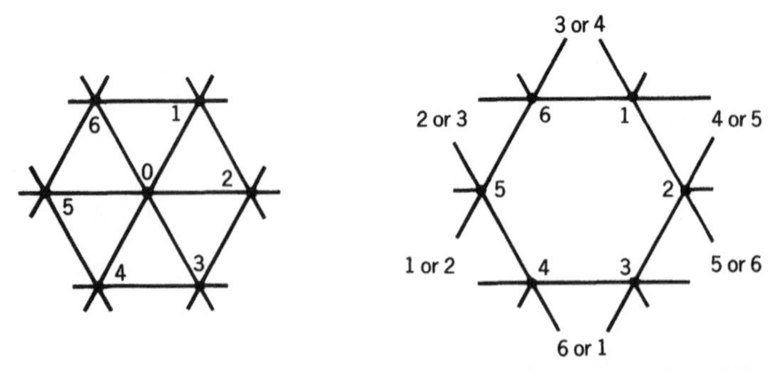

Figure 3.32. Partial rotation projections for putative triangular imbeddings of K_7.

Similarly, each of the edges 23, 34, 45, 56, and 61 must have type 0, as indicated. We shall see that there are but two ways to extend this partial system to a rotation system that has only three-sided faces, and that all of the other edges in the complete system also have type 0, thereby proving that the surface is orientable.

The edge 16 must lie in a second triangle besides 016. The third vertex of this triangle must be 3 or 4, since the edges 56 and 12 are already in use. By similar reasoning applied to the five edges 12, 23, 34, 45, and 56, we arrive at the partial rotation projection shown on the right of Figure 3.32, where the vertex 0 and its incident edges are now omitted, to simplify the picture.

Suppose that 3 is the third vertex of the second triangle containing the edge 16. Then the third vertex of the second triangle containing the edge 56 must be 2. Continuing in this manner around the figure, we obtain the partial rotation projection shown at the left of Figure 3.33. The remaining free edge at each vertex is now determined, thereby yielding the following complete rotation system, whose projection is illustrated at the right of Figure 3.33:

0. 1 2 3 4 5 6
1. 2 0 6 3 5 4
2. 3 0 1 4 6 5
3. 4 0 2 5 1 6
4. 5 0 3 6 2 1
5. 6 0 4 1 3 2
6. 1 0 5 2 4 3

If the edge 13 had type 1, then the successive edges 61 and 13 would be followed by the edge 35 in the Face Tracing Algorithm execution, which blocks the completion of a three-sided face. Thus, the edge 13 must have type 0. Similarly, the edges 24, 35, 46, 51, and 62 must have type 0. The only

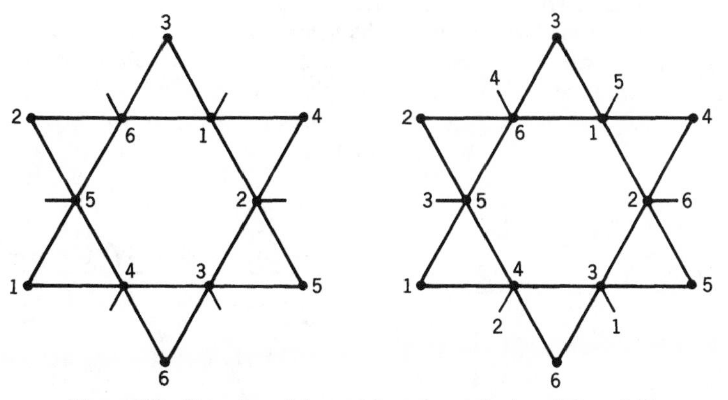

Figure 3.33. Extensions of the partial rotation projection of Figure 3.32.

remaining edges of undetermined type are now 36, 25, and 14. However, in order for the face tracing sequence 61, 13, 36 to continue with 61, rather than 64, it is necessary that the edge 36 have type 0. Similarly, the edges 25 and 14 have type 0.

If the third vertex of the second triangle containing the edge 16 were 4, instead of 3, then the same arguments would again yield a pure rotation system. It follows that the complete graph K_7 has no imbedding in the Klein bottle, and that $\tilde{\gamma}(K_7) = 3$. □

3.4.5. Some Graphs of Crosscap Number 1 but Arbitrarily Large Genus

In view of Theorem 3.4.5, one might expect to find a general upper bound on the genus of a graph, based on its crosscap number. However, an example due to Auslander et al. (1963) shows that no such upper bound exists. The key to understanding this example is the idea of cutting a surface apart along a nonseparating cycle of an imbedded graph, which we have already encountered in the proof of the invariance of Euler characteristic.

Let $G \to S$ be a graph imbedding, let B be a (nonreduced) band decomposition for $G \to S$, and let C be the cycle

$$v_1, e_1, v_2, e_2, \ldots, v_n, e_n, v_1$$

in G. Then the bands for the vertices and edges of C form a topological neighborhood $N(C)$ that is homeomorphic to an annulus, if C is orientation preserving, or to a Möbius band, if C is orientation reversing. In either case, the cycle C is a topological loop that runs around the middle of the neighborhood.

Suppose that we cut the surface S along the cycle C, that we regard the 0-band for each vertex v_i as having been split into a v_i'-band and v_i''-band, and that we regard the 1-band for each edge e_j as having been split into an e_j'-band and an e_j''-band. This is illustrated in Figure 3.34.

Suppose further that we cap off the hole or holes created by cutting on C with one or two new 2-bands, as required. The graph imbedding $G' \to S'$ consistent with the resulting band decomposition is said to have been obtained by "splitting" the imbedding $G \to S$ "along the cycle C". An easy counting

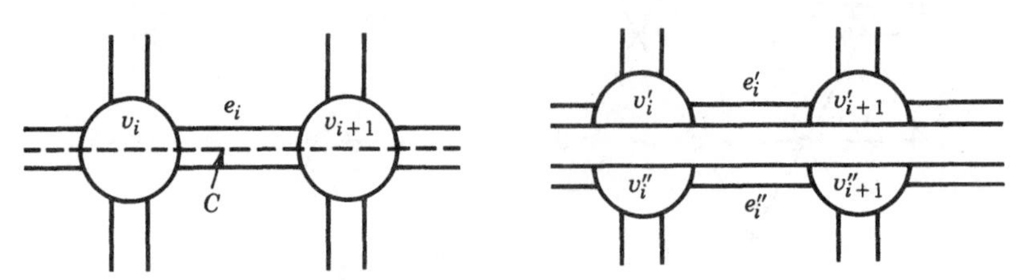

Figure 3.34. Splitting an imbedding along a cycle.

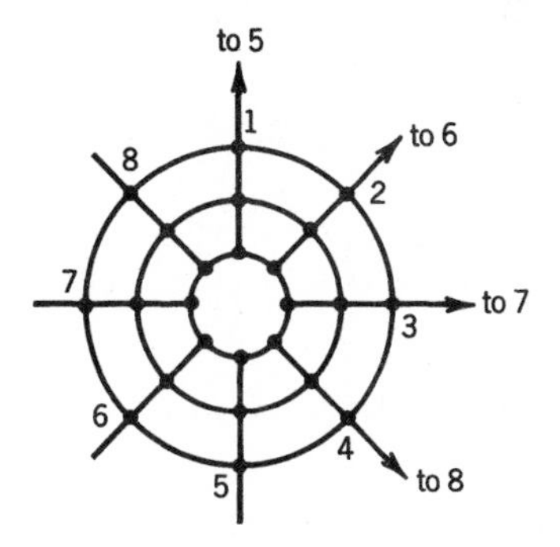

Figure 3.35. The graph A_n (for $n = 2$).

argument on vertices, edges, and faces, as given in the proofs of Theorems 3.3.7 and 3.3.9, yields the following theorem.

Theorem 3.4.7. *Let $G \to S$ be a cellular graph imbedding, let C be a cycle of G, and let $G' \to S'$ be the imbedding obtained by splitting $G \to S$ along C. If C is orientation preserving, then $\chi(S') = \chi(S) + 2$. If C is orientation reversing, then $\chi(S') = \chi(S) + 1$.*

Example 3.4.2 (Auslander et al. 1963). *For each positive integer n, the graph A_n is formed from $n + 1$ concentric cycles, each of length $4n$, plus some additional edges, as shown in Figure 3.35. There are $4n^2$ "inner" edges that connect the $n + 1$ cycles to each other and $2n$ "outer" edges that adjoin antipodal vertices on the outermost cycle.*

Suppose that the $n + 1$ concentric cycles of the graph A_n and the inner edges are imbedded on a disk so that the vertices of the outermost cycle appear equally spaced on the boundary of the disk. Suppose also that a rectangular sheet with $2n$ lines from top to bottom has its left and right sides identified so that it becomes a Möbius band. If the boundary of the Möbius band is identified to the boundary of the disk so that the $2n$ lines become images of the outer edges of A_n, then the result is an imbedding $A_n \to N_1$. Figure 3.36 shows that the graph A_1 contains a subdivision of $K_{3,3}$. Since A_n contains a subdivision of A_1, for all $n \geq 1$, it follows that $\tilde{\gamma}(A_n) = 1$.

It is clear from Figure 3.36 that $\gamma(A_1) = 1$. Suppose, by way of induction, that $\gamma(A_{n-1}) \geq n - 1$. Then let C be the innermost cycle of the graph A_n, as it appears in Figure 3.35, and let $A_n \to S$ be an orientable cellular imbedding.

First, assume that the cycle does not separate the surface S. Then let the imbedding $A'_n \to S'$ be the result of splitting the imbedding $A_n \to S$ along the cycle C. By Theorem 3.4.7, $\gamma(S) = \gamma(S') + 1$. Since the graph A_{n-1} is a subdivision of a subgraph of the graph A_n, it follows from the induction hypothesis that $\gamma(S') \geq n - 1$. Thus $\gamma(S) \geq n$.

On the other hand, one might assume that the cycle C does separate the surface S. Since the graph $A_n - C$ is connected, the cycle C must bound a

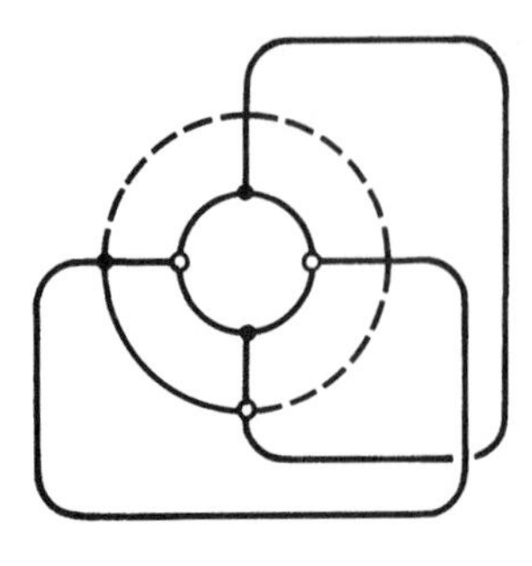

Figure 3.36. A subdivision of the Kuratowski graph $K_{3,3}$ in the graph A_1.

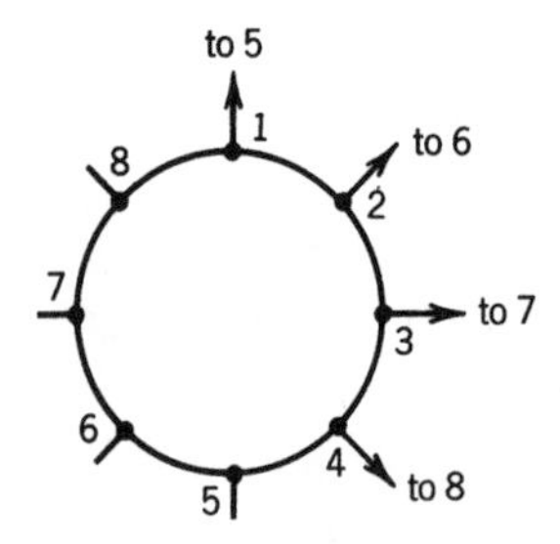

Figure 3.37. A rotation projection for the imbedding $Y_n \to T$.

face of the cellular imbedding $A_n \to S$. Now suppose that images of all the edges on the n other concentric cycles of the graph A_n are deleted, and that every resulting vertex of valence 2 is smoothed over. The result is a possibly noncellular imbedding $Y_n \to S$. If every noncellular face of $Y_n \to S$ is replaced by a 2-cell, then one obtains the imbedding $Y_n \to T$ whose rotation projection is shown in Figure 3.37. Obviously, $\gamma(T) \leq \gamma(S)$.

An elementary face-tracing argument shows that the imbedding $Y_n \to T$ has only two faces, so that $\chi(T) = 4n - 6n + 2 = 2 - 2n$, which implies that $\gamma(T) = n$. It follows that $\gamma(S) \geq n$.

We conclude from this two-pronged inductive argument that $\gamma(A_n) \geq n$. Since there is only one outer face in the imbedding of A_n given by the rotation projection of Figure 3.35, it follows from a calculation of Euler characteristic that the imbedding surface has genus n. Thus $\gamma(A_n) = n$.

3.4.6. Maximum Genus

The simple formula $\bar{\gamma}_M(G) = \beta(G)$ for maximum crosscap number is established by Theorem 3.4.3. The problem of calculating the maximum genus $\gamma_M(G)$ is not so easily solved, but N. H. Xuong has demonstrated that it is still much easier than computing the genus $\gamma(G)$. The immediate goal is to determine which graphs have orientable one-face imbeddings. A survey of results on maximum genus up to the time of Xuong's characterization is given by Ringeisen (1978).

Two edges are called "adjacent" if they have a common endpoint.

Lemma 3.4.8. *Let d and e be adjacent edges in a connected graph G such that $G - d - e$ is a connected graph having an orientable one-face imbedding. Then the graph G has a one-face orientable imbedding.*

Proof. Let $V(d) = \{u, v\}$, and let $V(e) = \{v, w\}$. First, extend the one-face imbedding $(G - d - e) \to S$ to a two-face imbedding $(G - e) \to S$ by placing the image of d across the single face. Of course, the vertex v lies on both faces. Thus, if one attaches a handle from one face of $(G - e) \to S$ to the other, one may then place the image of edge e so that it runs across the handle, thereby creating a one-face imbedding $G \to S'$. □

Example 3.4.3. *Consider the complete graph K_5 as a Cayley graph with vertices* $0, 1, 2, 3, 4$. *Then the spanning tree T with edges* 01, 12, 23, *and* 34 *has a one-face imbedding in the sphere. By Lemma* 3.4.8, *the graph* $T' = T + 04 + 42$ *also has a one-face imbedding, as does the graph* $T'' = T' + 20 + 03$, *as does* $K_5 = T'' + 31 + 14$. *Thus* $\gamma_M(K_5) = 3$.

Lemma 3.4.9. *Let G be a connected graph such that every vertex has valence at least* 3, *and let G have a one-face orientable imbedding $G \to S$. Then there exist adjacent edges d and e in G such that $G - d - e$ has a one-face orientable imbedding.*

Proof. Let d be an edge of G whose two occurrences in the single boundary walk of the imbedding $G \to S$ are the closest together, among all edges of G. Then the boundary walk can be written in the form dAd^-B, where no edge appears twice in the subwalk A. Since the graph G has no vertex of valence 1, the subwalk A is nonempty, so that it has a first edge e. By case ii of Theorem 3.3.5, edge-deletion surgery on edge d in the imbedding $G \to S$ yields a two-face imbedding $(G - d) \to S'$. The boundary walks of the two faces are A and B, and the edge e appears in both A and B. Thus, by case i of Theorem 3.3.5, the result of edge-deletion surgery on e in the imbedding $(G - d) \to S'$ is a one-face imbedding of $G - d - e$. □

The "deficiency $\xi(G, T)$ of a spanning tree" T for a connected graph G is defined to be the number of components of $G - T$ that have an odd number of edges. The "deficiency $\xi(G)$ of the graph" G is defined to be the minimum of $\xi(G, T)$ over all spanning trees T.

Example 3.4.4. *On the left in Figure* 3.38 *is a spanning tree T for a graph G such that* $\xi(G, T) = 3$. *On the right is a spanning tree T' for the same graph G such that* $\xi(G, T') = 1$. *Since the edge complement of any spanning tree for this graph has* $1 - 7 + 11$ *edges, an odd number, the deficiency of any spanning tree must be at least one. Therefore* $\xi(G) = \xi(G, T') = 1$.

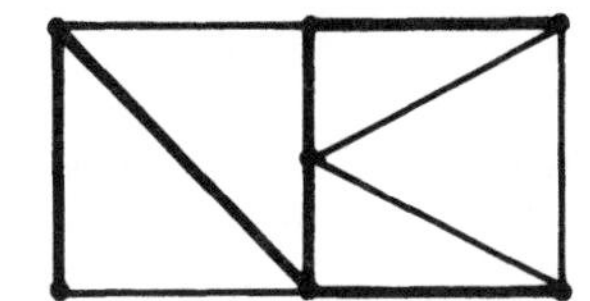
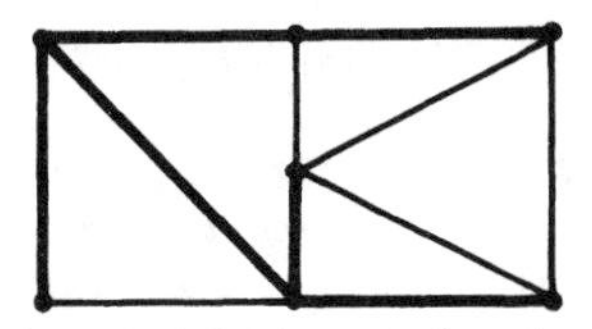

Figure 3.38. Two spanning trees for the same graph, one of deficiency 3, the other of deficiency 1.

Lemma 3.4.10. *Let T be a spanning tree for a graph G, and let d and e be a pair of adjacent edges in $G - T$. If $\xi(G - d - e, T) = 0$, then $\xi(G, T) = 0$.*

Proof. Every component of $G - d - e - T$ that meets either of the edges d or e has an even number of edges, since $\xi(G - d - e, T) = 0$. The number of edges in the component of $G - T$ that contains the edges d and e is two plus the sum of these even numbers. All the other components of $G - T$ have evenly many edges, as in $G - d - e - T$. Thus, $\xi(G, T) = 0$. □

Lemma 3.4.11. *Let G be a graph other than a tree, and let T be a spanning tree such that $\xi(G, T) = 0$. Then there are adjacent edges d and e in $G - T$ such that $\xi(G - d - e, T) = 0$.*

Proof. Let H be a nontrivial component of $G - T$. Since H is connected and has at least two edges, there are adjacent edges d and e in H such that $H - d - e$ has at most one nontrivial component. (See Exercise 16.) Thus, $\xi(G - d - e, T) = 0$. □

Theorem 3.4.12 (Xuong, 1979). *Let G be a connected graph. Then G has a one-face orientable imbedding if and only if $\xi(G) = 0$.*

Proof. As the basis for an induction, one observes that if $\#E_G = 0$, then both clauses of the conclusion are trivially true. Next, assume that the conclusion holds for any graph with n or fewer edges, and let G be a graph with $n + 1$ edges.

As a preliminary, suppose that G has a vertex v of valence 1 or 2, and let G' be the graph obtained by contracting an edge incident on vertex v. Then obviously, the graph G has a one-face orientable imbedding if and only if the graph G' does. Also, $\xi(G) = 0$ if and only if $\xi(G') = 0$. Since the graph G' has one edge less than the graph G, it follows from the induction hypothesis that G' has a one-face orientable imbedding if and only if $\xi(G') = 0$. The conclusion follows immediately.

In the main case, every vertex of G has valence 3 or more. Suppose first that G has a one-face orientable imbedding. By Lemma 3.4.9, there exist adjacent edges d and e in G such that $G - d - e$ has a one-face orientable imbedding. It follows from the induction hypothesis that $\xi(G - d - e) = 0$,

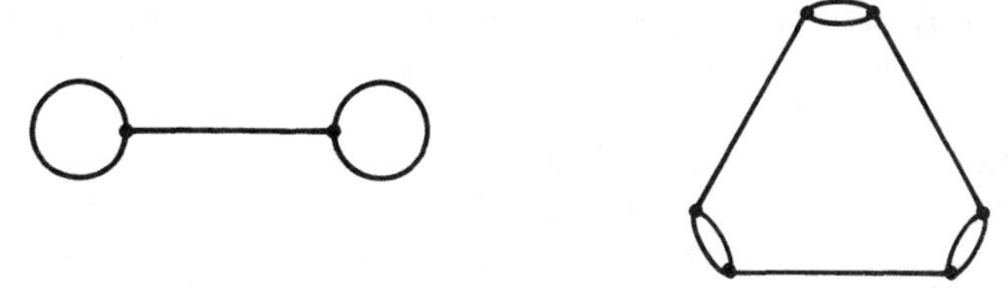

Figure 3.39. Two graphs that have deficiency 2.

so there is a spanning tree T in $G - d - e$ such that $\xi(G - d - e, T) = 0$. Of course, the tree T also spans G. By Lemma 3.4.10, it follows that $\xi(G, T) = 0$, which implies that $\xi(G) = 0$.

Conversely in the main case, one may suppose that $\xi(G) = 0$, so that there is a spanning tree T such that $\xi(G, T) = 0$. Since G has no vertex of valence 1, it is not a tree. By Lemma 3.4.11, it follows that there are adjacent edges d and e in $G - T$ such that $\xi(G - d - e, T) = 0$. Thus, $\xi(G - d - e) = 0$. By the induction hypothesis, the graph $G - d - e$ has a one-face orientable imbedding. By Lemma 3.4.8, so does the graph G. □

Example 3.4.5. *The dumbbell graph G on the left of Figure 3.39 has only one spanning tree, from which it follows that $\xi(G) = 2$. It is easily verified that the removal of any pair of adjacent edges from the graph G' on the right of Figure 3.39, followed by a sequence of edge contractions to eliminate vertices of valence 1 or 2, yields the dumbbell graph. Therefore by Lemma 3.4.9, it follows that $\beta(G') = 2$ also.*

A graph is "k-edge-connected" if the removal of fewer than k edges from G still leaves a connected graph. The two graphs in Example 3.4.5 can be generalized to obtain 1-edge-connected and 2-edge-connected planar graphs with arbitrarily large deficiency. On the other hand, Kundu (1974) and Jaeger (1976) have shown that every 4-edge-connected graph G has a spanning tree T whose edge complement $G - T$ is connected. Thus if G is 4-edge-connected, then $\xi(G) = 0$ or 1, according to whether $\beta(G)$ is even or odd.

Theorem 3.4.13 (Xuong, 1979). *Let G be a connected graph. Then the minimum number of faces in any orientable imbedding of G is exactly $\xi(G) + 1$.*

Proof. An equivalence to the conclusion is the statement that the graph G has an orientable imbedding with $n + 1$ or fewer faces if and only if $\xi(G) \leq n$. The proof of this equivalent statement is by induction on the number n. It holds for $n = 0$, by Theorem 3.4.12, and we now assume that it holds for all values of k less than n, where $n > 0$.

First, we suppose that $G \to S$ is an orientable imbedding with $\#F_G = n + 1$. Then perform edge-deletion surgery on an edge e common to two faces of the imbedding $G \to S$. By case i of Theorem 3.3.5, the resulting imbedding $(G - e) \to S'$ has n faces. From the induction hypothesis, it follows that $\xi(G - e) \leq n - 1$. Therefore, $\xi(G) \leq n$.

Conversely, one may suppose that $\xi(G) = n$, so that there is a spanning tree T in G such that $\xi(G, T) = n$. Let H be a component of $G - T$ with an odd number of edges. Certainly the subgraph H has an edge e that does not disconnect H or such that one endpoint of e has valence 1 in H. Accordingly, $\xi(G - e, T) = n - 1$. By the induction hypothesis, the graph $G - e$ has an orientable imbedding with at most n faces. Therefore the graph G has an orientable imbedding with at most $n + 1$ faces. □

Corollary (Xuong, 1979). *Let g be a connected graph. Then $\gamma_M(G) = \frac{1}{2}(\beta(G) - \xi(G))$.*

Proof. Let $g = \gamma_M(G)$. Then $2 - 2g = \#V - \#E + (\xi(G) + 1)$, by Theorem 3.4.13. It follows that $g = \frac{1}{2}(\beta(G) - \xi(G))$. □

Example 3.4.6. *Let T be the tree in the complete graph K_n consisting of all edges incident on a particular vertex. Then*

$$\xi(K_n, T) = \begin{cases} 0 & \text{if } \binom{n-1}{2} \text{ is even} \\ 1 & \text{if } \binom{n-1}{2} \text{ is odd} \end{cases}$$

or equivalently,

$$\xi(K_n, T) = \begin{cases} 0 & \text{if } n \equiv 1 \text{ or } 2 \text{ modulo } 4 \\ 1 & \text{if } n \equiv 0 \text{ or } 3 \text{ modulo } 4 \end{cases}$$

It follows that $\gamma_M(K_n) = \lfloor \beta(K_n)/2 \rfloor$.

The obvious computational problem presented by Theorem 3.4.13 is to calculate the deficiency of a graph. The number of spanning trees is exponential, and no polynomial-time algorithm was found in the immediate years after Xuong published his characterization. Ultimately, Furst, Gross, and McGeoch (1985a) developed a polynomial-time algorithm involving a reduction to matroid parity.

3.4.7. Distribution of Genus and Face Sizes

Suppose that a graph G has vertices $v_1, \ldots, v_n$ of respective valences $d_1, \ldots, d_n$. Then the total number of orientable imbeddings is

$$\prod_{i=1}^{n} (d_i - 1)!$$

Gross and Furst (1985) raise the problem of determining exactly how many of these imbeddings lie in each surface of the genus range. From such a determination, one would immediately know the minimum genus and the maximum genus. Furst, Gross, and Statman (1985b) combined topological methods and enumerative techniques to obtain a genus distribution formula for two infinite classes of graphs with expanding genus ranges. Gross, Robbins, and Tucker (1986) calculated the genus distributions of bouquets of circles, with the aid of a formula of Jackson (1986).

Another problem raised by Gross and Furst (1985) is to compute the distribution of face sizes over all imbeddings of a graph, which appears to be harder than computing genus distributions. They introduce an extensive hierarchy of invariants of the imbedding distribution of a graph, in which the genus distribution and the face-size distribution lie near the bottom. Higher invariants keep track of the distribution of face sizes within each imbedding and the relationships between the faces.

The set of imbeddings of a graph has various statistical properties of interest. For instance, all the genus distributions mentioned above are strongly unimodal. Also, Gross and Klein (1986) classified the graphs of average genus less than or equal to 1.

Little else is known about the number of imbeddings of a given graph in a given surface, except when the imbedding surface has low genus. The proof of Theorem 3.4.6 shows that the complete graph K_7 has $2 \cdot 5!$ different imbeddings in the torus (once a rotation at vertex number 1 is fixed, there are only two choices for the imbedding). By Whitney's theorem (Theorem 1.6.6), 3-connected planar graphs have "only one" equivalence class of imbeddings in the plane. Negami (1983, 1985) has investigated the uniqueness of imbeddings in the torus and the projective plane.

3.4.8. Exercises

1. Show that the sequence of rotation projections for the complete graph K_5 given in Figure 3.40 are adjacent in pairs from left to right and that the associated surfaces have genus 1, 2, and 3 (again from left to right).

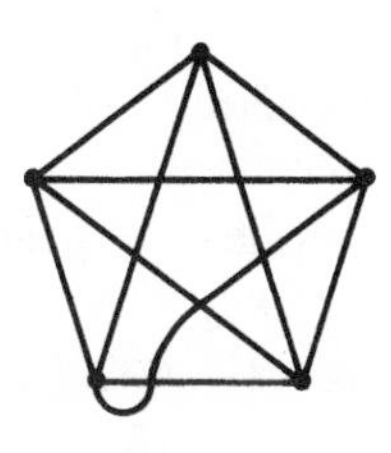

Figure 3.40. Three rotation projections of the complete graph K_5.

2. Draw a sequence of rotation projections for K_5 in surfaces of crosscap number $1, 2, 3, 4, 5, 6$ such that each imbedding is obtained from the previous imbedding by edge-twisting surgery.
3. Draw a sequence of rotation projections for adjacent imbeddings of K_6 in orientable surfaces of genus $1, 2, 3, 4, 5$.
4. Write a sequence of rotation systems in list format for adjacent imbeddings of the octahedral graph O_3 in surfaces of genus $0, 1, 2, 3$.
5. The "Euler characteristic range" $\mathrm{ER}(G)$ of a connected graph G is defined to be the set of numbers c such that G has a cellular imbedding in a surface of Euler characteristic c. Prove that $\mathrm{ER}(G)$ is an interval.
6. Prove that $\mathrm{ER}(G) = [c, c]$ if and only if $\beta(G) \leq 1$. Conclude that if $\mathrm{ER}(G) = [c, c]$, then $c = 2$.
7. Give a genus-1 imbedding of the graph Y_n shown in Figure 3.37.
8. Let G be connected and not a tree, and let k be the smallest integer such that G imbeds, cellularly or not, in the nonorientable surface N_k. Prove that G has a cellular imbedding in N_k.
9. Prove that the maximum genus of the complete bipartite graph $K_{m,n}$ is $\lfloor \beta(K_{m,n})/2 \rfloor$. (Ringeisen, 1972)
10. Find the maximum genus of the octahedral graph O_n, for all $n > 0$.
11. Let e be an edge of the graph G such that $G - e$ has two components G_1 and G_2. Prove that $\gamma(G) = \gamma(G_1) + \gamma(G_2)$ and that $\gamma_M(G) = \gamma_M(G_1) + \gamma_M(G_2)$. (*Hint:* Use Exercise 3.3.4 and Theorem 3.3.3.)
12. Find a graph that has genus range $[0, 1]$. Find a graph that has genus range $[1, 2]$. Use Exercise 11 to prove that for any $m \geq 0$ and any $n \geq 2m$, there is a graph with genus range $[m, n]$.
13. Prove that $\mathrm{GR}(G) = [m, m]$, $m \geq 0$, if and only if no vertex of the graph G lies on two distinct cycles. (*Hint:* Exercise 7 might help.)
14. Use Exercise 13 to prove the result of Nordhaus et al. (1972) that if $\mathrm{GR}(G) = [m, m]$, then $m = 0$.
15. By Exercise 12, there is a graph with genus range $[2, n]$, for any $n \geq 4$. By Exercise 14, there is no graph with genus range $[2, 2]$. Is there a graph with genus range $[2, 3]$? In general, find a graph of genus range $[m, n]$ for some $m > 0$, $m < n < 2m$.
16. Prove that if H is a connected graph with more than one edge, then there are adjacent edges d and e such that $H - d - e$ has only one nontrivial component.
17. Calculate the genus distribution of the following graphs:
    (a) K_4.
    (b) $K_{3,3}$.
    (c) $K_2 \times K_3$.

18. Calculate the face-size distribution of the following graphs:
(a) K_4.
(b) $K_{3,3}$.
(c) $K_2 \times K_3$.

3.5. ALGORITHMS AND FORMULAS FOR MINIMUM IMBEDDINGS

Historically, most of the early emphasis in topological graph theory was to determine the genus and crosscap number of a graph. Since this problem is inherently finite, one might look for algorithms to compute them. One possibility, which is incorporated into the Heffter–Edmonds algorithm, requires an examination of all the rotation systems, an unfeasibly lengthy search.

The most frequently adopted approach has been to calculate genus and crosscap number formulas for interesting families of graphs and for the effect of various graph operations. Most of the derivations of such formulas have used covering space constructions, either in explicit topological form or in an equivalent combinatorial guise. However, before extending voltage graph techniques to include a lift of the imbedding—the main theme of Chapter 4—it is instructive to study two operations whose imbeddings have not required covering space constructions: cartesian products and one-vertex amalgamations.

3.5.1. Rotation-System Algorithms

In its least complicated form, the Heffter–Edmonds algorithm for minimum genus is as follows: First, list all the pure rotation systems of a graph. Next, compute the number of faces for each rotation system. Then, choose one having the most faces. The obstruction to usefulness is that the number of rotation systems is exponential in the number of vertices. For instance, a regular $(r+1)$-valent graph with n vertices has $(r!)^n$ pure rotation systems.

Sampling from the set of all pure rotation systems in order to approximate the genus of a given graph seems a plausible approach. However, not much is known about the distribution of genera within a rotation system, except for a few special cases. It seems that relatively few imbeddings are minimum genus imbeddings. For example, the genus range of the complete graph K_7 is $[1, 7]$, and by details in the proof of Theorem 3.4.6, we see that only $2 \cdot 5!$ of $(5!)^7$ possible pure rotation systems give a genus-1 imbedding, or about one in 10^{12}. A heuristic explanation of this phenomenon is given in Exercises 2–4.

A less exhaustive use of rotation systems might be to hill-climb by moving from one imbedding of the given graph to an adjacent imbedding (as defined in Section 3.4), while always trying to increase the number of faces at each step. Unfortunately, as the following theorem shows, "local genus minima" abound in the collection of orientable imbeddings of a graph.

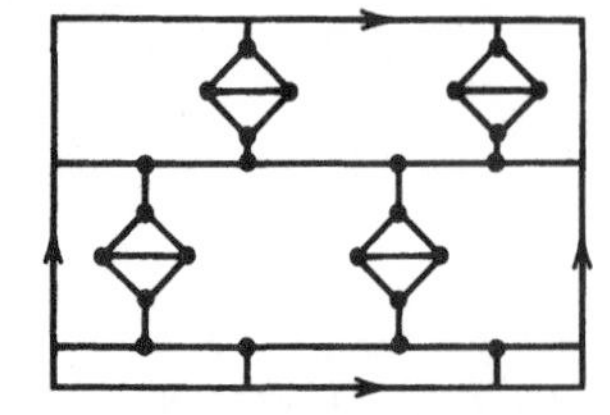

Figure 3.41. Two imbeddings of a graph.

Theorem 3.5.1. *Let $G \to S$ be an imbedding such that every boundary walk is a cycle, and such that the only boundary walks passing through both endpoints of an edge are the two boundary walks containing that edge. Then every adjacent imbedding has larger genus.*

Proof. Let e be any edge of the graph G, and let $G' \to S'$ be the imbedding obtained by deletion surgery on edge e. This imbedding has one less face than the original imbedding, since the two faces containing edge e have been joined into one. Moreover, no face of the imbedding $G' \to S'$ passes through both endpoints of edge e, except the joined face. Therefore, when the edge e is reattached in a different way, it must form a bridge joining two more faces. Hence, any adjacent imbedding has fewer faces and larger genus. □

Example 3.5.1 (Gross and Tucker, 1979b). *It is easy to verify that the toroidal imbedding on the left of Figure* 3.41 *satisfies the hypothesis of Theorem* 3.5.1. *Thus, every adjacent imbedding has larger genus. However, the given graph has the planar imbedding given on the right. It follows that the imbedding at the left is a local minimum that is not a global minimum.*

One might try to salvage an algorithm by defining a different kind of adjacency for graph imbeddings. One obvious candidate is "vertex-adjacency": two imbeddings are vertex-adjacent if their rotation systems agree at every vertex except one. However, a regular graph of valence 3, such as that in Example 3.5.1, has only two rotations at any vertex, so that vertex-adjacent imbeddings of these graphs would also be adjacent, presenting us with the same problem of local genus minima. In general, algorithms based on rotation systems have not been sufficiently fast.

3.5.2. Genus of an Amalgamation

Suppose that H and K are graphs imbeddable in surfaces S_h and S_k, respectively, and that G is a graph obtained by amalgamating H and K at a vertex. Then there is no difficulty in constructing an imbedding of G into a surface of genus $h + k$. First, one chooses a disk on S_h, whose interior is disjoint from the imbedded graph H and whose boundary intersects H only at

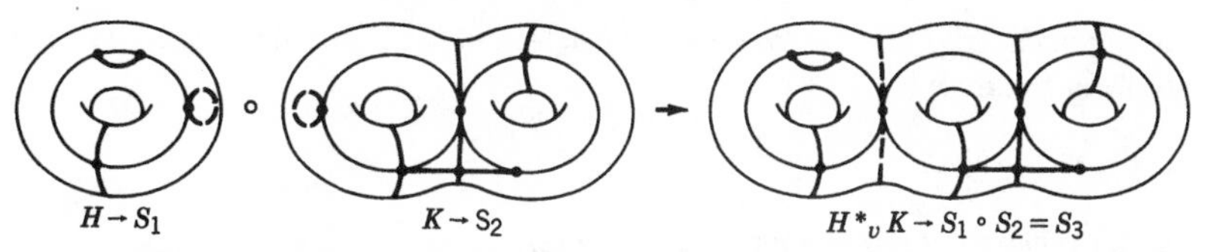

Figure 3.42. Any amalgamation of two graphs is imbeddable into a disk sum of any two of their respective imbedding surfaces.

the amalgamation vertex. Next, one chooses an analogous disk on the surface S_k. Then, one deletes the interiors of both disks and identifies the resulting surface boundaries so that the graphs H and K are amalgamated, as illustrated in Figure 3.42. As we have mentioned in Exercise 3.3.11, this composition of surfaces is called a "disk sum".

For this simplified type of graph amalgamation, we often use the notation $H *_v K$, where v is the vertex of amalgamation, to represent the resulting graph. From the above discussion, it is obvious that

$$\gamma(H *_v K) \leq \gamma(H) + \gamma(K)$$

The so-called *BHKY* theorem (for Battle, Harary, Kodama, and Youngs) asserts the nonobvious fact that

$$\gamma(H *_v K) = \gamma(H) + \gamma(K)$$

By way of contrast, the analogous assertion, is not true for maximum genus (see Exercise 5), and it is not true for crosscap number (see Theorem 3.5.5).

The straightforward way to try to prove the *BHKY* theorem is to attempt to show that an amalgamated graph $H *_v K$ must have a minimum imbedding that permits reversal of the combined disk sum and graph amalgamation operation. That is, one hopes for an imbedded circle in the surface that intersects $H *_v K$ only at the amalgamation vertex v and that separates the imbedding surface into two parts, one of which contains H and the other of which contains K. Necessarily, all the H-edges would have to be grouped contiguously in the rotation at v (as would the K-edges). If one started with a minimum imbedding of $H *_v K$ in which the H-edges were not continuously grouped, then one would wish to regroup, without increasing the genus of the imbedding surface. The following lemma, whose application is not restricted to imbeddings of amalgamated graphs, is what facilitates regrouping.

Lemma 3.5.2. *Let*

$$v. \quad ab \cdots cd$$

be the rotation at a vertex v of an orientable imbedding $G \to S$ such that the corners ab and cd lie on the same face. Let $G \to S'$ be the adjacent imbedding *such that the edge a is inserted between edges c and d, so that vertex v has the resulting rotation*

$$v. \quad b \ldots c\,a\,d$$

and all other rotations remain the same. Then $\gamma(S') \leq \gamma(S)$.

Proof. Let $G \to S''$ be the imbedding obtained from $G \to S$ by deletion surgery on edge a. We shall use F, F', and F'' to denote the face sets of the respective imbeddings $G \to S$, $G \to S'$, and $G \to S''$. If the edge a appears twice on the same face boundary of the imbedding $G \to S$, then $\#F'' = \#F + 1$, from which it follows that $\#F' \geq \#F'' - 1 = \#F$. Otherwise, the deletion surgery on edge a in the imbedding $G \to S$ unites the face having the corners ab and cd with another face, and the resulting "new" face in $G \to S''$ contains the corner cd plus another occurrence of v across the "new" face to the corner cd, thereby resplitting the "new" face. Thus, $\#F' = \#F$. □

Theorem 3.5.3 [The *BHKY* Theorem (Battle, Harary, Kodama, and Youngs, 1963)]. $\gamma(H *_v K) = \gamma(H) + \gamma(K)$.

Proof. In any imbedding of $H *_v K$, the rotation at vertex v contains sequences of edges from the graph H interspersed with sequences from K. Maximal sequences from H and K are called "H-segments" and "K-segments", respectively, relative to that imbedding. Let $H *_v K \to S$ be a minimum-genus imbedding that maximizes the sum of the squares of the segment lengths at v, taken over all imbeddings of $H *_v K$ into the surface S. We assert that for this imbedding, there is but one H-segment and one K-segment.

In order to verify the assertion, we observe that in any imbedding of $H *_v K$, there exists an edge d from H and an edge e from K such that d immediately precedes e in the rotation at v. Moreover, whatever face boundary contains the corner de must also contain a corner $e'd'$ where e' is from K and d' from H. Since edges of H and K meet only at v, it follows that the rotation v has the form

$$v. \quad d\,e \ldots e'\,d' \ldots$$

We claim that the edges d and d' must lie in the same H-segment. If not, then suppose that the lengths of the H-segments containing d and d' are n and n', respectively. Without loss of generality, we shall assume that $n \leq n'$. Apply Lemma 3.5.2, with edges d, e, e', and d' filling the roles of a, b, c, and d, respectively. Evidently, the adjacent imbedding with the modified rotation

$$v. \quad e \ldots e'\,d\,d'$$

(and no other changes) would have the same imbedding surface—lower genus is impossible, since S is a minimum-genus imbedding surface—but a larger sum of squares of segment lengths, since

$$(n-1)^2 + (n'+1)^2 > n^2 + (n')^2$$

because $n \leq n'$. Thus, the edges d and d' are in the same H-segment. Similarly, the edges e and e' are in the same K-segment. Therefore, we have established an assertion that there is only one H-segment and only one K-segment.

We must still prove that a closed arc C through vertex v from corner de to corner $e'd'$ in the interior of their common face must separate the imbedding surface. However, if not, then closed arc C would lie on a handle of surface S. If the surface S were cut open along C and reclosed with two disks, the result would be an imbedding of the disjoint union of the graphs H and K on a surface S' of genus $\gamma(S) - 1$. Such an imbedding would be noncellular, since any region that contained both a closed H-walk and a closed K-walk in its boundary would fail to be simply connected. By the construction of S', no such region could arise. It follows that the closed arc C separates the surface S. Since graph H lies on one side of the separation, that side must have genus at least $\gamma(H)$. Similarly, the other side must have genus at least $\gamma(K)$. It follows that

$$\gamma(H *_v K) \geq \gamma(H) + \gamma(K)$$

Since we have previously established the opposite direction of this inequality, the *BHKY* theorem is proved. □

A maximal 2-connected subgraph of a graph G is called a "block" of the graph G. For example, the graph in Figure 3.43 has four blocks. A slight generalization of Theorem 3.5.3 is the following.

Theorem 3.5.4. *If the blocks of graph G are $G_1, \ldots, G_n$, then*

$$\gamma(G) = \gamma(G_1) + \cdots + \gamma(G_n).$$

Proof. One uses a simple induction on the number of vertices of the graph G, while applying the fact that a subgraph H is a block graph G if and only if $G = H *_v K$ for some vertex v and some subgraph K (of course, K itself may have several blocks). □

Figure 3.43. A graph having four blocks.

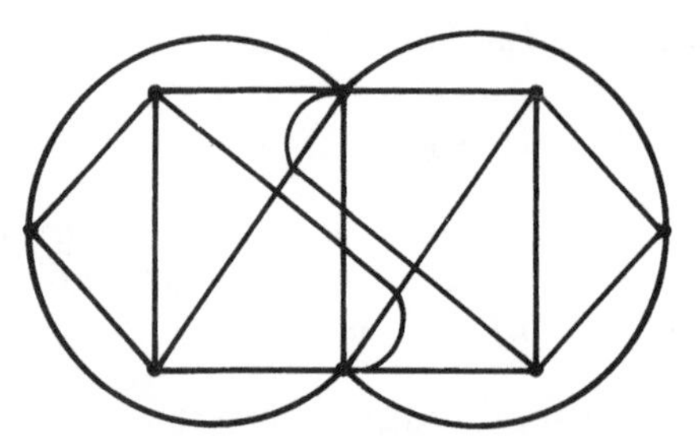

Figure 3.44. A rotation projection of $K_5 *_e K_5$.

More general types of amalgamations have also been studied. The situation here is much more complicated, as the following example illustrates.

Example 3.5.2. *The rotation projection shown in Figure* 3.44 *gives an imbedding in a torus of an amalgamation of* K_5 *with* K_5 *along an edge. Thus the genus of this graph is* 1, *not the seemingly obvious value of* 2.

The results known for other amalgamations are all of the form $\gamma(H *_f K) = \gamma(H) + \gamma(K) + \delta$, where the parameter δ is given a possible range of values, and each of these values does occur for some pair of graphs H and K. Alpert (1973) shows that when the pasting function f identifies an edge of H with an edge of K, then $\delta = 0$ or -1. Stahl (1980) and Decker et al. (1981) prove that $\delta = 1, 0$, or -1 when f identifies copies of $K_2{}^c$. Stahl also shows in the same paper that $\delta = 2, 1, 0, -1$, or -2, when the pasting function f amalgamates two copies of $K_3{}^c$.

In general amalgamations, the range of δ values is much greater and determining when δ takes on a particular value is quite complicated [see, for example, Decker et al. (1985)]. The best general theorem would be a bound on δ that depends only on the number k of vertices being identified by the amalgamation. An upper bound for δ if $k = 1$ is easily obtained and is achieved for some amalgamations when $k = 2$. Archdeacon (1986) has shown, however, that there is no lower bound for δ even when $k = 3$; he constructs a sequence of n-vertex graphs G_n and 3-vertex amalgamations H_n of G_n with itself such that $\gamma(H_n) = 2\gamma(G_n) + n$.

3.5.3. Crosscap Number of an Amalgamation

Computation of the crosscap number of a one-vertex amalgamation of two graphs is complicated by the fact that a nonorientable surface can be represented as the disk sum of a nonorientable surface and an orientable surface.

Example 3.5.3. *The crosscap number of the complete graph* K_7 *is* 3, *by Theorem* 3.4.6. *However, since* K_7 *is imbeddable in a torus, the amalgamation*

$K_7 *_v K_7$ has an imbedding in the nonorientable surface N_5 obtained from a disk sum of imbeddings of K_7 in a torus and in the surface N_3. Thus, in general, $\bar{\gamma}(H *_v K) \neq \bar{\gamma}(H) + \bar{\gamma}(K)$.

Theorem 3.5.5 (Stahl and Beineke, 1977). *$\bar{\gamma}(H *_v K) = \bar{\gamma}(H) + \bar{\gamma}(K) + \delta$ where $\delta = -1$ if either $\bar{\gamma}(H) > 2\gamma(H)$ or $\bar{\gamma}(K) > 2\gamma(K)$, and $\delta = 0$ otherwise.*

Proof. Let g denote the value of the right-hand side of the equality and let $G = H *_v K$. To see that $\bar{\gamma}(G) \leq g$, use disk sums and the general fact that $\bar{\gamma}(H) \leq 2\gamma(H) + 1$ (and similarly for K).

Conversely, to show that $\bar{\gamma}(G) \geq g$, it suffices to prove, as in Theorem 3.5.3, that there is an imbedding of the graph G in the nonorientable surface of minimum crosscap number such that the rotation at vertex v has one H-segment and one K-segment. The details of the proof of Theorem 3.5.3 can be used again. The only difficulty lies in the nonorientable version of Lemma 3.5.2. The type of the edge d might have to be changed when it is reattached at vertex v, in order not to decrease the number of faces. (Certainly one of the two possible types for edge d will be wrong: we want the sides of edge d in the new imbedding to belong to separate faces, yet an extra twist in an edge whose sides belong to separate faces joins the faces.) One might fear that changing the type of edge d could make the new imbedding surface orientable. However, in the present context, the sides of the edge d belongs to distinct faces in the original imbedding. As we observed in Section 3.4 in the discussion of extra twists, changing the type of edge under such a circumstance always yields a nonorientable surface. □

3.5.4. The White–Pisanski Imbedding of a Cartesian Product

The cartesian product of two graphs contains many cycles of length 4, because each instance of the product $K_2 \times K_2$ is a cycle of length 4. One might hope, therefore, that a cartesian product $G \times H$ has imbeddings in which all the faces are quadrilaterals. If the graphs G and H both have girth greater than 3, then the resulting imbeddings would be minimal genus or crosscap number imbeddings. Indeed, it might be much easier to compute the genus of a product $G \times H$ than to compute the genus of either G or H. White (1970) was the first to exploit this idea. Our presentation includes generalizations due to Pisanski (1980).

Define a "patchwork" of an imbedding $G \to S$ to be a collection P of faces such that the union of all the boundary walks of the faces in the collection P forms a 2-factor of the graph G. A set of patchworks $P_1, P_2, \ldots, P_r$ is "disjoint" if they are pairwise as collections of faces, that is, no face in P_i is also a face in P_j unless $i = j$. (However, the boundary walk of a face in P_i may share an edge with a boundary walk of a face in P_j.) The faces of an imbedding $G \to S$ that are not in a given set of patchworks are called

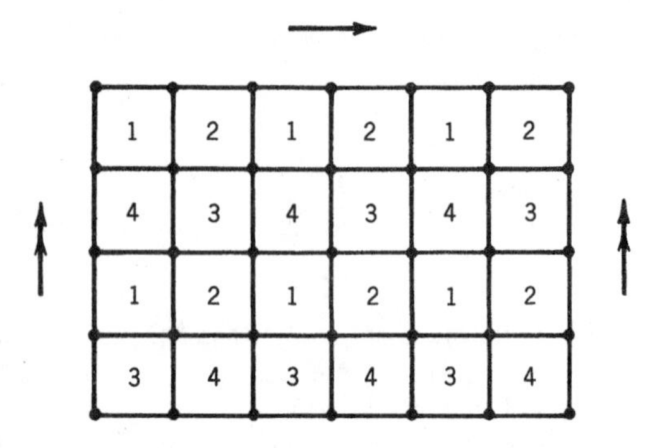

Figure 3.45. Some disjoint patchworks in a toroidal imbedding of $C_4 \times C_6$.

"residual". A patchwork is called "even" if all of its faces have even size, and it is called "quadrilateral" if all of its faces have size 4.

Example 3.5.4. *Let P_i be the collection of faces labeled i in the toroidal imbedding given in Figure* 3.45. *Then $P_1, \ldots, P_4$ are four disjoint quadrilateral patchworks; this set of patchworks has no residual faces. The set P_1, P_2 of patchworks has* 12 *residual faces.*

An "edge-coloring" of a graph G is an assignment of a "color", usually denoted by an integer, to each edge such that edges incident to the same vertex have different colors. A color i used in an edge-coloring of the graph G is "saturated" if there is an i-colored edge incident to every vertex of G. Thus, the set of edges of a saturated color in a given edge-coloring of the graph G form a 1-factor of the graph. A regular graph of valence p has an edge-coloring with p colors if and only if it is 1-factorable.

Example 3.5.5. *On the left of Figure* 3.46 *is an edge-coloring of the cycle graph C_6 using two colors, both of which are saturated. On the right is an edge-coloring of the graph C_5 using three colors, none of which is saturated.*

Theorem 3.5.6 (White, 1970; Pisanski, 1982). *Let $G \to S$ be an imbedding that has r disjoint, even patchworks such that all residual faces are quadrilaterals and such that at least q of the patchworks are quadrilateral. Let H be a graph that has an edge-coloring with r colors, at least $r - q$ of which are saturated. Then the cartesian product $G \times H$ has an imbedding with at least $2r - q$ disjoint patchworks, such that all faces of the imbedding are quadrilaterals. The imbedding is orientable if and only if the imbedding $G \to S$ is orientable and the graph H is bipartite.*

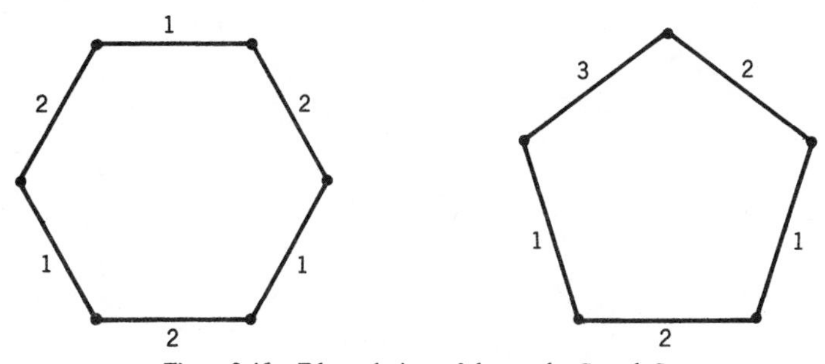

Figure 3.46. Edge-colorings of the graphs C_6 and C_5.

Proof. Let $1, 2, \ldots, r$ denote the colors of the edge-coloring of the graph H with the colors that are not saturated listed first. Let $P_1, P_2, \ldots, P_r$ be the disjoint even patchworks of the imbedding $G \to S$ with the quadrilateral patchworks listed first. We construct the desired imbedding for the graph $G \times H$ by giving its rotation at each vertex (u, v). Recall that edges incident to vertex (u, v) are of the form (d, v) where d is an edge of the graph G incident on vertex u or of the form (u, e) where e is an edge of the graph H incident on vertex v. For convenience, we call the former G-edges and the latter H-edges. We call the edges d and e their projected edges. The G-edges are to appear in the rotation at vertex (u, v) in the same order as their projected edges in the rotation at vertex u for the imbedding $G \to S$. The H-edges are then interspersed as follows: if the projection of an H-edge is colored i in the edge-coloring of the graph H, then that edge is placed between the two G-edges whose projections form the corner of the face from patchwork P_i at vertex u. The type of every G-edge is the same as its projected edge while every H-edge is given type 1. It is best to view this imbedding as $\#V(H)$ copies of the imbedding $G \to S$ placed at different levels with "tubes" joining corresponding faces at different levels where needed for H-edges. Figure 3.47 shows how a pair of H-edges having the same projection fit into four corners of corresponding faces from patchwork P_i. A pair of corresponding faces of

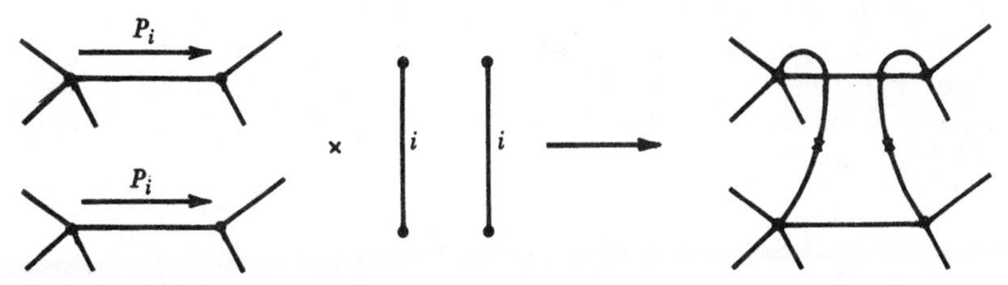

Figure 3.47. Part of a rotation projection of the imbedding for $G \times H$; on the left, before H-edges are added and, on the right, after.

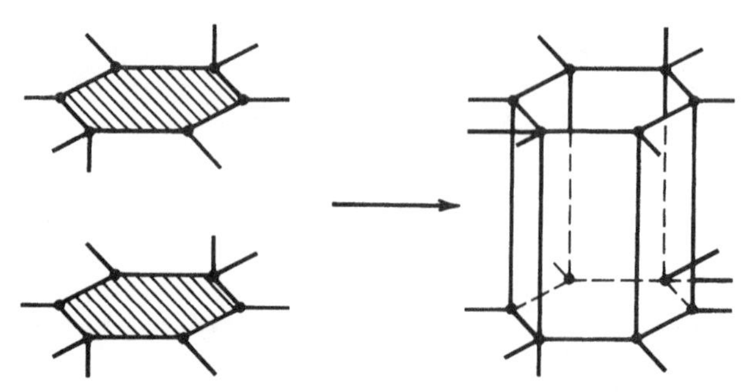

Figure 3.48. A pair of corresponding faces joined by a tube.

the imbedding $G \to S$ is thereby transformed into a tube of quadrilaterals (see Fig. 3.48).

If color i is not saturated, then the patchwork P_i, which is quadrilateral by the choice of listing of the patchworks, generates a quadrilateral patchwork consisting of alternate quadrilaterals around each tube between levels joined by H-edges colored i and faces of P_i itself at levels where no H-edge colored i is incident. If color i is saturated, then patchwork P_i generates a pair of disjoint quadrilateral patchworks formed by alternate quadrilaterals around each tube joining a pair of corresponding faces of P_i. Thus the resulting imbedding has at least $2r - q$ disjoint patchworks, as claimed. It is clear that every face of the imbedding is quadrilateral, by construction. Finally, the imbedding surface is orientable if and only if the imbedding $G \to S$ is orientable (since G-edges have the same type as their projections), and the graph H has no cycle of odd length (since every H-edge has type 1). □

The important part of the conclusion of Theorem 3.5.6 is that the constructed imbedding has all faces quadrilateral. The number of disjoint patchworks in the imbedding is useful for repeated cartesian products.

3.5.5. Genus and Crosscap Number of Cartesian Products

The White–Pisanski imbedding enables us on compute the genus and crosscap number of a variety of cartesian products. In all the applications that follow, the minimal genus or crosscap imbedding has all faces quadrilateral. Thus the number of faces is half the number of edges. In particular, if the graph is regular of valence t and has n vertices, the genus of the imbedding surface S, if it is orientable, is

$$1 - \chi(S)/2 = 1 - (n - nt/2 + nt/4)/2 = 1 + n(t - 4)/8$$

If the imbedding surface is nonorientable, its crosscap number is twice this.

Theorem 3.5.7 (White, 1970). *Let* $G = C_{n_1} \times C_{n_2} \times \cdots \times C_{n_r}$ *where* $r > 1$ *and* $n_i > 3$ *for all* i. *Then*

a. $\gamma(G) = 1 + \#V(G)(r-2)/4$ *if each* n_i *is even.*
b. $\bar{\gamma}(G) = 2 + \#V(G)(r-2)/2$ *if* $r > 2$, n_1 *and* n_2 *are even, and* n_3 *is odd.*

Proof. The graph $C_{n_1} \times C_{n_2}$ has an imbedding in the torus that has all faces quadrilateral and four disjoint patchworks (see Example 3.5.4), whenever n_1 and n_2 are even. Repeated application of Theorem 3.5.6 gives an all-quadrilateral imbedding of the graph G, orientable in case a and nonorientable in case b. Since the girth of the graph G is greater than three, the equations follow from Euler's equation. □

Theorem 3.5.8 (Ringel, 1955). *The genus of the n-cube graph* Q_n *is* $1 + 2^{n-3}(n-4)$.

Proof. The proof is like that for Theorem 3.5.7; use the quadrilateral imbedding of $Q_2 = C_4$ to get started. □

Theorem 3.5.9 (Pisanski, 1980). *Let G and H be* 1*-factorable, regular r-valent graphs with girth greater than three. If both G and H are bipartite, then*

$$\gamma(G \times H) = 1 + mn(r-2)/4$$

and if G and H are not both bipartite, then

$$\bar{\gamma}(G \times H) = 2 + mn(r-2)/2$$

where $m = \#V(G)$ *and* $n = \#V(H)$.

Proof. First, we construct an imbedding $G \to S$ that has r disjoint even partitions (and hence no residual faces). Choose any cyclic ordering of the 1-factors of a 1-factorization of the graph G, and let the rotation at each vertex be given by this single ordering. Let every edge have type 1. Then each face of the resulting imbedding consists of edges alternating between two 1-factors. Thus, the imbedding has r disjoint even patchworks. The imbedding surface is orientable if and only if the graph G is bipartite. The theorem then follows from Theorem 3.5.6, since the graph H has an edge-coloring with r colors all of which are saturated. □

Theorem 3.5.10 (Jungerman, 1978a). $\bar{\gamma}(Q_n) = 2\gamma(Q_n)$ *for* $n \neq 4, 5$ *and* $\bar{\gamma}(Q_n) = 2\gamma(Q_n) + 1$ *for* $n = 4, 5$.

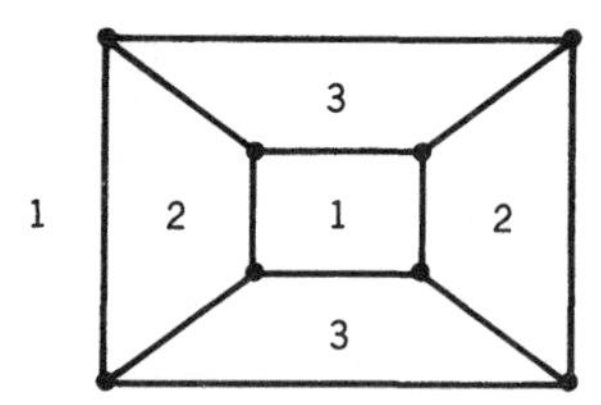

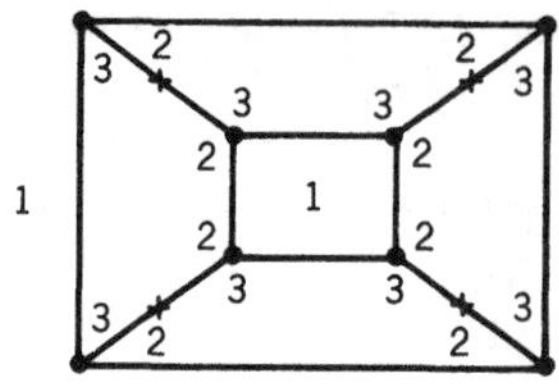

Figure 3.49. Rotation projections for two imbeddings of Q_3.

Proof. Since $\gamma(Q_n) = 0$ for $n < 4$, there is nothing to be proved there. Jungerman shows that neither Q_4 nor Q_5 has a nonorientable quadrilateral imbedding, but they do have orientable ones, hence the special equation for $n = 4, 5$. We shall show that Q_4 has a nonorientable imbedding with two disjoint patchworks and all residual faces quadrilateral. Theorem 3.5.6 then provides a nonorientable, quadrilateral imbedding of $Q_4 \times Q_2 = Q_6$ with four disjoint patchworks, from which one easily obtains quadrilateral imbeddings of Q_n for all $n > 6$, again by Theorem 3.5.6.

Consider the two imbeddings of Q_3 whose rotation projections are given in Figure 3.49. Both imbeddings have three disjoint patchworks consisting each of faces with the same label; the imbedding on the right has two faces of size 8, each of which forms a patchwork by itself. We have labeled the corners of those faces. Now perform the White–Pisanski construction on $Q_3 \times Q_1 = Q_4$, only using the left imbedding of Q_3 on one level and the right imbedding on the other level and putting Q_1-edges into the corners of faces labeled 1. Faces labeled 2 then form a patchwork in the resulting imbedding, while those labeled 3 form another. Residual faces are the quadrilaterals on the tubes joining faces labeled 1. Finally, the reader can easily verify that the resulting imbedding has a cycle with an odd number of type-1 edges. □

Clearly an edge-coloring of a graph with maximum valence r requires at least r colors. The following converse has been established by Vizing (1964).

Theorem 3.5.11 (Vizing). *Any graph with maximum valence r can be edge-colored with $r + 1$ colors or less.*

Proof. Omitted. □

3.5.6. Exercises

1. Prove that given any integer m, there is a graph having an edge whose contraction reduces the genus of the graph by m. (*Hint:* Any n-vertex graph reduces to a one-vertex graph by a sequence of $n - 1$ edge contractions.)

2. Given a pure rotation system for a graph G, let ρ be the permutation of the edge set DE(G) such that for each vertex v the set of directed edges with initial vertex v, ordered by the given rotation system, forms a cycle of ρ. Let ι be the involution of DE(G) given by $\iota(e^{\sigma}) = e^{-\sigma}$, $\sigma = \pm$. Show that the cycles of $\rho \circ \iota$ are the directed boundary walks of the associated orientable imbedding.
3. For each permutation $\pi \in \mathscr{S}_n$, define $t(\pi)$ to be the permutation obtained by erasing the symbol n in the cycle notation for π. For example, if $\pi = (1\,3\,4\,6\,2)(5)$ then $t(\pi) = (1\,3\,4\,2)(5)$. Use t to prove that the expected number of cycles in an element of $\mathscr{S}_n$ is $1 + 1/2 + 1/3 + \cdots + 1/n$ [see Knuth (1969) for more on cycle structure of random permutations].
4. Use Exercises 2 and 3 to argue that one should expect the number of faces of a random orientable imbedding of an m-edge graph to be around $ln(2m)$. Contrast this with the number of faces of a "good" (e.g., triangular) minimal-genus imbedding. What should one expect for the genus of a random imbedding of K_7?
5. Prove that $\gamma_M(K_7 *_v K_7) = \gamma_M(K_7) + \gamma_M(K_7) + 1$.
6. Find graphs H and K such that $\gamma_M(H *_v K) = \gamma_M(H) + \gamma_M(K)$.
7. Prove that $\gamma_M(H *_v K) = \gamma_M(H) + \gamma_M(K) + \delta$, where $\delta = 0, 1$.
8. For each possible value of δ, give an example of a pair of graphs H and K and a function f identifying copies of K_2 in the graphs such that $\gamma(H *_f K) = \gamma(H) + \gamma(K) + \delta$.
9. Prove that if G and H are regular bipartite graphs of the same valence r having m and n vertices, respectively, then $\gamma(G \times H) = 1 + mn(r-2)/4$.
10. Let H be any bipartite graph with maximum valence r and girth greater than 3. Use Vizing's theorem and Theorem 3.5.6 to compute the genus of $Q_n \times H$ for all $n > r$.
11. Prove that if H is any graph with maximum valence $r > 3$, then $Q_n \times H$ has a nonorientable quadrilateral imbedding for all $n > r + 2$. (*Hint:* The proof of Theorem 3.5.10 provides nonorientable quadrilateral imbeddings of Q_n with many disjoint patchworks for $n > 5$.)
12. Compute $\gamma(K_{3,3} \times Q_n)$ for all n.

4

Imbedded Voltage Graphs and Current Graphs

Suppose that a graph G is simultaneously given a voltage assignment α and an imbedding in some surface S. Then there is a natural way to construct a derived surface S^α and an imbedding of the derived graph G^α into S^α; the key idea is to "insert a region" into every closed walk of G^α that covers a boundary walk of the "base" imbedding $G \to S$. Moreover, there is a surface map $S^\alpha \to S$ that is topologically a "branched covering" and whose restriction to the derived graph G^α is the natural covering projection $p: G^\alpha \to G$. This form of the voltage graph construction makes it easy to prove various theoretical results, including Alexander's theorem that every closed orientable surface is a branched covering space over the sphere. Furthermore, imbedded voltage graphs and their duals, which are known as "current graphs", are of primary importance in many calculations of the genus of graphs, because they provide a powerful technique for generating complicated imbeddings from simple ones. Finally, imbedded voltage graphs can be used to understand "surface symmetry", by which we mean group actions on surfaces.

4.1. THE DERIVED IMBEDDING

Let $G \to S$ be an imbedding of a graph G in a surface S, and let α be a voltage assignment in a group $\mathscr{A}$. We call the pair $\langle G \to S, \alpha \rangle$ an "imbedded voltage graph". The surface S may be orientable or not, and the voltages may be ordinary or permutation voltages, although one usually employs the subscripted notation $\langle G \to S, \alpha \rangle_n$ to indicate permutation voltages in $\mathscr{S}_n$. Then there is a natural way to define a "derived surface" S^α and a "derived imbedding" $G^\alpha \to S^\alpha$, either in terms of a lifted rotation system or in terms of lifted faces.

4.1.1. Lifting Rotation Systems

Suppose first that α is an ordinary voltage assignment in a group $\mathscr{A}$, and let $p: G^\alpha \to G$ be the natural projection of the derived graph onto the base graph. We call the given imbedding $G \to S$ of the base graph the "base imbedding".

We shall initially assume that the base imbedding is specified by a rotation system for the base graph G, and specify how to construct the "derived rotation system".

Suppose that the vertex u of the base graph G has the rotation

$$u.\quad c\,d \ldots e$$

Then a vertex u_a in the fiber over u has the rotation

$$u_a.\quad c_a\,d_a \ldots e_a$$

If some edge incident on u has type 1 in the base imbedding (i.e., if the local orientations at its endpoints are inconsistent), then the corresponding edge in the derived graph is also assigned type 1. The surface S^α induced by this rotation system on the derived graph is called the "derived surface", and the imbedding $G^\alpha \to S^\alpha$ is called the "derived imbedding".

This definition of the derived imbedding translates immediately into a band decomposition. Each 0-band of the base imbedding lifts to $\#\mathscr{A}$ different 0-bands in the derived imbedding, each of which has the same instructions (up to $\mathscr{A}$-coordinates) for attaching 1-bands. Moreover, each 1-band of the base imbedding lifts to $\#\mathscr{A}$ different 1-bands, each with the same "twist" as the base 1-band.

Example 4.1.1. *Consider the graph imbedding $G \to S$ corresponding to the rotation system*

$$\begin{array}{lllll} u. & a^1 & c & d & \\ v. & b & d & c & \\ w. & b & e^1 & a^1 & e^1 \end{array}$$

with voltages in $\mathscr{Z}_2$, as indicated by the labeled rotation projection at the bottom of Figure 4.1. *Then the derived rotation system is*

$$\begin{array}{lllll} u_0. & a_0^1 & c_0 & d_0 & \\ u_1. & a_1^1 & c_1 & d_1 & \\ v_0. & b_1 & d_0 & c_1 & \\ v_1. & b_0 & d_1 & c_0 & \\ w_0. & b_0 & e_1^1 & a_0^1 & e_0^1 \\ w_1. & b_1 & e_0^1 & a_1^1 & e_1^1 \end{array}$$

It is illustrated by the rotation projection at the top of Figure 4.1.

4.1.2. Lifting Faces

The base imbedding in Example 4.1.1 has three faces. The boundary walks of these faces and their net voltages are as follows:

$$\begin{array}{ll} a\,e\,b\,c^- & 0+1+1-1 \equiv 1 \bmod 2 \\ d\,b^-e\,a^- & 0-1+1-0 \equiv 0 \bmod 2 \\ c\,d^- & \qquad\quad 1-0 \equiv 1 \bmod 2 \end{array}$$

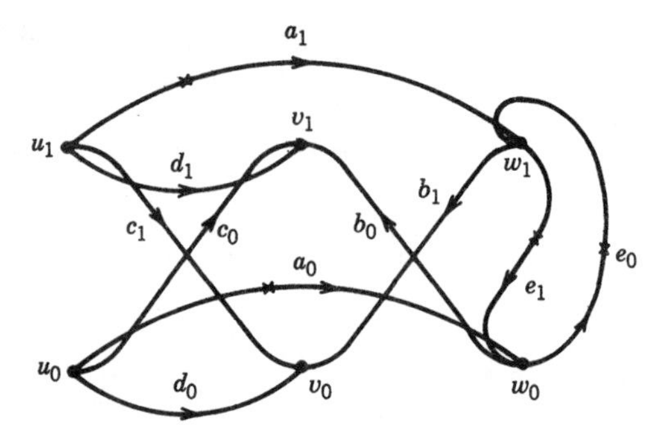

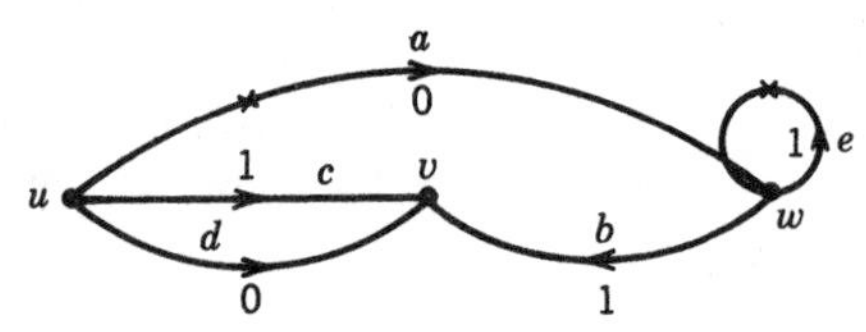

Figure 4.1. Rotation projections for an imbedded voltage graph (bottom) and its derived imbedding (top).

The derived imbedding has four faces and these boundary walks:

$$\begin{array}{llllllll} a_0 & e_0 & b_1 & c_1^- & a_1 & e_1 & b_0 & c_0^- \\ d_0 & b_1^- & e_1 & a_0^- \\ d_1 & b_0^- & e_0 & a_1^- \\ c_0 & d_1^- & c_1 & d_0^- \end{array}$$

Obviously, this information might be obtained by means of the Face Tracing Algorithm, an arduous task if the derived graph is large. Fortunately, however, the number and sizes of the faces of the derived imbedding are always computed using general properties of the voltage graph construction, properties which are apparent in the pattern of boundary walks just given for Example 4.1.1.

Since rotations in the derived imbedding $G^\alpha \to S^\alpha$ are the same as those in the base imbedding $G \to S$, up to subscripts in the voltage group $\mathscr{A}$, the faces of the derived imbedding are determined by the faces of the base imbedding. In particular, the sequence of edges in a boundary walk of $G^\alpha \to S^\alpha$ corresponds exactly (after erasure of subscripts) to one or more iterations of the edge sequence in a boundary walk of the base imbedding. The iterations continue until the first edge would once again have the same voltage subscript it started with. In terms of the natural projection $p: G^\alpha \to G$, the boundary walks of the derived imbedding are precisely the components of lifts of the boundary walks of the base imbedding. Thus, the following theorem is essentially a corollary of Theorem 2.1.3.

Theorem 4.1.1 (Gross and Alpert, 1974; Gross, 1974). *Let C be the boundary walk of a face of size k in the imbedded voltage graph $\langle G \to S, \alpha \rangle$. If the net voltage on the closed walk C has order n in the voltage group $\mathscr{A}$, then there are $\#\mathscr{A}/n$ faces of the derived imbedding $G^\alpha \to S^\alpha$ corresponding to the region bounded by C, each with kn sides.* □

In Example 4.1.1 the base region $a\,e\,b\,c^-$ has length 4 and net boundary voltage of order 2, so it lifts to one eight-sided face $a_0\,e_0\,b_0\,c_1{}^-\,a_1\,e_1\,b_1\,c_0{}^-$, whose boundary walk covers the boundary walk of $a\,e\,b\,c^-$ twice. The base region $d\,b^-\,e\,a^-$ has length 4 and net boundary voltage of order 1, so it lifts to two four-sided faces, $d_0\,b_1{}^-\,e_1\,a_0{}^-$ and $d_1\,b_0{}^-\,e_0\,a_1^-$, each of whose boundary walks covers the boundary walk of $d\,b^-\,e\,a^-$ once. The base digon $c\,d^-$ has net boundary voltage of order 2, so it lifts to one four-sided face $c_0\,d_1{}^-\,c_1\,d_0{}^-$.

We emphasize the principle that faces of the derived imbedding are computed in practice from the net voltages on boundary walks of the base imbedding, not by the lengthy task of face tracing in the derived rotation system. Another example further illustrates this point.

Example 4.1.2. *Let $\langle G \to S_1, \alpha \rangle$ be the imbedded voltage graph shown in Figure 4.2, where voltages are in the cyclic group $\mathscr{Z}_6$. The four boundary walks have net voltages (listed clockwise starting from the upper left) 0, 5, 4, and 3. Therefore the derived imbedding has six faces of size 4, one face of size 24, two faces of size 12, and three faces of size 8. Since the base graph has four vertices and eight edges, the derived surface has Euler characteristic*

$$\chi(S^\alpha) = 6 \cdot 4 - 6 \cdot 8 + (6 + 1 + 2 + 3) = -12$$

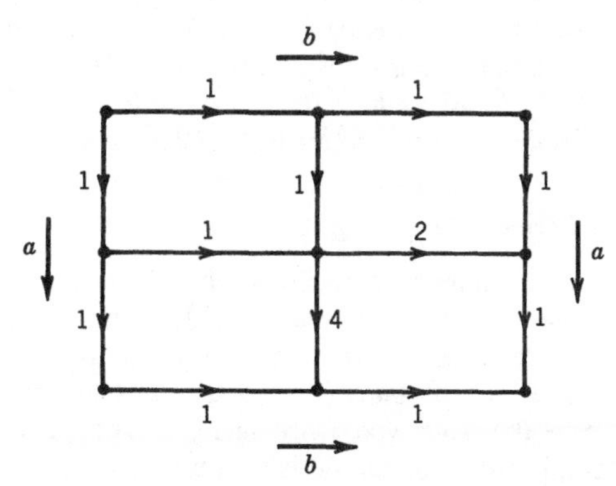

Figure 4.2. A voltage graph imbedded in the torus, with voltages in the cyclic group $\mathscr{Z}_6$.

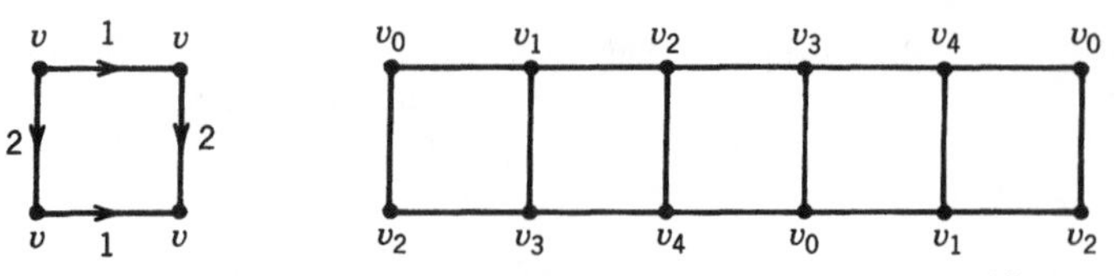

Figure 4.3. An imbedding $B_2 \to S_1$ with voltages in $\mathscr{Z}_5$ and the derived imbedding $K_5 \to S_1$.

4.1.3. The Kirchhoff Voltage Law

A face of an imbedded voltage graph is said to satisfy the Kirchhoff voltage law (abbreviated KVL) if the net voltage on its boundary walk is the identity. Such faces are especially important in constructing minimal imbeddings, since the lifts of a face satisfying KVL all have the same number of sides as the face itself. For example, a triangular base imbedding with all faces satisfying KVL gives rise to a triangular derived imbedding. Computation of the Euler characteristic of the derived surface is also simpler when faces satisfy KVL.

Theorem 4.1.2. *Let $\langle G \to S, \alpha \rangle$ be an imbedded voltage graph such that every face satisfies KVL. Then $\chi(S^\alpha) = n\chi(S)$, where n is the order of the voltage group.*

Proof. By Theorem 4.1.1, each face of the base imbedding lifts to n faces of the derived imbedding. The derived graph always has n times as many vertices and edges as the base graph. Thus

$$\chi(S^\alpha) = n\#V_G - n\#E_G + n\#F_G = n\chi(S) \quad \square$$

Example 4.1.3. *Let $G \to S$ be the imbedding of a bouquet of two circles in the torus with voltages in $\mathscr{Z}_5$ as shown on the left in Figure 4.3. Since KVL holds for the one and only face of the base imbedding, it follows that $\chi(S^\alpha) = 5\chi(S) = 0$. Every edge of the derived imbedding has type 0, because every edge of the base imbedding does. Therefore, the derived surface S^α is orientable, and from its Euler characteristic, we infer that it must be the torus. The derived graph G^α is the complete graph K_5. Thus, we have found a quadrilateral imbedding of K_5 in the torus. This imbedding is shown on the right of Figure 4.3.*

4.1.4. Imbedded Permutation Voltage Graphs

Let $\langle G \to S, \alpha \rangle_n$ be an imbedded permutation voltage graph. The "derived rotation system" and the resulting "derived imbedding" $G^\alpha \to S^\alpha$ and "derived surface" S^α are defined just as for ordinary voltages: rotations at corresponding vertices are the same (up to erasure of subscripts), and corresponding edges have the same type. Once again, it is easier to calculate faces of the derived graph not from the derived rotation system, but by directly lifting faces of the base graph.

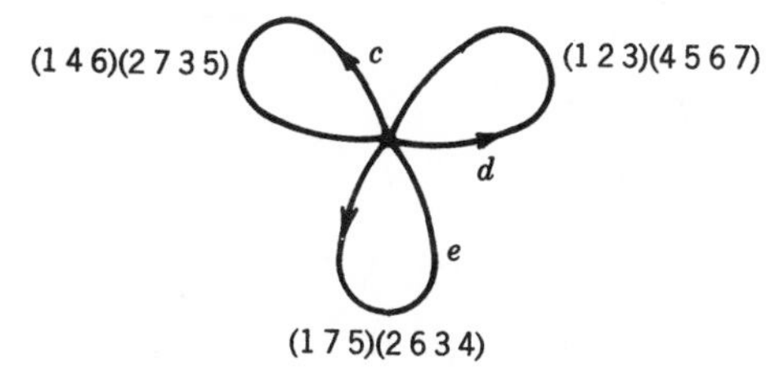

Figure 4.4. A permutation voltage graph imbedded in the sphere.

Theorem 4.1.3. *Let C be the boundary walk of a k-sided face in the imbedded permutation voltage graph $\langle G \to S, \alpha \rangle_n$. Suppose that the net permutation voltage on C has cycle structure $(c_1, c_2, \ldots, c_n)$. Then there are $c_1 + c_2 + \cdots + c_n$ faces of the derived imbedding $G^\alpha \to S^\alpha$ corresponding to the face C, including for $j = 1, 2, \ldots, n$ exactly c_j faces with kj sides.*

Proof. This is an easy corollary of Theorem 2.4.3. □

Example 4.1.4. *Consider the imbedded permutation voltage graph $\langle B_3 \to S_0, \alpha \rangle_7$ shown in Figure 4.4. The derived graph has seven vertices, 21 edges, and no multiple edges or loops, which implies that it is the complete graph K_7. The voltage on each of the three monogons of the bouquet B_3 has cycle structure $(0, 0, 1, 1, 0, 0, 0)$, so it gives rise to one quadrilateral and one triangle in the derived imbedding. The three-sided outside face of the base imbedding has a net voltage of $(15732)(46)$; thus it lifts to one 15-sided face and one 6-sided face in the derived imbedding. The derived surface S^α is orientable of genus 4, since*

$$\chi(S^\alpha) = 7 - 21 + 8 = -6$$

4.1.5. Orientability

Let $\langle G \to S, \alpha \rangle$ be an imbedded voltage graph. If the surface S is orientable, then the orientability algorithm of Section 3.2 enables us to construct an equivalent imbedding in which every edge of G has type 0. Therefore, we may assume that every edge of the derived imbedding also has type 0, from which it follows that the derived surface is orientable. If the base surface S is nonorientable, the situation is not so clear. On the one hand, the derived surface in Example 4.1.1 is nonorientable, since the closed walk $a_0 b_0 c_0^-$ is type 1. On the other hand, the derived surface might be orientable, as in the following example.

Example 4.1.5. *Let $\langle G \to S, \alpha \rangle$ be the imbedded voltage graph with $\mathscr{Z}_4$-voltages, as shown on the left in Figure 4.5. The imbedding has only one face, so we know from its Euler characteristic that the base surface S is the projective plane.*

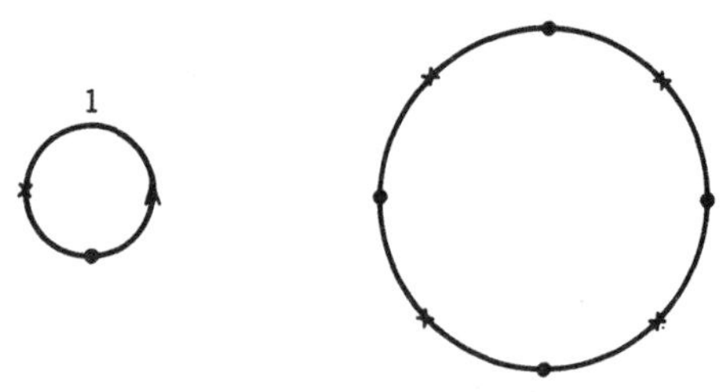

Figure 4.5. Rotation projections of a nonorientable imbedded $\mathscr{Z}_4$-voltage graph and its orientable derived imbedding.

The derived imbedding, shown on the right, has two four-sided faces, and its only cycle consists of four edges of type 1. *Thus the derived surface is the sphere; in particular, it is orientable. If the voltages shown were in $\mathscr{Z}_3$ instead, then the derived surface would have had a cycle with an odd number of type-*1 *edges and would have been nonorientable.*

From this example, one may correctly suspect that the derived surface of an imbedded voltage graph $\langle G \to S, \alpha \rangle$ is nonorientable, if and only if some type-1 closed walk in G has a net voltage of odd order. In fact, one only need check for net voltages of order 1.

Theorem 4.1.4. *Let $\langle G \to S, \alpha \rangle$ be an imbedded voltage graph. Then the derived surface S is nonorientable if and only if the net voltage on some type-*1 *closed walk in the base imbedding $G \to S$ equals the identity.*

Proof. Let W' be a type-1 closed walk in the derived surface S^α, and let $p: G^\alpha \to G$ be the natural projection. Then the image $W = p(W')$ of the sequence of edges in W is a type-1 closed walk, and, by Theorem 2.1.2, the net voltage on the walk W is the identity. Conversely, let W be a type-1 closed walk in the base imbedding $G \to S$ whose net voltage equals the identity. Then the lift $p^{-1}(W)$ consists of $\#\mathscr{A}$ copies of the closed walk W, each of which is type 1. □

Clearly one cannot check the net voltage of every one of the infinite number of type-1 closed walks to see if it equals the identity. Some finite collection of "basic" cycles is needed, as in the orientability algorithm of Section 3.2, or as in the method described in Section 2.5 for finding generators of the local group $\mathscr{A}(u)$. The difficulty is that the voltages assigned to basic type-0 cycles have just as much influence over the orientability of the derived imbedding as the voltages assigned to basic type-1 cycles. The globalness is illustrated by the following example.

Example 4.1.6. *Let $B_2 \to S$ be any imbedding of the bouquet of two circles such that one loop d of B_2 is type* 1 *and the other loop e is type* 0. *Let α be the $\mathscr{Z}_2$-voltage assignment given by $\alpha(d) = \alpha(e) = 1$. Since the only "basic" type-*1 *cycle d does not have a net voltage of* 0, *it might appear that the derived surface* S^α *is orientable. However, the closed walk d e is also type* 1, *and its net voltage is* 0, *since* $\alpha(e) = 1$. *Thus the derived surface is, in fact, nonorientable.*

Example 4.1.6 indicates that the derived surface is orientable when the net voltages on type-1 closed walks are in some way "disjoint" from the net voltages on type-0 closed walks. To make this idea more precise, a base vertex u is needed so that the local voltage group $\mathscr{A}(u)$ can be used. Let $\mathscr{A}^0(u)$ denote the collection of net voltages on all type-0 closed walks based at vertex u. Since the set of such walks is closed under the product of walks and the reversing of walks, the collection $\mathscr{A}^0(u)$ forms a subgroup of the group $\mathscr{A}(u)$; this subgroup is called the "type-0 local (voltage) group".

Theorem 4.1.5. *Let* $\langle G \to S, \alpha \rangle$ *be an imbedded voltage graph with voltage group* $\mathscr{A}$, *and let u be any vertex of G. Then the type-*0 *local group* $\mathscr{A}^0(u)$ *has index* 1 *or* 2 *in the local group* $\mathscr{A}(u)$. *Moreover, the following are equivalent:*

- **i.** $\mathscr{A}^0(u)$ has index 2 in $\mathscr{A}(u)$.
- **ii.** No type-1, u-based closed walk has its net voltage in $\mathscr{A}^0(u)$.
- **iii.** The derived surface S^α is orientable.

Proof. We will first prove that i and ii are equivalent. To this end, let C be any type-1, u-based closed walk, and suppose its net voltage is b. We claim that $\mathscr{A}(u) = \mathscr{A}^0(u) \cup b\mathscr{A}^0(u)$. In support of this claim, suppose that $a \in \mathscr{A}(u)$, so that a is the net voltage on some u-based closed walk D. If D is type 0, then $a \in \mathscr{A}^0(u)$ by definition. If D is type 1, then $C^- D$ is type 0, so its net voltage $b^{-1}a$ is in $\mathscr{A}^0(u)$. Therefore, $a \in b\mathscr{A}^0(u)$, and the claim is proved. It follows that $\mathscr{A}^0(u)$ has index 1 or 2 in $\mathscr{A}(u)$. Since the type-1 closed walk C is arbitrary, this also shows that $\mathscr{A}^0(u)$ has index 2 if and only if no type-1, u-based closed walk has its net voltage in $\mathscr{A}^0(u)$. Thus, i and ii are equivalent.

Now we will show that ii and iii are equivalent. Suppose that the derived surface S^α is nonorientable. Then, by Theorem 4.1.4, there is a type-1 closed walk C in the base graph G with net voltage equal to the identity. If C is not already based at the vertex u, then the closed walk DCD^-, where D is a path from u to the initial vertex of C, is based at u and still is type-1 with net voltage equal to the identity. Of course, the identity is in $\mathscr{A}^0(u)$. Thus, we have proved that ii implies iii. Conversely, suppose that some type-1, u-based closed walk C has its net voltage in $\mathscr{A}^0(u)$. By the definition of the subgroup $\mathscr{A}^0(u)$, some type-0, u-based closed walk D has the same net voltage as C. Thus, CD^- is a type-1 closed walk in the base graph, whose net voltage equals the identity. It follows from Theorem 4.1.4 that the derived surface S^α is nonorientable. Thus iii implies ii. □

Example 4.1.7. *Let* $B_3 \to S$ *be any imbedding of a bouquet of three loops c, d, and e, such that each loop is type* 1. *Let* α *be the unique voltage assignment in the group* $\mathscr{Z}_2 \times \mathscr{Z}_2$ *such that* $\alpha(c) = (1,0)$, $\alpha(d) = (0,1)$, *and* $\alpha(e) = (1,1)$.

Since each of the three possible index-2 subgroups of $\mathscr{Z}_2 \times \mathscr{Z}_2$ contains the net voltage of a type-1 loop, it follows from Theorem 4.1.5 that the derived surface is nonorientable.

4.1.6. An Orientability Test for Derived Surfaces

We wish to develop a finite test for the orientability of a derived surface based on Theorem 4.1.5, in which the only closed walks to be checked are in the "basic" set of cycles given by a spanning tree. Unfortunately, the type-0 local group might not be generated by the net voltages on the basic type-0 cycles. For instance, in returning to Example 4.1.7, we see that there are no type-0 cycles at all, yet the type-0 local group is not trivial. (In fact, it is the whole voltage group.) Thus, the test is not quite so direct as one might hope. Before giving the test, we shall prove a useful algebraic lemma about index-2 subgroups.

Theorem 4.1.6 (An Algebraic Lemma). *Let $\mathscr{B}$ be a subgroup of index 2 in a group $\mathscr{A}$. Then $\mathscr{B}$ is the kernel of a homeomorphism $\phi: \mathscr{A} \to \mathscr{Z}_2$. In particular, a product $a_1 \cdots a_n$ is in the subgroup $\mathscr{B}$ if and only if the number of elements $a_i \notin \mathscr{B}$ is even.*

Proof. It suffices to show that the subgroup $\mathscr{B}$ is normal in the group $\mathscr{A}$. This follows immediately from the observation that, if $a \notin \mathscr{B}$, then $a\mathscr{B} = \mathscr{B}a = \mathscr{A} - \mathscr{B}$. □

Orientability Test for the Derived Surface. *Let $\langle G \to S, \alpha \rangle$ be a connected, nonorientable imbedded voltage graph with voltage group $\mathscr{A}$. First, select a spanning tree T and a base vertex u for the graph G, and change the local orientation where necessary (as in the orientability algorithm of Section 3.2) so that each edge of the tree T is type 0. Next, compute the T-voltage assignment α_T and the local group $\mathscr{A}(u)$, as in Section 2.5. Then the derived surface S^α is orientable if and only if this condition holds:*

> *There is a subgroup $\mathscr{B}$ of index 2 in the local group $\mathscr{A}(u)$ such that $\alpha_T(e) \in \mathscr{B}$ when e is a type-0 edge and $\alpha_T(e) \notin \mathscr{B}$ when e is a type-1 edge.*

This algorithmizable test was devised by Gross and Tucker (1979a). It is related to criteria given by Stahl and White (1976).

To verify that this orientability test works, we first recall that the derived surfaces for voltage assignments α and α_T are the same. If their common derived surface is orientable, then by Theorem 4.1.5, the type-0 local group $\mathscr{A}^0(u)$ has index 2 in the local group $\mathscr{A}(u)$, and no type-1 closed walk has a

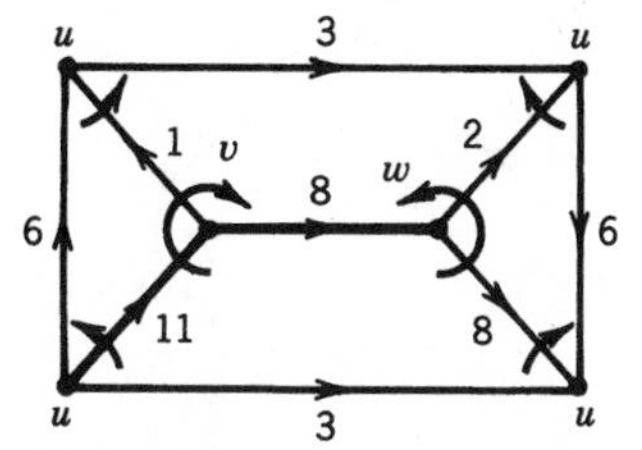

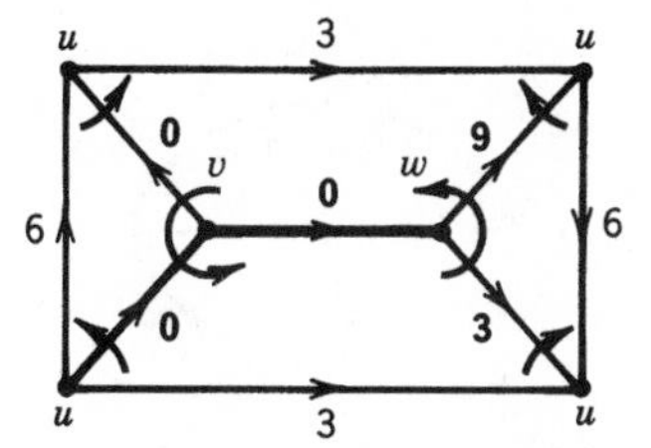

Figure 4.6. A voltage graph imbedded in the Klein bottle (the original voltages are given on the left, the T-voltages on the right).

net voltage in $\mathscr{A}^0(u)$. Because every edge of the tree T is type 0 and has T-voltage equal to the identity, it follows that every edge e not in the tree T corresponds to a unique u-based closed walk having the same type as e and net voltage equal to $\alpha_T(e)$. Therefore, $\mathscr{A}^0(u)$ is the desired subgroup $\mathscr{B}$.

Conversely, suppose that the subgroup $\mathscr{B}$ exists. Then α_T assigns every type-1 edge a voltage not in $\mathscr{B}$. By the algebraic lemma (Theorem 4.1.6), it follows that every type-1 closed walk has a net voltage not in $\mathscr{B}$. In particular, no type-1 closed walk has net voltage equal to the identity. Therefore, by Theorem 4.1.4, the derived surface is orientable.

Example 4.1.8. *Let $\langle G \to S, \alpha \rangle$ be the imbedded voltage graph given on the left in Figure 4.6 where voltages are in the cyclic group $\mathscr{Z}_{12}$. The base imbedding surface S is a Klein bottle. The graph G is the same as that of Example 2.5.3, and the tree T and root vertex u selected are again the same as in Example 2.5.3. Local orientations are indicated by arrows around vertices. On the right of Figure 4.6 is the same imbedding with the orientation at vertex v reversed so that each edge in the tree T is type 0. In addition the T-voltages have been given as computed in Example 2.5.3. The local group $\mathscr{A}(u)$ is $\{0,3,6,9\}$. The only possible 2-index subgroup is $\{0,6\}$. It is easily verified that $\alpha_T(e) \in \{0,6\}$ if and only if the edge e is type 0. We conclude that the derived surface S^α is orientable. We also observe that the derived surface S^α has three components. Finally, since two of the boundary walks of the imbedding $G \to S$ have net voltage 0 and the other two have net voltage 6, it follows that the derived imbedding has 36 faces. Therefore, $\chi(S^\alpha) = 36 - 84 + 36 = -12$, from which it follows that the derived surface S^α consists of three orientable components of genus 3.*

It should be noted that the existence of the index-2 subgroup $\mathscr{B}$ in the test described above is equivalent, by Theorem 4.1.6, to the existence of a homeomorphism $\phi\colon \mathscr{A}(u) \to \mathscr{Z}_2$ such that $\phi(\alpha_T(e)) = 0$ when e is a type-0 edge and $\phi(\alpha_T(e)) = 1$ when e is a type-1 edge. This viewpoint is especially convenient if the local group $\mathscr{A}(u)$ is given by a presentation with generating set $\{\alpha_T(e) \mid e \in E_G - T\}$. There really is no choice for the desired homeomor-

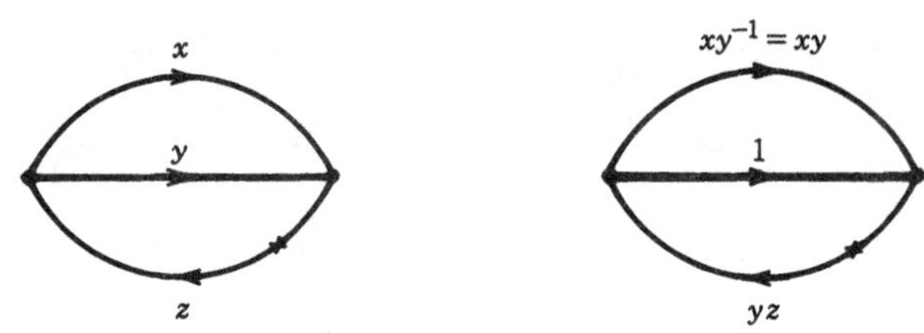

Figure 4.7. A rotation projection for an imbedded voltage graph and a corresponding set of T-voltages.

phism ϕ, since its values on the generating set are determined. It is necessary to check only that this assignment of 0s and 1s to the generators of $\mathscr{A}(u)$ does define a homeomorphism; that is, each relator for the given presentation must have the value 0.

Example 4.1.9. *Let $\langle G \to S, \alpha \rangle$ be the imbedded voltage graph given on the left in Figure 4.7. The voltage group $\mathscr{A}$ is a finite group with a presentation of the form*

$$\langle x, y, z: x^2 = y^2 = z^2 = 1, (xy)^4 = (yz)^4 = (xz)^3 = 1, \ldots \rangle$$

On the right in Figure 4.7 are the T-voltages for the spanning tree shown in boldface. If we let $s = xy$ and $t = yz$, then the local group $\mathscr{A}(u)$ has a presentation of the form

$$\langle s, t: s^4 = t^4 = (st)^3 = 1, \ldots \rangle$$

Then the desired homeomorphism $\phi: \mathscr{A}(u) \to \mathscr{Z}_2$ must have $\phi(s) = 0$ and $\phi(t) = 1$. But then $\phi((st)^3) = 3\phi(s) + 3\phi(t) \neq 0$. Therefore the derived surface is nonorientable.

4.1.7. Exercises

1. Let $\langle G \to S, \alpha \rangle$ be the imbedded voltage graph whose rotation projection appears in Figure 4.8*a*. Find the number of faces of each size in the derived imbedding and identify the derived surface.
2. Verify that the derived imbedding of the imbedded permutation voltage graph given in Figure 4.8*b* is an imbedding of the complete bipartite graph $K_{4,4}$ in which every face is a quadrilateral.
3. Construct an imbedding of the complete graph K_6 that has two triangles, two 6-gons, and three quadrilaterals. (*Hint:* Begin with a bouquet of three circles imbedded in the sphere. Multiple edges in the derived graph must be identified.)

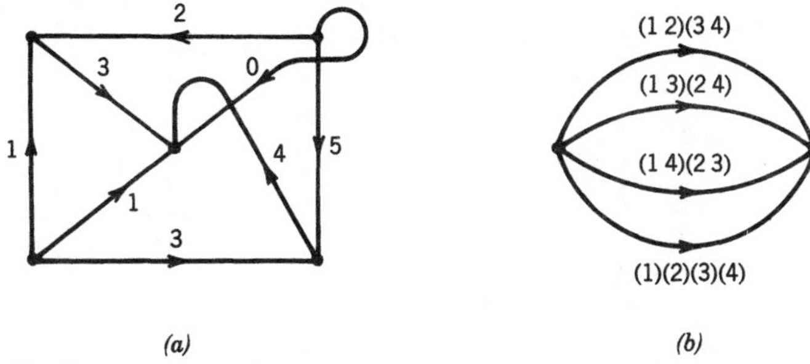

Figure 4.8. Two rotation projections of imbedded voltage graphs. (a) Ordinary voltages in $\mathscr{Z}_6$, (b) permutation voltages in $\mathscr{S}_4$.

4. Derive a one-face imbedding for K_5, starting with a bouquet of two circles imbedded in the Klein bottle.
5. Find an orientable imbedding of the octahedron graph O_3 in which all the faces are quadrilaterals.
6. Suppose that the abelian group $\mathscr{A}$ has a generating set X that contains no elements of order 2. If $\#X = n$ is even, show that the Cayley graph $C(\mathscr{A}, X)$ has an orientable imbedding in which every face is a $2n$-gon.
7. Find an imbedding in the torus of the 4-regular graph $(C_3 \cup C_4)^c$, which was previously considered in Example 2.3.7.
8. Use Example 2.1.6 to construct an orientable imbedding of the 3-cube graph Q_3 in which all faces are 6-gons.
9. Let $\langle G \to S, \alpha \rangle$ be an imbedded voltage graph where the voltage group $\mathscr{A}$ is nontrivial. Prove that if $\chi(S) < 0$, then $\chi(S^\alpha) < \chi(S)$, and that if $\chi(S) = 0$, then $\chi(S^\alpha) \leq 0$.
10. Let $\mathscr{A}$ be a nontrivial finite group with presentation $\langle x, y{:}\ x^3 = y^5 = (xy)^2 = 1, \ldots \rangle$. Prove that the Cayley graph $C(\mathscr{A}, \{x, y\})$ has an imbedding in a connected, orientable surface of Euler characteristic $\#\mathscr{A}/30$, and that therefore $\#\mathscr{A} = 60$. Conclude that the group $\mathscr{A}$ is simple. (*Hint:* Show that any nontrivial quotient of $\mathscr{A}$ has a presentation of the same form as that of $\mathscr{A}$ and hence has order 60.)
11. Let $\mathscr{A}$ be a finite group with presentation $\langle x, y{:}\ x^3 = y^6 = (xy)^2 = 1, \ldots \rangle$. Prove that the genus of the Cayley graph $C(\mathscr{A}, \{x, y\})$ is at most 1.
12. Use the orientability test for derived surfaces to show that the derived surface in Example 4.1.1 is nonorientable.
13. Let T be a spanning tree and u a root vertex for a connected, imbedded voltage graph $\langle G \to S, \alpha \rangle$. Let $\mathscr{B}'$ be the subgroup of the local group generated by $\{\alpha_T(e) \mid e \text{ is a type-0 edge}\}$. Show by an example that the derived surface S may be nonorientable even if $\alpha_T(e) \in \mathscr{B}'$ when the edge e is type 0 and $\alpha_T(e) \notin \mathscr{B}'$ when the edge is type 1.

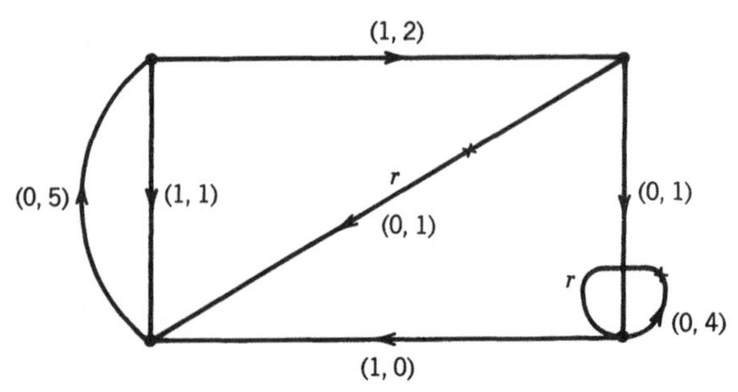

Figure 4.9. A rotation projection of an imbedded voltage graph, with voltages in $\mathscr{Z}_2 \times \mathscr{Z}_8$.

14. Let $\langle G \to S, \alpha \rangle$ be the imbedded voltage graph given by the rotation projection in Figure 4.9 where the voltage group is $\mathscr{Z}_2 \times \mathscr{Z}_8$. Use the methods of Example 4.1.8 to determine the topological type of the derived surface S^α.

15. Let $\mathscr{A}$ be a finite group with a presentation of the form $\langle x, y: xyx^{-1}y = 1, \ldots \rangle$. Prove that the subgroup $\mathscr{B}$ generated by x^2 and y is normal in $\mathscr{A}$ and has index at most 2. Show that if $\mathscr{B}$ has index 2 in $\mathscr{A}$, then the Cayley graph $C(\mathscr{A}, \{x, y\})$ has an imbedding in the torus. Conclude that in any case, the genus of the group $\mathscr{A}$ is at most 1.

4.2. BRANCHED COVERINGS OF SURFACES

For any voltage graph $\langle G, \alpha \rangle$, the natural graph projection $p: G^\alpha \to G$ is a covering projection in the topological sense; that is, every point $y \in G$ has a neighborhood N_y such that each component of the lift $p^{-1}(N_y)$ is mapped homeomorphically by p onto N_y, as described in Section 2.3. For any imbedded voltage graph $\langle G \to S, \alpha \rangle$ one might expect that the natural projection $p: G^\alpha \to G$ extends to a map $\bar{p}: S^\alpha \to S$ with properties similar to those of p. This is indeed the case, for there is an extension of p that is "almost" a covering map.

4.2.1. Riemann Surfaces

Let $p: \tilde{S} \to S$ be a continuous function between closed surfaces. Then $p: \tilde{S} \to S$ is called a "branched covering" if there is a finite set of points Y in S such that p restricted to the punctured surface $\tilde{S} - p^{-1}(Y)$ is a covering onto the punctured surface $S - Y$. Since $\tilde{S}$ and S are compact, it is equivalent to specify that the function p is a local homeomorphism except possibly at the points of $p^{-1}(Y)$. Points in the exceptional set Y are called "branch points", and points in its lift $p^{-1}(Y)$ are called "prebranch points".

The "order" of a prebranch point x is r if the function p is r-to-1 in a small neighborhood of x. If the surface S is connected, an elementary topological argument shows that the fiber size $\#p^{-1}(z)$ is the same for all nonbranch points $z \in S$. This common value is called the "number of sheets" of the branched covering. The "deficiency" of the branch point y, denoted $\text{def}(y)$, is $n - \#p^{-1}(y)$, where n is the number of sheets of the branched covering.

Example 4.2.1. *Let $\tilde{S}$ and S both be the extended complex plane, that is, the complex plane with a point at ∞ adjoined (the resulting surface is homeomorphic, by stereographic projection, to the sphere). Let $p: \tilde{S} \to S$ be the function $p(z) = z^n$, where $n \geq 1$ is an integer. Since p simply multiplies the argument of each complex number by n, the effect of p is to wrap the complex plane n times around the origin. The restricted map $p: \tilde{S} - \{0, \infty\} \to S - \{0, \infty\}$ is clearly a covering, so $p: \tilde{S} \to S$ is a branched covering. The number of sheets is n, since any complex number other than 0 and ∞ has n different nth roots. Both 0 and ∞ are branch points of deficiency $n - 1$, and the corresponding prebranch points, again 0 and ∞, have order n.*

Example 4.2.2. *Let S be the extended complex plane, and let $\tilde{S}$ be the set of points $\{(z, w) \in C \times C | w^2 = z(z-1)(z-2)\}$ with a point at ∞ adjoined. It can be shown that every point of $\tilde{S}$ has a neighborhood homeomorphic to the unit disk, so that $\tilde{S}$ is a surface. Let $p: \tilde{S} \to S$ be the projection $p(z, w) = z$. Then p fails to be a local homeomorphism only at the points $(0,0)$, $(1,0)$, $(2,0)$ and ∞. Therefore p is a branched covering with branch points at 0, 1, 2, and ∞. We observe that p is two-sheeted, since the equation $w^2 = z(z-1)(z-2)$ has two solutions for w for each value of z other than 0, 1, 2, or ∞.*

Example 4.2.1 gives a typical view of a branched covering near a prebranch point, and the surface $\tilde{S}$ of Example 4.2.2 together with the map $p: \tilde{S} \to S$ is the Riemann surface associated with the equation $w^2 = z(z-1)(z-2)$. The theory of Riemann surfaces has had a long and powerful influence in low-dimensional topology. A small amount of this theory, including some classical results of the 19th and early 20th centuries, is developed here in graph-theoretical terms. For more about Riemann surfaces from an elementary viewpoint, see Springer (1957).

The idea of a branched covering generalizes to higher dimensions. A simplicial map $p: M \to N$, where M and N are n-dimensional complexes is a branched covering if there is a subcomplex Y in N of dimension at most $n - 2$ such that p restricted to $M - p^{-1}(Y)$ is a covering onto $N - Y$, that is, each point in $M - p^{-1}(Y)$ has an open neighborhood mapped homeomorphically by p onto an open set of $N - Y$ (see Fox, 1957). If instead the subcomplex Y has dimension $n - 1$, then the map p is said to be a "folded covering". Such maps were first studied by Tucker (1936), but they have received little attention. Bouchet's (1982a) work on lifting triangulations is one of the few uses of folded surface coverings in topological graph theory.

4.2.2. Extension of the Natural Covering Projection

The source of the branched coverings that we study is imbedded voltage graphs. The following is the fundamental theorem relating branched coverings to voltage graphs.

Theorem 4.2.1 (Gross, 1974; Gross and Tucker, 1977). *Let $\langle G \to S, \alpha \rangle$ be an imbedded voltage graph, with either ordinary or permutation voltages. Then the natural projection $p: G^\alpha \to G$ on the graphs extends to a branched covering $p: S^\alpha \to S$ on their imbedding surfaces. If α is a permutation assignment in $\mathscr{S}_n$, and if the net voltage on face f has cycle structure $(c_1, \ldots, c_n)$, then the projection p has a branch point inside face f with exactly c_j prebranch points of order j, for $j = 1, \ldots, n$. If α is an ordinary assignment in the group $\mathscr{A}$ and the net voltage on face f has order r, then p has a branch point inside face f with $\#\mathscr{A}/r$ prebranch points, each of order r.*

Proof. The natural projection $p: G^\alpha \to G$ is already known to be a local homeomorphism, because it is a covering projection. We first extend p to the 0-bands of the imbedding $G^\alpha \to S^\alpha$ so that p remains a local homeomorphism. This is possible because the rotation at any vertex v_i of G is lifted from the rotation at $v = p(v_i)$, that is, the rotation at the vertex v_i is the same as that at the vertex v up to erasure of subscripts. Next we extend p to the 1-bands, still keeping p a local homeomorphism. Since the type of any edge e_i in the edge-fiber $p^{-1}(e)$ is defined to be the same as the base edge e, and since p maps the interior of the edge e_i homeomorphically onto the interior of the edge e, it follows that there is a homeomorphism of the 1-band for e_i onto the 1-band for e, which can be used to extend p as desired.

Finally, in order to extend the map p to the 2-bands, we suppose that f is a face of the imbedding $G \to S$, and that the net permutation voltage on the boundary walk C has cycle structure $(c_1, \ldots, c_n)$. Then, by Theorem 4.1.3, there are c_j faces of the derived imbedding $G^\alpha \to S^\alpha$, each of whose boundary walks is wrapped by p exactly j times about the boundary walk C. The map p can then be extended inside the 2-band for each of these faces so that its restriction to any one of these 2-bands looks like the map $z \to z^j$ on the unit disk in the complex plane. In particular, the branch point inside the face f will have exactly c_j prebranch points of order j, for $j = 1, \ldots, n$. If α is an ordinary voltage assignment, then essentially the same proof holds, only using Theorem 4.1.1 instead of Theorem 4.1.3. □

Corollary. *Let $\langle G \to S, \alpha \rangle_n$ be an imbedded voltage graph satisfying KVL on every face. Then a natural projection $p: S^\alpha \to S$ is an unbranched covering.* □

We should like to call the branched covering constructed in Theorem 4.2.1 "the" natural projection of the derived surface S^α onto the base surface S. Certainly in terms of the combinatorial structure of vertices, edges, and faces

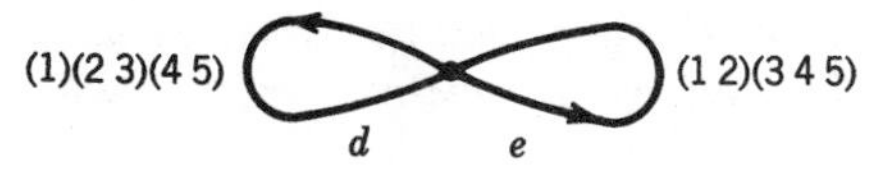

Figure 4.10. A rotation projection for an imbedded permutation voltage graph.

for these surfaces, the extension of the natural graph projection $p: G^\alpha \to G$ is unique. However, it is not at all unique topologically. For example, let us try to be more precise about how p extends to the 2-bands so that it "looks like" the map $z \to z^j$. If B is a 2-band in the derived surface and B is the corresponding 2-band in the base surface, we first need homeomorphisms $\tilde{h}: \tilde{B} \to D$ and $h: B \to D$, where D is the unit disk in the complex plane. Then p is defined to be $h^{-1}g\tilde{h}$ on $\tilde{B}$, where $g(z) = z^j$. Thus the extension of p depends upon the choice of homeomorphisms h and $\tilde{h}$. In fact, the extension of p to the 0-bands and 1-bands also depends on choices of homeomorphisms.

We could try to get around this difficulty by defining "equivalent" branched coverings just as we defined equivalent imbeddings in Section 3.1. Rather than become any more deeply involved with topological considerations, however, we simply define a "natural (surface) projection" for an imbedded voltage graph $\langle G \to S, \alpha \rangle$ to be any branched covering $q: S^\alpha \to S$ that extends the natural graph projection $p: G^\alpha \to G$ and has at most one branch point in each face of the base imbedding.

Example 4.2.3. *Let $\langle G \to S, \alpha \rangle_5$ be the imbedded permutation voltage graph given by the rotation projection in Figure* 4.10. *By Theorem* 4.2.1, *there is a natural surface projection $p: S^\alpha \to S$ that is an extension of the natural graph projection $p: G^\alpha \to G$. The face bounded by the loop d has a branch point of deficiency* $5 - 3 = 2$ *and prebranch points of orders* 1, 2, *and* 2. *The face bounded by e has a branch point of deficiency* $5 - 2 = 3$ *and prebranch points of orders* 2 *and* 3. *By computing its net boundary voltage, we determine that the outside face has a branch point of deficiency* 3 *and prebranch points of order* 1 *and* 4. *Note that some prebranch points might have order* 1, *in which case the map p is still a local homeomorphism there.*

4.2.3. Which Branched Coverings Come from Voltage Graphs?

It might appear that branched coverings arising from imbedded permutation voltage graphs are special in some way, but the following theorem shows this not to be the case. Just as every graph covering can be obtained by the permutation voltage graph construction, every branched covering of surfaces can be obtained by extending the natural projection of a permutation voltage graph.

Theorem 4.2.2 (Gross and Tucker, 1977). *Let $q: \tilde{S} \to S$ be an n-sheeted branched covering. Let $G \to S$ be any cellular graph imbedding such that no* *branch point of q lies on G and at most one branch point lies in any face. Then there is a permutation voltage assignment α in $\mathscr{S}_n$ such that q is a natural projection $S^\alpha \to S$.*

Proof. By hypothesis, the restriction of the covering q to the graph preimage $q^{-1}(G)$ is a local homeomorphism, and hence a covering. By Theorem 2.4.4, this restriction of q must be the natural projection of a permutation voltage graph $\langle G, \alpha \rangle_n$. In order to show that the rotation system for the imbedding $q^{-1}(G) \to \tilde{S}$ is the same as the derived rotation system for the derived imbedding $G^\alpha \to S^\alpha$, let v be a vertex of the imbedding $q^{-1}(G) \to \tilde{S}$. Since the covering q is a local homeomorphism at vertex v, it takes a small disk D around vertex v homeomorphically onto a small disk $q(D)$ around the vertex $q(v)$ in the imbedding $G \to S$. Thus, if the rotation at vertex v—which is determined by the disk D—is $e_1 \cdots e_m$, then the rotation at vertex $q(v)$—which is determined by the disk $q(D)$—must be $q(e_1) \cdots q(e_n)$. That is, the rotation at the vertex v in the covering space S is the derived rotation. A similar argument shows that each edge e of the imbedding $q^{-1}(G) \to \tilde{S}$ has the same type as its image edge $q(e)$. □

Example 4.2.1 Revisited. *The branched covering of the extended complex plane given by $p(z) = z^n$ could be obtained by extending the natural projection of the permutation voltage graph $\langle G \to S, \alpha \rangle_n$ where G consists of a single loop d imbedded in the sphere and $\alpha(d) = (1\,2\,3 \cdots n)$.*

Example 4.2.2 Revisited. *The branched covering of the Riemann surface for $w^2 = z(z-1)(z-2)$ could be obtained from the imbedded voltage graph $\langle B_3 \to S_0, \alpha \rangle$ given in Figure 4.11. The voltage group is $\mathscr{Z}_2$.*

Theorem 4.2.2 can be used to obtain nontrivial information about branched coverings. As one example, we show that any connected covering of the sphere $p: S \to S_0$ must have at least two branch points if it is nontrivial (more than one sheet). Indeed, if p has one branch point or none, the graph G of Theorem 4.2.2 can be chosen to consist of a single loop in the sphere enclosing the branch point of p if it has one. The voltage assigned to the loop cannot be the identity since the covering space is connected and has more than one sheet. However, a nonidentity voltage on the loop yields two branch points, one

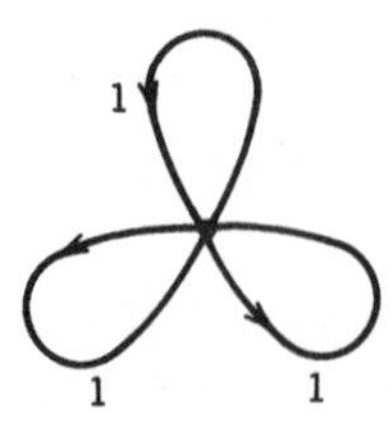

Figure 4.11. An imbedded voltage graph for the Riemann surface of Example 4.2.2.

inside the loop and another outside the loop. The case of two branch points in the sphere is considered in Exercise 13.

4.2.4. The Riemann–Hurwitz Equation

Given a branched covering $p: \tilde{S} \to S$, it is useful to express the Euler characteristic $\chi(\tilde{S})$ in terms of $\chi(S)$ and the deficiencies of the branch points of p. The following theorem establishes the Riemann–Hurwitz equation, which does exactly that.

Theorem 4.2.3. *Let $p: \tilde{S} \to S$ be an n-sheeted covering and let Y be the set of branch points of p. Then*

$$\chi(\tilde{S}) = n\chi(S) - \sum_{y \in Y} \operatorname{def}(y)$$

Proof. By Theorem 4.2.2, the map p is a natural surface projection for an imbedded permutation voltage graph $\langle G \to S, \alpha \rangle_n$. Since $\#V(G^\alpha) = n\#V(G)$ and $\#E(G^\alpha) = n\#E(G)$, we see that all that is needed to compute the Euler characteristic $\chi(\tilde{S}) = \chi(S^\alpha)$ of the derived surface is the number of faces of the derived imbedding $G^\alpha \to S^\alpha$. Each face of the imbedding $G \to S$ lifts to n faces if the face contains no branch point and to $n - \operatorname{def}(y)$ faces if the face contains the branch point y. Therefore, $\#F(G^\alpha) = n\#F(G) - \sum_{y \in Y} \operatorname{def}(y)$, from which it follows that

$$\begin{aligned} \chi(S^\alpha) &= n\#V(G) - n\#E(G) + n\#F(G) - \sum_{y \in Y} \operatorname{def}(y) \\ &= n\chi(S) - \sum_{y \in Y} \operatorname{def}(y) \quad \square \end{aligned}$$

Example 4.2.4. *The Riemann surface for the equation $w^3 = z(z-1)(z-2)$ has branch points at $z = 0, 1, 2, \infty$, each of deficiency 2. By the Riemann–Hurwitz equation, the Euler characteristic of the Riemann surface is $3\chi(S_0) - 4 \cdot 2 = -2$, and we infer that the surface has genus 2.*

4.2.5. Alexander's Theorem

One of the more startling results in the theory of manifolds is Alexander's theorem: any closed, orientable, triangulated manifold of dimension n is a branched covering of the sphere of dimension n. Although the surface version of this theory was known before 1920, the importance of Alexander's generalization to higher dimensions, especially dimensions 3 and 4, was not fully realized for another half-century. The proof given here for surfaces is an easy application of voltage graphs.

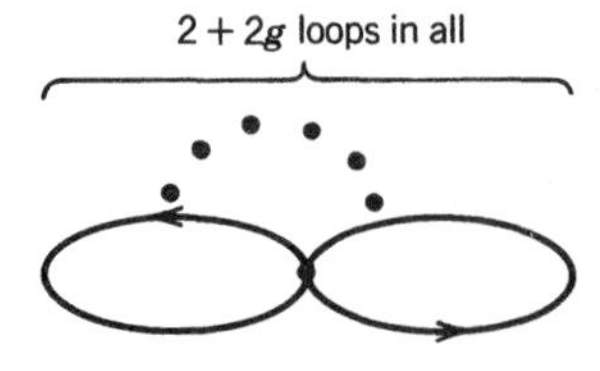

Figure 4.12. An imbedded voltage graph.

Theorem 4.2.4 (Alexander, 1920). *Every closed orientable surface S is a branched covering space over the sphere S_0. In fact, S_g is a two-sheeted covering of S_0 with $2 + 2g$ branch points.*

Proof. Imbed the bouquet B_{2+2g} in the sphere S_0 as shown in Figure 4.12. Let α assign to each loop the voltage 1 in the group $\mathscr{Z}_2$. Then the net voltage on the exterior face is 0, and on each of the $2 + 2g$ monogons it is 1. Therefore, a natural surface projection $p\colon S^\alpha \to S_0$ has $2 + 2g$ branch points, each of deficiency 1. From the Riemann–Hurwitz equation,

$$\chi(S^\alpha) = 2\chi(S_0) - (2 + 2g) = 2 - 2g$$

it follows that the derived surface has genus g. □

The sharpened conclusion given here, that the branched covering can be chosen to be two-sheeted, does not hold in higher dimensions. It is known, however, that every closed, orientable 3-manifold is a three-sheeted branched covering of the 3-sphere (Hilden, 1976; Montesinos, 1976). The following nonorientable version of Alexander's theorem is also not known in higher dimensions.

Theorem 4.2.5. *Every closed nonorientable surface N_k is a branched covering space over the projective plane N_1. In fact, there is a branched covering $p\colon N_k \to N_1$ having two sheets and k branch points if k is even, or three sheets and $(k + 1)/2$ branch points if k is odd.*

Proof. For an even crosscap number k, let $B_k \to N_1$ be the imbedding obtained by giving a twist to one loop of the imbedding in Figure 4.12. Again let α assign to each loop the voltage 1 in the group $\mathscr{Z}_2$. Then the base imbedding has k faces, and each has net voltage 1, including the "exterior" face. Thus, a natural projection from the derived surface has k branch points, each of deficiency 1. By the Riemann–Hurwitz equation, the derived surface has Euler characteristic $2\chi(N_1) - k = 2 - k$. Since an orientability test for the derived surface shows it is nonorientable, the derived surface must be N_k.

For an odd crosscap number k, we imbed the bouquet $B_{(k+1)/2}$ in the projective plane in the same way as in the even case. This time, however, the

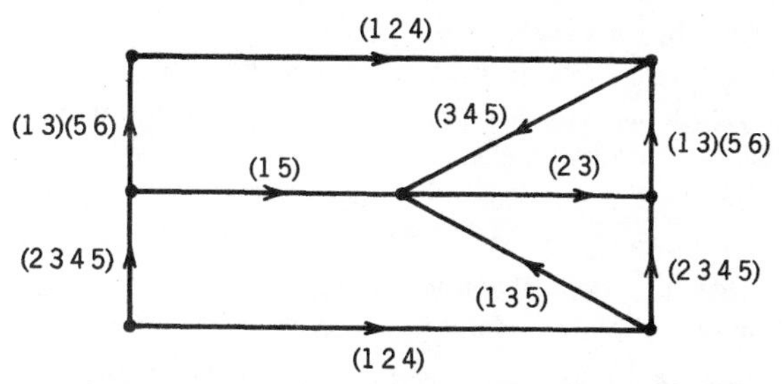

Figure 4.13. A permutation voltage graph imbedded in the torus.

type-0 loops are assigned the voltage 1 in $\mathscr{Z}_3$ and the single type-1 loop is assigned a voltage that gives the exterior face a nonzero net voltage. Then there are $(k + 1)/2$ branch points, each of deficiency 2. By the Riemann–Hurwitz equation, the derived surface has Euler characteristic $3\chi(N_1) - 2(k + 1)/2 = 2 - k$. Since the number $2 - k$ is odd, the derived surface must be nonorientable, from which we conclude that it is homeomorphic to N_k. □

It is natural to ask whether the anomalous use of a three-sheeted branched covering for odd k is really necessary. In fact, no two-sheeted covering space, branched or unbranched, can have odd Euler characteristic (see Exercises 11 and 12).

4.2.6. Exercises

1. Let $\langle G \to S, \alpha \rangle_6$ be the permutation voltage graph imbedding in the torus given by Figure 4.13. Describe the branch and prebranch points of a natural projection $p\colon S^\alpha \to S$. Compute $\chi(S^\alpha)$ first by the Riemann–Hurwitz equation, then by simply counting the number of vertices, edges, and faces of the derived imbedding.
2. Find the branch points and prebranch points of the Riemann surface for the equation $w^4 = z^3 - z$. What is its genus?
3. Show that the torus is a two-sheeted unbranched covering space over the Klein bottle.
4. Generalize Exercise 3 by proving that the orientable surface S_g is a two-sheeted unbranched covering of the nonorientable surface N_{g+1} for all $g \geq 0$. (*Hint:* Represent the surface N_{g+1} by a bouquet of $g + 1$ type-1 loops.)
5. Use Theorem 3.4.5 and Exercise 4 for $g = 0$ to show that there are graphs of arbitrarily large genus having two-sheeted planar coverings.
6. Give an imbedding of the Petersen graph in the projective plane such that every face has size 5. Conclude that the 1-skeleton of the dodecahedron is a two-sheeted covering of the Petersen graph.

7. Prove that the only surfaces that are unbranched coverings of themselves (more than one-sheeted) are the torus and Klein bottle. Prove that the only surfaces that are coverings of themselves with branch points are the sphere and projective plane.
8. Prove that if S is an orientable surface, then S is a branched covering over the sphere with exactly three branch points.
9. Prove that the nonorientable surface N_k, $k > 3$, is an unbranched covering over the nonorientable surface N_3.
10. Prove that the orientable surface S_g, $g > 2$, is an unbranched covering over the orientable surface S_2.
11. Show that for any imbedded voltage graph with voltage group $\mathscr{Z}_2$ the sum of the net boundary voltages over all faces is 0. Conclude that any two-sheeted branched covering must have an even number of branch points.
12. Use Exercise 11 to prove that no surface of odd Euler characteristic is a two-sheeted branched covering space over any surface.
13. Let $p: S \to S_0$ be a branched covering with two branch points in the sphere S_0. Prove that if S is connected, it must be the sphere as well.

4.3. REGULAR BRANCHED COVERINGS AND GROUP ACTIONS

The natural projection of an ordinary voltage graph is more than a covering; it is also the quotient map of the free action of the voltage group on the derived graph. In the same way, a natural surface projection of an imbedded ordinary voltage graph is not just a branched covering, but also the quotient map of a group action on a surface.

4.3.1. Groups Acting on Surfaces

An "action of a group $\mathscr{A}$ on a surface S" is given by associating to each group element $a \in \mathscr{A}$ a homeomorphism ϕ_a on the surface S such that

1. $\phi_{ab} = \phi_a \circ \phi_b$ for all $a, b \in \mathscr{A}$.
2. ϕ_a is the identity homeomorphism if and only if a is the identity element of the group $\mathscr{A}$.

A "fixed point" of the action is a point $x \in S$ such that $\phi_a(x) = x$ for some group element $a \neq 1$. An action with no fixed points is called "free" (or "fixed-point free") and an action with a finite number of fixed points is called "pseudofree".

Example 4.3.1. *When the surface S_2 is imbedded in 3-space as shown in Figure 4.14a, there is an action of the Klein group $\mathscr{Z}_2 \times \mathscr{Z}_2$ on S_2 whose nonidentity elements are* 180° *rotations around the x, y, or z axis (the composition of any*

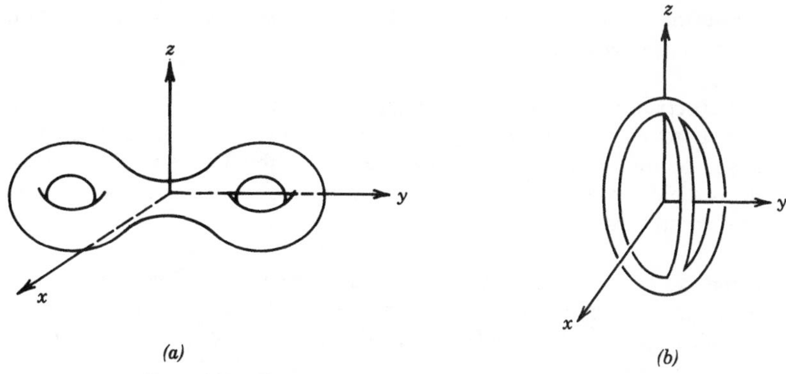

Figure 4.14. Two placements of a surface of genus 2 in 3-space.

two of these is the third). This action on S_2 clearly is pseudofree, but not free; in particular, a $180°$ *rotation about the y axis has six fixed points. A free action of $\mathscr{Z}_2$ on the surface S_2 is generated by the antipodal homeomorphism $h(x, y, z) = (-x, -y, -z)$. The action of the group $\mathscr{Z}_2 \times \mathscr{Z}_2 \times \mathscr{Z}_2$ generated by the three reflections through the coordinate planes is neither free nor pseudofree, since it has entire circles of fixed points; however, this action contains both the two previously described actions. Finally, the 3-space imbedding of the surface S_2 depicted in Figure* 4.14*b reveals an action of the cyclic group $\mathscr{Z}_3$ generated by a* $120°$ *rotation about the z axis.*

Example 4.3.2. *In the graph imbedding whose band decomposition is illustrated in Figure* 4.15, *there are three faces: an inner* 5*-gon, an outer* 5*-gon, and a* 20*-gon in between. There is an obvious action of the cyclic group $\mathscr{Z}_5$ on that band decomposition. Since this action takes each boundary walk to itself, the action can be extended radially inside each* 2*-band to obtain an action of $\mathscr{Z}_5$ on the closed surface. Each nonidentity element of the $\mathscr{Z}_5$ action has three fixed points —one at the center of each face. Note that the closed surface is orientable (just reverse the rotation at each vertex on one of the pentagons) and that its Euler*

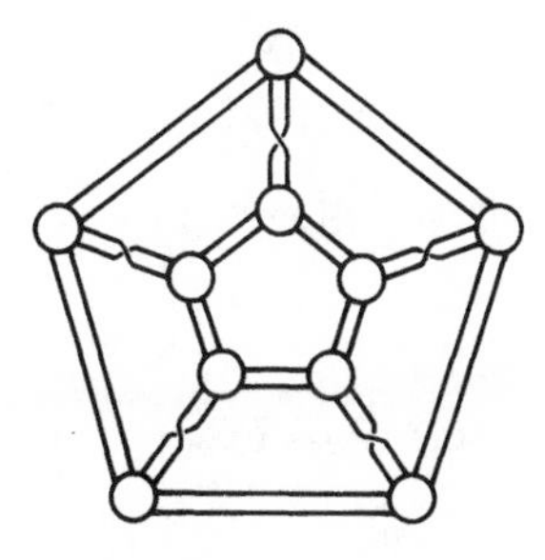

Figure 4.15. A band decomposition of a surface of genus 2.

characteristic equals $10 - 15 + 3 = -2$. *Thus, we have an action of* $\mathscr{Z}_5$ *on a closed surface of genus* 2.

We observe that the various group actions in Example 4.3.1 could each be obtained from an action on a graph at the "core" of the solid in 3-space bounded by the surface. For instance, the core of the double torus in Figure 4.14a is a bouquet of two circles that run inside the handles. This observation can be generalized.

Theorem 4.3.1. *Let* $\mathscr{A}$ *be a group that acts on the graph* G. *Then* $\mathscr{A}$ *also acts on a surface* S *whose genus is the cycle rank* $\beta(G) = \#E - \#V + 1$. *If the action of* $\mathscr{A}$ *on the graph* G *is free, then so is the action of* $\mathscr{A}$ *on the surface* S.

Proof. Imbed the graph G in 3-space and thicken it to form a solid, and let S be the boundary of that solid. Obviously, the action of the group $\mathscr{A}$ on the graph G extends to an action on the solid, from which we can extract an action on the surface S. The action on the solid and on its boundary S is free if the action on the graph G is free. If the graph G is a tree, then the surface S is a sphere bounding a misshapen 3-cell. If T is a spanning tree from G, then each edge not in T adds a handle to the surface S. Therefore, the genus of S is $\beta(G)$, the number of edges not in the tree T. □

4.3.2. Graph Automorphisms and Rotation Systems

In Example 4.3.2, an action of a group on a graph is extended to an imbedding surface for the graph. In general, to extend an automorphism of a graph to a homeomorphism of an imbedding surface, there must be a vertex-orbit consistency within the rotation system corresponding to the imbedding, and an edge-type consistency.

Suppose that the graph imbedding $G \to S$ has a rotation

$$v. \quad e_1 \, e_2 \, \ldots \, e_n$$

at the vertex v, and that h is an automorphism of the graph G. We say that the automorphism h "respects the rotation at the vertex v" if the rotation at the image vertex $h(v)$ is

$$h(v). \quad h(e_1)\, h(e_2) \, \ldots \, h(e_n)$$

We say that h "reverses the rotation at v" if the rotation at $h(v)$ is

$$h(v). \quad h(e_n)\, h(e_{n-1}) \, \ldots \, h(e_1)$$

We say that the graph automorphism h "preserves the type of edge e" if the image edge $h(e)$ has the same type as the edge e.

Finally, we say that the graph automorphism h "respects the rotation system" if h preserves the type of each edge and respects the rotation at each vertex. Similarly, we say that h "reverses the rotation system" if h preserves the type of each edge and reverses the rotation at each vertex.

Theorem 4.3.2. *Let $G \to S$ be an imbedding, and let $h: G \to G$ be a graph automorphism such that h either respects or reverses the rotation system. Then h extends to a surface homeomorphism $\bar{h}: S \to S$. Any face of the imbedding $G \to S$ whose boundary walk is invariant under the graph automorphism h contains a fixed point of the extension $\bar{h}$. Moreover, if the graph automorphism h has no fixed points, then its extension $\bar{h}$ can be chosen to have at most one fixed point in any face.*

Proof. Topologically, the condition that the graph automorphism h respects or reverses the rotation system simply ensures that h can be extended to the 0-bands and 1-bands of the imbedding $G \to S$; the extension to the whole surface S can then be easily accomplished, say radially, inside each 0-band. Combinatorially, what that condition ensures is that each boundary walk of a face is mapped by h to a boundary walk of a face, since boundary walks are determined completely by the rotation system. Any boundary walk mapped to itself by h must bound a face mapped to itself by the extension $\bar{h}$. By the Brouwer fixed-point theorem [e.g., see Milnor (1965b) for a concise, self-contained proof], the continuous function $\bar{h}$ must have at least one fixed point in that face. If the graph automorphism h has no fixed points, then no boundary walk will have any fixed points, in which case a radial extension of the graph automorphism $\bar{h}$ inside each face will have the property that there is at most one fixed point for the extension h inside any face. □

Corollary. *Let $G \to S$ be a graph imbedding, and let $\mathscr{A}$ be a group of automorphisms acting on the graph G. If every element of $\mathscr{A}$ respects or reverses the rotation system, then the action of $\mathscr{A}$ can be extended to an action on the imbedding surface S. Moreover, if $\mathscr{A}$ acts freely on the graph G, then the extended action can be chosen to have at most one fixed point inside any face.* □

It should be noted that the extended homeomorphism of Theorem 4.3.2 might have fixed circles, as in the reflections in Example 4.3.1; this observation will prove useful in Chapter 6. The currently important application of Theorem 4.3.2 is for imbedded voltage graphs.

Theorem 4.3.3. *Let $\langle G \to S, \alpha \rangle$ be an imbedded voltage graph with voltage group $\mathscr{A}$. Then the natural action of $\mathscr{A}$ on the derived graph G^α extends to an action of $\mathscr{A}$ on the derived surface S^α, having at most one fixed point inside any face.*

Proof. It follows immediately from the definitions of the derived rotation system and the natural action of the voltage group that every element of the

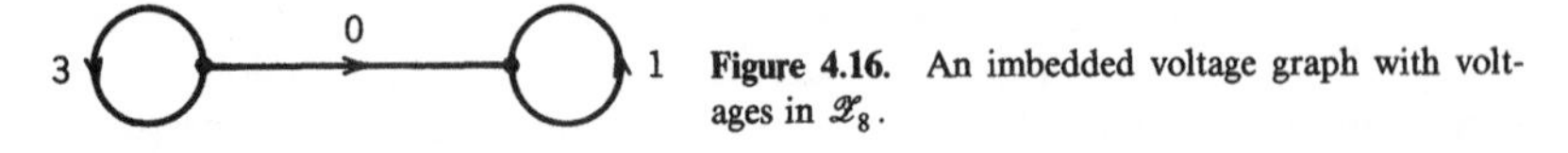

Figure 4.16. An imbedded voltage graph with voltages in $\mathscr{Z}_8$.

natural action respects the derived rotation system and that the natural action is free. By the corollary to Theorem 4.3.2, the conclusion of Theorem 4.3.3 holds. □

As with the natural projection, it would be convenient to be able to call the extended action of the voltage group $\mathscr{A}$ on the derived surface S^α "the" natural (surface) action. However, the extension is not unique topologically, even if it is unique combinatorially. Therefore we define a "natural (surface) action" of the voltage group $\mathscr{A}$ on the derived surface S^α to be any action of $\mathscr{A}$ on S^α that extends the natural action of $\mathscr{A}$ on the derived graph G^α and that has at most one fixed point inside any face.

Example 4.3.3. *Let $\langle G \to S_0, \alpha \rangle$ be the imbedded voltage graph shown in Figure* 4.16 *whose voltages are in $\mathscr{Z}_8$. The net voltages on the boundary walks are* 3, 4, *and* 1, *from which it follows that surface S^α has six faces, and that its Euler characteristic $\chi(S^\alpha) = 16 - 24 + 6 = -2$. Since S^α is evidently connected and orientable, it must be the surface of genus* 2. *Therefore, by Theorem* 4.3.3, *we have established that the cyclic group $\mathscr{Z}_8$ acts on the surface of genus* 2. (*In fact, it can be shown that the full automorphism group of the derived graph G^α acts on S^α, but we postpone discussion of this until Chapter* 6.)

4.3.3. Regular Branched Coverings and Ordinary Imbedded Voltage Graphs

We have seen that a regular covering of a graph corresponds to a group action and to ordinary voltages. Similarly, we shall now see that it is possible to define a regular branched covering of a surface that corresponds to a group action on that surface and to an imbedding of an ordinary voltage graph.

Let the group $\mathscr{A}$ act on a closed surface S. By the "orbit" of a point $x \in S$, denoted $[x]$, we mean the set $\{\phi_a(x): a \in \mathscr{A}\}$.

Let $S/\mathscr{A}$ denote the set of all orbits, and let $p: S \to S/\mathscr{A}$ denote the natural projection $p(x) = [x]$. The set $S/\mathscr{A}$ is given the strongest topology in which the natural projection p is continuous. That is, a subset $U \subset S/\mathscr{A}$ is open if and only if its inverse image $p^{-1}(U)$ is open. With such a specification of open sets, the resulting topological space is called the "quotient space" of the action of the group $\mathscr{A}$ on the surface S, and the map p is called the "quotient map".

It can be shown that if the group $\mathscr{A}$ acts pseudofreely, then the quotient space $S/\mathscr{A}$ is, in fact, a surface. Moreover, if a point $x \in S$ is not left fixed by a nonidentity element of $\mathscr{A}$, then some neighborhood of x is moved off itself by every nonidentity element of $\mathscr{A}$. This ensures that the quotient map

$p\colon S \to S/\mathscr{A}$ is locally one-to-one, except at fixed points. Since the domain is compact, the quotient map is a branched covering. A branched covering obtained this way is called "regular", and the group $\mathscr{A}$ is called the group of "covering transformations". Note that $p \circ \phi_a = p$ for any $a \in \mathscr{A}$.

Example 4.2.1 Revisited. *Once again we consider the complex analytic function $p(z) = z^n$. If $p(z_1) = p(z_2)$, then the points z_1 and z_2 have the same modulus, and their arguments differ by an integer multiple of $2\pi/n$. Thus, they are in the same orbit of the pseudofree action of the cyclic group $\mathscr{Z}_n$ on the extended complex plane generated by a rotation about the origin through an angle of $2\pi/n$. It follows that the function p can be viewed as the quotient map of this action, and accordingly, p is a regular branched covering.*

This example indicates a general procedure for showing that a given branched covering $p\colon \tilde{S} \to S$ is regular. It suffices to find a group $\mathscr{A}$ acting pseudofreely on the covering space $\tilde{S}$ and satisfying this condition for every pair of points x_1 and x_2 in $\tilde{S}$: $p(x_1) = p(x_2)$ if and only if x_1 and x_2 are in the same orbit of the action of $\mathscr{A}$.

The following extension theorem for regularity in the branched covering is a sharpening of Theorem 4.2.1 for ordinary voltages. We omit the proof, since it is routinely related to the proofs of Theorems 4.2.1 and 4.3.3.

Theorem 4.3.4. *Let $\langle G \to S, \alpha\rangle$ be an imbedded ordinary voltage graph. The natural projection $p\colon G^\alpha \to G$ extends to a regular branched covering $p\colon S^\alpha \to S$ that has at most one branch point inside each face. If the net voltage on a face has order r in the voltage group $\mathscr{A}$, then that face has a branch point of deficiency $\#\mathscr{A} - \#\mathscr{A}/r$, each of whose prebranch points has order r.* □

Corollary. *Let $\langle G \to S, \alpha\rangle$ be an imbedded ordinary voltage graph satisfying KVL on every face. Then the natural projection $p\colon G^\alpha \to G$ extends to an (unbranched) regular covering.* □

4.3.4. Which Regular Branched Coverings Come from Voltage Graphs?

As the following theorem shows, regular surface coverings arising from ordinary voltage graphs are not unusual. The proof is a routine sharpening of the proof of Theorem 4.2.2 and is omitted.

Theorem 4.3.5. *Let $p\colon \tilde{S} \to S$ be a regular branched covering of surfaces, and let $G \to S$ be any cellular imbedding in the base surface having at most one branch point in any face and no branch points in G. Then there is an ordinary voltage assignment α in the group of covering transformations of p such that p is equivalent to a natural surface projection $S^\alpha \to S$.* □

Theorem 4.3.5 is an important aid to understanding pseudofree group actions and regular branched coverings for surfaces. It implies that any regular branched covering behaves around branch points the same way that a natural projection of an imbedded voltage graph behaves on faces. Thus we have the following information about branch points.

Corollary. *Let $p: \tilde{S} \to S$ be a regular branched covering with covering transformation group $\mathscr{A}$. Then for each branch point y of the base surface, there is an element $b \in \mathscr{A}$ such that, if r is the order of b in the group $\mathscr{A}$, then the number of prebranch points of y is $\#\mathscr{A}/r$, and each has order r.*

Proof. By Theorem 4.3.5, the covering p is a natural surface projection $S^\alpha \to S$ for some imbedded voltage graph $\langle G \to S, \alpha \rangle$. The prebranch points of the branch point y correspond to the faces in the derived imbedding that cover the base face containing y. Thus, the conclusion of the corollary follows directly from an application of Theorem 4.1.1 in which the group element b equals the net voltage on the face containing the branch point y. □

This corollary provides most of the information we need for later use, but much more can be said. For instance, the subgroup of the covering transformation group $\mathscr{A}$ that leaves fixed a particular prebranch point of y is a cyclic group, and it is generated by a conjugate of the element b (see Exercise 13).

Many consequences of Theorem 4.3.5 result from the freedom allowed by this theorem in the choice of the base imbedding $G \to S$. Two such consequences are now derived. The first is of general importance, especially for the study of the genus of groups in Chapter 6. The second is narrower, but deeper, and it draws upon a variety of results developed so far.

Theorem 4.3.6. *If the group $\mathscr{A}$ acts pseudofreely on the connected surface $\tilde{S}$, then there is a Cayley graph for $\mathscr{A}$ that is imbeddable in $\tilde{S}$.*

Proof. Just let a bouquet of circles assume the role of the graph G of Theorem 4.3.5. By Theorem 2.2.3, the derived graph G^α is a Cayley graph for $\mathscr{A}$. □

Theorem 4.3.7. *No finite group $\mathscr{A}$ acting on an orientable surface $\tilde{S}$ can have exactly one fixed point.*

Proof. Since any element of the group $\mathscr{A}$ leaving a point fixed generates a cyclic subgroup leaving that point fixed, it suffices to prove the theorem for the case in which the group $\mathscr{A}$ is cyclic. Under that additional condition, let $S = \tilde{S}/\mathscr{A}$ be the quotient surface, and let $p: \tilde{S} \to S$ be the quotient map. If the action of $\mathscr{A}$ has only one fixed point, then the regular branched covering $p: \tilde{S} \to S$ has exactly one prebranch point, and its order must be $\#\mathscr{A}$.

Suppose first that the quotient surface S is not the sphere. Then let $G \to S$ be a one-face imbedding of a bouquet of circles containing the single branch point of the quotient map p in its only face. The boundary walk of that face is of the form

$$a_1 b_1 a_1^{-1} b_1^{-1} a_2 b_2 a_2^{-1} b_2^{-1} \dots \quad \text{if } S \text{ is orientable}$$
$$c_1 c_1 c_2 c_2 c_3 c_3 \dots \quad \text{if } S \text{ is nonorientable}$$

By Theorem 4.3.5, there is a voltage assignment α such that the quotient map p is equivalent to a natural projection $S^\alpha \to S$. If the quotient surface S is orientable, then the net voltage on the only boundary walk is the identity, since the group $\mathscr{A}$ is abelian. This would contradict the existence of a branch point of p. On the other hand, if the quotient surface S is nonorientable, then the net voltage is of the form $2a$ for some $a \in \mathscr{A}$. It follows that $\#\mathscr{A}$ cannot be even, since then the order of the net voltage $2a$ and the associated prebranch point would be at most $\#\mathscr{A}/2$, that is, not equal to $\#\mathscr{A}$. However if $\#\mathscr{A}$ is odd, then the derived surface would be nonorientable, by Theorem 4.1.5. Thus, in either case, there is a contradiction.

Finally, if the quotient surface S is a sphere, then let $G \to S$ be an imbedding of a single loop containing the branch point of p in one of its two faces. However, the other face has the same boundary walk and net voltage, so it must also contain a branch point. □

4.3.5. Applications to Group Actions on the Surface S_2

The Riemann–Hurwitz equation is a surprisingly powerful tool for determining which groups act pseudofreely on which surfaces. As an example of its power, we shall show how it can be used to prove that there is no pseudofree action of the cyclic group $\mathscr{Z}_n$ on the surface S_2 of genus 2, for any prime number $n > 5$. Since Examples 4.3.1 and 4.3.2 exhibit actions of $\mathscr{Z}_n$ on the surface S_2 for $n = 2, 3$, and 5, this demonstration completes a classification of the prime, cyclic, pseudofree actions on S_2.

Let $n > 3$ be prime, and suppose there is a pseudofree action of $\mathscr{Z}_n$ on the surface S_2. Then the quotient map $p: S_2 \to S_2/\mathscr{Z}_n$ is a regular n-sheeted branched covering. By the corollary to Theorem 4.3.5, it follows that each branch point has deficiency of the form $n - n/r$, where r is the order of a voltage. However, since n is prime, we infer that the only possible deficiency is $n - 1$. Therefore, if $\chi(S_2/\mathscr{Z}_n) = k$, and if the number of branch points is t, it follows from the Riemann–Hurwitz equation that

$$-2 = \chi(S_2) = nk - t(n-1)$$

Since $n > 2$, it is impossible that $t = 0$. Moreover, since $n - 1 > 2$, it follows that k must be positive. Since the group order n is odd and $n - 1$ is even, it also follows that k must be even. Since k is the Euler characteristic of a surface, the only possibility is that $k = 2$. Thus

$$-2 = 2n - t(n-1)$$

Solving this equation for n, we obtain $n = 1 + 4/(t-2)$, which is at most 5.

Information about the number of fixed points in a group action can also be obtained directly from the Riemann–Hurwitz equation. In the preceding computations, if $n = 5$, then the number t of branch points equals 3, and the quotient surface is the sphere. Since each branch point has deficiency 4, there are also only three prebranch points in all. Prebranch points of a regular covering are just the fixed points of the group of covering transformations. Therefore, any pseudofree action of $\mathscr{Z}_5$ on S_2 has exactly three fixed points, as is the case in Example 4.3.2. A similar inspection of the preceding computations shows that any pseudofree action of $\mathscr{Z}_3$ on S_2 having the sphere as quotient space has four fixed points, as is the case in Example 4.3.1 (the possibility of the torus as a quotient is considered in Exercise 7). Deeper applications of the Riemann–Hurwitz equation are deferred until Chapter 6.

4.3.6. Exercises

1. Show that every finite group acts freely on some surface.
2. Draw a band decomposition for the surface S_3 that reveals an action of $\mathscr{Z}_7$.
3. Prove that the cyclic group $\mathscr{Z}_n$ acts freely on the orientable surface S if and only if n divides $\chi(S)$.
4. Generalize Exercise 3 to arbitrary finite abelian groups.
5. Use the Riemann–Hurwitz equation to prove that any pseudofree action on the projective plane has exactly one fixed point.
6. Prove that no group of even order acts pseudofreely on N_k for k odd. (*Hint:* Use Exercise 4.2.12 and the fact that any group of even order has an element of order 2.)
7. Show that every pseudofree action of $\mathscr{Z}_3$ on the surface S_2 has exactly four fixed points.
8. Let $p: \tilde{S} \to S$ be a regular branched covering such that the surface S is orientable, and suppose that the covering transformation group is abelian. Prove that p cannot have only one branch point. Show by examples that p can have exactly one branch point if $\mathscr{A}$ is nonabelian or if S is nonorientable. Does this contradict Theorem 4.3.7?
9. Prove that $\mathscr{Z}_{12}$ does not act pseudofreely on the surface S_2. (This is like showing that $\mathscr{Z}_{16}$ does not act pseudofreely on S_2, but there are more cases to consider.)
10. Complete the classification of cyclic pseudofree actions on S_2 by showing that $\mathscr{Z}_n$ acts pseudofreely on S_2 if and only if $n = 2, 3, 4, 5, 6, 8, 10$.
11. Prove for prime numbers $n > 2$ that $\mathscr{Z}_n$ acts pseudofreely on the surface S_g if and only if n is of the form $(2g - 2 + m)/(m - k)$, where k is even, $k \leq 2$, and $m > 1$.
12. Use Exercise 11 to find for which primes n there is a pseudofree action of $\mathscr{Z}_n$ on S_5.

13. Prove in regard to the corollary to Theorem 4.3.5 that the subgroup of the covering transformation group $\mathscr{A}$ that leaves fixed a particular prebranch point of x must be cyclic and generated by a conjugate of the element b.

4.4. CURRENT GRAPHS

The historical origins of topological graph theory lie largely in map-coloring problems concerned with the relationships between regions on a surface. From our present perspective, it should not be surprising that duality plays a central role. In fact, the main method used in the original solution of the Heawood map-coloring problem is a dual form of the voltage graph called a "current graph". The precursors of this technique can be traced back through Ringel's work to Heffter (1891).

4.4.1. Ringel's Generating Rows for Heffter's Schemes

In trying to establish that the Heawood number for the surface S_g is the chromatic number, and not just an upper bound, Heffter (1891) sought the minimal-genus surface consisting of n regions (faces), each of which is a neighbor of all the others. Heffter did consider the problem simultaneously in primal and dual forms, and he also wished to compute the genus of the complete graph K_n. To describe the surface, he started with n $(n-1)$-sided polygons, and he labeled them $1, 2, \ldots, n$. The sides of polygon i were then labeled by

$$1, 2, \ldots, i-1, i+1, \ldots, n$$

to indicate where the neighboring polygons should be attached. Figure 4.17, which is taken from Heffter's paper, illustrates this procedure for a surface with five regions. The identifications along the sides of polygon 1 have already been made. When the remaining sides are pasted together, the segments AB and DC are identified as are the segments AD and BC, thereby yielding a torus.

For the most part, Heffter did not depend on pictures and, instead, gave a "table" of n rows, in which row i lists in cyclic order the labels for the sides of polygon i. The table he gave for the preceding example of five regions was

$$\begin{array}{ll} (1) & 3\,2\,4\,5 \\ (2) & 4\,3\,5\,1 \\ (3) & 5\,4\,1\,2 \\ (4) & 1\,5\,2\,3 \\ (5) & 2\,1\,3\,4 \end{array}$$

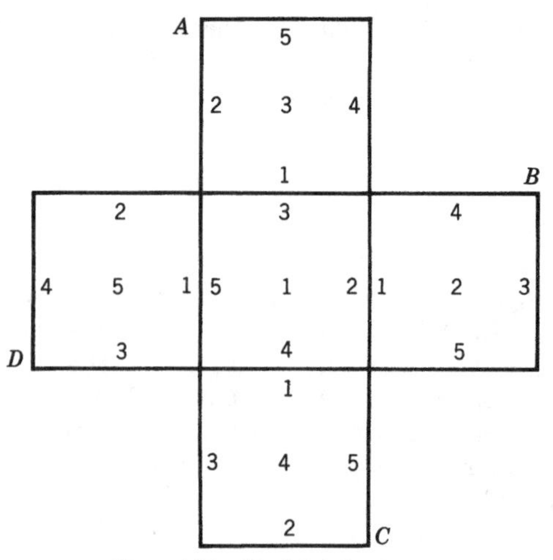

Figure 4.17. Five labeled polygons.

We now recognize such a table as a rotation system for an imbedding of the complete graph K_5, to which the mapping of Figure 4.17 is a dual. Heffter showed how the cycle of polygons meeting at a vertex can be recovered from the table; that is, he gave a face tracing algorithm for the imbedding of K_n. Having thus realized the "purely arithmetical nature of the problem", he constructed tables yielding minimal imbeddings of K_n for all $n \leq 12$. Although Ringel was well aware of Heffter's methods, their present popularity evidently rose because of the abstract of Edmonds (1960).

In addition to the use of duality and the introduction of rotation systems, it is especially noteworthy that Heffter gave tables for K_n, $n = 5, 6, 7$, in which row $i + 1$ was obtained by adding 1 modulo n to the numbers of row i. Heffter proved that such conveniently generated tables exist for general $n > 6$ only if n is of the form $12s + 7$, and he constructed them in the cases in which $4s + 3$ is prime and 2^k is not congruent to ± 1 modulo $4s + 3$ for $0 < k < 12s + 1$. (It is still not known whether there are infinitely many such s.) These tables generated from a single row using the cyclic group $\mathscr{Z}_n$ have ultimately evolved into voltage graphs. For example, the table given above for $n = 5$ is precisely the derived rotation system for the voltage graph of Example 4.1.3.

When Ringel began his work on the genus of complete graphs, he also chose to describe surfaces with tables like Heffter's, calling them "schemes". In an important generalization of Heffter's work, he designed the strategy of letting most of the rows of the scheme be generated from a small number of rows. Unlike Heffter, however, he had spectacular success with this technique. For instance, Ringel (1961) obtained the minimal imbedding of K_n for all n of the form $12s + 7$, by constructing a single generating row for the scheme that Heffter had been unable to obtain without restricting s. Despite Ringel's

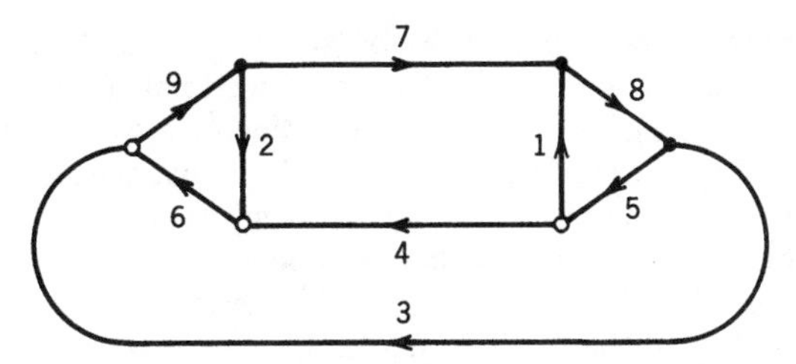

Figure 4.18. A current graph with currents in $\mathscr{Z}_{19}$.

particular brilliance in constructing such generating rows, a general method to find them was still lacking. Such a method was ultimately provided in rudimentary form by Gustin (1963).

4.4.2. Gustin's Combinatorial Current Graphs

Gustin began with a planar drawing of a graph G. At each vertex he assigned a direction, clockwise or counterclockwise. This led to what Gustin thought of as flows within the graph, which we now recognize as face boundaries. Each edge was assigned a direction and a "current", by which he meant an element of some group $\mathscr{A}$.

Example 4.4.1. *The current graph in Figure* 4.18 *leads to a single generating row for a triangular imbedding of the complete graph* K_{19} *in the surface* S_{20}. *To get that row (in our present notation), write the vertex* v_0 *at the left, which is to show which vertex has the rotation specified by the row. Then proceed as follows.*

Start at any directed edge, for instance, the one labeled with current 9 *modulo* 19. *Write that current as the subscript on the first vertex in the row. (Since the resulting graph will have no multiple edges, the entries in the rows of the scheme can be vertices, rather than directed edges.)*

At the end of that first directed edge is a filled vertex, indicating that the second edge is to be obtained by moving in a clockwise direction. That is, the directed edge with current 7 *is next, so write* v_7 *next in the generating row. Continue in this manner, going clockwise at every solid vertex* (•) *and counterclockwise at every hollow vertex* (○), *until every edge has been traversed in both directions. When proceeding in the minus direction on an edge, one writes the additive inverse modulo* 19 *of the current shown on the plus direction. The resulting generating row is*

$$v_0. \quad v_9\, v_7\, v_8\, v_3\, v_{13}\, v_{15}\, v_{14}\, v_{11}\, v_{18}\, v_4\, v_{17}\, v_{10}\, v_{16}\, v_5\, v_1\, v_{12}\, v_2\, v_6$$

The next row is

$$v_1. \quad v_{10}\, v_8\, v_9\, v_4\, v_{14}\, v_{16}\, v_{15}\, v_{12}\, v_0\, v_5\, v_{18}\, v_{11}\, v_{17}\, v_6\, v_2\, v_{13}\, v_3\, v_7$$

In general, the subscript on the jth *entry in row* v_i *is obtained by adding* i *modulo* 19 *to the subscript on the* jth *entry of row* v_0.

Gustin imposed a number of rules on the graph, on the assignment of rotations, and on the assignment of currents, whose total effect is to guarantee that the resulting scheme described an imbedding of a Cayley graph for the current group. The Kirchhoff current law (KCL) is that the sum (or product, for nonabelian currents) of the currents leaving each vertex is the group identity. Gustin's requirement that the graph be regular and that every vertex satisfy KCL has the effect of ensuring that every face in the resulting imbedding has the same number of sides as the valence of every vertex in the current graph.

Gustin's current graphs were regarded as computational tools ("nomograms" in Youngs's terminology) to aid in the construction of generating rows for schemes. Progressively more constructive power was obtained over the next few years as Gustin's original rules were relaxed. Very importantly, Youngs (1967) explained how to use current graphs with excess currents at some vertices. Whereas Gustin (1963) discussed only the cases of one, two, or three generating rows, Jacques (1969) gave a general exposition, still restricted to Cayley graphs, but covering the possibility of arbitrarily many generating rows. Finally, Gross and Alpert (1973, 1974) showed how a topological viewpoint makes it possible to eliminate all of Gustin's restrictions, and they unified all previously defined kinds of current graphs into the one now described.

4.4.3. Orientable Topological Current Graphs

Let $G \to S$ be an imbedding in an oriented surface. An "(ordinary) current assignment" is a function β from the set of directed edges of G into a group $\mathscr{B}$ such that $\beta(e^-) = \beta(e^+)^{-1}$ for every edge e. The values of β are called "currents" and $\mathscr{B}$ is called the "current group". The pair $\langle G \to S, \beta \rangle$ is called a current graph. If the clockwise and counterclockwise directions used by Gustin to order the entries in a generating row are reinterpreted as vertex rotations, the result is a legitimate imbedding. Hence, Gustin's combinatorial current graphs are properly considered a special case of a topological covering space construction.

The point of assigning currents to an imbedding is to obtain a "derived graph G_β" and "derived imbedding $G_\beta \to S_\beta$". The vertex set of G_β is the cartesian product $F_G \times \mathscr{B}$, and the edge set is the product $E_G \times \mathscr{B}$. Thus, vertices of the derived graph correspond to faces of the current graph, as in Gustin's construction. Endpoints and plus directions of derived edges are given as follows. The plus direction of the derived edge (e, b) has initial vertex (f, b) and terminal vertex (g, bc), where the rotation at the initial vertex of the directed edge e^+ carries face f to face g and $\beta(e^+) = c$. Thus, the directed edge $(e, b)^-$ goes from the vertex (g, bc) to the vertex (f, b). It will be notationally convenient to refer sometimes to the directed edge $(e, bc^{-1})^-$, which goes from vertex (g, b) to vertex (f, bc^{-1}), as (e^-, b); also, we let (e^+, b) stand for $(e, b)^+$. The rotation system for the derived imbedding is obtained by lifting boundary walks of the imbedding $G \to S$, as follows. If the directed boundary walk of face f is

$$e_1^{\epsilon_1} e_2^{\epsilon_2} \cdots e_n^{\epsilon_n}$$

then the rotation at vertex (f, b) is

$$(e_1, b_1)^{\epsilon_1} (e_2, b_2)^{\epsilon_2} \cdots (e_n, b_n)^{\epsilon_n}$$

where $b_i = b$ if $\epsilon_i = +$, and $b_i = b\beta(e_i^-)$ if $\epsilon_i = -$. This rotation is more easily denoted

$$(e_1^{\epsilon_1}, b)(e_2^{\epsilon_2}, b) \cdots (e_n^{\epsilon_n}, b).$$

Observe that the rotation at vertex (f, ab) is obtained from the rotation at vertex (f, b) by multiplying all $\mathscr{B}$ coordinates on the left by a.

Example 4.4.1 Revisited. *For $j = 1, \ldots, 9$, let e_j be the edge whose plus direction carries the current j. Gustin's method of obtaining a generating row is clearly no more than an application of the Face Tracing Algorithm. Thus the base imbedding $G \to S$ has one face whose directed boundary walk is*

$$e_9\, e_7\, e_8\, e_3\, e_6^-\, e_4^- \ldots$$

Denote this face by v. For $j = 1, \ldots, 9$ and $i = 1, \ldots, 19$, the derived edge $(e_j, i)^+$ runs from vertex (v, i) to vertex $(v, i + j)$. Thus, the derived graph is the complete graph K_{19}. The rotation at vertex $(v, 0)$ is

$$(v,0).\quad (e_9,0)(e_7,0)(e_8,0)(e_3,0)(e_6,13)^-\ (e_4,15)^- \ \ldots$$

Since the derived graph is simplicial, this rotation can also be given in vertex form. If we let $v_i = (v, i)$, the result is

$$v_0.\quad v_9\, v_7\, v_8\, v_3\, v_{13}\, v_{15} \cdots$$

which is the generating row given by Gustin's construction. Moreover, the full rotation system is also the same as Gustin's scheme, since the rotation at vertex v_i is obtained from that at vertex v_0 by adding i modulo 19 to all $\mathscr{B}$ coordinates.

Example 4.4.2 (Jungerman, 1975). *Consider the current graph $\langle G \to S, \beta \rangle$ given by Figure 4.19, where the current group is $\mathscr{Z}_{16}$. An application of the Face Tracing Algorithm shows that the base imbedding has two faces, which we denote u and v, and that vertices u_0 and v_0 of the derived imbedding have the following rotations:*

$$u_0.\quad v_3\, u_2\, v_{15}\, u_6\, v_7\, v_5\, v_{11}\, v_{13}\, u_{14}\, v_1\, u_{10}\, v_9$$
$$v_0.\quad u_3\, v_{14}\, u_9\, u_{15}\, u_{13}\, v_6\, u_{11}\, v_2\, u_5\, v_{10}\, u_7\, u_1$$

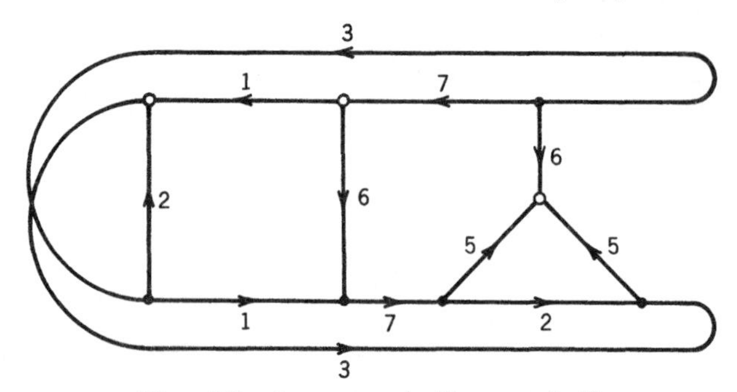

Figure 4.19. A current graph with currents in $\mathscr{Z}_{16}$.

An inspection of these two rows reveals that the vertex u_i is adjacent to the vertex v_j if and only if $j - i$ is odd, and that u_i is adjacent to u_j (similarly v_i and v_j) if and only if $j - i \equiv 2 \bmod 4$. It follows that $\{u_i: i \text{ even}\} \cup \{v_j: j \text{ odd}\}$ is the vertex set for one component of the derived graph. In this component, every pair of vertices is adjacent except u_i and u_j, where $j - i \equiv 0 \bmod 4$ and v_i and v_j, where $j - i \equiv 0 \bmod 4$. Thus, this component is isomorphic to the complete multipartite graph $K_{4,4,4,4}$. The other 16 vertices of the derived graph lie in a second copy of $K_{4,4,4,4}$.

4.4.4. Faces of the Derived Graph

Just as the vertices of a derived graph correspond to the faces of its current graph, faces of a derived graph correspond to vertices of its current graph. The proof of the following theorem is postponed to the next section, where it is shown to be a simple consequence of the duality between current and voltage graphs. Originally, the theorem was proved by an application of the Face Tracing Algorithm to the derived imbedding.

Theorem 4.4.1 (Gross and Alpert, 1974). *Let $e_1^{\epsilon_1} \ldots e_n^{\epsilon_n}$ be the rotation at vertex v of the current graph $\langle G \to S, \beta \rangle$, and let c_i be the current carried by the direction of the edge e_i that has v as its initial vertex. Let $c = c_1 \ldots c_n$, and let r be the order of c in the current group $\mathscr{B}$. Then the derived imbedding has $\#\mathscr{B}/r$ faces corresponding to vertex v, each of size rn, and each of the form*

$$\begin{aligned}
&(e_1^{\epsilon_1}, b), (e_2^{\epsilon_2}, bc_1), (e_3^{\epsilon_3}, bc_1c_2) \ldots (e_n^{\epsilon_n}, bc_1c_2 \ldots c_{n-1})\\
&(e_1^{\epsilon_1}, bc), (e_2^{\epsilon_2}, bcc_1), (e_3^{\epsilon_3}, bcc_1c_2) \ldots\\
&\vdots\\
&(e_1^{\epsilon_1}, bc^r) = (e_1^{\epsilon_1}, b) \quad \square
\end{aligned}$$

The product $c_1 \ldots c_n$ in Theorem 4.4.1 is called the excess current at vertex v. The Kirchhoff current law (KCL) holds at vertex v if the excess current is the identity.

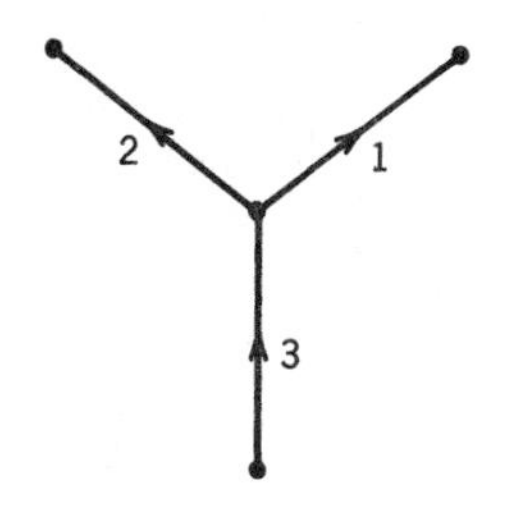

Figure 4.20. A current graph with currents in $\mathscr{Z}_6$.

Corollary. *Let $\langle G \to S, \beta \rangle$ be a current graph such that KCL holds at every vertex and every vertex has valence three. Then the derived imbedding is triangular.* □

Since KCL holds at every vertex in both Example 4.4.1 and Example 4.4.2, both derived imbeddings are triangular. Thus the first gives a minimal-genus imbedding of K_{19}, whereas the second gives a minimal-genus imbedding of $K_{4,4,4,4}$.

Example 4.4.3. *The current graph given in Figure 4.20 with currents in $\mathscr{Z}_6$ has one face, denoted v. By face tracing, we see that the rotation at vertex v_0 of the derived imbedding is*

$$v_0. \quad v_2\, v_4\, v_1\, v_5\, v_3\, v_3$$

The order of the excess current at the bottom vertex of the current graph is 2. If we excise the resulting three digons of the derived imbedding and reclose the surface by identifying opposite sides, we thereby identify three pairs of multiple edges from v_i to v_{i+3}, for $i = 1, 2, 3$. The final product is an imbedding of the complete graph K_6. The 3-valent vertex of the current graph satisfies KCL and it generates six triangles, and the two remaining 1-valent vertices together yield one hexagon and two triangles. Thus, the Euler characteristic of the derived surface is $6 - 15 + 9 = 0$. The scheme whose generating row is the given rotation for v_0 with the extra v_3 deleted is precisely the table Heffter gives for his minimal-genus imbedding of K_6.

Example 4.4.4. *Consider the current graph given in Figure 4.21 with currents in the cyclic group $\mathscr{Z}_{100}$. There are two faces, which we denote u and v. In the derived graph, there is an edge between u_i and v_{i+1} for all i, and there are no other edges. Therefore the derived graph has two components; the vertex set of one component is*

$$\{u_i\colon i \text{ even}\} \cup \{v_j\colon j \text{ odd}\}$$

Each vertex of the current graph has excess current ± 2 and valence 2, and thereby generates a pair of 100-gons; each of these 100-gons passes through every vertex in its component. Add a vertex inside each of the six 100-gons of one

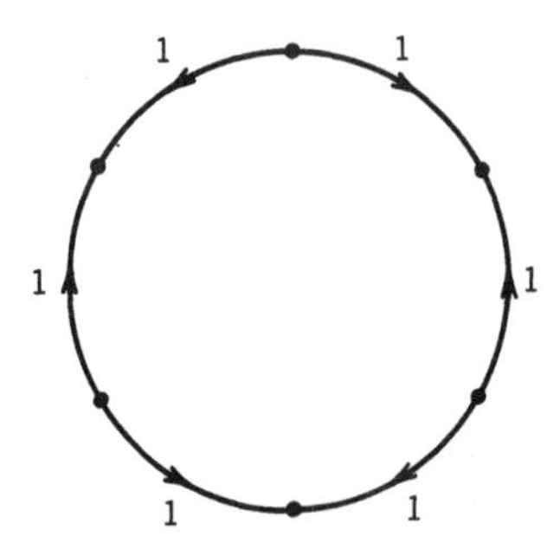

Figure 4.21. A current graph with currents in $\mathscr{Z}_{100}$.

component, and add edges joining that vertex to the 100 *original vertices. Then delete all the original edges. The resulting graph is the complete bipartite graph* $K_{6,100}$, *and every face of the resulting imbedding is a quadrilateral. Hence the imbedding has minimal genus.*

Clearly, Example 4.4.4 generalizes to arbitrary even integers m and n, yielding quadrilateral and hence minimal-genus imbeddings of $K_{m,n}$ for these cases. Exercise 11 shows how quadrilateral imbeddings of $K_{m,n}$ can be obtained when m is odd and $n \equiv 2 \bmod 4$.

4.4.5. Nonorientable Current Graphs

It is possible to consider current graphs in nonorientable surfaces as well. The difficulty is that vertex rotations are reversed by type-1 edges during the Face Tracing Algorithm, so care must be taken in using vertex rotations to define the derived graph. For instance, suppose that e is a type-1 edge lying between faces f and g, as illustrated in Figure 4.22. Unlike the orientable case, the rotation at either endpoint of edge e sends face f to face g. If e^+ carries current c and e^- were to carry current c^{-1}, then in the derived graph, there would be edges corresponding to e running from vertex (f, b) to both of the vertices (g, bc) and (g, bc^{-1}). Therefore, both directions of a type-1 edge must be assigned the same current. Also, in tracing faces of the derived graph, the edge e is traversed in the same direction twice; it follows that in computing the excess currents at the endpoints of edge e, the current c must be used twice. Thus type-1 edges in nonorientable current graphs ("cascades" in Youngs's terminology) appear in print with arrows in both directions.

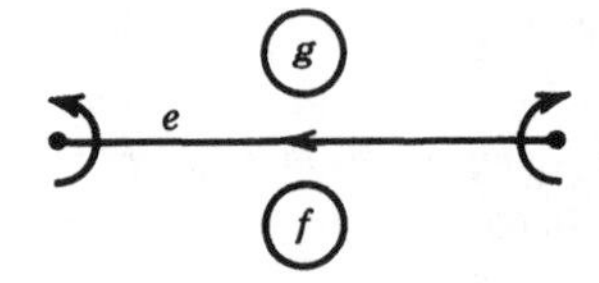

Figure 4.22. A type-1 edge.

We conclude that a current graph $\langle G \to S, \beta \rangle$ in a nonorientable surface must satisfy the rule that

$$\beta(e^-) = \begin{cases} \beta(e^+)^{-1} & \text{if } e \text{ is type } 0 \\ \beta(e^+) & \text{if } e \text{ is type } 1 \end{cases}$$

Note the implication that the vertex rotations and edge types of the imbedding $G \to S$ must be specified before currents are assigned (even in the orientable case, an orientation of the imbedding surface must be chosen to specify vertex rotations). The derived graph G_β is then defined as before.

To define the derived imbedding $G_\beta \to S_\beta$, a direction for each boundary walk must also be specified, since the Face Tracing Algorithm does not determine a unique direction in the nonorientable case. It is also necessary to specify at which corners a vertex rotation is reversed while tracing a face; such a corner is called "reversed". Now suppose that the directed boundary of face f is $e_1^{\epsilon_1} \ldots e_n^{\epsilon_n}$. The rotation at vertex (f, b) in the derived imbedding is then

$$(e_1, b_1)^{\epsilon_1} \ldots (e_n, b_n)^{\epsilon_n}$$

where $b_i = b$ if $\epsilon_i = +$ and the corner $e_{i-1}e_i$ is not reversed or if $\epsilon_i = -$ and the corner e_ie_{i+1} is reversed, and where $b_i = b\beta(e^+)^{-1}$ otherwise. Edge types of the derived imbedding are defined so that if both directions of edge e appear in the directed boundary walks of the imbedding $G \to S$, then edge (e, b) is type 0 for all $b \in \mathscr{B}$, and otherwise edge (e, b) is type 1 for all $b \in \mathscr{B}$.

Example 4.4.5 (Youngs, 1968a). *Let $\langle G \to S, \beta \rangle$ be the current graph illustrated in Figure 4.23 with currents in $\mathscr{Z}_{15}$. As before, a filled vertex indicates a clockwise rotation and unfilled, counterclockwise. Youngs's convention of drawing type-1 edges broken in the middle with arrows in both directions, suggestive of the usual "×", is used. The imbedding has one face, which we shall call v. If we let e_i denote the edge carrying current i, then the directed boundary of face v is*

$$e_1\, e_6\, e_4^-\, e_3\, e_7\, e_3\, e_5\, e_6\, e_2^-\, e_2\, e_4^-\, e_7^-\, e_5\, e_1^-$$

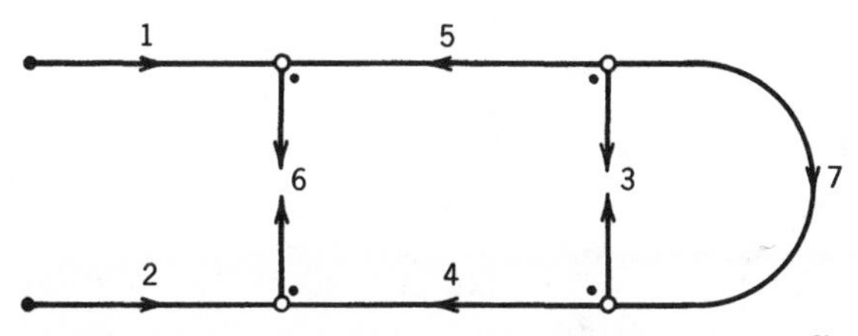

Figure 4.23. A nonorientable current graph with currents in $\mathscr{Z}_{15}$.

Again following Youngs, reversed corners are indicated by dots. The rotation in the derived imbedding at vertex $(v, 0)$, *in edge form, is*

$$(v,0).\quad (e_1,0)\,(e_6,0)^1\,(e_4,0)^{1-}\;(e_3,12)^1\,(e_7,0)\,(e_3,0)\,(e_5,10)^1\,\ldots$$

*where type-*1 *edges are given the superscript* 1, *as usual. The vertex form of the rotation with* $(v_i, 0) = v_i$ *is*

$$v_0.\quad v_1\,v_6{}^1\,v_4{}^1\,v_{12}{}^1\,v_7\,v_3\,v_{10}{}^1\,v_9{}^1\,v_{13}\,v_2\,v_{11}{}^1\,v_8\,v_5{}^1\,v_{14}$$

The rotation at vertex v_i *is obtained by adding* i *modulo* 15 *to the subscripts in the rotation at* v_0. *The derived graph* G_β *is clearly the complete graph* K_{15}.

Faces of the derived graph are determined exactly as in the orientable case. In Example 4.4.5 each of the 3-valent vertices satisfies *KCL* and, accordingly, each generates 15 triangles in the derived graph. Each of the two 1-valent vertices has an excess current of order 15 and induces in the derived imbedding a 15-gon whose boundary passes through each vertex. If an extra vertex is placed in the center of each 15-gon and joined by edges to the original vertices, the result is a triangular imbedding of the graph $K_{17} - K_2$. An extra twist in an edge common to the two original 15-gons allows the extra two vertices to be joined by an edge, thereby yielding a minimal nonorientable imbedding of K_{17}.

Example 4.4.6. *The current graph given by Figure* 4.24 *with currents in* $\mathscr{Z}_{13}$ *has a single face* v. *Since there is an edge in the derived graph between vertices* v_i *and* v_{i+j}, *for every* i *and for* $j = 1, 2, 3, 4, 5, 6$, *the derived graph is isomorphic to the complete graph* K_{13}. *Since KCL holds at every vertex, the derived imbedding is triangular.*

The derived imbedding of a nonorientable current graph can be orientable (see Exercise 12). A criterion for deciding when the derived imbedding is orientable is best understood in terms of dual voltage graphs and is considered

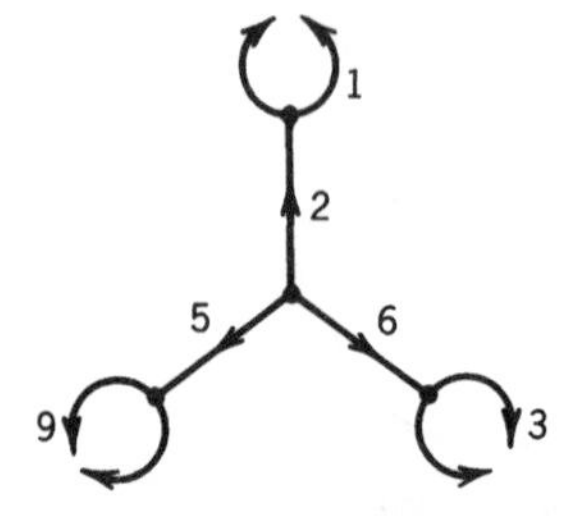

Figure 4.24. A nonorientable current graph with currents in $\mathscr{Z}_{13}$.

in Section 4.5. For present purposes, it suffices to know that a current graph in a nonorientable surface has a nonorientable derived graph if the current group has odd order.

4.4.6. Exercises

1. Give a current graph whose derived imbedding is a quadrilateral imbedding of the complete graph K_5.
2. Draw a current graph that yields a triangular imbedding of the complete graph K_7. Give the resulting rotation system (scheme) in vertex form.
3. Find the first two rows in vertex form, of the rotation system for the derived imbedding of Example 4.4.6.
4. Describe the derived imbedding for the current graph given in Figure 4.25*a*; the current group is $\mathscr{Z}_{21}$.
5. Show how a triangular imbedding of the complete 4-partite graph $K_{5,5,5,5}$ can be obtained from the current graph in Figure 4.25*b*. The current group is $\mathscr{Z}_{10} \times \mathscr{Z}_2$ (Jungerman, 1975).
6. Obtain a triangular imbedding of $K_{10} - K_3$ by changing the current group in Figure 4.20.
7. Change the current group in Exercise 2 to get a triangular imbedding of the octahedral graph O_4.
8. Assign currents in $\mathscr{Z}_{13}$ to the nonorientable imbedding given in Figure 4.26*a* so that the derived imbedding can be used to obtain a nonorientable, triangular imbedding of $K_{16} - K_3$.
9. Complete the current assignment in $\mathscr{Z}_{25}$ for the nonorientable imbedding given in Figure 4.26*b* in order to yield a nonorientable triangular imbedding for K_{25}.
10. Find the derived graph for the current graph illustrated in Figure 4.27 with current group $\mathscr{Z}_{15}$. (*Hint:* Show that the current graph has three faces u, v, and w such that one component of the derived graph has vertex set $\{(u, i), (v, j), (w, k) \mid i \equiv 0 \bmod 3,\ j \equiv 1 \bmod 3,\ k \equiv 2 \bmod 3.\}$

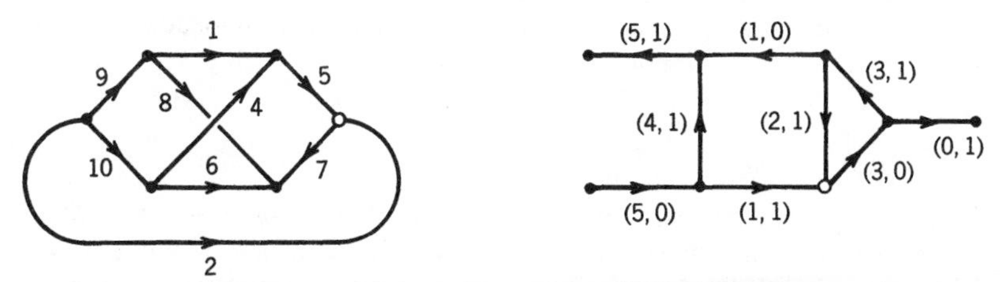

Figure 4.25. Two current graphs; the current groups are, on the left, $\mathscr{Z}_{21}$ and, on the right, $\mathscr{Z}_{10} \times \mathscr{Z}_2$.

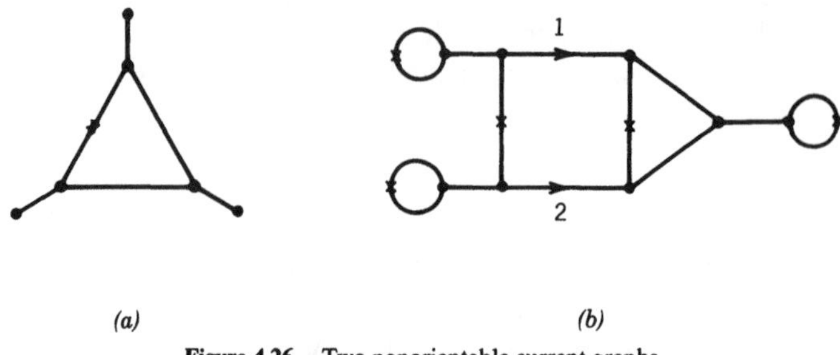

(a) (b)

Figure 4.26. Two nonorientable current graphs.

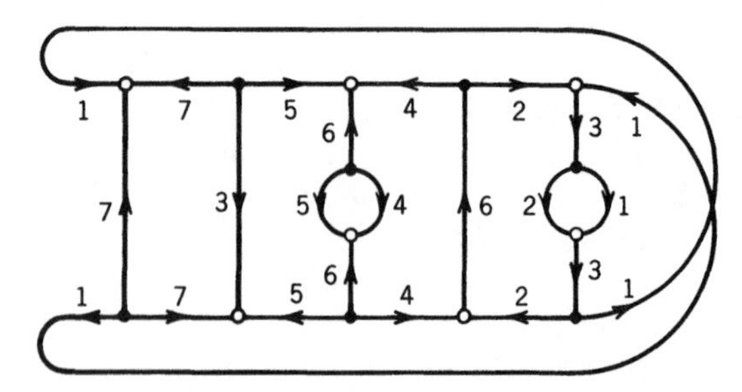

Figure 4.27. A current graph with currents in $\mathscr{Z}_{15}$.

11. Suppose that m is odd and $n \equiv 2 \bmod 4$, where $m > 2$ and $n > 2$. Assign currents in $\mathscr{Z}_n$ to an m-cycle so that the derived graph can be used to construct a quadrilateral imbedding of $K_{m,n}$. (One current will be 3, and all others will be 1.)

12. Give a nonorientable current graph whose derived graph is orientable.

4.5. VOLTAGE–CURRENT DUALITY

The voltage and current graph constructions have obvious similarities, with faces of a current graph playing a role of vertices of an imbedded voltage graph and vice versa. We now give details of the formal duality.

4.5.1. Dual Directions

Since both voltage and current assignments work with directed edges, it is necessary to have a convention to specify the plus direction of a dual edge, given the plus direction of the corresponding primal edge. Let $G \to S$ be a locally oriented graph imbedding with plus directions specified for all edges. The "dual plus direction" of the edge e^*, denoted e^{+*}, is the one crossing the

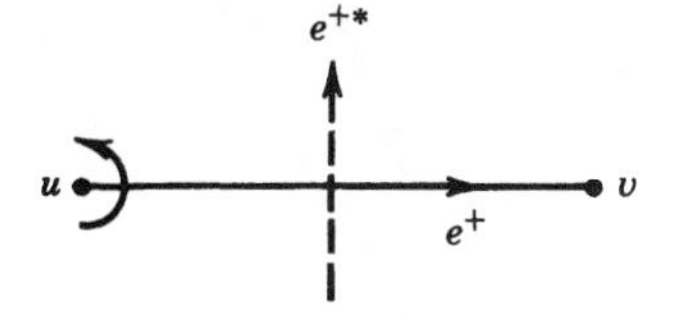

Figure 4.28. The dual plus direction.

primal edge e in the same direction as the orientation at the initial vertex of e^+ (see Fig. 4.28).

It is important to note that the dual plus direction of e^* depends as much on the choice of orientation at the initial vertex of e^+ as on the choice of plus direction for the edge e. Indeed, changing the orientation at the initial vertex of e^+ always changes the dual plus direction, but changing the plus direction of edge e does not change the dual plus direction if the edge is type 1. One unfortunate consequence of this dependence on local orientations is that the dual of the dual plus direction is not necessarily the primal plus direction. This is partly because there is no natural way to orient dual vertices; the Face Tracing Algorithm can give different directions to a boundary walk around a primal face depending on the choice of initial vertex for the walk. In some cases, however, there is no way at all of orienting dual vertices that avoids this problem.

Example 4.5.1. *Figure 4.29 shows part of a locally oriented imbedding with dual edges indicated by broken lines. The edge e is type 1 and the edge d is type 0. If a clockwise orientation is chosen for the dual vertex f^*, then $d^{+**} = d^-$; if a counterclockwise orientation for the face f is chosen instead, then $e^{+**} = e^-$.*

On the other hand, suppose that S is an orientable surface and the imbedding $G \to S$ has been locally oriented by an orientation of the surface S. Then the Face Tracing Algorithm gives every boundary walk a direction consistent with the opposite orientation of the surface. Thus the natural local orientation of the dual graph is given by the opposite orientation of the surface. In this case, the dual of the dual plus direction is the primal plus direction (see Fig. 4.30). Hence, for an orientable surface, dual directions behave nicely as long as the primal graph is locally oriented by one orientation of the surface and the dual graph by the opposite.

It is still necessary to specify a orientation of the surface for the primal imbedding, but this is done automatically by a pure rotation system for the

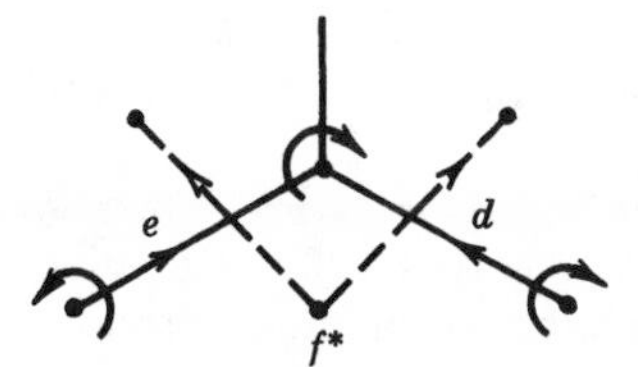

Figure 4.29. A locally oriented imbedding and its dual.

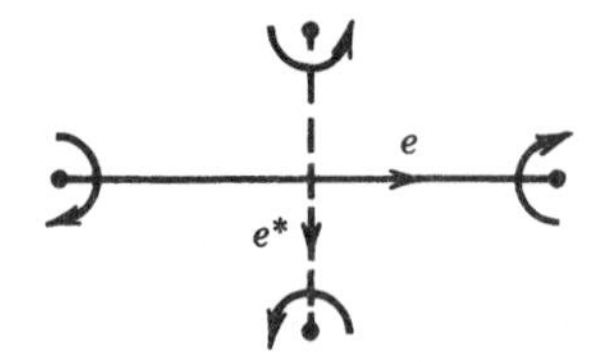

Figure 4.30. A directed edge and its dual.

imbedded graph. If the imbedding is given by a rotation projection, the convention that all rotations are clockwise specifies an orientation. Only if the imbedding is given by drawing the graph upon a polygon with edges to be identified must care be taken to indicate the chosen orientation (by a circular arrow, for example).

Warning. *Drawing a dual imbedding is confusing if the primal imbedding is given by a rotation projection, but relatively easy if the primal imbedding is drawn on a polygon with edges to be identified. It frequently pays to convert a presentation of the former type to one of the latter type before attempting to find the dual. An alternative is first to list the directed boundary walks of the primal imbedding and then to use the resulting rotation system for the dual to draw a picture of the dual. In fact, dual imbeddings in orientable surfaces can be computed in terms of pure rotation systems simply by composing two permutations (see Exercise* 4); *this method is obviously the best suited for computers.*

4.5.2. The Voltage Graph Dual to a Current Graph

Let $\langle G \to S, \beta\rangle$ be a locally oriented current graph, and let $G^* \to S$ be the dual imbedding with dual plus directions. The "voltage assignment dual to β", denoted β^*, is the one satisfying the rule $\beta^*(e^{+*}) = \beta(e^+)$. The resulting imbedded voltage graph $\langle G^* \to S, \beta^*\rangle$ is called the "imbedded voltage graph dual to current graph $\langle G \to S, \beta\rangle$".

Example 4.4.1 Revisited. *Paying heed to our previous warning, we first draw the given current graph on a polygon with edges to be identified. It is readily verified that the rotation at each vertex in Figure* 4.31*a is the same as that at the corresponding vertex of Figure* 4.18 (*the specified orientation of the surface* S_2 *is clockwise*). *Therefore, the imbedding given in Figure* 4.31*a is equivalent to the original imbedding. Its dual imbedded voltage graph is given in Figure* 4.31*b. The corners of the octagon represent a single vertex of the dual graph, and edges of the polygon are also edges of the dual graph. It is instructive to check that the rotation at the dual vertex, when read counterclockwise, agrees with the first row of the rotation system for the derived graph given in Example* 4.4.1, *and that all dual plus directions are correctly given.*

The dual relationship between the current and voltage graph constructions can now be given explicitly. In its original form, before the development of

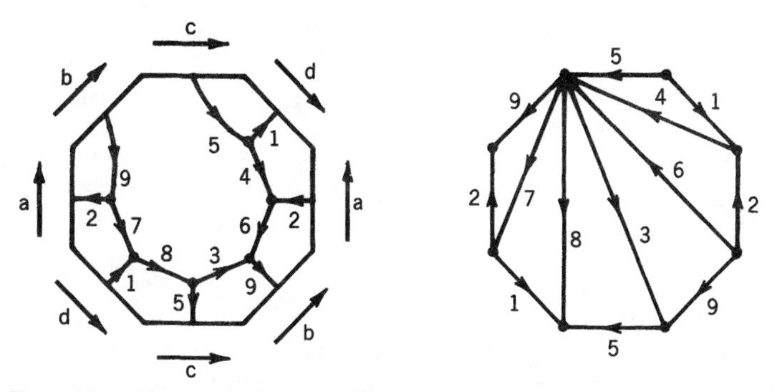

Figure 4.31. The current graph of Example 4.4.1 and its dual imbedded voltage graph.

voltage graphs, the following theorem stated that the derived imbedding for any current graph $\langle H \to S, \beta \rangle$ is a branched covering of the dual of the imbedding $H \to S$. In its present form, the proof is a routine verification that the definitions are satisfied, and it is omitted.

Theorem 4.5.1 (Gross and Alpert, 1974). *Let $\langle H \to S, \beta \rangle$ be a locally oriented current graph and let $\langle G \to S, \alpha \rangle$ be its dual imbedded voltage graph. Then the derived imbeddings $H_\beta \to S_\beta$ and $G^\alpha \to S^\alpha$ are identical.*

Theorem 4.5.1 explains many aspects of current graphs that were not considered in the previous section, such as orientability and connectivity tests for current-derived imbeddings, or were considered only briefly and without justification, such as face tracing (Theorem 4.4.1) and excess current. These aspects are much easier to study with voltage graphs.

Example 4.5.2 (Alpert). *The nonorientable $\mathscr{Z}_{28}$ current graph given in Figure 4.32 has one face. The derived graph is the Cayley graph for the cyclic group $\mathscr{Z}_{28}$ with generating set $\{1, 2, 3, \ldots, 13\}$. Thus, it is isomorphic to the octahedral graph O_{14}. All faces of the derived imbedding are triangles, by KCL, except for two 28-gons corresponding to the two 1-valent vertices. If an extra vertex is placed at the center of each 28-gon and is joined by edges to each of the original 28 vertices, then the result is a triangular imbedding of O_{15}. But what is the*

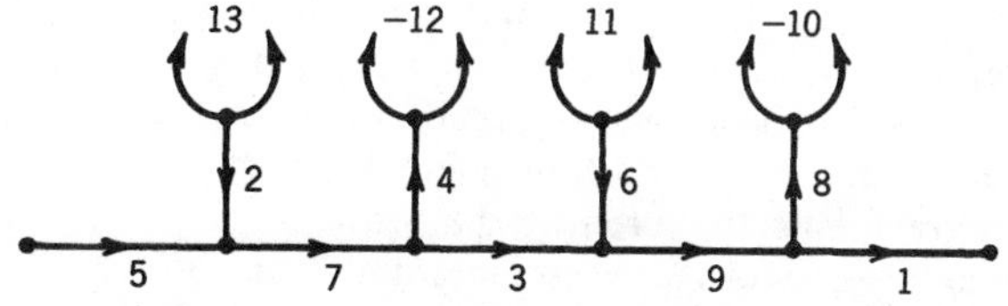

Figure 4.32. A nonorientable current graph with currents in $\mathscr{Z}_{28}$.

orientability type of the surface? The edges carrying currents $-10, 11, -12$, *and* 13 *are traversed twice in the same direction in tracing the face of the current graph imbedding. Therefore, the corresponding loops in the dual imbedded voltage graph are type* 1. *Since the voltages* -12 *and* -10 *are in the only index-2 subgroup of* $\mathscr{Z}_{28}$, *the derived imbedding is nonorientable. This current graph can be generalized to yield nonorientable triangular imbeddings for all octahedral graphs* O_n, $n \equiv 0 \bmod 3$, $n > 3$ (*see Exercise* 10).

A word of caution about local orientation is in order at this point. Changing the local orientation at a vertex of a current graph reverses the plus direction of some dual edges. In effect, this simply inverts the value of the dual voltage assignment on some edges. It should not be surprising that, even in the orientable case, the two possible orientations of a surface can induce nonisomorphic derived graphs for the same current graph! This is because there are voltage graphs such that inverting the voltage assignment to each edge changes the isomorphism type of the derived graph (see Exercise 8). This phenomenon can only happen, however, if the current group is nonabelian (Exercise 9). Therefore, for most applications, it is not a matter of concern.

4.5.3. The Dual Derived Graph

Let $\langle H \to S, \beta\rangle$ be a locally oriented graph, and let $\langle G \to S, \alpha\rangle$ be its dual imbedded voltage graph. There is another graph imbedding that arises naturally in the current-graph construction and has not yet been discussed: the dual $H_\beta{}^* \to S_\beta$ of the derived imbedding $H_\beta \to S_\beta$. It is best to think of this imbedding as $p^{-1}(H) \to S_\beta$, where p is a natural surface projection $S_\beta \to S$, since then p restricts to a graph map $H_\beta{}^* \to H$. The imbedding $H_\beta{}^* \to S_\beta$ is called the "dual derived imbedding", the graph $H_\beta{}^*$ is called the "dual derived graph", and the map $p\colon H_\beta{}^* \to H$ is called the "natural projection" for the dual derived graph.

Example 4.5.3. *At the bottom left of Figure* 4.33 *is a current graph in the sphere with current group* $\mathscr{Z}_4$. *At the bottom right is the dual imbedded voltage graph. At the upper right is the derived imbedding, the* 3-*cube. At the upper left is the dual derived imbedding, the* 3-*octahedron. Both base graphs can be viewed as quotient graphs under the* $\mathscr{Z}_4$ *action generated by a* 90° *rotation about the vertical axis. For the cube* (*derived graph*) *the action is free*; *for the octahedron* (*dual derived graph*) *the action has fixed points at the top and bottom vertex.*

Let $\langle H \to S, \beta\rangle$ be a current graph, and let $\langle G \to S, \alpha\rangle$ be its dual imbedded voltage graph. Since the surface projection $p\colon S^\alpha \to S$ may wrap a face of the derived imbedding many times around a face of the imbedded voltage graph, as in Example 4.5.3, the natural projection $p\colon H_\beta{}^* \to H$ of the dual derived graph onto the current graph is not, in general, a covering map. Indeed, if r is the order of the excess current at vertex v of the current graph H, then $p\colon H_\beta{}^* \to H$ is locally r-to-1 near each vertex of $p^{-1}(v)$.

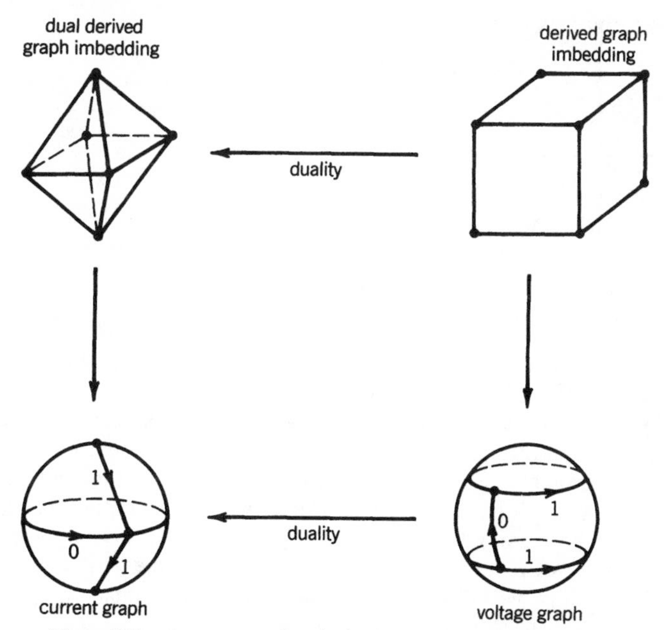

Figure 4.33. A current graph, a dual voltage graph, and derived graphs.

The graph map $p\colon H_\beta^* \to H$ is not a branched covering, but rather a folded covering, because the set of points where the map fails to be a local homeomorphism has dimension not 2, but 1, less than the complex H. Parsons et al. (1980) and Jackson et al. (1981) [also Gross et al. (1982)] use the term "wrapped quasi-covering" in their papers that analyze the construction, and they develop it for direct use in graph-imbedding problems. We shall call it simply a "wrapped covering".

Even if KCL holds at very vertex of a current graph so that the dual derived graph does not cover the current graph, it is still not easy to determine the structure of the dual derived graph. The following example illustrates the problem in one case and presents an ad hoc solution.

Example 4.5.4 (White, 1974). *Suppose that $\mathscr{A}$ is an abelian group and $X = \{x_1, \ldots, x_{2m}\}$ is a set of distinct generators for $\mathscr{A}$ such that $x_i \neq x_j^{-1}$ for all i and j. Let $H \to S$ be the imbedding of a bouquet of $2m$ circles in the surface of genus m given by the rotation*

$$v.\quad e_1\, e_2\, e_1^-\, e_2^-\, e_3\, e_4\, e_3^-\, e_4^- \;\ldots$$

Let β be the current assignment such that $\beta(e_i^+) = x_i$, for $i = 1, \ldots, 2m$. Figure 4.34 illustrates the case $m = 2$, where the edges of the octagon form the dual voltage graph.

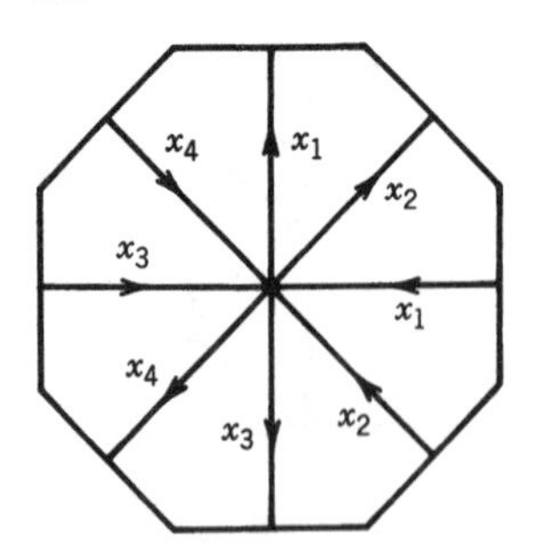

Figure 4.34. A current graph.

Clearly the derived graph is the Cayley graph $C(\mathscr{A}, X)$. We claim this imbedding is self-dual; that is, the dual derived graph $H_\beta{}^*$ is also the Cayley graph $C(\mathscr{A}, X)$. Since *KCL* holds at the one vertex of the current graph H, the graph $H_\beta{}^*$ is a regular covering of H, and accordingly, it must be a Cayley graph for the group $\mathscr{A}$. However, it is not at all obvious that the generating set for this Cayley graph is X. To show that it is, we give an explicit labeling of vertices and coloring of edges that makes $H_\beta{}^*$ the desired Cayley graph.

By Theorem 4.4.1, for each $a \in \mathscr{A}$, exactly one of the $\#\mathscr{A}$ faces of the derived imbedding, which we label a, has a directed boundary walk of the form

$$(e_1, a)(e_2, ax_1)(e_1{}^-, ax_1x_2)(e_2{}^-, ax_1x_2x_1{}^{-1})(e_3, ax_1x_2x_1{}^{-1}x_2{}^{-1}) \cdots$$

This can be simplified, since the group $\mathscr{A}$ is abelian, to

$$(e_1, a)(e_2, ax_1)(e_1{}^{-1}, ax_1x_2)(e_2{}^{-1}, ax_2)(e_3, a)(e_4, ax_3) \ldots$$

Using the relation $(e_1{}^{-1}, ax_1x_2) = (e_1, ax_1x_2x_1{}^{-1})^- = (e_1, ax_2)^-$, we further simplify to

$$(e_1, a)(e_2, ax_1)(e_1, ax_2)^- \ (e_2, a)^- \ (e_3, a)(e_4, ax_3) \ldots$$

The directed edge $(e_1, a)^-$ then forms part of the boundary walk of the face labeled $ax_2{}^{-1}$, and the edge $(e_2, a)^+$ lies on the directed boundary of the face labeled $ax_1{}^{-1}$. If the vertex dual to face a is also labeled a, then the dual edge $(e_1, a)^*$ has its plus direction from dual vertex a to dual vertex ax_2, and the plus direction of $(e_2, a)^*$ goes from dual vertex $ax_1{}^-$ to dual vertex a. The other pairs of edges e_{2i-1} and e_{2i}, for $i > 1$, behave similarly. Thus, if the plus direction of the dual edge $(e_{2i-1}, a)^*$ is colored x_{2i} and the minus direction of the dual edge $(e_{2i}, a)^*$ is colored x_{2i-1}, then the result is a Cayley color graph for the group $\mathscr{A}$ and generating set X.

If the abelian group $\mathscr{A}$ in Example 4.5.4 is the cyclic group $\mathscr{Z}_{4m+1}$ and $X = \{1, 2, \ldots, 2m\}$, then the result is a self-dual imbedding of the complete graph K_{4m+1}. The graph K_n has a self-dual imbedding only if $n \equiv 0, 1 \bmod 4$

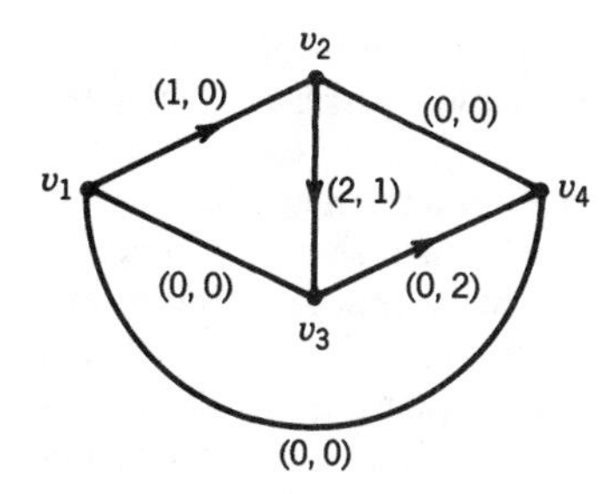

Figure 4.35. A $\mathscr{Z}_n \times \mathscr{Z}_n$-current graph.

(Exercise 3); Stahl (1979) and Pengelley (1975) have obtained such self-dual imbeddings of K_{4m+1} (see Exercise 11). If the group $\mathscr{A}$ is $\mathscr{Z}_{4m+2}$ and $X = \{1, 2, \ldots, 2m\}$ again, then the result is a self-dual imbedding of the octahedron graph O_{2m+1}. Gross (1978) uses a variation of this construction to compute the least number of edge crossings of certain octahedron graphs imbedded in certain surfaces (Exercises 12–14).

Although it is difficult to determine the structure of the dual derived graph, it is possible to give a readily applied criterion to decide when a dual derived graph of a simplicial current graph is simplicial. The proof of the following theorem is a straightforward application of Theorem 4.4.1 and is omitted.

Theorem 4.5.2. *Let $\langle H \to S, \beta \rangle$ be a current graph without multiple edges or loops. Then the dual derived graph has a multiple edge or loop if and only if there are adjacent vertices with excess currents b and c such that $b^s = c^t \neq 1$ for some s and t.*

Example 4.5.5. *Let $\langle H \to S, \beta \rangle$ be the current graph given in Figure 4.35 with the current group $\mathscr{Z}_n \times \mathscr{Z}_n$. The excess currents at vertices v_1, v_2, v_3, v_4 are $(1,0)$, $(1,1)$, $(-2,1)$, and $(0,-2)$, respectively. As long as the number n is odd, no multiple (consider the group operation to be addition) of $(1,1)$ or $(-2,1)$ has a zero coordinate. If $s(1,1) = t(-2,1)$, then $-2t \equiv t \bmod n$, so $3t \equiv 0 \bmod n$. Thus, if $n \equiv 1$ or $5 \bmod 6$, then the adjacent vertices of H satisfy the condition of Theorem 4.5.2. Therefore the dual derived graph is simplicial. Since each vertex of the graph H has excess current of order n, there are n vertices of the dual derived graph corresponding to each v_i, $i = 1, \ldots, 4$. Since no vertex corresponding to v_i is adjacent to another corresponding to v_i and each dual derived vertex has valence $3n$, the dual derived graph is the complete multipartite graph $K_{n,n,n,n}$. The imbedding is triangular and, therefore, minimal, because the dual voltage graph is regular of valence 3.*

Example 4.5.5 can be generalized to any triangular imbedding $H \to S$, where the vertices of graph H can be covered by a disjoint union of paths of length 3. The derived graph H_β has n vertices $v_1, \ldots, v_n$ corresponding to each vertex v in H and edges from u_i to v_j if and only if there is an edge from u to

v in H. Such a graph is denoted $H_{(n)}$ (it is the "composition" of the graph H by a set of n independent vertices). If H is the complete graph K_m, then $H_{(n)}$ is the complete multipartite graph $K_{n,n,\ldots,n} = K_{m(n)}$. A number of authors have sought minimal imbeddings for $H_{(n)}$ given a minimal imbedding, usually triangular or quadrilateral, for H. Bouchet (1982b) has shown, through an ingenious use of "nowhere-zero flows", that if H has a triangular imbedding, so does $H_{(n)}$ for any n not divisible by 2, 3, or 5. Jackson et al. (1981) and Parsons et al. (1980) use wrapped coverings to show that a triangular imbedding for H lifts to one for $H_{(n)}$ for all n relatively prime to a number M that depends only on the chromatic number of H. Bouchet (1982a) proves by means of folded coverings that if H has a eulerian circuit, then any triangular imbedding for H lifts to a triangular imbedding of $H_{(n)}$, for all n. Bouchet (1983) and Abu-Sbeih and Parsons (1983) obtain similar results for quadrilateral imbeddings of bipartite graphs.

4.5.4. The Genus of the Complete Bipartite Graph $K_{m,n}$

In Section 4.4, current graphs were used to construct orientable quadrilateral imbeddings for the complete bipartite graph $K_{m,n}$ when m and n are both even (Example 4.4.4) or when m is odd and $n \equiv 2 \bmod 4$ (Exercise 4.4.11). It is instructive to reconsider these imbeddings in terms of both voltage graphs and dual derived graphs.

Suppose that m is odd and that $n \equiv 2 \bmod 4$, where $m > 2$ and $n > 2$. The left side of Figure 4.36 illustrates, for $m = 7$, the $\mathscr{Z}_n$-current graph that was suggested by Exercise 4.4.11. The vertex having excess current 2 generates an n-gon in the derived graph, just as the other vertices do, because $n/2$ is odd. Accordingly, a quadrilateral imbedding can be constructed from the derived graph, by adding a vertex inside each face, as in Example 4.4.4. In the middle of Figure 4.36 is the voltage graph dual to the current graph shown on the left. Clearly, the derived graph has n vertices and m faces, each bounded by a hamiltonian cycle.

Since the desired imbedding of $K_{m,n}$ is obtained from the derived imbedding by adding a vertex inside each face, one might suspect that a quadrilateral imbedding of $K_{m,n}$ can also be constructed directly as a dual derived

Figure 4.36. A voltage graph and two current graphs.

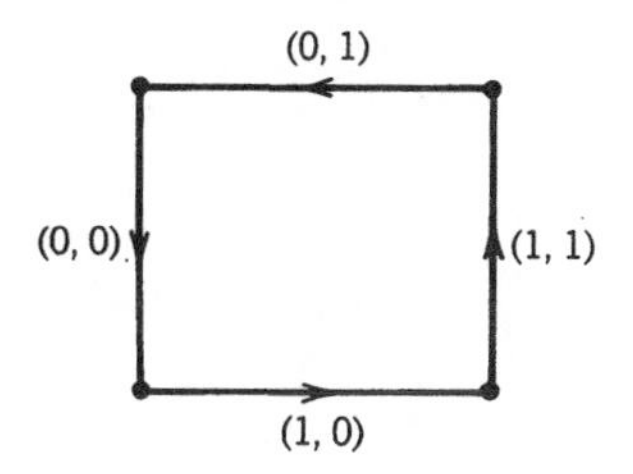

Figure 4.37. A $\mathscr{Z}_{m/2} \times \mathscr{Z}_{n/2}$-current graph.

graph. On the right of Figure 4.36 is another current graph with currents in $\mathscr{Z}_{n/2}$, illustrated for the case $m = 7$. By Theorem 4.5.2, the dual derived graph has no multiple edges or loops. Since the top and bottom vertices satisfy KCL, they generate $n/2 + n/2$ vertices, each of valence m. Thus, they form one part of the complete bipartite graph $K_{m,n}$. The excess current at each of the 2-valent vertices has order $n/2$ in $\mathscr{Z}_{n/2}$. Therefore, these vertices generate m vertices of valence n. Since every face of the current graph imbedding is a quadrilateral, so is every face of the derived imbedding.

A quadrilateral imbedding of $K_{m,n}$ for m and n both even can also be given as a dual derived graph. In this case, the construction is especially elegant. The current group is $\mathscr{Z}_{m/2} \times \mathscr{Z}_{n/2}$, and the current graph is given in Figure 4.37. It is left to the reader to verify that the dual derived graph is $K_{m,n}$.

Since $K_{m,n}$ has no cycles of length less than 4, it follows that

$$\gamma(K_{m,n}) \geq 1 - (m + n - mn + mn/2)/2 = (m-2)(n-2)/4.$$

The last quantity is an integer if and only if m and n are both even, or one or the other is congruent to 2 mod 4. Quadrilateral imbeddings have already been constructed for these cases. To finish the computation of $\gamma(K_{m,n})$ for all m, n, an ad hoc construction is used for the nonquadrilateral imbeddings.

Theorem 4.5.3 (Ringel, 1965). $\gamma(K_{m,n}) = \lceil (m-2)(n-2)/4 \rceil$.

Proof. Consider first the irregular case $m \equiv 3 \bmod 4$, $n \equiv 1 \bmod 4$. Then $K_{m-1,n}$ has a quadrilateral imbedding, which for $n = 9$ looks like Figure 4.38 (left) near a vertex u in the $(m-1)$ part of $K_{m-1,n}$. Split the vertex u into two vertices v and w and join each of these vertices to the nearest "half" of the surrounding n vertices as shown in the center of Figure 4.38. Extra edges from v and w are needed to obtain $K_{m,n}$; this in turn will necessitate some extra handles. The imbedding surface for $K_{m-1,n}$ has genus $(m-3)(n-2)/4$, whereas the desired imbedding for $K_{m,n}$ has genus $\lceil (m-2)(n-2)/4 \rceil$, which for $m \equiv 3 \bmod 4$ and $n \equiv 1 \bmod 4$ is the same as $((m-2)(n-2)+1)/4$. Thus, there is a leeway of $(n-1)/4$ handles that can be used to put in the extra edges at vertices v and w. The right side of Figure 4.38 shows how to add the handles: each labeled face at vertex v is joined by a tube to the

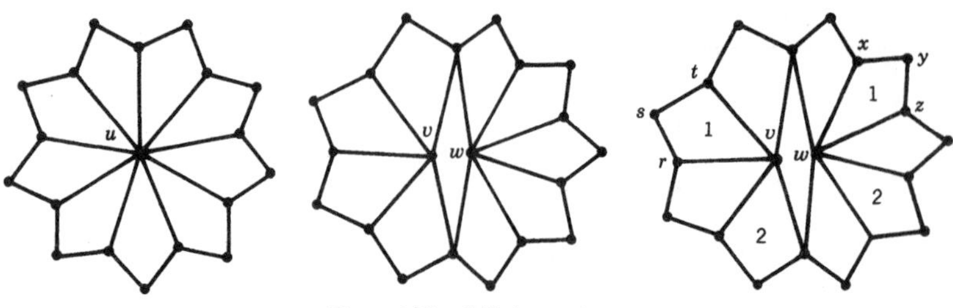

Figure 4.38. Splitting vertex u.

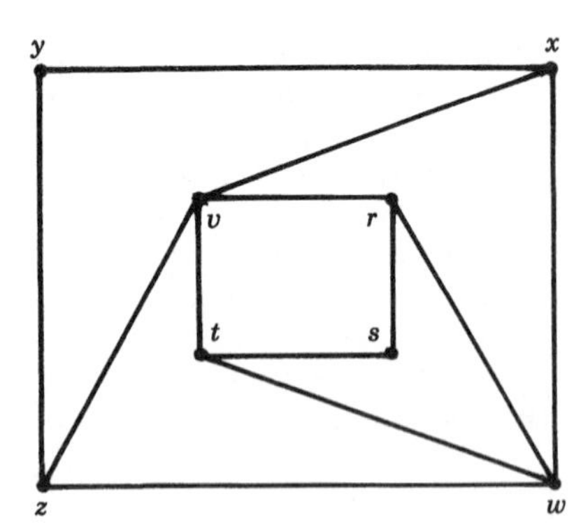

Figure 4.39. The insides of tube 1.

like-labeled face at vertex w. Four edges are then drawn on tube 1, two from v and two from w, as illustrated in Figure 4.39. Three edges are drawn on tube 2, two from v and one from w. For general $n \equiv 1 \bmod 4$, each tube carries four edges except the bottom tube, which carries three.

Similar constructions work for $m \equiv 3 \bmod 4$ and $n \equiv 0$ or $3 \bmod 4$; $n/4$ handles are added for $n \equiv 3 \bmod 4$. Again, the same construction works for $m \equiv 1 \bmod 4$, $n \equiv 0 \bmod 4$. Finally, for $m \equiv 1 \bmod 4$ and $n \equiv 1 \bmod 4$, a quadrilateral imbedding for $K_{m-3,n}$ is obtained first. Three vertices are split, and $(n-1)/4$ handles are added near each split vertex, as before. Since $\lceil (m-2)(n-2)/4 \rceil = (m-5)(n-2)/4 + 3(n-1)/4$ for $m \equiv n \equiv 1 \bmod 4$, this yields the desired imbedding of $K_{m,n}$. All other combinations of congruence classes mod 4 are covered by interchanging m and n. □

4.5.5. Exercises

1. Draw the current graph whose dual voltage graph is given on the left in Figure 4.6.
2. Draw the current graph whose dual voltage graph is given in Figure 4.1.
3. Prove that if K_n has a self-dual imbedding, then $n \equiv 0$ or $1 \bmod 4$.

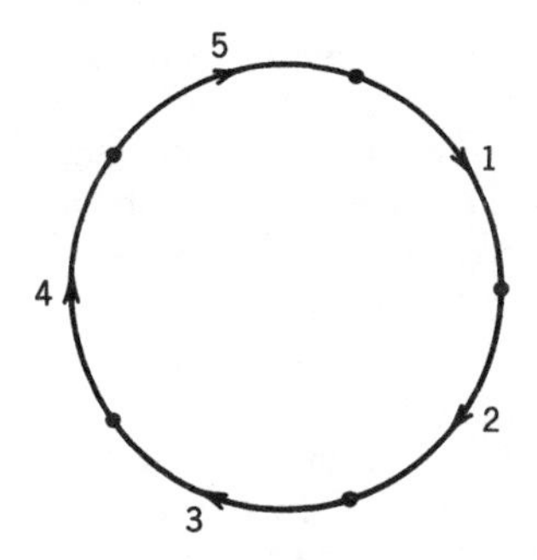

Figure 4.40. A $\mathscr{Z}_5$-current graph.

4. In Exercise 3.5.2 it was shown how to represent an orientable imbedding $G \to S$ by a single permutation ρ of the directed edge set DE(G). Show that the dual imbedding $G^* \to S$ is represented by $\rho \circ \iota$, where ι is the involution of DE(G) given by $\iota(e^{\epsilon}) = e^{-\epsilon}$. Prove that the dual of the dual imbedding is the primal imbedding, using this viewpoint.
5. Name the edges of the imbedded voltage graph given in Exercise 4.1.1. Represent this imbedding with a permutation ρ of the directed edge set, as in Exercise 4. Then use the dual permutation $\rho \circ \iota$ to draw the current graph dual to the original imbedded voltage graph.
6. Obtain a triangular imbedding of $K_{5,5,5}$ from the $\mathscr{Z}_5$ current graph shown in Figure 4.40. Generalize this construction to get triangular imbeddings of $K_{mn,n,n}$ for all $m, n > 0$.
7. Obtain a triangular imbedding of $K_{n,n,n}$ as a dual derived imbedding by assigning currents in $\mathscr{Z}_n \times \mathscr{Z}_n$ to a triangle.
8. Let $\mathscr{A}$ be a group, and let $X = \mathscr{A} - \{1_{\mathscr{A}}\}$. Prove that if there is a direction-preserving automorphism of the Cayley color graph $C(\mathscr{A}, X)$ taking each edge labeled x to the edge labeled x^{-1}, then $\mathscr{A}$ is abelian. Then use the trick of Frucht's theorem, Section 2.2, to show that if $\mathscr{A}$ is nonabelian, then there is an $\mathscr{A}$-voltage graph $\langle G, \beta\rangle$ such that the derived graphs G^{α} and G^{β} are not isomorphic, where β is the voltage assignment $\beta(e) = \alpha(e)^{-1}$. Conclude that for an orientable, nonabelian current graph, it is possible to generate nonisomorphic derived graphs by choosing different orientations of the current-graph surface.
9. Prove that if the group $\mathscr{A}$ is abelian, then the two possible derived graphs of an orientable $\mathscr{A}$ current graph $\langle G \to S, \alpha\rangle$ corresponding to the two orientations of S are isomorphic (see Exercise 8).
10. Generalize Example 4.5.2 to arbitrary n, $n \equiv 0 \bmod 3$, $n > 3$.
11. A self-dual imbedding of K_n, $n \equiv 0 \bmod 4$, can be constructed as follows (Pengelley, 1975). Suppose $n = 2^r t$, $r > 1$. Let $\mathscr{A}$ be the group $\mathscr{Z}_2 \times \cdots \times \mathscr{Z}_2 \times \mathscr{Z}_t$ having r factors of $\mathscr{Z}_2$. Let X be the set of elements of $\mathscr{A}$ having order greater than 2 and Y the set of elements of

order 2. Assign half of the elements of X as voltages to a $\#X$-sided polygon, as in Example 4.5.4 (check first that $\#X \equiv 0 \bmod 4$). Then attach a bouquet of $\#Y$ circles at one vertex of the polygon and assign the elements of Y as voltages (do not nest the circles). Prove that the derived imbedding, after identifying the sides of digons, is a self-dual imbedding of K_n.

12. Let p be a prime, $p \equiv 1 \bmod 4$. The multiplicative group of non-zero elements of $\mathscr{Z}_p$ is cyclic generated by some element q. Let $G \to S$ be the imbedding of a bouquet of $(p-1)/2$ circles in the surface of genus $(p-1)/4$ such that the rotation at the only vertex is $e_1\, e_2\, e_3 \ldots e_1^-\, e_2^-\, e_3^- \ldots$. Let α be the voltage assignment given by $\alpha(e_i) = q^i - q^{i-1}$. Prove that the derived imbedding is a self-dual imbedding of the complete graph K_p in the surface of genus $(p-1)(p-4)/4$ such that any face is a simple closed walk.

13. Let G be the graph constructed from the union of primal and dual complete graphs K_p in Exercise 12 by adding edges from each dual vertex to each primal vertex on the face containing the dual vertex. Show that the graph G is the octahedron graph O_p.

14. Suppose that a graph G is drawn on the surface S_g so that there are c points where edges cross. Prove that $2 - 2g \leq (\#V + c) - (\#E + 2c)/3$. Use this inequality and Exercise 13 to conclude that the minimum number of edge crossings in a drawing of the octahedron graph O_p on the surface $S_{(p-1)(p-4)/4}$ is $p(p-1)/2$, where p is a prime, $p \equiv 1 \bmod 4$ (Gross, 1978).

5

Map Colorings

The chromatic number of a surface S is equal to the maximum of the set of chromatic numbers of simplicial graphs that can be imbedded in S, as we recall from Section 1.5. Heawood (1890) showed that there is a finite maximum, even though there is no limit to the number of vertices of a graph that can be imbedded in S. In particular, if the surface S has Euler characteristic $c \leq 1$, then

$$\text{chr}(S) \leq \left\lfloor \frac{7 + \sqrt{49 - 24c}}{2} \right\rfloor$$

which is now known as the "Heawood inequality". The value of the expression on the right-hand side is called the "Heawood number" of the surface S and is denoted $H(S)$.

The determination of the chromatic numbers of the surfaces other than the sphere is called the "Heawood problem". Its solution, mainly by Ringel and Youngs (1968), gave topological graph theory the critical momentum to develop into an independent research area. The solution is that, except for the Klein bottle, which has chromatic number 6, the chromatic number of every surface equals the Heawood number. For example, the projective plane has chromatic number 6 and the torus has chromatic number 7, exactly their Heawood numbers. The idea of the proof is to imbed in each surface a complete graph whose number of vertices equals the Heawood number of the surface.

Since the sphere has Euler characteristic $c = 2$, its Heawood number is 4. However, Heawood's argument for other surfaces cannot be used to establish four as an upper bound for the chromatic number of the sphere. Although the sphere is the least complicated closed surface, and although some of the most distinguished mathematicians attempted to solve the problem, the chromatic number of the sphere was the last to be known. This last case was resolved when Appel and Haken (1976) established that $\text{chr}(S_0) = H(S_0) = 4$, which is called the Four-Color Theorem. Since the proof of the Four-Color Theorem is fully explained elsewhere, quite lengthy, and not topological in character, we confine our attention to the Heawood problem.

The original solution to the Heawood problem occupies about 300 journal pages, spread over numerous separate articles, and it requires several different kinds of current graphs, whose properties are individually derived. Ringel (1974) has condensed this proof somewhat. Although Ringel considers every case, it remains useful to refer to the original papers for complete details of some of the more difficult cases (such as "orientable case 6").

The introduction of voltage graphs and topological current graphs unifies and simplifies the geometric part of the solution. However, the construction of appropriate assignments of voltages or currents remains at about the same level of difficulty as originally. The present review of the Heawood problem concentrates on a representative sample of the cases whose solutions are most readily generalizable to other imbedding problems.

5.1. THE HEAWOOD UPPER BOUND

The first step in calculating the chromatic numbers of all the closed surfaces except the sphere is to derive the Heawood inequality. The second step is to use Heawood's inequality to reduce the Heawood problem to finding the genus and the crosscap number of every complete graph. A computation of the genus of each complete graph K_n such that $n \equiv 7 \text{ modulo } 12$ illustrates the basic approach to completing the second step.

5.1.1. Average Valence

In order to establish an upper bound for the possible chromatic numbers of graphs that can be imbedded in a surface S, the main concept needed is average valence. By using the Euler characteristic, it is possible to show that the average valence is bounded.

Theorem 5.1.1. *Let S be a closed surface of Euler characteristic c, and let G be a simplicial graph imbedded in S. Then*

$$\text{average valence}(G) \leq 6 - \frac{6c}{\#V}$$

Proof. Whether or not the imbedding is a 2-cell imbedding, we know that

$$\#V - \#E + \#F \geq c$$

For a 2-cell imbedding, we have equality. Otherwise, we observe that all the nonsimply connected regions could be subdivided into cellular regions by adding edges to the graph G, thereby increasing $\#E$ without increasing $\#F$. From the edge-region inequality $2\#E \geq 3\#F$, established in Theorem 1.4.2, we obtain an upper bound

$$\#F \leq \tfrac{2}{3}\#E$$

for $\#F$, which we substitute into the previous inequality. This yields the inequality

$$\#V - \tfrac{1}{3}\#E \geq c$$

and its consequence

$$\#E \leq 3\#V - 3c$$

By Theorem 1.1.1, the sum of the valences is equal to $2\#E$. Thus, the average valence is $2\#E/\#V$. Substituting the upper bound $3\#V - 3c$ for $\#E$, we conclude that

$$\text{average valence}(G) \leq 6 - \frac{6c}{\#V} \quad \square$$

5.1.2. Chromatically Critical Graphs

A graph G is called "chromatically critical" if, no matter what edge is removed, the chromatic number is decreased. Given any graph imbedded in S whose chromatic number is that of the surface S, one can successively delete edges until a chromatically critical graph imbedded in S is obtained. Obviously, every chromatically critical graph is simplicial and connected.

Theorem 5.1.2. *Let S be a closed surface, and let G be a chromatically critical graph such that $chr(G) = chr(S)$. Then for every vertex v of G, $chr(S) - 1 \leq$ valence (v).*

Proof. Suppose that v is a vertex of G with fewer than $\text{chr}(S) - 1$ neighbors. Since G is chromatically critical, its subgraph $G - v$ can be colored with $\text{chr}(S) - 1$ colors. At most $\text{chr}(S) - 2$ of these are assigned to neighbors of v. This leaves one of those $\text{chr}(S) - 1$ colors available to color v, thereby contradicting the fact that $\text{chr}(G) = \text{chr}(S)$. $\square$

Example 5.1.1. *Since the projective plane N_1 has Euler characteristic $c = 1$, it follows from Theorem 5.1.1 that any graph G imbedded in N_1 has a vertex of valence five or less. From Theorem 5.1.2 it then follows that $\text{chr}(N_1) \leq 6$. Figure 5.1 shows an imbedding of the complete graph K_6 in the projective plane. Since $\text{chr}(K_6) = 6$, it follows that $\text{chr}(N_1) = 6$, as first proved by Tietze (1910).*

Example 5.1.2. *The torus S_1 has Euler characteristic $c = 0$. Thus, by Theorem 5.1.1, the average valence of a simplicial graph G imbeddable in S_1 is less than or equal to 6. By Theorem 5.1.2, it follows that $\text{chr}(S_1) \leq 7$. Figure 5.2 shows an imbedding of the complete graph K_7 in S_1. Since $\text{chr}(K_7) = 7$, it follows that $\text{chr}(S_1) = 7$, which was first proved by Heffter (1891).*

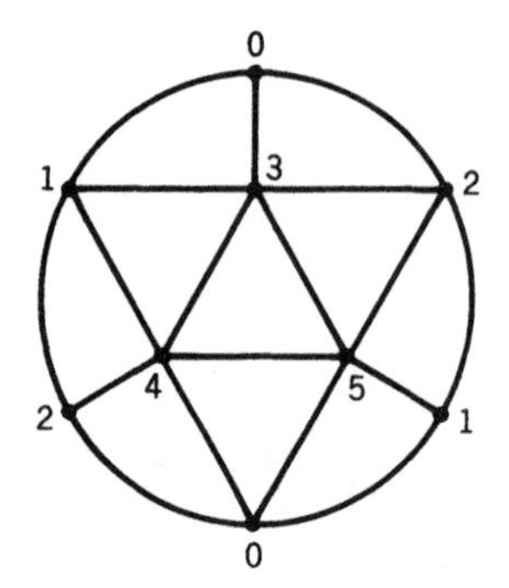

Figure 5.1. An imbedding of the complete graph K_6 in the projective plane N_1.

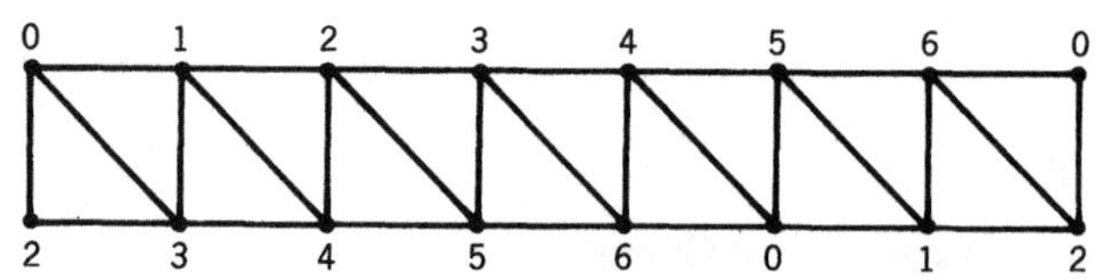

Figure 5.2. An imbedding $K_7 \to S_1$ is obtained by pasting the left side to the right side and then pasting the top to the bottom with a 2/7 twist.

Example 5.1.3. *Like the torus, the Klein bottle N_2 has Euler characteristic $c = 0$. By the same argument as for the torus it is proved that* $\text{chr}(N_2) \leq 7$. *Unlike the torus, however, the Klein bottle has chromatic number* 6, *one less than its Heawood number, because—as we established in Theorem* 3.4.6—*the crosscap number of K_7 is* 3. *This anomaly was discovered by Franklin* (1934).

Theorem 5.1.3 (Heawood, 1890). *Let S be a closed surface with Euler characteristic $c \leq 1$. Then*

$$\text{chr}(S) \leq \left\lfloor \frac{7 + \sqrt{49 - 24c}}{2} \right\rfloor$$

Proof. If $c = 1$, then the expression on the right-hand side of the inequality has the value 6, and the surface S is a projective plane, so that $\text{chr}(S) = 6$, as we saw in Example 5.1.1. Thus the inequality holds.

If $c \leq 0$, our immediate objective is to prove that the quadratic expression

$$\text{chr}(S)^2 - 7\,\text{chr}(S) + 6c$$

is nonpositive. To this end, let G be a graph imbedded in S such that $\text{chr}(G) = \text{chr}(S)$, and such that G is chromatically critical. From Theorem 5.1.1, it follows that

$$\text{average valence}(G) \leq 6 - \frac{6c}{\#V}$$

From Theorem 5.1.2, it follows that

$$\text{average valence}(G) \geq \text{chr}(S) - 1$$

Combining these two inequalities, we obtain the inequality

$$\text{chr}(S) - 1 \leq 6 - \frac{6c}{\#V}$$

Since $c \leq 0$, we have $-6c/\#V \leq -6c/\text{chr}(S)$, because $\#V \geq \text{chr}(S)$. Thus, we infer

$$\text{chr}(S) - 1 \leq 6 - \frac{6c}{\text{chr}(S)}$$

from which we immediately produce the inequality

$$\text{chr}(S)^2 - 7\,\text{chr}(S) + 6c \leq 0$$

By factoring the quadratic expression on the left-hand side we obtain

$$\left(\text{chr}(S) - \frac{7 - \sqrt{49 - 24c}}{2}\right)\left(\text{chr}(S) - \frac{7 + \sqrt{49 - 24c}}{2}\right) \leq 0$$

For $c \leq 0$, the value of the expression $7 - \sqrt{49 - 24c}$ is less than or equal to 0. Thus, the value of the first factor is positive. It follows that the value of the second factor is nonpositive. Since $\text{chr}(S)$ is an integer, the conclusion follows. □

5.1.3. The Five-Color Theorem

If one substitutes $c = 2$ into the Heawood inequality, one obtains the result $\text{chr}(S_0) \leq 4$. Even though Appel and Haken have subsequently proved this, it does not follow from Heawood's argument. Indeed, one of the purposes of Heawood's paper (1890) was to show the error in a purported proof by Kempe (1879) that $\text{chr}(S_0) \leq 4$. What Heawood was able to prove about the sphere is the following theorem.

Theorem 5.1.4 (Heawood, 1890). *The chromatic number of the sphere S_0 is at most 5.*

Proof. Let G be a graph imbedded in S_0 such that $\text{chr}(G) = \text{chr}(S_0)$ and that G is chromatically critical. By Theorem 5.1.1, the average valence of G is less than 6, so that G must have a vertex v of valence less than or equal to 5. By Theorem 5.1.2, the chromatic number of G is at most 6. Since G is

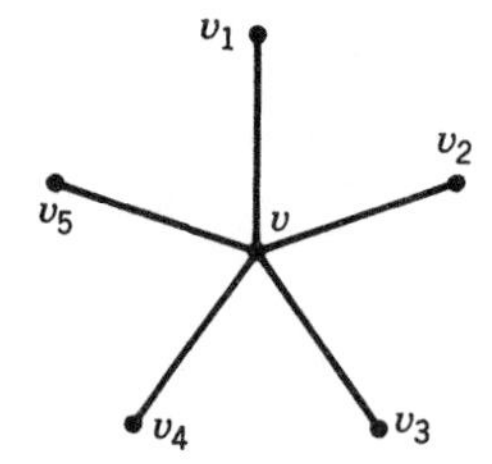

Figure 5.3. The five neighbors of the vertex v.

chromatically critical, we can color the graph $G - v$ with five colors, say the colors $c_1, \ldots, c_5$. If any two of the five neighbors of v had the same color, then some color c_i would not be applied to any neighbor of v. Accordingly we might extend the coloring of $G - v$ to a coloring of G by assigning color c_i to v. Thus, we suppose that the five neighbors $v_1, \ldots, v_5$ of v are assigned the colors $c_1, \ldots, c_5$, respectively, corresponding to the order in which they are encountered in a clockwise traversal around v. This is illustrated in Figure 5.3

For any two colors c_i and c_j, consider the subgraph of $G - v$ containing every vertex that is assigned color c_i or c_j. Every component of this subgraph is called a "c_i-c_j Kempe chain".

First, suppose that the vertex v_3 is not contained in the c_1-c_3 Kempe chain that contains the vertex v_1. One might recolor the vertices of this c_1-c_3 Kempe chain by reversing the roles of c_1 and c_3, that is, by recoloring every c_1-colored vertex c_3 and recoloring every c_3-colored vertex c_1. This results in a proper coloring of the graph $G - v$ such that v_1 and v_3 are both assigned color c_1, while vertices v_2, v_4, and v_5 are still assigned colors c_2, c_4, and c_5 respectively. We can extend this revised coloring of $G - v$ to a coloring of G by assigning color c_3 to v.

Alternatively, if the vertex v_3 is in the same c_1-c_3 Kempe chain as v_1, then v_2 and v_4 cannot be in the same c_2-c_4 Kempe chain, because of the Jordan curve theorem. Thus, we might revise the coloring of $G - v$ by reversing the assignment of c_2 and c_4 to the vertices in the c_2-c_4 Kempe chain that contains v_4. Then color c_4 would be assigned to no neighbor of v, and so it could be assigned to v, yielding a 5-coloring of G. □

5.1.4. The Complete-Graph Imbedding Problem

The Euler characteristic of the orientable surface S_g is $2 - 2g$. Thus,

$$H(S_g) = \left\lfloor \frac{7 + \sqrt{49 - 24(2 - 2g)}}{2} \right\rfloor = \left\lfloor \frac{7 + \sqrt{1 + 48g}}{2} \right\rfloor$$

The Euler characteristic of the nonorientable surface N_k is $2 - k$. Thus,

$$H(N_k) = \left\lfloor \frac{7 + \sqrt{49 - 24(2 - k)}}{2} \right\rfloor = \left\lfloor \frac{7 + \sqrt{1 + 24k}}{2} \right\rfloor$$

Table 5.1 Heawood number of S_g, for $g = 1, \ldots, 23$

genus g	1	2	3	4	5	6	7	8	9	10	11	12	13	14	15	16	17	18	19	20	21	22	23
$H(S_g)$	7	8	9	10	11	12	12	13	13	14	15	15	16	16	16	17	17	18	18	19	19	19	20

Table 5.2 Heawood number of N_k, for $k = 1, \ldots, 23$

crosscap number k	1	2	3	4	5	6	7	8	9	10	11	12	13	14	15	16	17	18	19	20	21	22	23
$H(N_k)$	6	7	7	8	9	9	10	10	10	11	11	12	12	12	13	13	13	13	14	14	14	15	15

Tables 5.1 and 5.2 show the calculated values of the Heawood number for some orientable surfaces and for some nonorientable surfaces, respectively.

From the formula for the Heawood number, one sees that it must increase as the genus or crosscap number increases, but at a rate proportional to the square root of the genus or crosscap number. Tables 5.1 and 5.2 confirm this. Thus, every sufficiently large integer is the chromatic number of at least one orientable surface and of at least one nonorientable surface. Moreover, the same chromatic number may serve for several different surfaces of consecutive genera or of consecutive crosscap numbers, and the number of consecutive surfaces that share the same chromatic number tends to increase along with the chromatic number.

One reads from Table 5.1 that $H(S_{20}) = H(S_{21}) = H(S_{22}) = 19$. Now suppose that there were a graph G imbeddable in S_{20} such that $\text{chr}(G) = 19$. It would follow that $\text{chr}(S_{20}) \geq 19$. Moreover, since any such graph that is imbeddable in S_{20} is also imbeddable in S_{21} and S_{22}, it would also follow that $\text{chr}(S_{21}) \geq 19$ and $\text{chr}(S_{22}) \geq 19$. By Theorem 5.1.3, the chromatic number of each of these three surfaces is also at most 19, their common Heawood number. Thus, they would all have chromatic number exactly 19. In fact, this is true, because the complete graph K_{19} is imbeddable in S_{20}, which is proved in Example 4.4.1. In general, complete graphs are used in this manner to show that the chromatic number of a surface is equal to its Heawood number. To formalize the process, we define the numbers

$$I(n) = \left\lceil \frac{(n-3)(n-4)}{12} \right\rceil \quad \text{and} \quad \bar{I}(n) = \left\lceil \frac{(n-3)(n-4)}{6} \right\rceil$$

Theorem 5.1.5. *Assume that $n \geq 7$ and that $\gamma(K_n) \leq I(n)$. Then for any orientable surface S_g such that $I(n) \leq g < I(n+1)$, the equation* $\text{chr}(S_g) = H(S_g) = n$ *holds.*

Proof. Substitution of $I(n)$ and $I(n+1)$ into the formula for the Heawood number establishes by direct calculation that $H(S_{I(n)}) = n$, that $H(S_{I(n+1)}) = n+1$, and that for a surface S_g such that $I(n) \leq g < I(n+1)$, $H(S_g) = n$. By Theorem 5.1.3, $\text{chr}(S_g) \leq H(S_g) = n$. Under the stated assumption, on the other hand, the complete graph K_n is imbeddable in every such surface S_g, so that $\text{chr}(S_g) \geq n = H(S_g)$. □

Theorem 5.1.6. *Assume that* $n > 8$ *and that* $\bar{\gamma}(K_n) \leq \bar{I}(n)$. *Then for any nonorientable surface* N_k *such that* $\bar{I}(n) \leq k < \bar{I}(n+1)$, *the equation* $\text{chr}(N_k) = H(N_k) = n$ *holds.*

The proof of Theorem 5.1.6 is simplistically analogous to the proof of Theorem 5.1.5, and so it is omitted. These two theorems together reduce the calculation of chromatic numbers to proving that the complete graph K_n can be imbedded in the surfaces $S_{I(n)}$ and $N_{\bar{I}(n)}$. In fact, no better imbeddings exist for K_n, as it is now proved.

Theorem 5.1.7 (The Complete-Graph Orientable-Imbedding Inequality).

$$\gamma(K_n) \geq \left\lceil \frac{(n-3)(n-4)}{12} \right\rceil$$

Proof. We begin with the Euler equation for the closed orientable surface of genus $\gamma(K_n)$, that is,

$$\#V - \#E + \#F = 2 - 2\gamma(K_n)$$

From the edge-region inequality $2\#E \geq 3\#F$, we infer that $\#F \leq \frac{2}{3}\#E$, and substitute for $\#F$ into the equation above to obtain

$$\#V - \tfrac{1}{3}\#E \geq 2 - 2\gamma(K_n)$$

from which we routinely calculate that

$$\gamma(K_n) \geq \frac{n^2 - 7n + 12}{12} = \frac{(n-3)(n-4)}{12}$$

Since $\gamma(K_n)$ is an integer, the conclusion follows. □

Theorem 5.1.8 (The Complete-Graph Nonorientable-Imbedding Inequality).

$$\bar{\gamma}(K_n) \geq \left\lceil \frac{(n-3)(n-4)}{6} \right\rceil$$

Proof. Proceeding exactly as in the proof of Theorem 5.1.7, we obtain the inequality

$$n - \frac{1}{3}\frac{n(n-1)}{2} \geq 2 - \bar{\gamma}(K_n)$$

from which we immediately derive

$$\bar{\gamma}(K_n) \geq \frac{n^2 - 7n + 12}{6}$$

The conclusion follows, since $\bar{\gamma}(K_n)$ is an integer. □

5.1.5. Triangulations of Surfaces by Complete Graphs

The main part of the solution to the Heawood problem is now before us: to construct minimum imbeddings of the complete graphs in both orientable and nonorientable surfaces. The general form of these imbeddings seems to depend strongly on the residue class of n modulo 12.

First, if $n \equiv 0$, 3, 4, or 7 modulo 12, then the value of the expression $(n-3)(n-4)/12$ is an integer. The value of the expression is not an integer for other values of n. It follows that

$$I(n) = \left\lceil \frac{(n-3)(n-4)}{12} \right\rceil = \frac{(n-3)(n-4)}{12} \quad \text{for } n \equiv 0, 3, 4, \text{ or } 7 \text{ modulo } 12$$

A review of the proof of Theorem 5.1.7 reveals that this equality is obtained if and only if $\#F = \frac{2}{3}\#E$, which happens, of course, if and only if the graph K_n triangulates the surface of genus $I(n)$.

In general, the quantity $I(n) - (n-3)(n-4)/12$ measures by how much an orientable imbedding $K_n \to S_{I(n)}$ would fail to be a triangulation, in the following sense. For each positive integer i, let F_i denote the set of i-sided faces of the imbedding. Then

$$I(n) - (n-3)(n-4)/12 = \frac{1}{6}\left(\sum_i \#F_i - 3\#F\right) = \frac{1}{6}(2\#E - 3\#F)$$

For instance, if $n = 1$, 6, 9, or 10 modulo 12, then $I(n) - (n-3)(n-4)/12 = 3/6$. The number of extra edges one would need to add to such an imbedding to obtain a triangulation is 3. Table 5.3 gives this quantity for every residue class.

Similarly if $n \equiv 0, 1, 3, 4, 6, 7, 9$, or 10 modulo 12, then $(n-3)(n-4)/6$ is an integer. It is not an integer for other values of n. Table 5.4 shows the values

Table 5.3 The values of $I(n) - (n-3)(n-4)/12$

n modulo 12	0	1	2	3	4	5	6	7	8	9	10	11
$I(n) - (n-3)(n-4)/12$	0	$\frac{3}{6}$	$\frac{5}{6}$	0	0	$\frac{5}{6}$	$\frac{3}{6}$	0	$\frac{2}{6}$	$\frac{3}{6}$	$\frac{3}{6}$	$\frac{2}{6}$

Table 5.4 The values of $\bar{I}(n) - (n-3)(n-4)/6$

n modulo 12	0	1	2	3	4	5	6	7	8	9	10	11
$\bar{I}(n) - (n-3)(n-4)/6$	0	0	$\frac{2}{3}$	0	0	$\frac{2}{3}$	0	0	$\frac{2}{3}$	0	0	$\frac{2}{3}$

of $\bar{I}(n) - (n-3)(n-4)/6$, which indicate by how much a nonorientable imbedding $K_n \to N_{\bar{I}(n)}$ would fail to be a triangulation.

Tables 5.3 and 5.4 suggest immediately that there might be at least four cases to consider for orientable imbeddings and at least two cases to consider for nonorientable imbeddings of K_n. Only upon deeper investigation does one discover the properties that lead to a natural split into 12 orientable cases and 12 nonorientable cases. These are commonly known as orientable cases $0, \ldots, 11$ and nonorientable cases $0, \ldots, 11$, corresponding to the residue of n modulo 12. Orientable cases 0, 3, 4, and 7 and nonorientable cases 0, 1, 3, 4, 6, 7, 9, and 10 are called "regular cases", because the imbedded complete graph actually triangulates the surface in those cases. All other cases are called "irregular cases".

5.1.6. Exercises

1. Let G be a 16-vertex graph with four vertices each of valences 4, 5, 6, and 7. Prove that the graph G is not planar.
2. Construct a 5-regular planar graph with as few vertices as possible.
3. Suppose that G is a connected chromatically critical graph such that $\mathrm{chr}(G) = 3$. Is G necessarily an odd cycle?
4. Construct a chromatically critical graph G such that $\gamma(G) = 1$ and $\mathrm{chr}(G) = 5$.
5. Prove that the removal of an edge from K_7 yields a graph of crosscap number 2.
6. Construct an imbedding of K_6 in a Möbius band.
7. Let G be the edge-complement in K_8 of a 1-factor. What is the genus of G?
8. What is the chromatic number of the graph G of Exercise 7?
9. Suppose that $\gamma(G) = 1$ and that $\mathrm{chr}(G) = 7$. Prove that G contains a subgraph isomorphic to K_7.

5.2. QUOTIENTS OF COMPLETE-GRAPH IMBEDDINGS AND SOME VARIATIONS

It would be very convenient if minimum orientable and nonorientable imbeddings of all the complete graphs could be immediately derived from imbedded voltage graphs. That would have greatly simplified the solution to the Heawood problem. However, only orientable case 7 admits such a direct construction, and even in that case, it requires considerable ingenuity to assign the voltages correctly to the imbedded base graph. Each of the other three regular orientable cases requires a few additional tactics beyond the basic strategy.

5.2.1. A Base Imbedding for Orientable Case 7

Suppose that $n = 12s + 7$. Then the complete graph K_n has $12s + 7$ vertices and $(12s + 7)(6s + 3)$ edges. Our objective is to construct an imbedding with $(12s + 7)(4s + 2)$ faces, every one a triangle. The greatest common divisor of these three numbers is $12s + 7$, so the obvious approach is to consider a base graph imbedding such that

$$\#V = 1 \quad \#E = 6s + 3 \quad \text{and} \quad \#F = 4s + 2$$

that is, an imbedding of the bouquet B_{6s+3} in the surface S_{s+1}.

There is little trouble in constructing candidates for a base imbedding, if one generalizes a construction for surfaces already considered in Figure 4.31. In that figure, opposite sides of an octagon are pasted to each other to produce the surface S_2. In general, the opposite sides of a $4g$-sided polygon can be pasted together to yield the surface S_g. To prove that this works, one considers the corners of the polygon to be vertices of a graph and the sides of the polygon to be edges. The result of the pasting leaves $2g$ edges and one face, obviously. It is not difficult to verify that it also leaves only one vertex. (See Exercises 1–3.) Thus the result is the bouquet B_{2g} imbedded in the orientable surface of Euler characteristic $1 - 2g + 1$, that is, the surface S_g.

To obtain an imbedding of the bouquet B_{6s+3} in the surface S_{s+1}, one begins with a $(4s + 4)$-sided polygon. When each side is pasted to the opposite side, the result is an imbedding $B_{2s+2} \to S_{s+1}$. If one now draws a line from one corner of the $(4s + 4)$-sided face to each of the other $4s + 1$ corners to which it is not adjacent, one obtains an imbedding $B_{6s+3} \to S_{s+1}$. The scallop-shell appearance of this imbedding is illustrated in Figure 5.4.

Ordinarily one tries to choose the voltage group to be cyclic, since cyclic groups are the least complicated. In this case, since the covering space to be constructed is $(12s + 7)$-sheeted, the obvious candidate for a voltage group is $\mathscr{Z}_{12s+7}$. Moreover, since numbers of the form $12s + 7$ can be prime, the only hope to lift some of the scallop-shell base imbeddings to triangular imbeddings of complete graphs is with cyclic voltage groups.

Any one-to-one assignment of the voltages $1, \ldots, 6s + 3$ modulo $12s + 7$ to the edges of the bouquet B_{6s+3} yields K_{12s+7} as the derived graph. However, they must be assigned so that the Kirchhoff voltage law holds globally, else the derived imbedding would not be a triangulation. There is no known general reason to suppose that a particular problem of this kind can be solved at all.

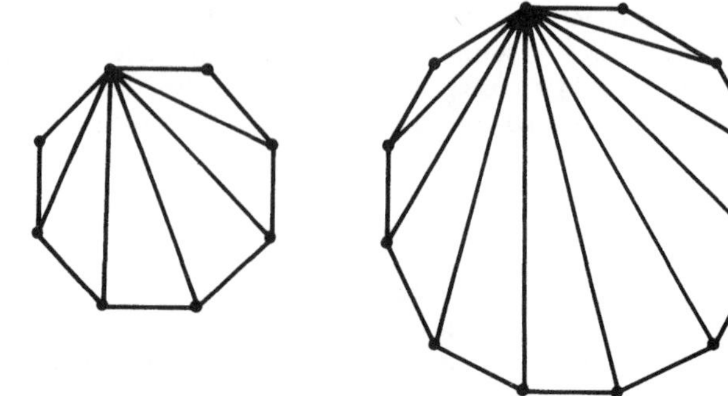

Figure 5.4. Imbeddings $B_9 \to S_2$ and $B_{15} \to S_3$. These are base graphs for imbeddings $K_{19} \to S_{20}$ and $K_{31} \to S_{64}$.

Alternatively, one may observe that the global Kirchhoff voltage law corresponds to a system of $4s + 2$ linear equations in $6s + 3$ unknown edge-voltages, which would seem to have many solutions. However, the requirement that each of the voltages $1, \ldots, 6s + 3$ be used exactly once corresponds to an inequality

$$\prod_{i,j} (x_i - x_j) \neq 0$$

of degree $\binom{6s+3}{2}$. Accordingly, one does not expect to find a solution by the routine application of general methods.

5.2.2. Using a Coil to Assign Voltages

To obtain a satisfactory voltage assignment for the scallop shell base imbedding, the voltages $1, \ldots, 6s + 3$ modulo $12s + 7$ are first partitioned into three intervals. Voltages $1, \ldots, 2s$ in the lower range are distributed along the rim of the shell in consecutive clockwise order, with plus edge directions assigned so they alternate. The voltage $2s + 1$ is assigned to the middle interior edge. Voltages $2s + 2, \ldots, 4s + 2$ in the middle range are assigned to the right half of the shell, and voltages $4s + 3, \ldots, 6s + 3$ are assigned to the left half. Figure 5.5 shows the way that they are assigned.

The relationship between the voltages $1, \ldots, 2s$ in the lower range and the voltages $2s + 2, \ldots, 4s + 2$ in the middle range is called a "coil". Note first that the voltage $3s + 2$ in the precise middle of the middle range is assigned to the top edge of the scallop shell. The first outer voltage to follow $3s + 2$ is the voltage 1, and the plus direction of its edge is opposite to that of the top edge. Thus, the top triangle will satisfy *KVL* if and only if $3s + 1$ is assigned to its

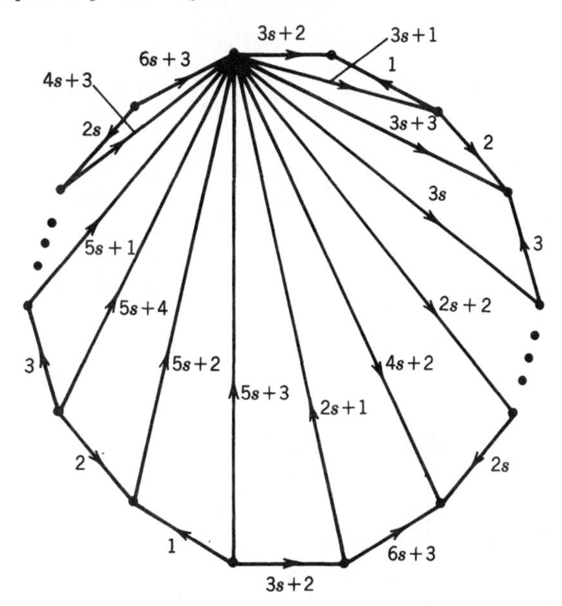

Figure 5.5. This assignment of voltages in $\mathscr{Z}_{12s+7}$ to the imbedded bouquet $B_{6s+3} \to S_{s+1}$ satisfies KVL and yields as a derived graph a triangular imbedding $K_{12s+7} \to S_{I(12s+7)}$.

other edge. Since the voltage 2 is assigned to the outer edge of the second triangle from the top, in the same direction as the voltage $3s + 1$, that triangle will satisfy KVL if its other edge is assigned the voltage $3s + 3$. In this manner the voltages $1, \ldots, 2s$ are used to coil from the middle voltage $3s + 2$ in the middle range outward to the limiting voltages $2s + 2$ and $4s + 2$. Figure 5.6 shows a combinatorial abstraction of the coil.

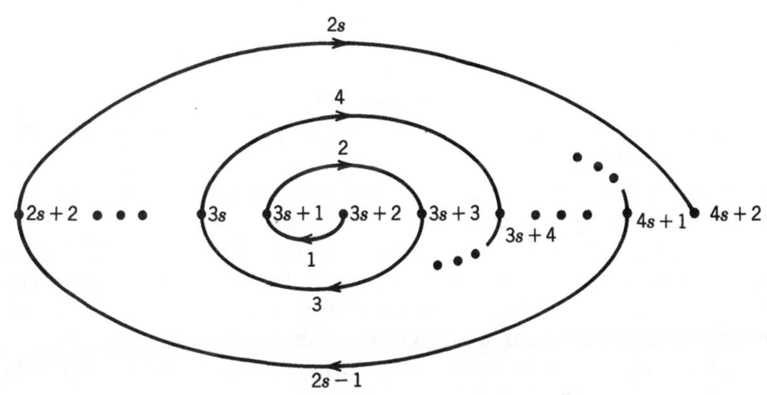

Figure 5.6. The voltages $1, \ldots, 2s$ are used to coil outward from $3s + 2$ toward $2s + 1$ and $4s + 2$.

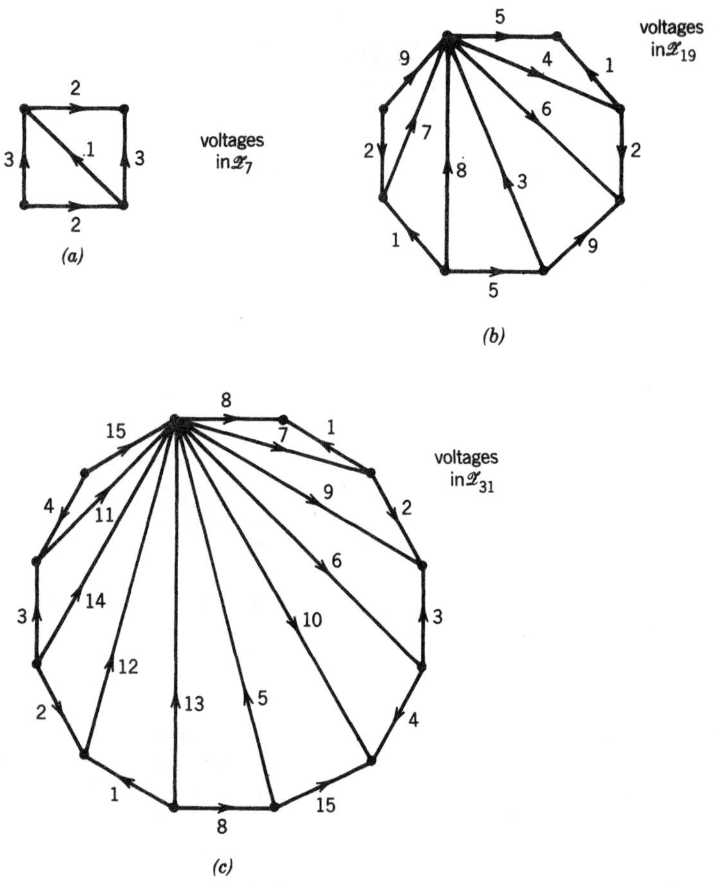

Figure 5.7. Imbedded voltage graphs that yield triangular imbeddings of (a) K_7, (b) K_{19}, and (c) K_{31}.

Because of the way the scallop shell is to be pasted together to form a closed surface, the lower voltages $1, \ldots, 2s$ also appear on the left rim. There they coil among the upper voltages $4s + 3, \ldots, 6s + 3$, starting from $5s + 3$ on the vertical interior edge and working outward. The result is that *KVL* also holds on the triangles on the left side of the shell.

The voltage $2s + 1$ on the middle interior edge welds the two coils together, so that *KVL* is also satisfied on the two triangular faces in which it lies. Thus, *KVL* holds globally, and the Heawood problem is solved for orientable case 7. Figure 5.7 shows the voltage assignments for imbeddings $B_3 \to S_1$, $B_9 \to S_2$, and $B_{15} \to S_3$ that have derived triangular imbeddings $K_7 \to S_1$, $K_{19} \to S_{20}$, and $K_{31} \to S_{63}$, respectively.

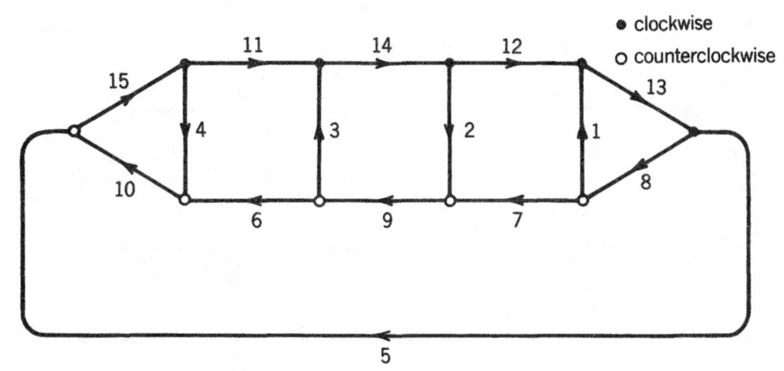

Figure 5.8. A Gustin current graph for K_{31}.

5.2.3. A Current-Graph Perspective on Case 7

If one now draws the dual of the imbedded voltage graph in Figure 5.7*b*, one obtains precisely the topological current graph already illustrated in Figure 4.31. The underlying graph is planar and 3-regular, so that it is also possible to draw an equivalent Gustin current graph in the plane, with no edge-crossings. Such a Gustin current graph appears in Figure 4.18. Figure 5.8 shows the Gustin current graph that is equivalent to the imbedded voltage graph in Figure 5.7*c*.

The coiling relationship among the currents is highly visible in this ladder-like Gustin current graph. It is also easy to see that *KCL* is satisfied at every vertex. This ease of visibility makes Gustin current graphs very useful in computations. On the other hand, while the base graph imbedding for the voltage graph solution is designed to have one vertex, it requires a certain amount of work to check the corresponding face in the Gustin current graph.

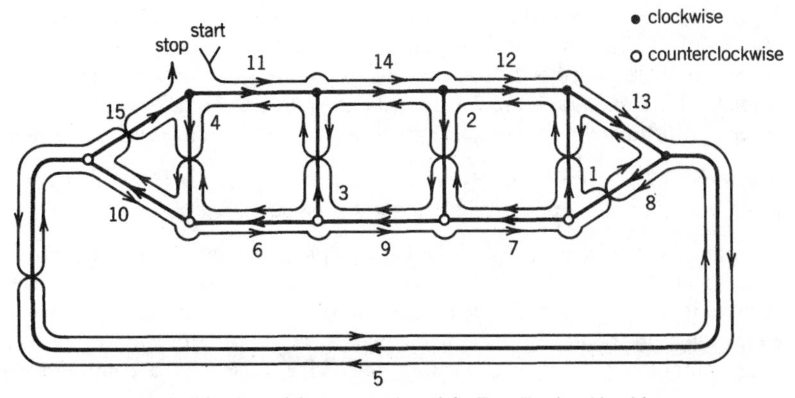

Figure 5.9. Pictorial representation of the Face Tracing Algorithm.

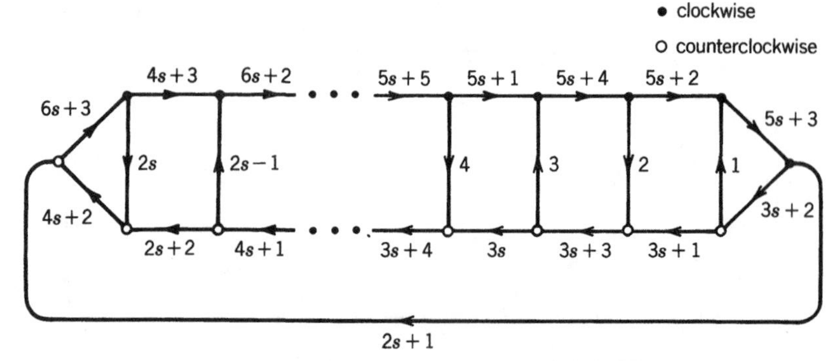

Figure 5.10. A general current graph for orientable case 7 of the Heawood problem.

Fortunately, there is a pictorial way to apply the Face Tracing Algorithm, illustrated in Figure 5.9.

One may start at any edge. In this example, we have started with the horizontal edge at the upper left, which carries the current 11. Since the terminal vertex of that edge is solid, the next edge we traverse is the one carrying the current 14. When this single edge orbit closes on itself, every edge has been traversed in both directions. Thus, the corresponding imbedding has only one face. The thin line that runs along the edges in Figure 5.9 traces out the boundary circuit of that face.

The general voltage graph for orientable case 7, shown in Figure 5.5, also has a dual that may be represented as a ladderlike Gustin current graph. This is shown in Figure 5.10. Ringel (1961) solved this case with purely combinatorial methods. The generating row he devised anticipates the technique of coiling.

5.2.4. Orientable Case 4: Doubling 1-Factors

If $n = 12s + 4$, then the complete graph K_n has $12s + 4$ vertices and $(6s + 2)(8s + 2)$ triangular faces. Since the greatest common divisor of these three numbers is $6s + 2$, the obvious properties one might want in an imbedded voltage graph are 2 vertices, $12s + 3$ edges, and $8s + 2$ triangular faces. Unfortunately, the complete graph K_{12s+4} cannot have such a quotient.

Theorem 5.2.1 (Gross and Tucker, 1974). *Let $p: K_n \to G$ be a covering projection. Then the cardinality of the fibers is an odd number.*

Proof. Let v be a vertex of the base graph G. Suppose that there are $2n$ vertices in the fiber over v. Then the subgraph of G spanned by these $2n$ vertices is isomorphic to K_{2n} and contains $n(2n - 1)$ edges. These $n(2n - 1)$ edges must be the union of the fibers of v-based loops. However, every edge fiber contains $2n$ edges, and $2n$ does not divide $n(2n - 1)$, a contradiction. □

It would not make sense to depend on the largest odd factor of $12s + 4$, since $12s + 4$ might be a power of 2, for example if $s = 5$ or 85. However, there is a tactic that does help, namely, to double some 1-factors of K_{12s+4}. "Doubling" a set of edges means to insert an additional adjacency along each edge in the set, that is, so that it has the same two endpoints.

If one 1-factor of K_{12s+4} were doubled, then the resulting graph would have $12s + 4$ vertices and $(6s + 2)(12s + 3) + (6s + 2)$ edges, that is, $(6s + 2)(12s + 4)$ edges. This looks promising, since assigning the voltages $1, \ldots, 6s + 2 \bmod (12s + 4)$ to the edges of the bouquet B_{6s+2} yields as a derived graph K_{12s+4} with a doubled 1-factor. The edge with voltage $6s + 2$ leads to the double edges in the derived graph, because it is a loop with a voltage of order 2.

One does not want a triangular imbedding of K_{12s+4} with a doubled 1-factor. Instead, the idea is to derive an imbedding so that each pair of doubled edges bounds a digon, but that all other faces are three-sided. Then each digon may be excised and its two sides sewn together, that is, identified. The result would be a triangular imbedding $K_{12s+4} \to S_{I(12s+4)}$. The number of digons in the derived imbedding would be $6s + 2$, and the number of triangular faces would be $(6s + 2)(8s + 2)$.

According to Theorem 4.1.1, such a derived imbedding could be obtained from an imbedded voltage graph with one vertex, $6s + 2$ edges, and $4s + 2$ faces. Of these faces, $4s + 1$ should be 3-sided and satisfy *KVL*. The other face should be a monogon whose net voltage has order two. An easy calculation reveals that the base surface would have Euler characteristic $1 - 2s$, an odd number, so the base surface would be nonorientable. Figure 5.11 shows such an imbedded voltage graph for K_4, when $s = 0$.

Some experimentation reveals that it is not easy to generalize Figure 5.11. One of the problems that would arise is the necessity of proving that the derived surface is orientable. Of course, this is obvious for Figure 5.11, because the derived surface has Euler characteristic 2. However, every nonpositive even Euler characteristic is shared by an orientable surface and a nonorientable surface.

These problems are completely circumvented if one doubles three 1-factors of K_{12s+4}, thereby obtaining a total of $(6s + 3)(12s + 4)$ edges. This time the derived imbedding has $18s + 6$ digons, to be excised, and $(6s + 2)(8s + 2)$ triangular faces. A plausible base imbedding for this configuration might have one vertex, $6s + 3$ edges, and $4s + 4$ faces. Three of these faces could be

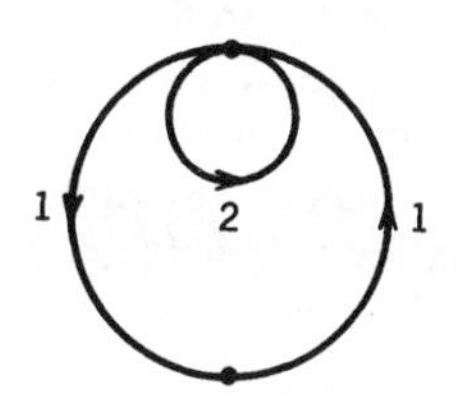

Figure 5.11. The bouquet B_2 imbedded in the projective plane N_1. Voltages are in the cyclic group $\mathscr{Z}_4$.

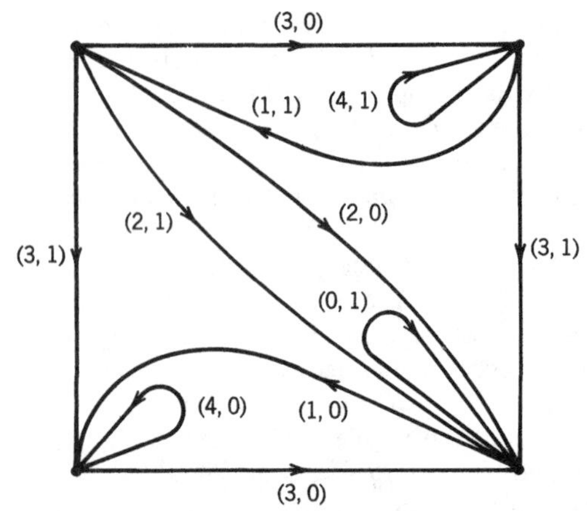

Figure 5.12. Voltages in $\mathscr{Z}_8 \times \mathscr{Z}_2$ for an imbedding $B_9 \to S_1$. The derived graph is K_{16} with three doubled 1-factors, and the imbedding surface is S_{13}.

monogons and the other $4s + 1$ could be three-sided. The Euler characteristic of the base surface would then be $2 - 2s$, so that it might be orientable.

However, this raises a new problem. The three monogons all need voltages of order 2 if they are to lift to digons, and these voltages must differ from each other if the derived graph is to be the one that is wanted. Therefore, the cyclic group $\mathscr{Z}_{12s+4}$ cannot be the voltage group for this kind of base imbedding.

The obvious voltage group to try is $\mathscr{Z}_{6s+2} \times \mathscr{Z}_2$, because it has the right order and it has three elements of order 2. For instance, one would try to obtain an imbedding of $K_{16} \to S_{13}$ by an assignment of voltages in $\mathscr{Z}_8 \times \mathscr{Z}_2$ to the imbedded bouquet $B_9 \to S_1$. Figure 5.12 shows an appropriate voltage assignment.

The three loops that are monogon boundaries get the voltages $(4, 0)$, $(4, 1)$, and $(0, 1)$ that have order 2. Thus, the derived imbedding has 24 digons, each of whose boundaries is a double adjacency. Each of these digons is excised and its two boundary edges are identified to eliminate a multiple adjacency. Since *KVL* holds on the boundary circuit of every base face except the monogons, and since these other faces are three-sided, all the faces in the derived imbedding except the digons are three-sided. Thus, excising the digons and closing up yields an imbedding $K_{16} \to S_{13}$.

The configuration illustrated in Figure 5.12 can be generalized. Figure 5.13 shows an imbedded bouquet $B_{21} \to S_3$ with voltages in $\mathscr{Z}_{20} \times \mathscr{Z}_2$ so that the derived imbedding is of K_{40} with three doubled 1-factors into S_{148}. Excising the digons and closing up would yield a triangular imbedding $K_{40} \to S_{148}$.

Once again, coiling is used to assign voltages so that *KVL* holds everywhere but on the three monogon boundaries. This time however, the voltages $(3, 0)$, $(3, 1)$, $(6, 0)$, $(6, 1)$, $(9, 0)$ and $(9, 1)$ are used on the rim of the polygon. The

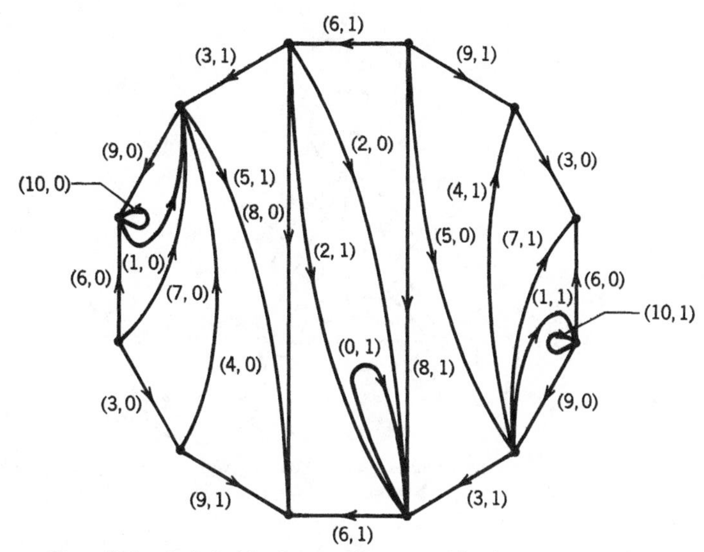

Figure 5.13. An imbedding bouquet $B_{21} \to S_3$ with voltages in $\mathscr{Z}_{20} \times \mathscr{Z}_2$.

voltages (3, 0), (6, 0) and (9, 0) coil among the voltages whose first coordinate is congruent to 1 modulo 3, while the voltages (3, 1) and (6, 1) coil among the voltages whose first coordinate is congruent of 2 modulo 3. The voltage (9, 1) is used to patch the coils together. The current graph in Figure 5.14 shows the coiling and patching clearly. The general solution to orientable case 4 was first achieved by Terry et al. (1970), and is left as an exercise. (See Exercise 11.)

5.2.5. About Orientable Cases 3 and 0

The two other regular orientable cases are much more difficult to solve than cases 7 and 4. What makes case 3 difficult is that the voltage graph for the solution has three vertices, rather than one. Nonabelian voltages are the complicating factor in case 0. Ringel (1974) gives a complete explanation of their solutions.

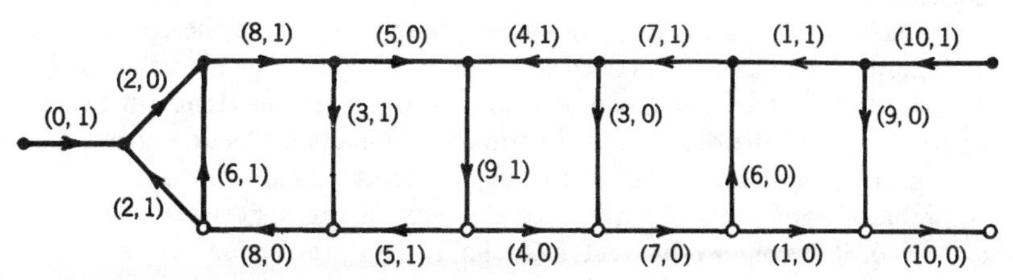

Figure 5.14. A current graph corresponding to the voltage graph in Figure 5.13.

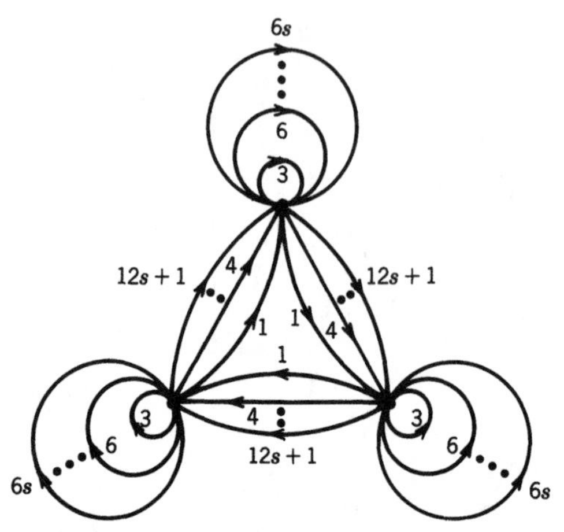

Figure 5.15. This base graph has $2s$ loops at each of its three vertices and $4s + 1$ edges between each pair of vertices. The derived graph consists of three isomorphic copies of K_{12s+3}.

A triangular imbedding of K_{12s+3} must have $12s + 3$ vertices, $(6s + 1)(12s + 3)$ edges, and $(6s + 1)(8s + 2)$ faces. In general, the largest common factor of these three numbers is $4s + 1$. Thus the smallest quotient graph must have at least three vertices and $18s + 3$ edges. Figure 5.15 shows an unimbedded voltage graph that has three isomorphic copies of K_{12s+3} as its derived graph. The solution to orientable case 3 is achieved by imbedding this voltage graph in the surface S_{3s} so that every face is three-sided and that *KVL* holds globally. The first solution to orientable case 3 was by Ringel (1961).

Case 3 perfectly illustrates the mathematical principle that a primal object and its dual can both be useful, without one entirely supplanting the other. Since the derived graph imbedding directly covers an imbedded voltage graph, theoretical investigations tend to concern themselves with voltage graphs, rather than with current graphs. On the other hand, for small-face imbeddings of high-valence graphs, it is often easier to assign appropriate currents than to assign voltages—when the work is by hand, but not necessarily for computer algorithms.

In particular, imagine assigning appropriate rotations to the voltage graph of Figure 5.15 so that the derived imbedding would be a genus imbedding for K_{12s+3}. It might be a tedious task to traverse face boundaries in order to verify *KVL*. On the other hand, Figure 5.16 contains a Gustin current graph that generalizes into a solution for all of orientable case 3. As soon as the combinatorial patterns of current assignments on various classes of edges are identified, it becomes clear that *KCL* holds. Note the coiled currents on the horizontal edges and on the globular rungs. An induction on the ladder length

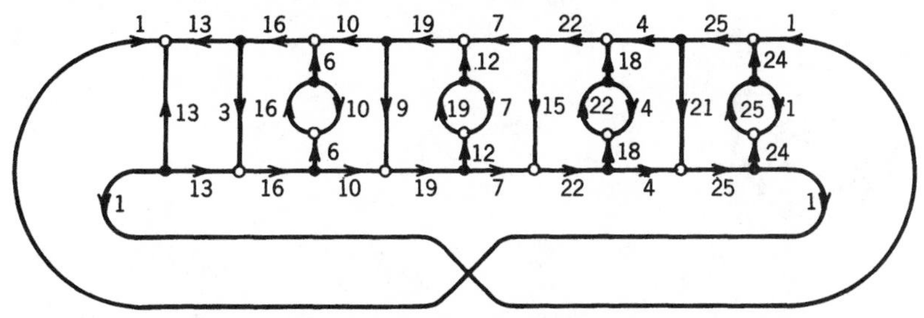

Figure 5.16. A three-orbit $\mathscr{Z}_{27}$-current graph that yields a triangular imbedding $K_{27} \to S_{46}$.

is used to show that the current graph has three edge orbits, each of which contains each nonzero current exactly once.

Orientable case 0 was first solved by Terry et al. (1967). The complication here is that if the number $12s$ has no large odd factors, then by Theorem 5.2.1, the complete graph K_{12s} has no small quotients. The solution here involves the doubling of many 1-factors and the use of nonabelian voltages or currents. In particular, if 2^k is the largest power of two that divides $12s$, then $2^k - 1$ 1-factors are doubled.

5.2.6. Exercises

1. Number the vertices of an eight-sided polygon in consecutive cyclic order from 0 to 7. For each pair of vertices that is identified in pasting opposite edges together, write a transposition. Note that for $i = 1, \ldots, 7$, there is a transposition $(i, i + 3 \text{ modulo } 8)$. How many vertex orbits are there in the permutation group generated by these transpositions?
2. Suppose that opposite sides of a $4n$-sided polygon are pasted together, and let the image of the polygon rim be regarded as a graph imbedded in the resulting surface. Prove that the graph has only one vertex. (*Hint:* See Exercise 1.)
3. Suppose that opposite sides of a $4n + 2$ sided polygon are pasted together as in Exercise 2. How many vertices does the graph have?
4. Try to complete the partial voltage assignment shown so that the derived imbedding is $K_{19} \to S_{20}$. Paste opposite sides together.

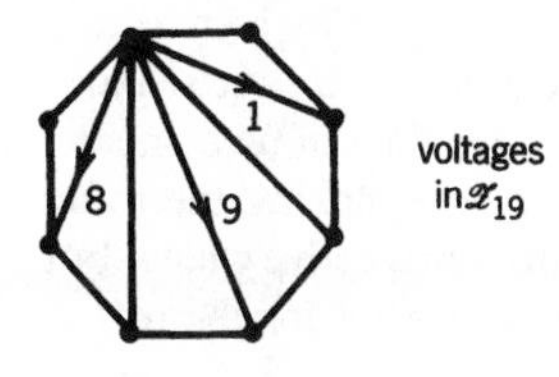

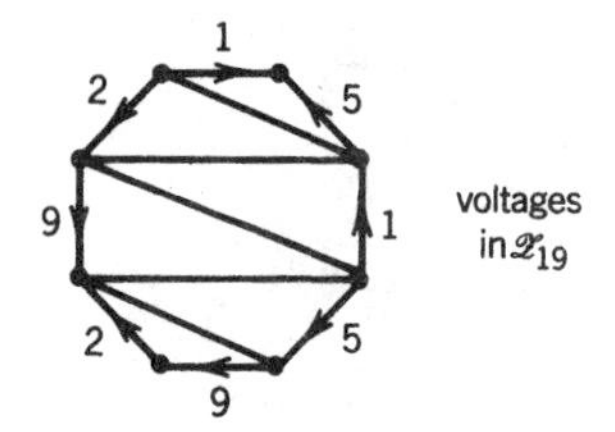

5. Complete the partial voltage assignment shown so that the derived imbedding is $K_{19} \to S_{20}$. The voltages shown imply the pasting rules.
6. The coil in Figure 5.6 might have been $3s + 2, 3s + 3, 3s + 1, \ldots,$ $4s + 2, 2s + 2$. Show how to use this in another solution to orientable case 4.
7. Draw the Gustin current graph for the imbedding $K_{43} \to S_{130}$.
8. Trace the face boundary in the current graph of Exercise 7.
9. Prove that if d is an odd divisor of n, then the complete graph K_n has a unique d-fold quotient (Gross and Tucker, 1974).
10. Classify the quotients of the complete bipartite graph $K_{m,n}$ (Sit, 1976).
11. Draw a general current graph solution for orientable case 4, modeled after Figure 5.14.

5.3. THE REGULAR NONORIENTABLE CASES

Perhaps surprisingly, the nonorientable cases of the Heawood problem seem to admit easier solutions than the orientable cases. Moreover, Ringel (1959) solved them all before the solution was known to any orientable case except case 5.

5.3.1. Some Additional Tactics

To solve the nonorientable cases of the Heawood problem, several interesting tactics are employed beyond the fundamental branched-covering-space strategy. Most of them are illustrated in Figure 5.17, which is the key to imbedding the complete graph K_{15} in the nonorientable surface $N_{\tilde{I}(15)}$.

Two of the regions in Figure 5.17 are special digons, marked by a cross in a circle. The interpretation of such a region is that the digon is to be excised and its two boundary edges matched together so that the plus directions coincide. This operation adds a crosscap to the surface. It is just like making the usual boundary-matching operation that converts a disk into a projective plane, but now viewed "inside-out". Thus the length of the boundary walk for the region

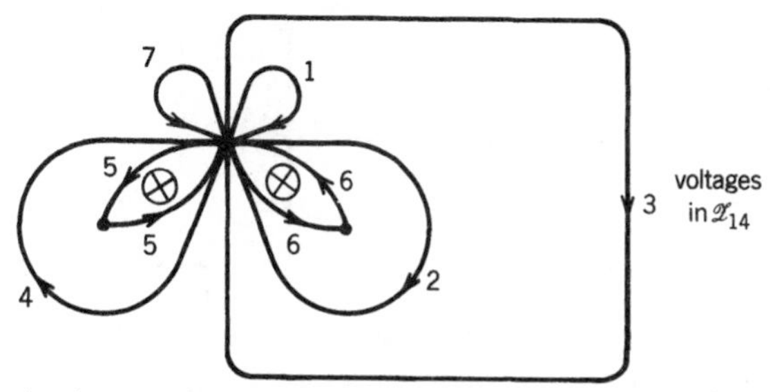

Figure 5.17. A voltage graph in N_2 that ultimately yields a triangular imbedding $K_{15} \to N_{22}$.

on the other side of the digon boundary remains the same as before the matching.

The two crosscaps in Figure 5.17 convert the apparently spherical imbedding surface in which the figure is drawn into the Klein bottle N_2. The edge identifications convert the graph into the bouquet B_7. Four of the regions are three-sided and satisfy *KVL* in the group $\mathscr{Z}_{14}$. In particular, $4 + 5 + 5 = 14$ and $2 + 6 + 6 = 14$. There are also two monogons, one with boundary voltage 7, the other with boundary voltage 1.

The derived graph is K_{14} with a doubled 1-factor, and the derived imbedding has $56 = 14 \cdot 4$ triangular faces, seven digons, and one 14-sided face. The Euler characteristic of the derived surface is $14 - 98 + 64 = -20$. Since the even numbers $0, 2, 4, 6, 8, 10, 12$ form the only subgroup of index 2 in $\mathscr{Z}_{14}$, and since the base edge carrying voltage 6 is orientation-reversing, the derived surface is nonorientable, by Theorem 4.1.4.

Excising the seven digons and closing up the holes converts the derived imbedding into an imbedding $K_{14} \to N_{22}$ such that every face is triangular, except for one 14-sided face whose boundary circuit is a Hamiltonian cycle. Insert a 15th vertex into the interior of that face and adjoin it to each of the other 14 vertices. The result is a triangular imbedding $K_{15} \to N_{22}$.

5.3.2. Nonorientable Current Graphs

In Section 4.4 we observed that assigning currents to a nonorientably imbedded graph does not automatically yield a well-defined derived imbedding. However, Youngs (1968a, b) established some combinatorial conventions that make it well-defined for certain restricted types. Youngs used these to give a current-graph explanation of Ringel's solution (1959) to the nonorientable cases. Figure 5.18 shows the Youngs nonorientable current graph that corresponds to the voltage graph in Figure 5.17.

The vertices of valence 2 in a Youngs nonorientable current graph are not topologically dual to a digon in the voltage graph. One of the edges with such an endpoint must have valence one at its other endpoint. That edge crosses the

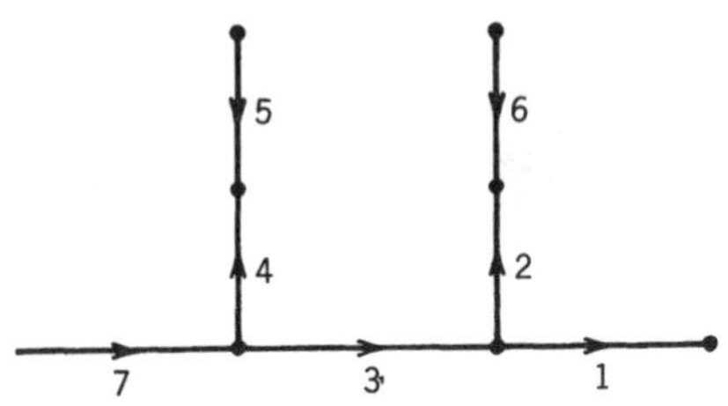

Figure 5.18. The Youngs nonorientable current graph to produce the imbedding $K_{15} \to N_{22}$.

"warp" in the surface indicated by an encircled cross in a drawing of a nonorientable voltage graph imbedding. The "warp" is where a digon was excised and the hole closed up with a crosscap matching.

Another convention used by Gustin, Ringel, and Youngs occurs in both orientable and nonorientable current graphs. If an edge has valence 1 at one endpoint and current of order 2, then the dot representing that endpoint is omitted. Under the topological interpretation of this combinatorial notation, the absent dot reminds the reader to excise all the digons that lie over the monogon dual to that endpoint, and to close up the holes in the surface by matching the two sides together.

5.3.3. Nonorientable Cases 3 and 7

The example in Figure 5.17 is easily generalized to a system of voltage graphs that solves nonorientable case 3. The triangular imbedding $K_{12s+3} \to N_{\bar{I}(12s+3)}$ is always obtained by first imbedding K_{12s+2} into $N_{\bar{I}(12s+3)}$ so that the boundary circuit of one face is a hamiltonian cycle and so that the other faces are all three-sided. The imbedding $K_{12s+2} \to N_{\bar{I}(12s+3)}$ is obtained by excising digons from an imbedding of K_{12s+2} with a doubled 1-factor in $N_{\bar{I}(12s+3)}$.

The imbedding of K_{12s+2} with a doubled 1-factor is derived from a voltage assignment in $\mathscr{Z}_{12s+2}$ to the bouquet B_{6s+1}. A monogon boundary loop carries the voltage $6s + 1$ of order 2, so that the monogon will lift to $6s + 1$ digons. Another monogon boundary carries the voltage 1, so that that monogon will lift to a $(12s + 2)$-sided face whose boundary circuit is hamiltonian. All of the other $4s$ faces are three-sided. The base surface is N_{2s}. Figure 5.19 shows the general Youngs current graph for nonorientable case 3.

The $2s$ vertices of valence 2 correspond to $2s$ crosscaps in the base surface. One observes that the odd currents $1, 3, \ldots, 4s - 1$ are assigned by coiling outward from $2s + 1$, from left to right in the row of horizontal edges.

The details of nonorientable case 7 are remarkably similar to the ones for nonorientable case 3. Figure 5.20 shows the general Youngs current graph. It cannot be used on the anomolous graph K_7, since $\bar{\gamma}(K_7) = 3 > \bar{I}(7) = 2$.

One difference in the solution to case 7 is that the vertical edge at the extreme left carries the current $4s + 2$, which has order 3 in $\mathscr{Z}_{12s+6}$. Thus, the monogon dual to its endpoint of valence 1 lifts to $4s + 2$ three-sided faces.

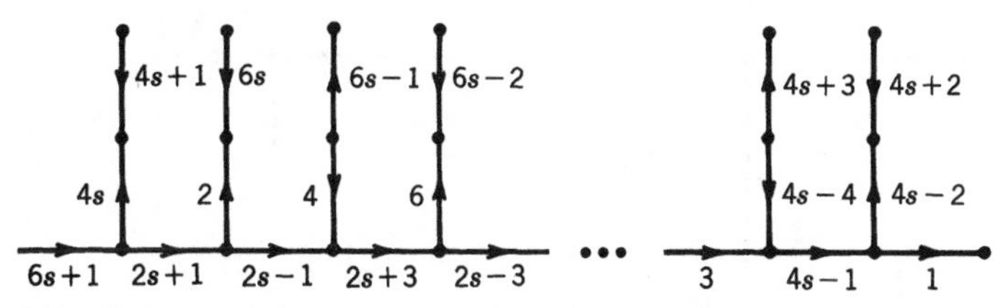

Figure 5.19. The Youngs nonorientable current graph that yields $K_{12s+3} \to N_{\bar{I}(12s+3)}$. The currents are in $\mathscr{Z}_{12s+2}$.

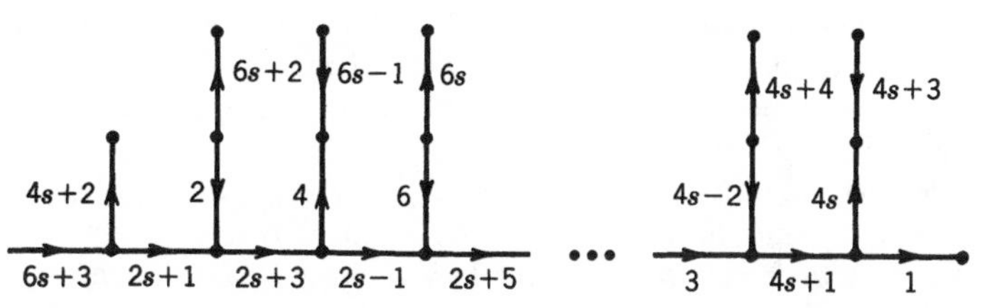

Figure 5.20. The Youngs current graph for nonorientable case 7. The current group is $\mathscr{Z}_{12s+6}$.

5.3.4. Nonorientable Case 0

To obtain an imbedding $K_{12s} \to N_{\bar{I}(12s)}$, one first derives an imbedding $K_{12s-1} \to N_{I(12s)}$ such that one face is $(12s - 1)$-sided and the others are three-sided. In that sense, nonorientable case 0 resembles nonorientable cases 3 and 7.

For instance, to imbed K_{24} into N_{70}, we start with K_{23}, which has 23 vertices and $23 \cdot 11$ edges. The kind of imbedding we seek in N_{70}, has one 23-sided face, whose boundary is a hamiltonian cycle, and $23 \cdot 7$ three-sided faces. Thus, we hope to derive it from a voltage assignment in $\mathscr{Z}_{23}$ on the bouquet B_{11}, imbedded in the nonorientable surface N_4. Figure 5.21 shows that this is possible.

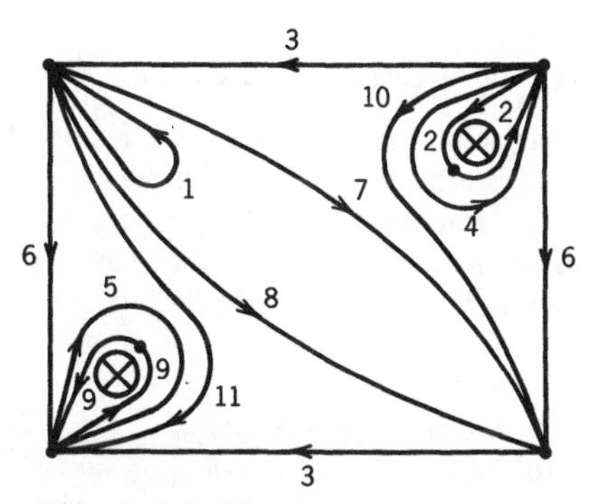

Figure 5.21. An imbedding $B_{11} \to N_4$ with voltages in $\mathscr{Z}_{23}$.

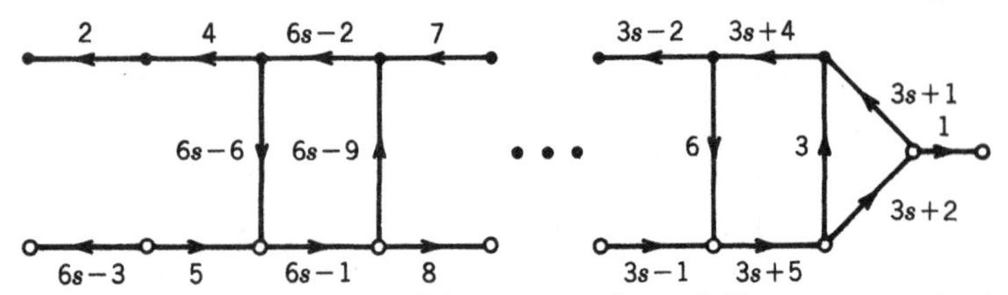

Figure 5.22. The Youngs current graph for nonorientable case 0. The current group is $\mathscr{Z}_{12s-1}$.

The rectangular sheet in Figure 5.21 is made into a torus in the usual manner, by matching opposite sides together. Then two crosscaps are inserted, to obtain N_4. The Kirchhoff voltage law holds on every face except the monogon.

In general, to derive the imbedding $K_{12s} \to N_{\bar{I}(12s)}$, the base graph is the bouquet B_{6s-1}, and the base surface is N_{2s}, formed by inserting two crosscaps into the orientable surface S_{s-1}. The base imbedding has one monogon and $4s-1$ triangular faces. The voltages $3, 6, \ldots, 6s-6$ coil among the voltages $4, 7, 10, \ldots, 6s-2$ and also among the voltages $5, 8, 11, \ldots, 6s-1$. The Youngs current graph is shown in Figure 5.22.

5.3.5. Nonorientable Case 4

Nonorientable case 4 is slightly different from nonorientable cases 0, 3, and 7, because the base surface has an odd number of crosscaps. Starting with the complete graph K_{12s+4}, one adds a 1-factor to create the possibility of a quotient graph with one vertex. With the 1-factor added, there are $(6s+2)(12s+4)$ edges, and the derived imbedding has $6s+2$ digons and $(6s+2)(8s+2)$ triangular faces. Taking a $(12s+4)$-fold quotient, the base graph has one vertex, $6s+2$ edges, one monogon, and $4s+1$ triangular faces, for an Euler characteristic of $1-2s$. Thus, the base surface is N_{2s+1}. Figure 5.23 shows the voltage graph for the case $s=2$.

5.3.6. About Nonorientable Cases 1, 6, 9, and 10

We omit details of the solutions for nonorientable cases 1, 6, 9, and 10. For nonorientable case 9, Ringel (1974) uses a three-orbit current graph, which means that, as in orientable case 3, the underlying voltage graph has three vertices. For nonorientable cases 1, 6, and 10, Ringel (1974) uses inductive constructions. Cases 6 and 10 are symbiotically intertwined.

5.3.7. Exercises

1. Using Figure 5.19, draw a voltage graph from which a triangular imbedding $K_{27} \to N_{92}$ can be obtained.

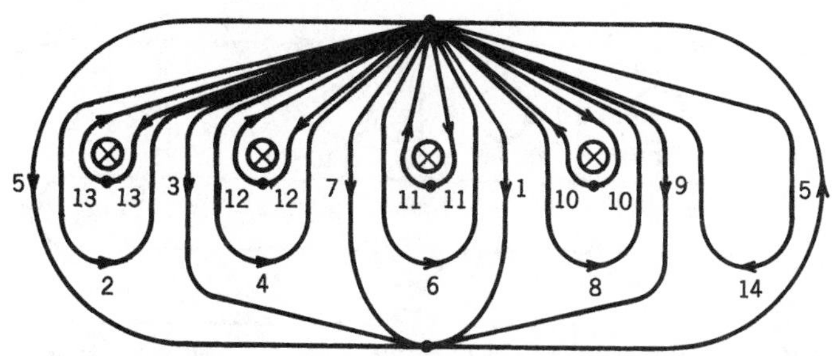

Figure 5.23. An imbedding $B_{14} \to N_5$ with voltages in $\mathscr{Z}_{28}$. The derived graph yields an imbedding $K_{28} \to N_{100}$.

2. For the current graph of Figure 5.20, calculate exactly how many faces of each size occur in the derived graph.

5.4. ADDITIONAL ADJACENCIES FOR IRREGULAR CASES

In orientable cases 1, 2, 5, 6, 8, 9, 10, and 11 and in nonorientable cases 2, 5, 8, and 11, the minimum imbedding cannot be a triangulation, as we recall from Tables 5.3 and 5.4. The solution strategy is the same for nearly all these cases. First, a slightly smaller graph than K_n is imbedded in a slightly less complicated surface than $S_{I(n)}$ or $S_{\bar{I}(n)}$. Then additional surface complications are added in order to permit the slightly smaller graph to be built up into K_n.

5.4.1. Orientable Case 5

The simplest example of an additional-adjacency solution occurs in case 5. As a preliminary, we observe that the complete graph K_5 has one too many edges to be imbedded in the sphere S_0. That is, the graph $K_5 - e$ already triangulates S_0, and there is no room for an additional edge between the two nonadjacent vertices. However, by adding a handle that runs from any face whose boundary contains one of those vertices to any face whose boundary contains the other, and by then running the missing edge across that handle, we achieve a genus imbedding $K_5 \to S_1$.

In general, we shall discover that the graph $K_{12s+5} - e$ has a triangular imbedding in the surface $S_{I(12s+5)-1}$. By consideration of fiber sizes and orbit sizes, we see immediately that the graph $K_{12s+5} - e$ has no quotients (see Exercise 1). It follows that we cannot hope to construct an imbedding

$$K_{12s+5} - e \to S_{I(12s+5)-1}$$

as a branched covering. However, consider what we would obtain if we deleted

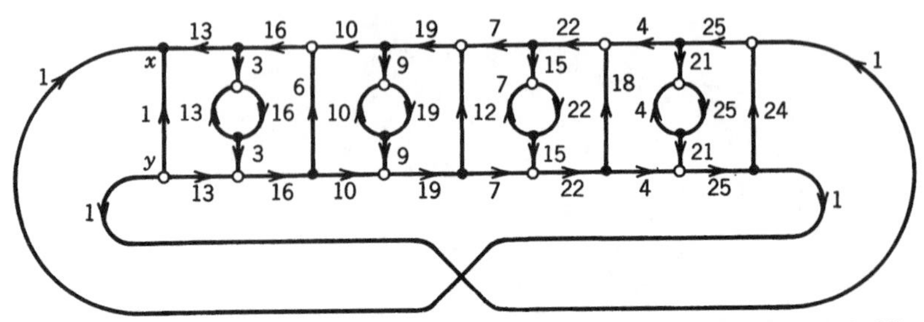

Figure 5.24. A three-orbit $\mathscr{Z}_{27}$-current graph that ultimately yields a triangular imbedding $K_{29} - K_2 \to S_{54}$.

the two $(12s + 3)$-valent vertices from such an imbedding—an imbedding of the complete graph K_{12s+3} into $S_{I(12s+5)-1}$ such that two face-boundaries are Hamiltonian circuits and all other faces are three-sided.

More specifically, such an imbedding

$$K_{12s+3} \to S_{I(12s+5)-1}$$

would be for a graph with $12s + 3$ vertices and $(12s + 3)(6s + 1)$ edges. There would be two $(12s + 3)$-gons and $(12s + 3) \cdot 4s$ triangles. One might hope that it covers an imbedded voltage graph with one vertex, $6s + 1$ edges, $4s$ triangles, and two monogons. Unfortunately, no such 1-vertex voltage graph or corresponding 1-orbit current graph for case 5 is known. The imbedding is achieved, instead, with a 3-orbit current graph closely related to case 3, as illustrated for K_{20} in Figure 5.24, in which x and y denote the vertices with excess current.

Once we have the imbedding $K_{12s+3} \to S_{I(12s+5)-1}$, we install a vertex in the interior of the two complete $(12s + 3)$-gons and adjoin it to each of the $12s + 3$ vertices on the region boundary. This yields an imbedding of $K_{12s+5} - e$ into $S_{I(12s+5)-1}$. An additional handle is added for the missing edge, thereby achieving our goal, an imbedding $K_{12s+5} \to S_{I(12s+5)}$.

5.4.2. Orientable Case 10

To solve orientable case 10 with additional adjacencies, we must go a little further out on a limb than in case 5, before we can crawl back. This time, as a preliminary, we notice that K_{10} has three too many edges to be imbedded in $S_{I(10)} = S_4$. If we could imbed $K_{10} - K_3$ into S_3 and then restore the three missing edges with only one handle, we would be done. Restoring all three edges of a missing 3-cycle with only one handle is not always possible, so special attention will be required.

The graph $K_{10} - K_3$ has 10 vertices and 42 edges. If it triangulated its imbedding surface, there would be 28 faces, from which it follows that the

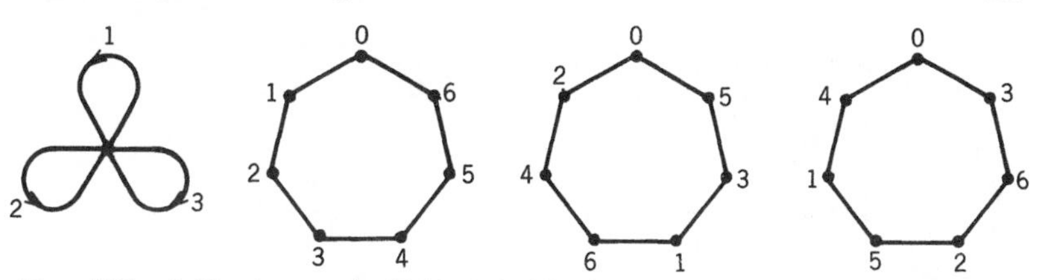

Figure 5.25. A $\mathscr{Z}_7$-voltage graph with base imbedding $B_3 \to S_0$ and derived imbedding $K_7 \to S_3$. The three complete 7-gons of the derived imbedding are shown at the right.

surface would have Euler characteristic $10 - 42 + 28 = -4$. If the imbedding surface were orientable, it would be S_3. Since the graph $K_{10} - K_3$ has no quotients (see Exercise 1), we delete the three 7-valent vertices from a hypothetical imbedding $K_{10} - K_3 \to S_3$ and thereby obtain an imbedding $K_7 \to S_3$ with three complete 7-gons and seven 3-gons. (A "complete" face boundary has one occurrence of every vertex.)

Such an imbedding of K_7 into S_3 can be derived from a voltage graph with one vertex, three edges, one 3-gon, and three 1-gons, as illustrated in Figure 5.25.

New vertices, labeled x, y, and z, are inserted into the interiors of the 7-gons covering the monogons with voltages 1, 2, and 3, respectively, and adjoined to the old vertices on their 7-gon boundaries. The result is a triangular imbedding of the graph $K_{10} - K_3$ into the surface S_3. Figure 5.26 illustrates the neighborhood of the vertex 0 in this imbedding.

Our objective is to add the 3-cycle x, y, z with the aid of only one additional handle. The handle itself is implemented by identifying the boundary of a hole in the 3-gon $01x$ to the boundary of a hole in the 3-gon 054. Next, the edges 04 and 05 are deleted, so that the vicinity of the vertex 0 becomes as illustrated in Figure 5.27. Figure 5.28 shows how to install the five edges xy, xz, yz, 04, and 05 so that the result is an imbedding $K_{10} \to S_4$.

In order to keep track of the four new edges that cross the hole boundary, we have marked them with the symbols

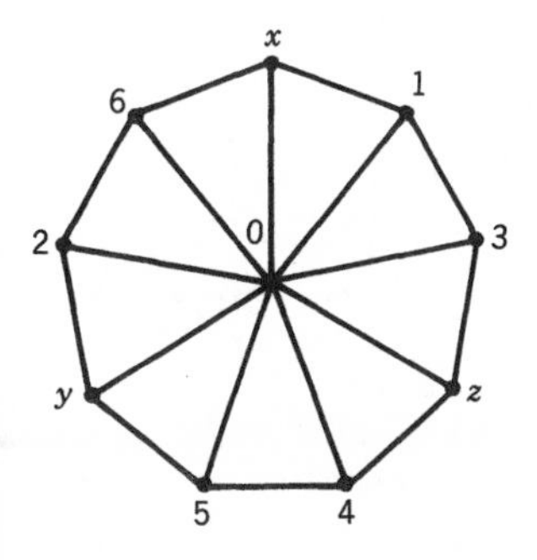

Figure 5.26. The vicinity of the vertex 0 in the imbedding $K_{10} - K_3 \to S_3$.

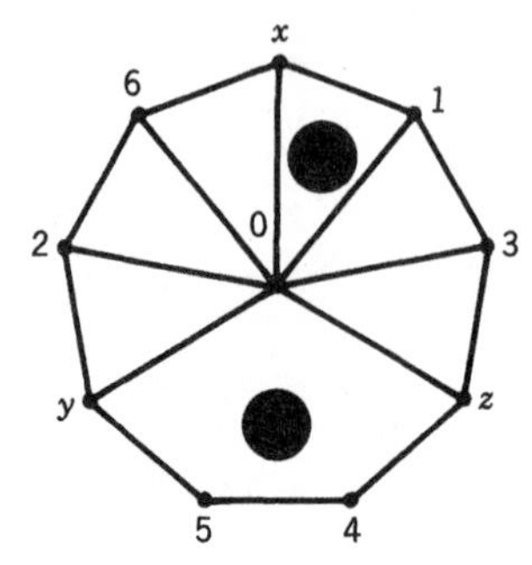

Figure 5.27. An imbedding of $(K_{10} - K_3) - \{04, 05\}$ into the surface S_4, in the vicinity of the vertex 0.

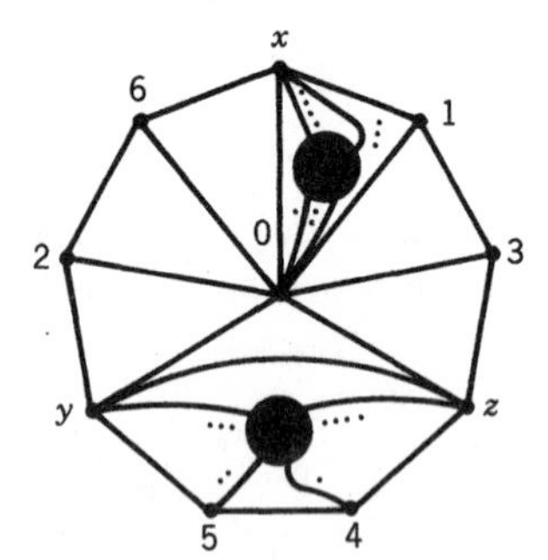

Figure 5.28. An imbedding $K_{10} \to S_4$, in the vicinity of the vertex 0.

Whereas this sequence of symbols appears in clockwise order on the lower hole in Figure 5.28, it appears in counterclockwise order on the upper hole. This corresponds to an identification of hole boundaries that ensures orientability of the resulting closed surface.

The general solution for orientable case 10 is very much like the solution for K_{10} itself. The strategy is first to imbed $K_{12s+10} - K_3$ into $S_{I(12s+10)-1}$ triangularly, and then to use a single new handle to create room to insert the edges of the missing 3-cycle. The derived imbedding associated with the current graph in Figure 5.29 is the graph K_{12s+7} in the surface $S_{I(12s+10)}$ with three complete $(12s + 7)$-gons and all other faces three-sided. We see that this current graph is closely related to the one used for orientable case 7.

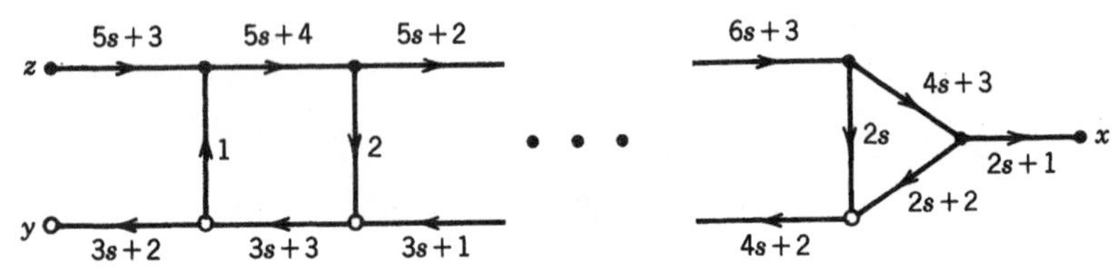

Figure 5.29. A $\mathscr{Z}_{12s+7}$-current graph used in the general solution to orientable case 10. The vertices with excess currents are labeled x, y, and z.

5.4.3. About the Other Orientable Cases

Although the original solution to case 9 was quite complicated, Jungerman (1974) ultimately designed an easier approach, which Ringel (1974) has reproduced. First, the complete graph K_{12s+8} is imbedded in the surface $S_{I(12s+9)-1}$ so that there are two $(6s+4)$-gons and all other faces are three-sided. The boundary circuits of the two $(6s+4)$-gons together form a 2-factor. Next, a handle is installed so that it runs from the interior of one $(6s+4)$-gon to the interior of the other, and a $(12s+9)$th vertex is placed on the handle. It is possible to adjoin the new vertex to every vertex of K_{12s+8}, which thereby creates an imbedding $K_{12s+9} \to S_{I(12s+9)}$.

Orientable case 1 is solved by combining the methods already used in cases 4 and 10. Using currents in $\mathscr{Z}_2 \times \mathscr{Z}_{6s-1}$, an imbedding of K_{12s-2} into $S_{I(12s+1)-1}$ is obtained, in which there are three $(12s-2)$-gons and all other faces are 3-sided. Edges are added to create an imbedding

$$K_{12s+1} - K_3 \to S_{I(12s+1)-1}$$

and then a handle is added to create room for the missing 3-cycle.

The rotation graph for orientable case 6 is a fascinating variation on the one for case 3, and the assignment of currents presents interesting new difficulties. Analogous to case 10, the near-final objective is a triangular imbedding

$$K_{12s+6} - K_3 \to S_{I(12s+6)-1}$$

Orientable cases 11, 2, and 8 require special ingenuity. In case 11, the branched covering space is a mostly triangular imbedding of K_{12s+6}, which has five exceptional faces, each a complete $(12s+6)$-gon. Figure 5.30 shows the current graph used when $s = 2$ and $12s + 11 = 35$. When new vertices are inserted into the $(12s+6)$-gons and adjoined to the vertices on the boundary cycles, the result is a triangular imbedding

$$K_{12s+11} - K_5 \to S_{I(12s+11)-2}$$

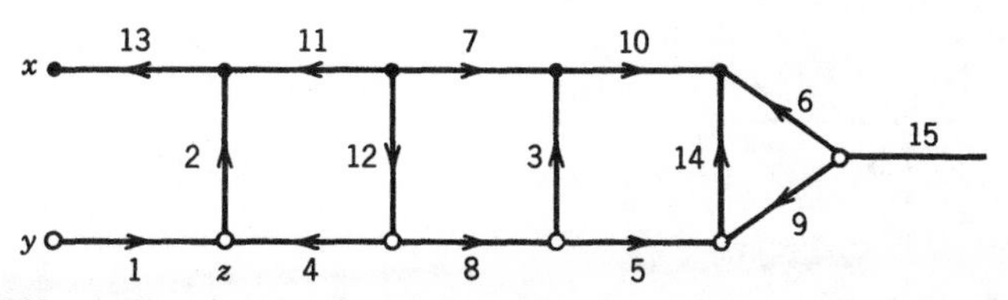

Figure 5.30. A $\mathscr{Z}_{30}$-current graph used to construct the imbedding $K_{35} \to S_{I(35)}$ from case 11. The vertices with excess currents are labelled x, y, and z.

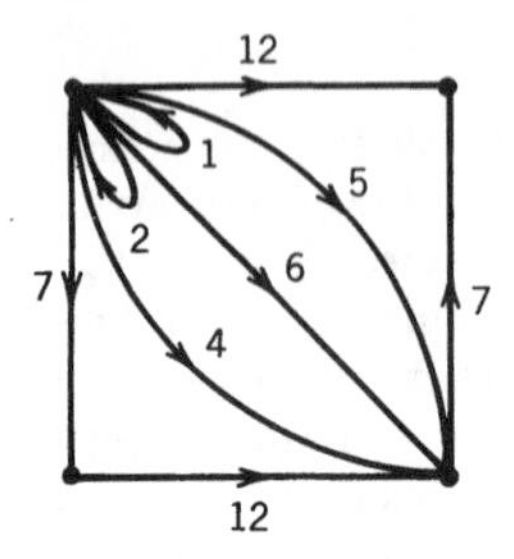

Figure 5.31. A $\mathscr{Z}_{15}$-voltage graph imbedding $B_7 \to N_2$ with derived imbedding $K_{15} \to N_{\bar{I}(17)-1}$. The two monogons are covered by complete 15-gons.

Jsing two handles to create room to add the missing ten edges of K_5 is ubstantially more difficult than the analogous problem of using one handle in ase 10 to add K_3. Cases 2 and 8 are like case 11 in some important ways, but ıey include complicated additional irregularities.

5.4.4. Nonorientable Case 5

Nonorientable case 5 provides a simple illustration of a nonorientable additional adjacency problem. Analogous to orientable case 5, the voltage graph problem is to derive an imbedding

$$K_{12s+3} \to N_{\bar{I}(12s+5)-1}$$

with two complete $(12s + 3)$-gons and all other faces three-sided. Figure 5.31 contains the voltage graph to be used when $s = 1$ and $K_{12s+5} = K_{17}$.

A new vertex is inserted into the interior of each of the two complete $(12s + 3)$-gons and adjoined to each vertex of the $(12s + 3)$-gon boundary. The result is a triangular imbedding

$$K_{12s+5} - K_2 \to N_{\bar{I}(12s+5)-1}$$

Figure 5.32 illustrates the vicinity of the vertex 0, where we have assigned x as the label on the new vertex installed into the 15-gon covering the monogon

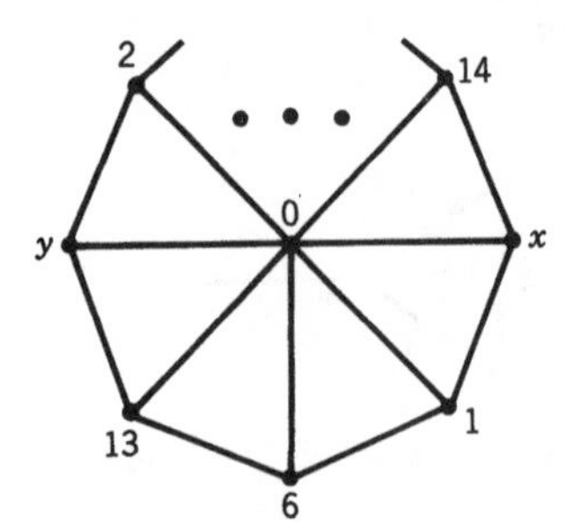

Figure 5.32. The vicinity of the vertex 0 in the imbedding $K_{17} - K_2 \to N_{\bar{I}(17)-1}$.

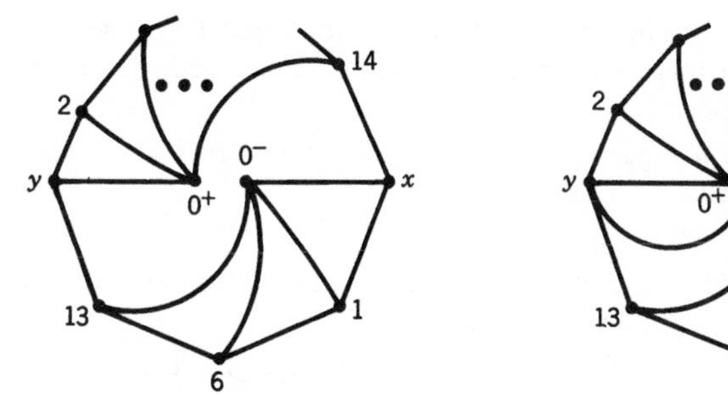

Figure 5.33. Since the vertices x and y have vertex 0 as a common neighbor, there is a way to adjoin them if only one more crosscap is added.

with voltage 1, and y as the label on the other new vertex, for the special case of K_{17}.

We shall now demonstrate that only one additional crosscap is needed to create room for the missing edge xy. Figure 5.33 illustrates the appropriate surgical procedure. First, split the vertex 0 into the two vertices 0^+ and 0^-, so that the vertex x and all the vertices on one side of the path $y0x$ are adjoined to 0^-, whereas the vertex y and all the vertices on the other side of the path $y0x$ are adjoined to 0^+.

Next, delete the interior of a disk whose boundary passes through both 0^+ and 0^- and parametrize that boundary by the unit circle so that 0^+ and 0^- are antipodal points. Also, draw edges from x and y, respectively, to another pair of antipodal points. Finally, reclose the surface by identifying all pairs of antipodal points on the parametrized boundary. The antipodal identification adds a crosscap, it completes an edge between x and y, and it reunites vertices 0^+ and 0^- into a single vertex 0 adjacent to every other vertex. Thus, the result is an imbedding $K_{17} \to N_{\bar{I}(17)}$. This technique easily generalizes to all of nonorientable case 5.

5.4.5. About Nonorientable Cases 11, 8, and 2

Nonorientable case 11 is quite similar to nonorientable case 5. Nonorientable case 8 is slightly more difficult, in that the derived imbedding contains two $(n/2)$-gons and one n-gon, rather than two n-gons. Nonorientable case 2 is solved inductively, rather like regular nonorientable cases 1, 6, and 10.

5.4.6. Exercises

1. Prove that graphs of the form $K_{12s+5} - e$ or of the form $K_{12s+10} - K_3$ have no nontrivial quotients.

2. Trace the edge orbits in Figure 5.24, in order to demonstrate that there are three orbits and that each nonzero current occurs exactly once on each orbit.
3. Use the pattern of Figure 5.24 to develop a current graph for the general solution of orientable case 5.
4. Draw the imbedded voltage graph used for K_{22}.
5. Use Figures 5.28 and 5.29 to construct a general solution to the additional-adjacency problem for case 10.
6. Describe the face-size distribution for the voltage graph dual to Figure 5.30 and the order of the net voltage on each face.

6

The Genus of a Group

In calculating the genus of the complete graph K_n, we represented K_n as a Cayley graph for the cyclic group $\mathscr{Z}_n$. By way of generalization, one might hope that similar methods can be used to compute the genus of Cayley graphs for other groups. As we saw in Chapter 5, the difficulties encountered even for cyclic groups were numerous. Thus, it is natural to suspect that calculating the genus of Cayley graphs for noncyclic groups might be an intractable problem, unless the choice of generating set is somehow restricted. Accordingly, in this chapter, we restrict our study to the case in which the generating set X is "irredundant", by which we mean that no proper subset of X also generates the group. In fact, to diminish the role of generating set even more, we shall focus on determining the minimum genus over all Cayley graphs for a given group $\mathscr{A}$, which White (1972) has called the genus of the group, and denoted by $\gamma(\mathscr{A})$.

One consequence of this concentration on the group itself rather than the generating set is that one is led inevitably to "highly symmetric" imbeddings, for which the group action on the Cayley graph extends to the imbedded surface. Surprising and intricate relationships are revealed between the genus of a group and the minimum genus surface on which the group acts. The result is a complex interplay of topology, combinatorics, and algebra.

Levinson (1970) has proved that the genus of an infinite group is either 0 or infinite (see Exercises 13 and 14). Accordingly, the emphasis is on the genus of finite groups.

6.1. THE GENUS OF ABELIAN GROUPS

Abelian groups are an obvious first target in a general attack on the genus of a group. Indeed, White (1972) initiated the modern study of the genus of a group by computing the genus of certain abelian groups, using methods discussed in Section 3.5. Jungerman and White (1980), whose work is now described, subsequently succeeded in determining the genus of "most" other abelian groups.

Before we proceed, a comment about generating sets is in order. A generating set X for a group $\mathscr{A}$ can be irredundant without being the

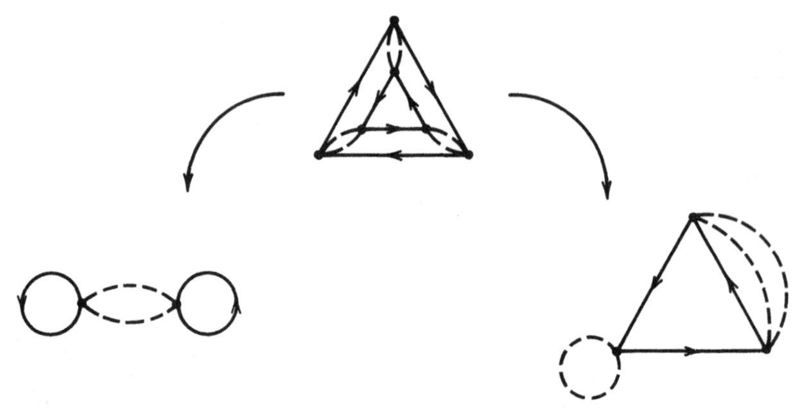

Figure 6.1. The Cayley graph $C(\mathcal{S}_3, X)$ and two quotients. Edges corresponding to generator (1 2 3) are solid, and generator (1 2), dashed.

minimum in size over all generating sets for $\mathscr{A}$. For example, consider the group $\mathscr{A} = \mathscr{Z}_6$ and the irredundant generating set $\{2, 3\}$. If the generating set X has "minimum" size, then the number $\#X$ is called the "rank" of the group $\mathscr{A}$.

In computing the genus of a group, one need consider only irredundant generating sets, because discarding extra generators serves only to delete edges from the associated Cayley graph. Obviously, deleting edges cannot increase the genus of a graph. On the other hand, as we shall see, it is possible that the genus of a group is not attainable by using any of its minimum-size generating sets. (See Exercise 6.4.16.) This was first observed by Sit (1980), in response to a problem of Gross and Harary (1980).

6.1.1. Recovering a Cayley Graph from Any of Its Quotients

For many applications, it is helpful to know how to assign voltages to a given quotient of a given Cayley graph, so as to recover the Cayley graph. For example, consider the full symmetric group $\mathcal{S}_3$ and let X be the generating set consisting of the permutations (1 2 3) and (1 2). The Cayley graph $C(\mathcal{S}_3, X)$ is depicted in the center of Figure 6.1. On the left is the quotient graph G under the cyclic action on $C(\mathcal{S}_3, X)$ generated by (1 2 3). On the right is the quotient graph H under the cyclic action on $C(\mathcal{S}_3, X)$ generated by (1 2). Observe that since the subgroup of $\mathcal{S}_3$ generated by (1 2 3) is normal, the graph G is itself a Cayley graph for the group $\mathscr{Z}_2$, with generating set $\phi(X)$, where $\phi\colon \mathcal{S}_3 \to \mathscr{Z}_2$ is the canonical quotient map. On the other hand, since the subgroup in $\mathcal{S}_3$ generated by (1 2) is not normal, the graph H is not a Cayley graph, but a Schreier coset graph. In either case, the problem is to assign voltages to the quotient so as to recover the original Cayley graph $C(\mathcal{S}_3, X)$.

It is tempting simply to assign the voltage x to each edge in the quotient graph that is the image under the quotient map of an edge colored x in the Cayley graph $C(\mathscr{A}, X)$. In the present example, the derived graph for such a voltage assignment on base graph G would have 12 vertices, and on base graph H, 18 vertices. If this assignment is to work, it is necessary that the derived graph consist of n copies of the original Cayley graph, where n is the number of vertices in the quotient graph. The following theorem establishes such a property. By the "natural" voltage assignment on the Schreier coset graph $S(\mathscr{A}: \mathscr{B}, X)$, we mean the map into the group $\mathscr{A}$ that assigns the voltage x to each directed edge colored x for every generator $x \in X$.

Theorem 6.1.1. *The natural voltage assignment α on the Schreier coset graph $G = S(\mathscr{A}: \mathscr{B}, X)$ yields a derived graph with $\#[\mathscr{A}: \mathscr{B}]$ components, each isomorphic to the Cayley graph $C(\mathscr{A}, X)$.*

Proof. Suppose that $\mathscr{B}_1 = \mathscr{B}, \mathscr{B}_2, \ldots, \mathscr{B}_n$ are the right cosets of $\mathscr{B}$. Then the $n\#\mathscr{A}$ vertices of the derived graph G^α are the pairs $(\mathscr{B}_i, a)$, where $a \in \mathscr{A}$ and $1 \leq i \leq n$. There is an edge from the vertex $(\mathscr{B}_i, a)$ to the vertex $(\mathscr{B}_j, b)$ if and only if there is a generator $x \in X$ such that $\mathscr{B}_j = \mathscr{B}_i x$ and $b = ax$. We define the vertex part of a graph map f from the Cayley graph $C(\mathscr{A}, X)$ to the derived graph G^α by the rule $f(a) = ([a], a)$, where $[a]$ denotes the right coset of $\mathscr{B}$ that contains a. Also, let f take the edge of $C(\mathscr{A}, X)$ from a to ax to the edge in G^α from $([a], a)$ to $([a]x, ax)$. Since $[a]x = [ax]$, we infer that the resulting map is a graph map. It clearly takes the set of directed edges originating at the vertex a of $C(\mathscr{A}, X)$ one-to-one onto the set of directed edges originating at the vertex $([a], a)$ of G^α. Therefore, f is a covering map from $C(\mathscr{A}, X)$ onto a component of G^α. Since f is also one-to-one on the whole vertex set of $C(\mathscr{A}, X)$, it follows that f is a graph isomorphism onto that component. The theorem follows, since the components of the derived graph of an ordinary voltage graph are mutually isomorphic. □

Corollary 1. *Let X be a generating set for a group $\mathscr{A}$, and ϕ a homeomorphism from $\mathscr{A}$ onto the group $\mathscr{Q}$. Let α be the voltage assignment for the Cayley graph $C(\mathscr{Q}, \phi(X))$ that gives a directed edge colored $\phi(x)$ the voltage x for each $x \in X$. Then the derived graph consists of $\#\mathscr{Q}$ copies of the Cayley graph $C(\mathscr{A}, X)$.*

Proof. Apply Theorem 6.1.1 with the kernel of ϕ as the subgroup $\mathscr{B}$, so that the Schreier coset graph $S(\mathscr{A}: \mathscr{B}, X)$ is isomorphic to the Cayley graph $C(\mathscr{Q}, \phi(X))$. □

Corollary 2. *Let X be a generating set for the group $\mathscr{A}$. Suppose that $\langle G \to S, \alpha \rangle$ is an oriented current graph with current group $\mathscr{A}$ satisfying these* *conditions*:

i. *In each directed face boundary there is exactly one edge assigned the current x and one the current x^{-1} for every generator $x \in X$, and all currents or their inverses are in X.*

ii. *There is a subgroup $\mathscr{B}$ contained in $\mathscr{A}$ and a one-to-one correspondence between the faces of the imbedding $G \to S$ and the cosets of $\mathscr{B}$, such that if the edge e lies in faces corresponding to cosets $\mathscr{B}_1$ and $\mathscr{B}_2$, then $\mathscr{B}_1\alpha(e^{\epsilon}) = \mathscr{B}_2$, where e^{ϵ} is the direction of e given by the directed boundry of the face for $\mathscr{B}_1$.*

Then each component of the derived graph G_α is isomorphic to the Cayley graph $C(\mathscr{A}, X)$.

Proof. Conditions i and ii imply that the dual graph for the imbedding $G \to S$ is the Schreier coset graph $S(\mathscr{A}: \mathscr{B}, X)$ and that the dual voltage assignment is the natural one. The corollary then follows from Theorem 6.1.1. □

The following two examples further develop the quotients of Figure 6.1 in the context of Theorem 6.1.1 and its corollaries. The first example concerns the left quotient, which is regular, and the second the right quotient, which is not.

Example 6.1.1. *On the left of Figure 6.2 is a voltage graph imbedded in the sphere with voltages in the symmetric group $\mathscr{S}_3$ generated by $X = \{x, y\}$, where $x = (1\,2\,3)$ and $y = (1\,2)$. The given voltage graph is the Schreier coset graph $S(\mathscr{S}_3: \mathscr{B}, X)$ where $\mathscr{B}$ is the subgroup generated by $(1\,2\,3)$. Alternatively, since $\mathscr{B}$ is normal in $\mathscr{S}_3$ with quotient map $\phi: \mathscr{S}_3 \to \mathscr{Z}_2$ such that $\phi(x) = 0$ and $\phi(y) = 1$, the voltage graph may be viewed as the Cayley graph $C(\mathscr{Z}_2, \{0, 1\})$. The vertices have been labeled by the cosets $\mathscr{B}$ and $\mathscr{B}y$ (rather than 0 and 1) to maintain the Schreier coset viewpoint. By Theorem 6.1.1, the derived imbedding for this voltage graph consists of two copies of the Cayley graph $C(\mathscr{S}_3, X)$. If one prefers to view the voltage graph as a Cayley graph, use the first corollary to Theorem 6.1.1 instead. On the right of Figure 6.2 is the dual current graph. The*

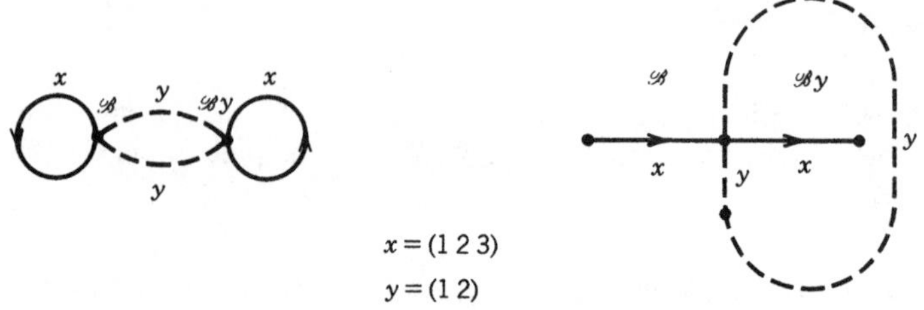

Figure 6.2. An imbedded voltage graph and its dual current graph.

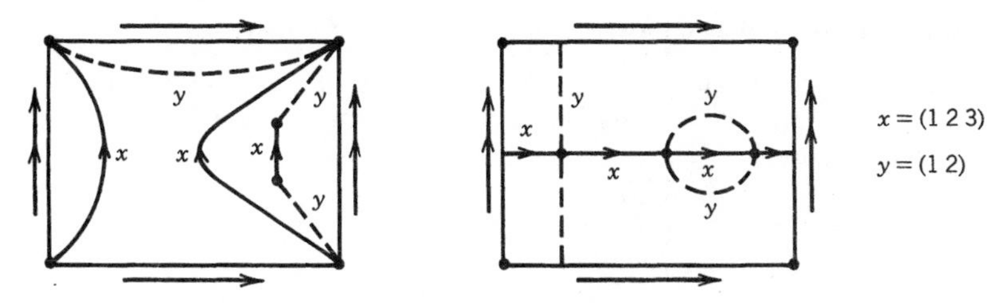

Figure 6.3. A current graph and its dual voltage graph.

reader should verify that the edge and face labels satisfy conditions i *and* ii *of the second corollary.*

Example 6.1.2. *On the left of Figure* 6.3 *is a current graph imbedded in the torus with currents in the symmetric group* $\mathscr{S}_3$ *generated by* $X = \{x, y\}$, *where* $x = (1\,2\,3)$ *and* $y = (1\,2)$. *Clearly condition* i *of the second corollary to Theorem* 6.1.1 *is satisfied. If the faces are labeled as illustrated, by cosets of the subgroup* $\mathscr{B}$ *generated by the permutation* $(1\,2)$, *then condition* ii *is also satisfied. Thus the derived graph for this current graph consists of two copies of the Cayley graph* $C(\mathscr{S}_3, X)$. *On the right of Figure* 6.3 *is the dual voltage-graph imbedding, which the reader should recognize as the right quotient in Figure* 6.1, *the Schreier coset graph* $S(\mathscr{S}_3\colon \mathscr{B}, X)$.

The information contained in an imbedded voltage graph or current graph like those in Examples 6.1.1 and 6.1.2 can be coded in tabular form. Since the voltages uniquely identify each directed edge incident on a given vertex, it follows that, to give the rotation system for such an imbedded voltage graph, it suffices to list the cyclic ordering of voltages at each vertex, where each vertex is identified by a coset of the subgroup $\mathscr{B}$. For example, the imbedded voltage graph of Example 6.1.2 is given by the following table, in which we adopt the notations $x = (1\,2\,3)$ and $y = (1\,2)$ of the figures:

$\mathscr{B}$	x	y	x^{-1}	y^{-1}
$\mathscr{B}x$	x	y	x^{-1}	y^{-1}
$\mathscr{B}x^2$	x	y^{-1}	x^{-1}	y

In terms of the dual current graph, this table gives the currents encountered in a trip around each face boundary. In the early literature, these are sometimes called the "logs of the circuits". In effect, the table of logs is just a condensed rotation system for the derived imbedding of the Cayley graph $C(\mathscr{A}, X)$. That is, edges incident on a vertex are identified by their "color" x or x^{-1}, $x \in X$, and the rotation at vertex a, where a is in coset $\mathscr{B}_i$, is simply the log for circuit $\mathscr{B}_i$.

6.1.2. A Lower Bound for the Genus of Most Abelian Groups

The classification theorem for finite abelian groups (e.g. see MacLane and Birkhoff (1967)) states that any finite abelian group $\mathscr{A}$ of rank r has a unique "canonical form"

$$\mathscr{Z}_{m_1} \times \cdots \times \mathscr{Z}_{m_r}$$

such that m_i divides m_{i+1} for $i = 1, \ldots, r-1$, and $m_1 > 1$. Each element of $\mathscr{A}$ is written as an r-tuple whose coordinates correspond to the canonical product form. The "canonical generating set" for $\mathscr{A}$ consists of the r elements that have 1 as one coordinate and 0 for all others.

Cyclic groups $\mathscr{Z}_n$ and abelian groups of the canonical form $\mathscr{Z}_2 \times \mathscr{Z}_m$ have planar Cayley graphs, and groups with canonical form $\mathscr{Z}_{m_1} \times \mathscr{Z}_{m_2}$, $m_1 > 2$, have genus 1 (see Exercise 6). Also, if the canonical form of an abelian group has any $\mathscr{Z}_2$-factors, then its genus is easily computed using methods of Section 3.5. Therefore, in the rest of this section, we consider only abelian groups of rank $r > 2$ having no $\mathscr{Z}_2$-factors in their canonical form.

The following theorem establishes a lower bound on the genus of such an abelian group that has no $\mathscr{Z}_3$-factors in its canonical form either. The bound given is simply the genus of an imbedding of $C(\mathscr{A}, X)$, where X is minimum size and all faces of the imbedding are quadrilaterals. At first glance this bound appears to be obvious, since any cycle of length 2 or 3 in a Cayley graph for an irredundant generating set must come from a generator of order 2 or 3, and the canonical generating set has no elements of order 2 or 3 under the given restrictions. However, there are other irredundant generating sets, and they can have elements of order 2 or 3. The argument needed to eliminate such generating sets is actually rather delicate.

Theorem 6.1.2 (Jungerman and White, 1980). *Let the abelian group $\mathscr{A}$ have the canonical form $\mathscr{Z}_{m_1} \times \cdots \times \mathscr{Z}_{m_r}$ such that $r > 1$ and $m_i > 3$ for all i. Then $\gamma(\mathscr{A}) \geq 1 + \#\mathscr{A} \cdot (r-2)/4$.*

Proof. Let X be a generating set for $\mathscr{A}$ such that $\gamma(C(\mathscr{A}, X)) = \gamma(\mathscr{A})$. As we have already observed, it may be assumed that X is irredundant. It cannot be assumed that X is canonical or even that $\#X = r$.

Let s be the number of generators in X of order at least 4, let t_2 be the number of order 2, and let t_3 be the number of order 3. We first claim that $s + t_2 \geq r$ and $s + t_3 \geq r$. To show that $s + t_3 \geq r$, let $\mathscr{B}$ be the subgroup of $\mathscr{A}$ consisting of all elements of order 2. Then $b \in \mathscr{B}$ if and only if the ith coordinate of b is 0, when m_i is odd, and either 0 or $m_i/2$, when m_i is even. Let $\phi: \mathscr{A} \to \mathscr{Q}$ be the natural quotient homomorphism whose kernel is $\mathscr{B}$. Clearly, the group $\mathscr{Q}$ is isomorphic to $\mathscr{Z}_{n_1} \times \cdots \times \mathscr{Z}_{n_r}$, where $n_i = m_i$ if m_i is odd and $n_i = m_i/2$ if m_i is even. In particular, the group $\mathscr{Q}$ has rank r, since $m_i > 2$ for all i. Therefore, the set $\phi(X)$ must contain at least r nonidentity elements, because $\phi(X)$ generates $\mathcal{Q}$. Since every element of order 2 is sent by ϕ to the identity, it follows that $s + t_3 \geq r$ as claimed. A similar argument using the subgroup of elements of order three establishes the other inequality $s + t_2 \geq r$.

Now suppose that $C(\mathcal{A}, X) \to S$ is an orientable imbedding. Let f_i be the number of i-sided faces. Then

$$f_2 \leq \#\mathcal{A} \cdot t_2/2 \quad \text{and} \quad f_3 \leq \#\mathcal{A} \cdot t_3/3$$

because digons and triangles arise only from elements of order 2 or 3 in the irredundant generating set X. Since $\Sigma i f_i = 2\#E$, it follows that

$$2f_2 + 3f_3 + 4(\#F - f_2 - f_3) \leq 2\#E$$

Therefore

$$4\#F \leq 2\#E + \#\mathcal{A}(t_2 + t_3/3)$$

Hence

$$\begin{aligned}
\chi(S) &= \#V - \#E + \#F \\
&\leq \#V - \#E/2 + \#\mathcal{A}(t_2 + t_3/3)/4 \\
&= \#\mathcal{A} - \#\mathcal{A}(s + t_2 + t_3)/2 + \#\mathcal{A}(t_2 + t_3/3)/4 \\
&\leq \#\mathcal{A}(1 - (2s + t_2 + t_3)/4)
\end{aligned}$$

From the inequalities $s + t_2 \geq r$ and $s + t_3 \geq r$, we obtain

$$\chi(S) \leq \#\mathcal{A}(1 - r/2)$$

and consequently

$$\gamma(S) \geq 1 + \#\mathcal{A}(r - 2)/4 \quad \square$$

6.1.3. Constructing Quadrilateral Imbeddings for Most Abelian Groups

The lower bound for $\gamma(\mathcal{A})$ in Theorem 6.1.2 could be realized by a quadrilateral, orientable imbedding of a Cayley graph $C(\mathcal{A}, X)$ with $\#X = r$. Since commutator relations in an abelian group provide an abundance of quadrilaterals, one expects that $\gamma(\mathcal{A}) = 1 + \#\mathcal{A}(r - 2)/4$ whenever this number is an integer and the canonical form for $\mathcal{A}$ contains no $\mathcal{Z}_3$-factors. This is precisely what Jungerman and White have shown, except when the group $\mathcal{A}$ has rank 3 and the first factor in the canonical form for $\mathcal{A}$ is odd.

Theorem 6.1.3 (Jungerman and White, 1980). *Let the abelian group $\mathcal{A}$ have the canonical form $\mathcal{Z}_{m_1} \times \cdots \times \mathcal{Z}_{m_r}$, where $r > 1$ and every $m_i > 3$, and such that either every m_i is even or $r > 3$. Then $\gamma(\mathcal{A}) = 1 + \#\mathcal{A}(r - 2)/4$, whenever the number on the right-hand side of this equation is an integer.*

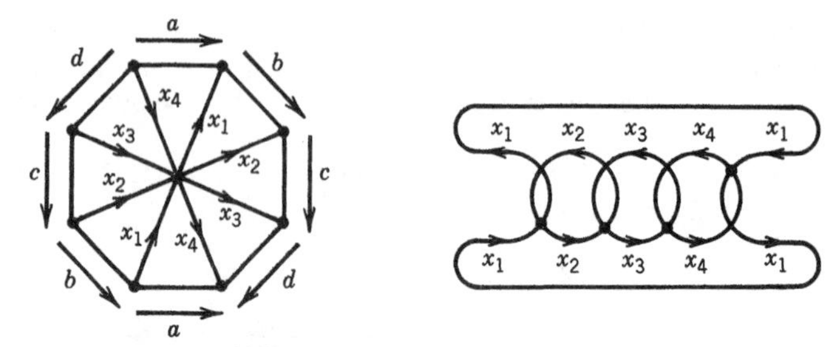

Figure 6.4. A voltage graph and its dual current graph.

Proof. The proof is broken down into various overlapping cases. In each case, a generating set X is given such that $\#X = r$ and a quadrilateral, orientable imbedding for the Cayley graph $C(\mathscr{A}, X)$ is constructed. In all but Case 1, the imbedding is obtained by a voltage–current–graph construction, using a quotient graph of $C(\mathscr{A}, X)$ corresponding to a certain subgroup $\mathscr{B}$.

Case 1. Every m_i is even.

Historically, this was the first case handled (White, 1972). It is a consequence of Theorem 3.5.7, where we take X to be the canonical generating set for the group $\mathscr{A}$. We observe that the Cayley graph $C(\mathscr{A}, X)$ is isomorphic to the cartesian-product graph $C_{m_1} \times \cdots \times C_{m_r}$. Although the proof of Theorem 3.5.7 does not involve voltage graphs, it is also possible to obtain this imbedding by a voltage-graph construction that has for its base graph a Cayley graph for the rank-r abelian 2-group $\mathscr{Z}_2 \times \cdots \times \mathscr{Z}_2$ (see Exercise 1).

Case 2. r even, m_r even.

Let x_i be the element of the group $\mathscr{A}$ having its rth and ith coordinates equal to 1 and all others 0. Let $\mathscr{B}$ be the subgroup consisting of all elements with even rth coordinate. Then the quotient group $\mathscr{A}/\mathscr{B}$ is isomorphic to $\mathscr{Z}_2$, and the Schreier coset graph $S(\mathscr{A}: \mathscr{B}, X)$ has two vertices and no loops. Consider for $r = 4$ the imbedded voltage graph given on the left in Figure 6.4. The base graph is $S(\mathscr{A}: \mathscr{B}, X)$, the base surface is S_2, and the voltage assignment is the natural one. On the right is the dual current graph (the reader should check that conditions i and ii of the second corollary to Theorem 6.1.1 both hold). The pattern for arbitrary even r should be obvious. It is easily verified that every face of the imbedded voltage graph is a quadrilateral with boundary walk of the form $x_i x_{i+1}^{\epsilon} x_i^{-1} x_{i+1}^{-\epsilon}$, $\epsilon = \pm 1$. Dually, KCL holds at every vertex of the current graph. Thus the derived graph is $C(\mathscr{A}, X)$, and every face of the derived imbedding is a quadrilateral.

Case 3. $r \equiv 2 \bmod 4$, m_1 odd.

Let X be the canonical generating set for $\mathscr{A}$. Let $\mathscr{B}$ be the subgroup containing every element of $\mathscr{A}$ whose sum of coordinates is congruent to $0 \bmod m_1$. Then the quotient group $\mathscr{A}/\mathscr{B}$ is isomorphic to $\mathscr{Z}_{m_1}$, and each directed edge of the Schreier coset graph $S(\mathscr{A}: \mathscr{B}, X)$ originating at vertex j

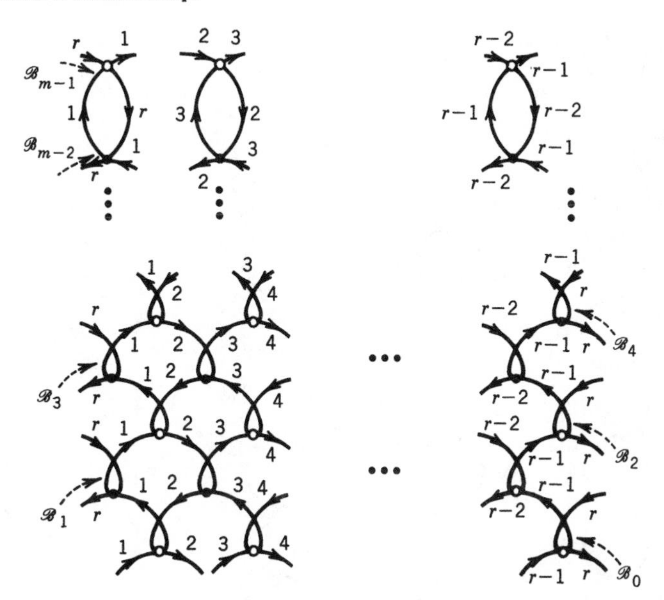

Figure 6.5. The current graph for Case 3. Hollow vertices have counterclockwise rotation and solid vertices clockwise rotation.

terminates at the vertex $j + 1 \bmod m_1$. Consider the current graph given in Figure 6.5. To avoid a blizzard of subscripts, we denote the canonical generator x_i simply by i, for $i = 1, \ldots, r$. An unfinished edge on the left of the diagram is to be identified with the unfinished edge on the right carrying the same current; similarly for unfinished edges at the top and bottom. To avoid double subscripting, let $m = m_1$. The correspondence between faces and cosets of $\mathscr{B}$ is given by the capital letters $\mathscr{B}_0, \ldots, \mathscr{B}_{m-1}$, where every element in $\mathscr{B}_j$ has a sum of coordinates congruent to $j \bmod m$.

The voltage graph dual to the current graph given in Figure 6.5 is more difficult to draw, because the valence, $2r$, of each vertex is too large. Instead, we give a rotation system for the imbedding of $S(\mathscr{A}: \mathscr{B}, X)$. That is, a table of logs for the current graph of Figure 6.5. As in the current graph, we simply denote generator x_i by i, and x_i^{-1} by $\bar{i}$. The vertices of $S(\mathscr{A}: \mathscr{B}, X)$ are the cosets $\mathscr{B}_0, \ldots, \mathscr{B}_{m-1}$, where again elements of $\mathscr{B}_i$ have sum of coordinates equal to $i \bmod m$. There are four kinds of rotation (logs): one for vertex $\mathscr{B}_{m-2}$, one for vertex $\mathscr{B}_{m-1}$, one for all other vertices $\mathscr{B}_j$ with j even, and one for all other vertices $\mathscr{B}_j$, with j odd, as given here for $r = 6$:

$$\mathscr{B}_j,\ j \text{ even:} \qquad \bar{1}\,2\,3\,\bar{4}\,\bar{5}\,6\,1\,\bar{2}\,\bar{3}\,4\,5\,\bar{6}$$

$$\mathscr{B}_j,\ j \text{ odd:} \qquad 6\,5\,\bar{4}\,\bar{3}\,2\,1\,\bar{6}\,\bar{5}\,4\,3\,\bar{2}\,\bar{1}$$

$$\mathscr{B}_{m-2}: \qquad \bar{1}\,6\,1\,\bar{6}\,\bar{5}\,4\,5\,\bar{4}\,\bar{3}\,2\,3\,\bar{2}$$

$$\mathscr{B}_{m-1}: \qquad 1\,\bar{6}\,\bar{1}\,6\,5\,\bar{4}\,\bar{5}\,4\,3\,\bar{2}\,\bar{3}\,2$$

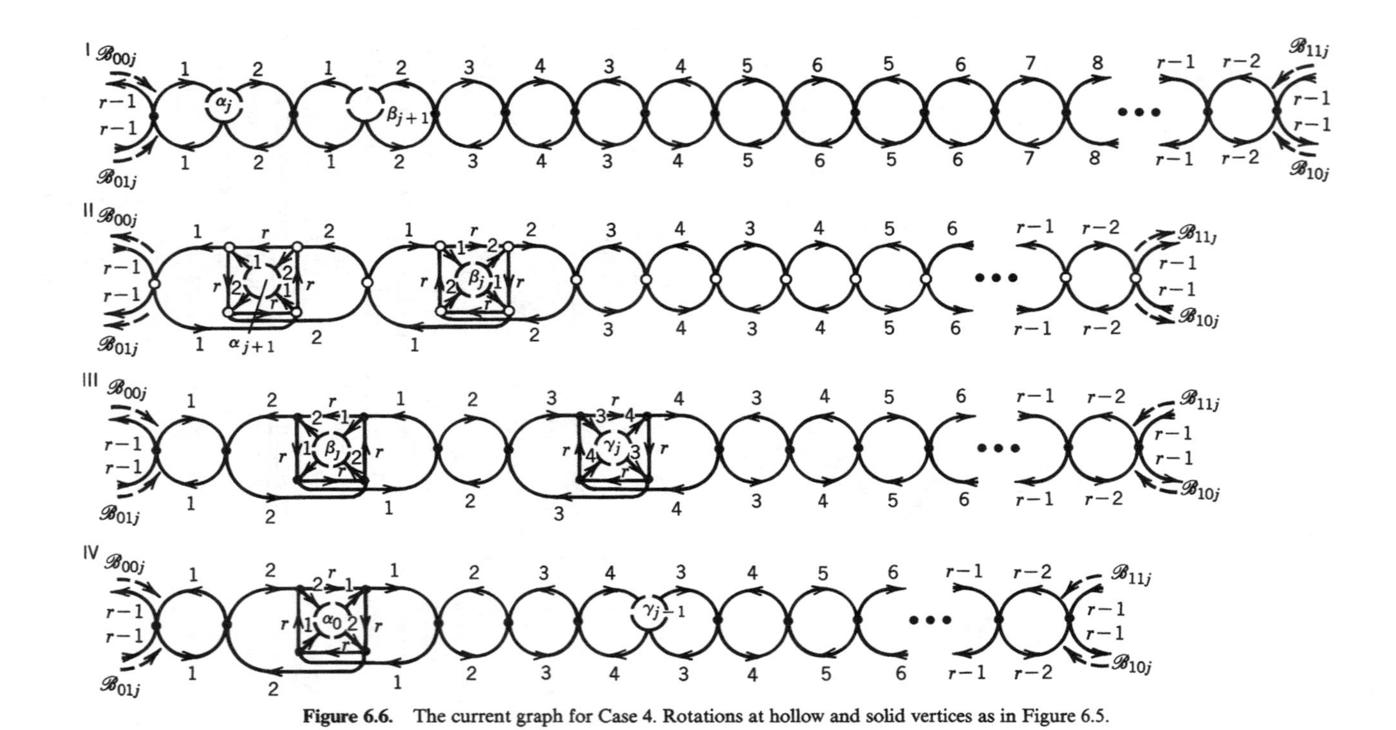

Figure 6.6. The current graph for Case 4. Rotations at hollow and solid vertices as in Figure 6.5.

It can be verified from this rotation scheme that all faces are quadrilaterals satisfying *KVL*. For example, start with edge 1 (i.e., x_1) at vertex $\mathcal{B}_{m-3}$. That edge terminates at vertex $\mathcal{B}_{m-2}$, where we obtain the next edge, 6. That edge takes us to vertex $\mathcal{B}_{m-1}$. There we obtain edge $\bar{1}$, which returns us to vertex $\mathcal{B}_{m-2}$, where we find edge $\bar{6}$ awaiting to return us to $\mathcal{B}_{m-3}$. The next edge is 1 again, so the face is complete (since $m - 3$ is even, the rotation for $\mathcal{B}_{m-3}$ is given by the first row). The face is a quadrilateral with net voltage $x_1 + x_6 - x_1 - x_6 = 0$.

Case 4. r odd, $r \geq 5$, m_{r-1} and m_r even, m_1 odd.

If i is even, let x_i have $(r - 1)$th, rth, and ith coordinate equal to 1 and all others equal to 0; if i is odd and $i \neq 1$, let x_i have rth and ith coordinate 1 and all others 0. Finally, x_1 has first coordinate 1 and all other coordinates 0. Then $X = \{x_i \mid 1 \leq i \leq r\}$ is a generating set for the group $\mathcal{A}$. Let $\mathcal{B}$ be the subgroup of all elements having first coordinate 0, and $(r - 1)$th and rth coordinates even. Then $\mathcal{A}/\mathcal{B} = \mathcal{Z}_{m_1} \times \mathcal{Z}_2 \times \mathcal{Z}_2$. The Schreier coset graph $S(\mathcal{A}: \mathcal{B}, X)$ is a Cayley graph for $\mathcal{A}/\mathcal{B}$ having edges labeled x_i from vertex (j, m, n) to vertex $(j, m + 1, n + 1)$ for i even, from (i, m, n) to $(i, m, n + 1)$ for i odd and $i \neq 1$, and from (j, m, n) to $(j + 1, m, n)$ for $i = 1$. The desired current graph is given in Figure 6.6. Unfinished edges on the left and right match up as in Case 3. Unfinished edges disappearing into the hole α_j (resp., β_j, γ_j) reappear in hole α_{j+1} (resp., β_{j+1}, γ_{j+1}). As in Case 3, we abbreviate i for x_i.

The corresponding directed rotation system for the dual voltage graph is as follows for $r = 7$.

j even, $j \neq m - 1$:

$(j,0,0).\quad 7\,\bar{1}\,6\,\bar{7}\,1\,\bar{6}\,5\,4\,\bar{5}\,\bar{4}\,3\,2\,\bar{3}\,\bar{2}$
$(j,0,1).\quad \bar{7}\,\bar{1}\,\bar{6}\,7\,1\,6\,\bar{5}\,\bar{4}\,5\,4\,\bar{3}\,\bar{2}\,3\,2$
$(j,1,0).\quad \bar{6}\,1\,7\,6\,\bar{1}\,\bar{7}\,\bar{2}\,3\,2\,\bar{3}\,\bar{4}\,5\,4\,\bar{5}$
$(j,1,1).\quad 6\,1\,\bar{7}\,\bar{6}\,\bar{1}\,7\,2\,\bar{3}\,\bar{2}\,3\,4\,\bar{5}\,\bar{4}\,5$

j odd, $j \neq m - 2$:

$(j,0,0).\quad \bar{6}\,\bar{1}\,\bar{7}\,6\,1\,7\,\bar{2}\,\bar{3}\,2\,3\,\bar{4}\,\bar{5}\,4\,5$
$(j,0,1).\quad 6\,\bar{1}\,7\,\bar{6}\,1\,\bar{7}\,2\,3\,\bar{2}\,\bar{3}\,4\,5\,\bar{4}\,\bar{5}$
$(j,1,0).\quad \bar{7}\,1\,6\,7\,\bar{1}\,\bar{6}\,\bar{5}\,4\,5\,\bar{4}\,\bar{3}\,2\,3\,\bar{2}$
$(j,1,1).\quad 7\,1\,\bar{6}\,\bar{7}\,\bar{1}\,6\,5\,\bar{4}\,\bar{5}\,4\,3\,\bar{2}\,\bar{3}\,2$

$j = m - 2$:

$(j,0,0).\quad 7\,\bar{6}\,\bar{1}\,\bar{7}\,6\,5\,1\,4\,\bar{5}\,\bar{4}\,3\,2\,\bar{3}\,\bar{2}$
$(j,0,1).\quad \bar{7}\,6\,\bar{1}\,7\,\bar{6}\,\bar{5}\,1\,\bar{4}\,5\,4\,\bar{3}\,\bar{2}\,3\,2$
$(j,1,0).\quad 1\,\bar{5}\,6\,7\,\bar{1}\,\bar{6}\,\bar{7}\,\bar{2}\,3\,2\,\bar{3}\,\bar{4}\,5\,4$
$(j,1,1).\quad 1\,5\,\bar{6}\,\bar{7}\,\bar{1}\,6\,7\,2\,\bar{3}\,\bar{2}\,3\,4\,\bar{5}\,\bar{4}$

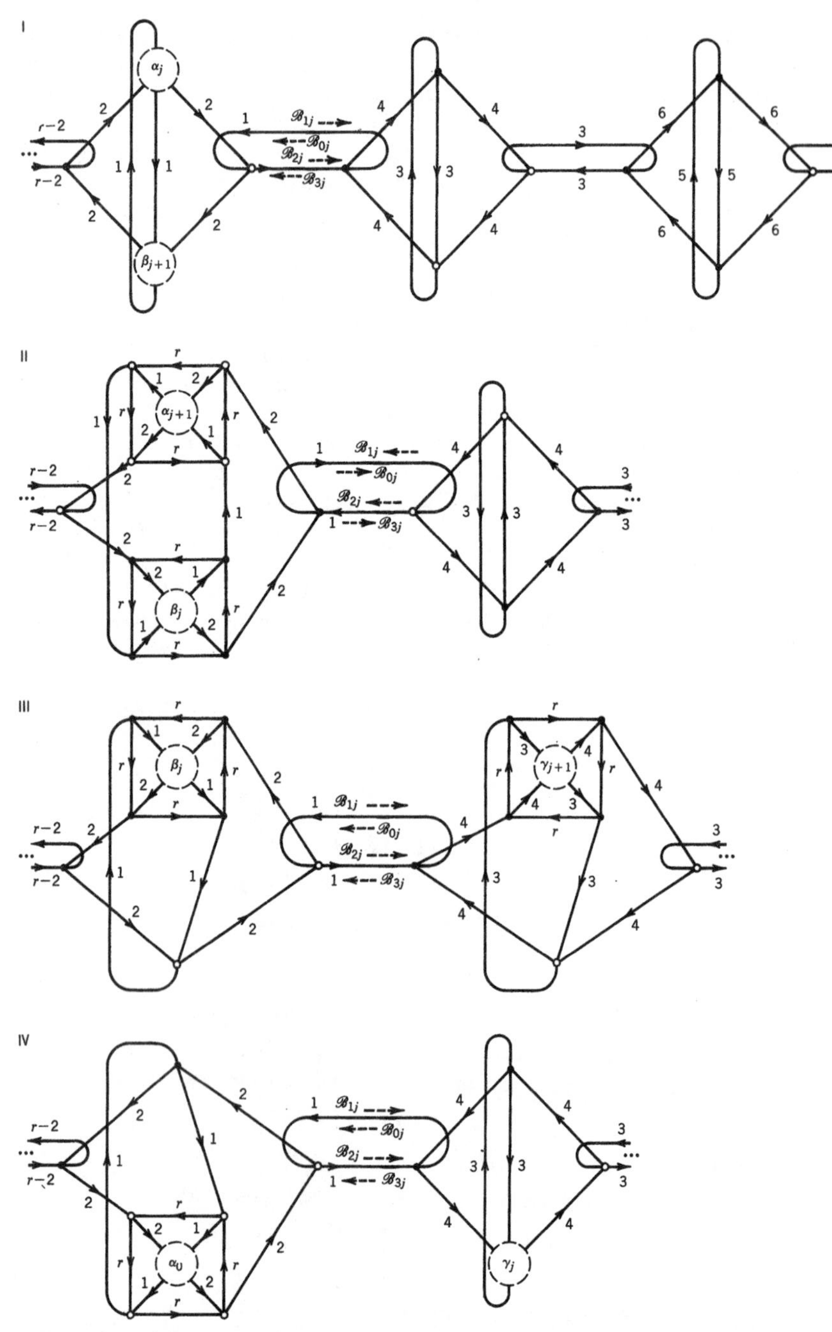

Figure 6.7. The current graph for Case 5. Rotations at hollow and solid vertices as in Figure 6.5.

$j = m - 1$:

$$(j,0,0).\quad \bar{7}\,6\,1\,7\,\bar{6}\,\bar{5}\,4\,\bar{1}\,5\,\bar{4}\,3\,2\,\bar{3}\,\bar{2}$$

$$(j,0,1).\quad 7\,\bar{6}\,1\,\bar{7}\,6\,5\,\bar{4}\,\bar{1}\,\bar{5}\,4\,\bar{3}\,\bar{2}\,3\,2$$

$$(j,1,0).\quad \bar{4}\,\bar{5}\,\bar{1}\,4\,5\,\bar{6}\,\bar{7}\,1\,6\,7\,\bar{2}\,3\,2\,\bar{3}$$

$$(j,1,1).\quad 4\,5\,\bar{1}\,\bar{4}\,\bar{5}\,6\,7\,1\,\bar{6}\,\bar{7}\,2\,\bar{3}\,\bar{2}\,\bar{3}$$

For arbitrary r, the generators $1, r-3, r-2, r-1, r$ behave like $1, 4, 5, 6, 7$, respectively, whereas all other generators follow, in pairs, the pattern of $2, 3$. For example, for $r = 9$, the second row under $j = m - 1$ would read:

$$(j,0,1).\quad 9\,\bar{8}\,\bar{1}\,9\,8\,7\,\bar{6}\,\bar{1}\,\bar{7}\,6\,\bar{5}\,\bar{4}\,5\,4\,\bar{3}\,\bar{2}\,3\,2$$

As in previous cases, it can be verified from the scheme above that every face is a quadrilateral satisfying *KVL*. For example, starting with edge 1 at vertex $(m-3,0,1)$, one obtains next the edge 7 at vertex $(m-3,1,1)$, then edge $\bar{1}$ at vertex $(m-2,1,1)$, then edge $\bar{7}$ at vertex $(m-3,1,1)$, then edge 1 at vertex $(m-3,0,1)$ again. We notice that $m-3$ is even, so that rotations come from the first and third groups.

Case 5. r odd, $r \geq 5$, $m_r \equiv 0 \bmod 4$, m_1 odd.

Let x have rth coordinate 2, ith coordinate 1, and all other coordinates 0, if i is even; let x_i have rth coordinate 1 and all other coordinates 0, if i is odd and $i \neq 1$. Finally, x_1 has first coordinate 1 and all other coordinates 0. Then $X = \{x_i \mid 1 \leq i \leq r\}$ is a generating set for the group $\mathscr{A}$. Let $\mathscr{B}$ be the subgroup of all elements having even $(r-1)$ and rth coordinates and first coordinate 0. Then $\mathscr{A}/\mathscr{B} = \mathscr{Z}_m \times \mathscr{Z}_4$, where again $m = m_1$. The Schreier coset graph $S(\mathscr{A}: \mathscr{B}, X)$ is a Cayley graph for the quotient group $\mathscr{A}/\mathscr{B}$ that has edges labeled x_i from vertex (j, m) to vertex $(j, m+2)$ for even i, from vertex (j, m) to vertex $(j, m+1)$ for odd i and $i \neq 1$, and from (j, m) to $(j+1, m)$ for $i = 1$. The desired current graph is given in Figure 6.7. Again, broken edges are identified by the same conventions as in Case 4.

The corresponding directed rotation system for the dual voltage graph is as follows for $r = 7$:

j even, $j \neq m - 1$:

$$(j,0).\quad 7\,6\,\bar{1}\,\bar{7}\,1\,\bar{6}\,3\,2\,\bar{3}\,\bar{2}\,5\,4\,\bar{5}\,\bar{4}$$

$$(j,1).\quad \bar{6}\,\bar{1}\,7\,1\,6\,\bar{7}\,\bar{4}\,5\,4\,\bar{5}\,\bar{2}\,3\,2\,\bar{3}$$

$$(j,2).\quad 6\,1\,\bar{7}\,\bar{1}\,\bar{6}\,7\,4\,\bar{5}\,\bar{4}\,5\,2\,\bar{3}\,\bar{2}\,3$$

$$(j,3).\quad \bar{7}\,\bar{6}\,1\,7\,\bar{1}\,6\,\bar{3}\,\bar{2}\,3\,2\,\bar{5}\,\bar{4}\,5\,4$$

j odd, $j \neq m - 3$:

$$(j,0).\quad \bar{6}\bar{1}\bar{7}167\bar{4}\bar{5}45\bar{2}\bar{3}23$$
$$(j,1).\quad \bar{7}6\bar{1}71\bar{6}\bar{3}23\bar{2}\bar{5}45\bar{4}$$
$$(j,2).\quad 7\bar{6}1\bar{7}\bar{1}63\bar{2}\bar{3}25\bar{4}\bar{5}4$$
$$(j,3).\quad 617\bar{1}\bar{6}\bar{7}45\bar{4}\bar{5}23\bar{2}\bar{3}$$

$j = m - 2$:

$$(j,0).\quad 54\bar{5}\bar{1}\bar{4}7\bar{6}\bar{7}1632\bar{3}\bar{2}$$
$$(j,1).\quad 67\bar{1}\bar{6}\bar{7}\bar{4}514\bar{5}\bar{2}32\bar{3}$$
$$(j,2).\quad \bar{6}\bar{1}\bar{7}6741\bar{5}\bar{4}52\bar{3}\bar{2}3$$
$$(j,3).\quad \bar{5}\bar{4}154\bar{7}6\bar{1}7\bar{6}\bar{3}\bar{2}32$$

$j = m - 1$:

$$(j,0).\quad 54\bar{5}1\bar{4}7\bar{6}\bar{1}\bar{7}632\bar{3}\bar{2}$$
$$(j,1).\quad 617\bar{6}\bar{7}4\bar{1}\bar{5}\bar{4}5\bar{2}32\bar{3}$$
$$(j,2).\quad \bar{6}\bar{7}167\bar{4}\bar{5}\bar{1}452\bar{3}\bar{2}3$$
$$(j,3).\quad \bar{5}45\bar{1}\bar{4}\bar{7}671\bar{6}\bar{3}\bar{2}32$$

As in Case 4, for arbitrary r, generators $1, r - 3, r - 2, r - 1, r$ behave like $1, 4, 5, 6, 7$, whereas all other generators behave in pairs like $2, 3$. For example, the vertex $(j, 2)$ for $j = m - 1$ has the following rotation for $r = 9$:

$$(j,2).\quad \bar{8}\bar{9}189\bar{7}\bar{8}\bar{1}782\bar{3}\bar{2}34\bar{5}\bar{4}5$$

As in the previous cases, it can be verified that every face is a quadrilateral satisfying *KVL*. □

The smallest abelian group whose genus is not ascertainable by application of Theorem 6.1.3 is $\mathscr{Z}_3 \times \mathscr{Z}_3 \times \mathscr{Z}_3$. For this group, we have the following special result, stated here without proof.

Theorem 6.1.4 (Mohar et al., 1985; Brin and Squier, 1986). *The genus of the group* $\mathscr{Z}_3 \times \mathscr{Z}_3 \times \mathscr{Z}_3$ *is* 7. □

It is not difficult (see Exercises 10 and 11) to prove that

$$5 \leq \gamma(\mathscr{Z}_3 \times \mathscr{Z}_3 \times \mathscr{Z}_3) \leq 10$$

and for about a decade, it was commonly suspected that the genus was 10, based on the reasoning that a lower genus could not be realized with a symmetric imbedding. After Mohar et al. (1985) produced a chaotic imbedding in the surface S_7, Brin and Squier (1986) responded with an ingenious case-by-case demonstration that no surface of smaller genus would be adequate.

Just to illustrate the difficulties posed by $\mathscr{Z}_3$-factors, although it is not hard to prove that $\gamma(\mathscr{Z}_3 \times \mathscr{Z}_3) = 1$ (see Exercise 6.4.14), there are many different toroidal imbeddings of the Cayley graph for $\mathscr{Z}_3 \times \mathscr{Z}_3$, each corresponding to an irredundant generating set. Beyond the obvious quadrilateral imbedding, there is an imbedding having six triangular faces and three hexagonal faces (Exercise 7), and there is another imbedding having six triangles, two quadrilaterals, and one face of size 10 (Exercise 8). In the same way, there are a variety of bizarre imbeddings of Cayley graphs for $\mathscr{Z}_3 \times \mathscr{Z}_3 \times \mathscr{Z}_3$ in the surface of genus 10 (see Exercises 9 and 10).

6.1.4. Exercises

1. Construct a voltage-graph imbedding (or current graph) for Case 1 of Theorem 6.1.3 using the quotient group $\mathscr{Z}_2 \times \cdots \times \mathscr{Z}_2$ of rank r.
2. Verify that conditions i and ii hold for the current graph in Case 2 of Theorem 6.1.3. Give the table of logs for this current graph.
3. Check that *KCL* holds at every vertex of the current graph given for Case 3 of Theorem 6.1.3. Verify at least three rows in the associated table of logs given in the text.
4. Same as Exercise 3, but for Case 4 of Theorem 6.1.3.
5. Same as Exercise 3, but for Case 5 of Theorem 6.1.3.
6. Prove that $\gamma(\mathscr{Z}_m \times \mathscr{Z}_n) = 1$, where m divides n and $m > 2$ (see also Exercise 6.4.14 for the case $m = n = 3$).
7. Use a one-vertex voltage graph imbedded in the sphere to obtain a toroidal imbedding of a Cayley graph for $\mathscr{Z}_3 \times \mathscr{Z}_3$.
8. Let $X = \{x, y\}$ be an irredundant generating set for $\mathscr{Z}_3 \times \mathscr{Z}_3$. Give a rotation scheme for an imbedding of $C(\mathscr{Z}_3 \times \mathscr{Z}_3, X)$ that has six triangles and at least one quadrilateral. (*Hint:* Draw a rectangle $xyx^{-1}y^{-1}$ bordered by a pair of x^3 triangles and a pair of y^3 triangles. This gives you the full rotation at the four vertices of the rectangle and part of the rotation at four more vertices. Now complete the rotation system so that the remaining x^3 and y^3 circuits are faces.) Given that $\gamma(\mathscr{Z}_3 \times \mathscr{Z}_3) = 1$, the imbedding constructed in this way must be toroidal. Why? What are the sizes of the faces?
9. Use one-vertex voltage graphs to construct three different imbeddings of a Cayley graph for $\mathscr{Z}_3 \times \mathscr{Z}_3 \times \mathscr{Z}_3$ in the surface S_{10}: one with 9 triangles, one with 18, and one with 27.
10. Let $X = \{x, y, z\}$ be an irredundant generating set for $\mathscr{Z}_3 \times \mathscr{Z}_3 \times \mathscr{Z}_3$. Let $\mathscr{B}$ be the subgroup generated by z. The Schreier coset graph

$S(\mathscr{A}: \mathscr{B}, X)$ is just a Cayley graph for $\mathscr{Z}_3 \times \mathscr{Z}_3$, having x and y as generators, with an additional loop for z attached at each vertex. Start with the usual quadrilateral imbedding of a Cayley graph for $\mathscr{Z}_3 \times \mathscr{Z}_3$ and then place z-loops inside some of the faces so that KVL holds around every face not enclosed by a z-loop. (The directions of the z-loops matter; one of the original quadrilateral faces contains three z-loops.) Show that the derived imbedding of $C(\mathscr{Z}_3 \times \mathscr{Z}_3 \times \mathscr{Z}_3, X)$ given by the natural voltage assignment has genus 10. Find the sizes of all faces of the derived imbedding.

11. Prove that $\gamma(\mathscr{Z}_3 \times \mathscr{Z}_3 \times \mathscr{Z}_3) \geq 5$.

12. Construct an imbedding of a Cayley graph for $\mathscr{Z}_3 \times \mathscr{Z}_3 \times \mathscr{Z}_3$ in the nonorientable surface of Euler characteristic $\chi = -15$. Notice that $\chi(S_{10}) = -18$. (A three-vertex voltage graph imbedded in the projective plane can be used. See also Exercise 6.4.15.)

13. Let $\mathscr{A}$ be an infinite group and $\mathscr{T}$ any finite subset of $\mathscr{A}$. First prove that there exists an element b in $\mathscr{A}$ such that $b\mathscr{T} \cap \mathscr{T}$ is empty. Next, suppose that a Cayley graph for the group $\mathscr{A}$ has a subgraph homeomorphic to K_5 or to $K_{3,3}$. Prove that $\mathscr{A}$ has infinitely many disjoint subgraphs homeomorphic to K_5 or to $K_{3,3}$.

14. Assume the infinite version of Kuratowski's theorem, due to Dirac and Schuster (1954), that an infinite graph can be imbedded in the plane if and only if it contains no subgraph homeomorphic to K_5 or to $K_{3,3}$. Use Exercise 13 to prove Levinson's theorem (1970) that the genus of an infinite group is either 0 or infinite.

6.2. THE SYMMETRIC GENUS

In attempting to determine the genus of a given Cayley graph $C(\mathscr{A}, X)$, one strategy is to suppose that there exists a minimum-genus imbedding that somehow reflects the symmetry of the graph. In particular, the natural action of the group $\mathscr{A}$ on the Cayley graph $C(\mathscr{A}, X)$ might extend to an action of $\mathscr{A}$ on the imbedding surface. If the natural action does so extend, then we call the imbedding "symmetric". Furthermore, if the natural action is orientation preserving, then we call the imbedding "strongly symmetric" and, otherwise, "weakly symmetric". The "symmetric genus" $\sigma(\mathscr{A})$ (resp., "strong symmetric genus" $\sigma^0(\mathscr{A})$) is then defined to be the smallest number n such that the surface S_n contains a symmetric (resp., strong symmetric) imbedding of a Cayley graph for the group $\mathscr{A}$. A main result of this section is that if the group $\mathscr{A}$ acts on an orientable surface S, then there is a symmetric imbedding in S of a Cayley graph for $\mathscr{A}$. It follows from this result that the symmetric genus could be defined equivalently without any reference at all to generating sets or Cayley graphs.

The concept of strong symmetric genus is actually the oldest version of a genus for groups. A form of it, where the quotient surface of the group action is restricted to be the sphere, was studied by Dyck (1882), in the context of automorphism of maps on surfaces. Burnside (1911) devotes Chapters 18 and 19 to this "graphical representation of a group" by map automorphisms and determines all groups of "genus" 0 and 1. To those working on conformal automorphisms of Riemann surfaces, "genus" means smallest genus surface, other than the sphere or torus, in which the group has a strongly symmetric imbedding. Maclachlin (1965) computes this genus for abelian groups. On the other hand, Tucker (1983) seems to have been first to introduce the symmetric genus, although Levinson and Maskit (1975) do discuss a related concept called the weak point-symmetric genus.

6.2.1. Rotation Systems and Symmetry

Suppose $C(\mathscr{A}, X) \to S$ is an imbedding of a Cayley graph in an oriented surface. Since the edges of $C(\mathscr{A}, X)$ are directed and labeled by elements of the generating set X, a rotation at a particular vertex can be represented as a cyclic ordering of the elements of X and their inverses. Thus, it makes sense to compare rotations at different vertices.

Example 6.2.1. *The dihedral group $\mathscr{D}_4$ is given by the presentation $\langle x, y: x^4 = y^2 = 1, yxy^{-1} = x^{-1}\rangle$, with generating set $X = \{x, y\}$. Consider the planar imbedding of the Cayley graph $C(\mathscr{D}_4, X)$ given on the left in Figure 6.8. Edges corresponding to x are solid, and edges corresponding to y are dashed. Every vertex has the (clockwise) rotation $xx^{-1}y^{-1}y$.*

Is the imbedding of Example 6.2.1 symmetric, or even strongly symmetric? Perhaps it is not exactly clear geometrically how the natural action of $\mathscr{D}_4$ on

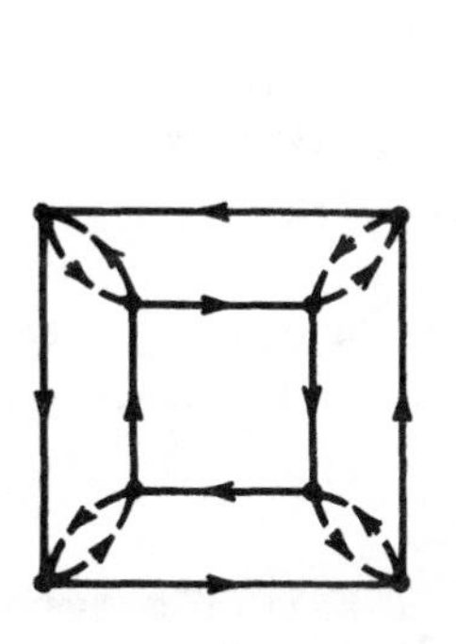

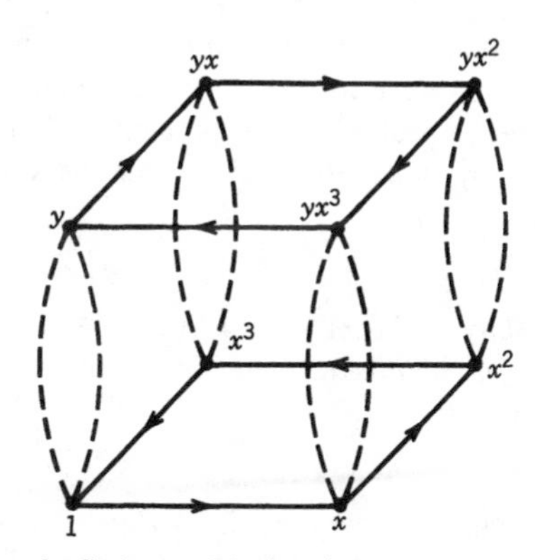

Figure 6.8. An imbedding of a Cayley graph in the sphere.

$C(\mathcal{D}_4, X)$ might extend to the sphere. On the right in Figure 6.8 is an equivalent imbedding, depicted as a cube. In the right-hand figure, the vertices of $C(\mathcal{D}_4, X)$ have been labeled by elements of $\mathcal{D}_4$. Left multiplication by x clearly rotates the bottom quadrilateral by 90° about the vertical line through its center, counterclockwise when looking from above. Left multiplication by x has the same effect on the top quadrilateral, since $xy = yx^{-1}$. Thus the natural action on $C(\mathscr{A}, X)$ given by left multiplication by x extends to a 90° rotation of the cube. Similarly, it is easily verified that left multiplication by y is equivalent to a 180° rotation about the line through the centers of the left front and right rear digons. Thus, the imbedding is symmetric in our precise sense. In fact, since the action on the sphere involves only rotations, and since rotations preserve orientation, the imbedding is strongly symmetric.

Although geometric visualization is important, one might also want a purely combinatorial way to show that a given imbedding is strongly symmetric. To discover such a way, we first observe that the natural action of a group $\mathscr{A}$ on the Cayley graph $C(\mathscr{A}, X)$ respects edge labels and directions. Accordingly, if the action of $\mathscr{A}$ extends to an imbedding surface, it must also respect the rotation system, in the sense of Section 4.3. That is, if the imbedding $C(\mathscr{A}, X) \to S$ is strongly symmetric, then the rotation at every vertex is the same, and if the imbedding is symmetric then the rotation at every vertex is the same or reversed. For strongly symmetric imbeddings, the converse is also true. If the rotation at every vertex is the same for the imbedding $C(\mathscr{A}, X) \to S$, then by Theorem 4.3.3, the natural action of $\mathscr{A}$ extends to an orientation-preserving action on the surface. Summarizing, we have the following result.

Theorem 6.2.1. *An imbedding $C(\mathscr{A}, X) \to S$ is strongly symmetric if and only if the rotation at every vertex is the same. If an imbedding $C(\mathscr{A}, X) \to S$ is symmetric, then the rotation at every vertex is the same or reversed.*

It might be helpful to restate Theorem 6.2.1 in terms of voltage graphs, at least for strongly symmetric imbeddings. Since the extended action can be chosen to have at most one fixed point in each face of the imbedding, the natural projection to the quotient surface is a regular branched covering, and hence by Theorem 4.3.5, it can be obtained by a regular voltage graph construction.

Corollary. *An imbedding $C(\mathscr{A}, X) \to S$ is strongly symmetric if and only if it it can be obtained as the derived imbedding for a one-vertex, regular voltage graph.* □

In this context, the Dyck–Burnside genus of a group $\mathscr{A}$ is simply the minimum genus obtained by a one-vertex voltage graph "in the sphere".

Maclachlin's genus is the minimum genus "greater than 1" obtained by a one-vertex voltage graph in any orientable surface.

The converse to the second statement of Theorem 6.2.1 is false, as readily illustrated by some nonsymmetric imbeddings for 2-generator presentations of odd-order cyclic groups. Since an orientation-reversing homeomorphism must reverse every rotation and the action of a group on its Cayley graph has no fixed points, exactly half the vertices in a weakly symmetric imbedding have one rotation and the other half have the reverse. In particular, no group of odd order can have a weakly symmetric Cayley graph imbedding. Necessary and sufficient conditions for a rotation system to yield a weakly symmetric imbedding are considered in Exercise 1.

One consequence of Theorem 6.2.1 should be mentioned: the number of symmetric imbeddings of a given Cayley graph is very small compared with the total number of imbeddings. For example, if $\mathscr{A} = \mathscr{Z}_n \times \mathscr{Z}_n \times \mathscr{Z}_n$, and if X is the canonical generating set for $\mathscr{A}$, then $C(\mathscr{A}, X)$ has $(5!)^{n^3}$ different imbeddings but only 5! different strongly symmetric imbeddings.

The situation for weakly symmetric imbeddings is greatly complicated by involutions. A variation on Example 6.2.1 will illuminate the problem. The group $\mathscr{Z}_2 \times \mathscr{Z}_4$ has the presentation $\langle x, y : x^4 = y^2 = 1, yxy^{-1} = x\rangle$, for which we let X denote the generating set $\{x, y\}$. Figure 6.8 also gives an imbedding of $C(\mathscr{Z}_2 \times \mathscr{Z}_4, X)$ in the sphere if the directions of the x-edges in the inside quadrilateral, on the left (or the top quadrilateral on the right) are all reversed. Although the underlying graph is isomorphic to the previously examined Cayley graph for the dihedral group $\mathscr{D}_4$, the group action on the graph has changed. Left multiplication by x is still accomplished by a 90° rotation of the cube. However, left multiplication by y now takes the vertices of the top quadrilateral to the vertices immediately below them. The only way this can be accomplished is by a reflection about the horizontal plane through the center of the cube. The trouble is that such a reflection takes each y-edge to itself, whereas left multiplication by y takes each edge labeled y to its "twin" edge.

Thus the imbedding of $C(\mathscr{Z}_2 \times \mathscr{Z}_4, X)$ is not even weakly symmetric, in our precise sense, because the natural action of $\mathscr{Z}_2 \times \mathscr{Z}_4$ on the Cayley graph cannot be extended to the imbedding. This could also be discovered from the rotation system of the imbedding. Every rotation is either $xx^{-1}y^{-1}y$ or $x^{-1}xy^{-1}y$, but these rotations are not the reverse of each other.

The difficulty here is really an artifact of representing involutions by doubled edges. Suppose we avoid double edges by using the alternative Cayley graph $C^1(\mathscr{Z}_2 \times \mathscr{Z}_4, X)$, and we tolerate the fact that the "natural action" on $C^1(\mathscr{Z}_2 \times \mathscr{Z}_4, X)$ fixes some edges. That is, left multiplication by y now simply reverses the undoubled y-edges in the same way as reflection in the horizontal plane through the cube's center. In terms of rotation systems, every rotation is either $xx^{-1}y$ or $x^{-1}xy$, which are respective reverses of each other.

Throughout the rest of this chapter we shall use only alternative Cayley graphs. This provides a minor variation on the voltage-graph viewpoint, since an alternative Cayley graph $C^1(\mathscr{A}, X)$ with an involution in its generating set is not the derived graph of a bouquet of circles.

6.2.2. Reflections

If the group $\mathscr{A}$ acts pseudofreely on the surface S, then the methods of Chapter 4 involving branched coverings and voltage graphs suffice to describe the action of $\mathscr{A}$ on S. However, some group actions on surfaces are not pseudofree. Under the most general conditions, the fixed point set of a surface homeomorphism can be very complicated, and one might fear that nonfree actions are completely intractable. Fortunately, the behavior of a surface homeomorphism of finite order is fairly simple, which is reassuring since our interest is in finite groups of homeomorphisms. For example, the main theorem of this section asserts that an orientation-preserving surface homeomorphism of finite order greater than 1 has only a finite number of fixed points.

We have already called attention, in passing, to some group actions that are not pseudofree, such as the action discussed just above of $\mathscr{Z}_2 \times \mathscr{Z}_4$ on the sphere, which involves reflection across the equator. For the sphere, reflection in the equator is, in effect, the only finite-order homeomorphism having an infinite number of fixed points. For a surface of higher genus, it is easy to visualize a similar reflection using a plane in 3-space to split the surface in "half". There are, however, other "reflections" that are not so easily visualized.

Example 6.2.2. *Let S be the torus given by the rectangle with sides to be identified, as in Figure 6.9. Note that after the top and bottom are identified, to form a cylinder, the left and right sides are identified by a half-twist. Let f be a reflection of the rectangle about the center broken line. Then f induces a homeomorphism h on the torus S. The broken line becomes a circle C left pointwise fixed by h, whereas the line $v\,u\,v$ becomes a circle C' that is taken to itself by h but is not left pointwise fixed—it is turned* 180°, *taking u to v and v to u. The circles C and C' split the torus S into halves that are interchanged by h. On the right of Figure* 6.9 *is the torus with edges identified. The symbols "F"*

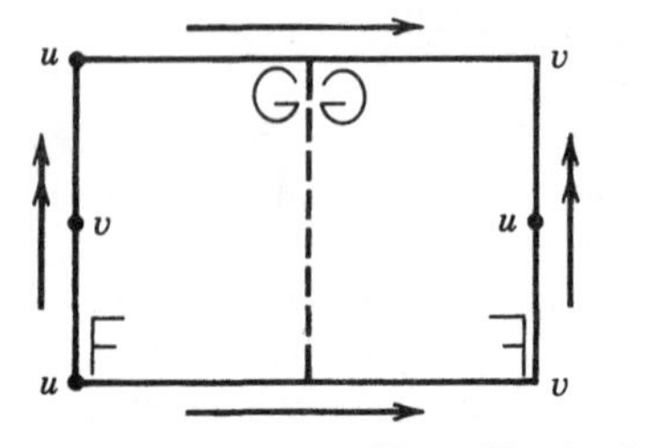

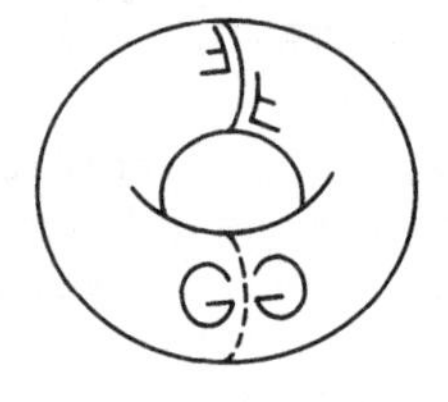

Figure 6.9. A reflection of the torus.

and "G" are given, together with their images under the reflection, to show how the homeomorphism h acts on the torus.

Up to this point the term "reflection", which is well defined in Euclidean geometry, has been used loosely to apply to certain types of surface homeomorphisms. With Example 6.2.1 in mind, we can now be more precise. A homeomorphism $h: S \to S$ on an orientable surface S is called a "reflection" if

- **i.** h is an orientation-reversing involution;
- **ii.** S is the union of two connected surfaces with boundary S_0 and S_1 such that $h(S_0) = S_1$ and $S_0 \cap S_1$ is a collection of disjoint circles imbedded in S, at least one of which is left pointwise fixed by h.

The surfaces S_0 and S_1 are called the "halves" of the reflection h, and the components of $S_0 \cap S_1$ are called the "dividing circles".

It can be shown (Exercise 3) that $\chi(S) = 2\chi(S_0)$. Since there are only finitely many surfaces with boundary having a given Euler characteristic, the choice of halves for reflections on a given surface is limited. For a torus, there are only two kinds of reflections: the one given in Example 6.2.2 and the obvious reflection that has the same halves as Example 6.2.2 but leaves both dividing circles pointwise fixed. Reflections for a higher-genus surface are considered in Exercise 4. The main result of this section can now be stated.

Theorem 6.2.2. *Any finite-order homeomorphism of an orientable surface either has a finite number of fixed points or is a reflection.*

Proof. The proof in the general topological case, due mostly to Kerekjarto (1921), with later corrections by Eilenberg (1934), is very difficult and is not considered here. If one assumes that the given homeomorphism is an automorphism of a graph imbedding that behaves "piecewise linearly" on the faces, then the proof follows from easily verified facts about linear transformations (see Exercises 5–10). □

Given a group $\mathscr{A}$ acting on an orientable surface S, the collection of orientation-preserving elements of $\mathscr{A}$ forms a subgroup of index at most 2. It is a subgroup because the composition of two orientation-preserving homeomorphisms is orientation-preserving. The index is at most 2, because the composition of two orientation-reversing homeomorphisms is orientation-preserving. This subgroup is called the "orientation-preserving subgroup" and is denoted $\mathscr{A}^0$. A given group can have many index-2 subgroups, and which one of these is the subgroup $\mathscr{A}^0$ depends, of course, on the given action.

Corollary. *Given an action of the group $\mathscr{A}$ on the orientable surface S, the orientation-preserving subgroup $\mathscr{A}^0$ acts pseudofreely.*

6.2.3. Quotient Group Actions on Quotient Surfaces

If the group $\mathscr{A}$ acts pseudofreely on the surface S, then by Theorem 4.3.6 there is a Cayley graph for $\mathscr{A}$ imbedded in S. We wish to prove that this result holds even if the action of $\mathscr{A}$ is not pseudofree. At the same time we should like to use voltage graphs and regular coverings as in Section 4.3 to obtain information about presentations for the group $\mathscr{A}$. Since the orientation-preserving subgroup $\mathscr{A}^0$ does act pseudofreely, the main difficulty is to understand the relation between the subgroup $\mathscr{A}^0$ and the reflections in $\mathscr{A}$. This is accomplished by studying the natural action of the quotient group $\mathscr{A}/\mathscr{A}^0$ on the quotient surface $S/\mathscr{A}^0$.

Let $\mathscr{A}$ act on the surface S, and let $\mathscr{B}$ be a normal subgroup of $\mathscr{A}$ that acts pseudofreely on S. Let $p: S \to S/\mathscr{B}$ be the natural projection of the regular branched covering associated with the action of $\mathscr{B}$. Given an element a in $\mathscr{A}$, let $\bar{a}$ denote the corresponding element of the quotient group $\mathscr{A}/\mathscr{B}$. Then an action of the group $\mathscr{A}/\mathscr{B}$ on the surface $S/\mathscr{B}$ can be given by associating to $\bar{a}$ the function $\psi_{\bar{a}}: S/\mathscr{B} \to S/\mathscr{B}$ defined by $\psi_{\bar{a}}(p(x)) = p(\phi_a(x))$, where ϕ_a is the homeomorphism of S for the element a given by the action of $\mathscr{A}$. This is called the "natural action" of the quotient group $\mathscr{A}/\mathscr{B}$ on the surface $S/\mathscr{B}$.

Of course it must be shown that $\psi_{\bar{a}}$ is well defined, that $\psi_{\bar{a}}$ is a homeomorphism, and that the association $\bar{a} \to \psi_{\bar{a}}$ does define an action, that is, $\psi_{\bar{a}\bar{b}} = \psi_{\bar{a}}\psi_{\bar{b}}$. First, suppose that $\bar{c} = \bar{a}$ and $p(y) = p(x)$. We must show that $\psi_{\bar{c}}(p(y)) = \psi_{\bar{a}}(p(x))$. By definition $\psi_{\bar{c}}p(y) = p\phi_c(y)$ and $\psi_{\bar{a}}(p(x)) = p\phi_a(x)$. Thus it suffices to show the orbit of $\phi_c(y)$ equals the orbit of $\phi_a(x)$ under the action of $\mathscr{B}$. Since we are given $p(y) = p(x)$, it follows that $y = \phi_b(x)$ for some $b \in \mathscr{B}$. Hence $\phi_c(y) = \phi_c\phi_b(x) = \phi_{cb}(x)$. Since $\overline{cb} = \bar{c}\bar{b} = \bar{a}$, $cb = b'a$ for some $b' \in \mathscr{B}$. Thus $\phi_{cb}(x) = \phi_{b'}\phi_a(x)$, which implies that $\phi_c(y) = \phi_{cb}(x)$ has the same orbit as $\phi_a(x)$, as desired. To show that $\psi_{\bar{a}}$ is a homeomorphism, we first observe that $\psi_{\bar{a}}p = p\phi_a$. Therefore, since p is a (possibly branched) covering and ϕ_a is a homeomorphism, it follows that $\psi_{\bar{a}}$ is a (possibly branched) covering. Moreover, since the number of sheets of the branched coverings $\psi_{\bar{a}}p$ and $p\phi_a$ must be equal, we infer that $\psi_{\bar{a}}$ is 1-sheeted, because ϕ_a is. Thus $\psi_{\bar{a}}$ is a homeomorphism. The proof that $\psi_{\bar{a}\bar{b}} = \psi_{\bar{a}}\psi_{\bar{b}}$ is left to the reader as Exercise 11.

Example 6.2.3. *Let S be the surface of genus 2, imbedded in 3-space as shown in Figure 6.10. Consider the action of the group $\mathscr{A} = \mathscr{Z}_2 \times \mathscr{Z}_2 \times \mathscr{Z}_2$ generated by reflections in the three coordinate planes:*

a: *reflection in the yz-plane,*
b: *reflection in the xz-plane,*
c: *reflection in the xy-plane.*

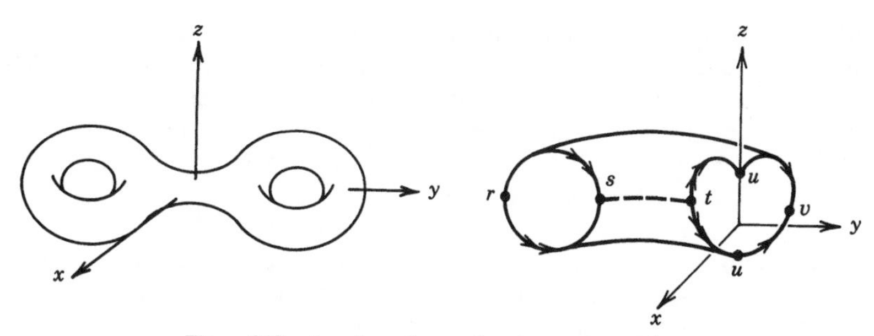

Figure 6.10. A surface of genus 2 and a quotient surface.

The orientation-preserving subgroup $\mathscr{A}^0$ *consists of the identity and the* 180° *rotations ab, ac, bc about the z, y, x axes, respectively.*

To visualize the quotient surface $S/\mathscr{A}^0$ and the quotient action of $\mathscr{A}/\mathscr{A}^0$, we proceed as follows. The surface S is divided into eight pieces by the three coordinate planes. We keep two adjacent pieces to form a surface S' (with boundary) and discard the rest of S. On the right of Figure 6.10 is shown the case in which S' consists of all points of S satisfying $x \leq 0$ and $y \leq 0$. The restriction of the quotient map $p: S \to S/\mathscr{A}^0$ to S' still maps onto $S/\mathscr{A}^0$, since the subsurface S' contains a point from every orbit of the action of $\mathscr{A}^0$. Moreover, the restriction of the covering projection p to the interior of S' is a homeomorphism, because that part of S' contains only one point from each orbit. On the other hand, the quotient map identifies pieces of the boundary of S'. Rotation ac about the y axis identifies the top and bottom halves of the circle in the yz-plane on the left and the semicircle in the yz-plane on the right, as indicated by double arrows. Rotation bc about the x axis identifies the top and bottom halves of the semicircle in the xz-plane, as indicated by single arrows.

Thus the quotient map closes off the holes in S' to form a sphere for the quotient surface $S/\mathscr{A}^0$. The branch points of the covering $p: S \to S/\mathscr{A}^0$ are indicated by heavy dots: r, s, t correspond to fixed points of rotation ac about the y axis, u to fixed points of rotation ab about the z axis, and v to fixed points of rotation bc about the x axis. The sphere $S/\mathscr{A}^0$ has an equator given by the arcs rs (semicircles identified), st (half of a dividing circle for reflection in the xy-plane), tu (quarter-circles identified), uv (quarter-circles identified), and vr (half of a dividing circle for reflection in the xy-plane). The action of the quotient group $\mathscr{A}/\mathscr{A}^0$ on the quotient surface $S/\mathscr{A}^0$ is simply reflection in this equator.

6.2.4. Alternative Cayley Graphs Revisited

As we have already indicated at the beginning of this section, reflections necessitate the use of alternative Cayley graphs when studying symmetric imbeddings. The difficulty with alternative Cayley graphs is that the natural action of a group on an alternative Cayley graph for the group is not necessarily free; if the given generating set contains an involution z, then left multiplication by z will "flip" the edge leading from the identity vertex to the vertex z, leaving the midpoint of the edge fixed. This implies that an alternative Cayley graph for a generating set with elements of order 2 is not a regular covering of a bouquet of circles.

There is, however, a characterization of alternative Cayley graphs that is analogous to the corollary to Theorem 2.2.3. Although the action of a group on an alternative Cayley graph for the group is not necessarily free, it is fixed-point free on the vertex set. The converse is also true:

Theorem 6.2.3 (Sabidussi, 1958). *Let the group $\mathscr{A}$ act on the graph G so that the action on the vertex set is free and transitive. Then G is a Cayley graph for $\mathscr{A}$, in which edges in the same orbit of the action correspond to the same generator of $\mathscr{A}$. Moreover, any edge left invariant by a nonidentity element of $\mathscr{A}$ corresponds to an involution in the generating set.*

Proof. Choose any vertex v, and label it by the group identity 1. For each $a \in \mathscr{A}$, label the vertex $\phi_a(v)$ by a. If $\phi_a(v) = \phi_b(v)$, then $\phi_{ab^{-1}}(v) = v$, which implies that $ab^{-1} = 1$, since $\mathscr{A}$ acts freely on the vertex set. Thus, there are no conflicts in the labeling. Moreover, every vertex receives a label, since the action of $\mathscr{A}$ is transitive on vertices.

Next, choose an edge e incident on the vertex v, and suppose that its other endpoint u is labeled with the group element x. Then assign the label x to every edge in the orbit of e. In addition, if $\phi_x(e) \neq e$, then for each $a \in \mathscr{A}$, direct the edge $\phi_a(e)$ so that its initial vertex is $\phi_a(v)$; if $\phi_x(e) = e$, do not assign any directions to the edges labeled x. Observe that the edge $\phi_a(e)$, which is labeled x, leads from vertex $\phi_a(v)$, which is labeled a, to the vertex $\phi_a(u) = \phi_a(\phi_x(v)) = \phi_{ax}(v)$, which is labeled ax. The only way a conflict in assigning directions could arise would be if there were a group element b such that $\phi_a(e) = \phi_b(e)$ and $\phi_a(v) \neq \phi_b(v)$. But then $\phi_{ab^{-1}}(e) = e$, which implies that $\phi_{ab^{-1}}(v) = u$, which in turn implies that $ab^{-1} = x$; and therefore, we have $\phi_x(e) = e$, which implies that no direction should be assigned, thereby ending the conflict. After edges in the orbit of the edge e are labeled and directed, choose another edge incident on v, and then label and direct the edges in its orbit. Continue in this manner until all edges incident on v have been labeled. Since the action of group $\mathscr{A}$ is transitive on vertices, it follows that all edges of the graph G have been labeled. As has already been observed, this labeling transforms G into a Cayley color graph for $\mathscr{A}$.

The given labeling certainly assigns the same generator to every edge in the same orbit. We still must show that if e is labeled x and if $\phi_a(e) = e$ for some nonidentity element a, then x is an involution. So suppose that the initial endpoint of e is labeled b. We may infer that $ab = bx$ and that $a(bx) = b$, since ϕ_a switches the endpoints of e. It follows that $b = a(bx) = (ab)x = bxx$, and therefore that $x^2 = 1$. □

6.2.5. Group Actions and Imbeddings

The following theorem is the fundamental result relating group actions on surfaces and imbeddings of Cayley graphs. The pseudofree case has already been considered in Theorem 4.3.6.

Theorem 6.2.4 (Tucker, 1983). *Let the finite group $\mathscr{A}$ act on an orientable surface S. Then there is a Cayley graph for $\mathscr{A}$ cellularly imbedded in S so that the natural action of $\mathscr{A}$ on the Cayley graph extends to the given action of $\mathscr{A}$ on the surface S.*

Proof. If there are no reflections in the action of $\mathscr{A}$ on S, then, by Theorem 6.2.2, the action is pseudofree, and according to Theorem 4.3.6, the conclusion holds. Assume, therefore, that the action of $\mathscr{A}$ on S contains reflections. Consider the action of the quotient group $\mathscr{A}/\mathscr{A}^0$ on the quotient surface $S/\mathscr{A}^0$, where, as usual, $\mathscr{A}^0$ denotes the index-2, orientation-preserving subgroup of the action $\mathscr{A}$. The only nonidentity element of the action of $\mathscr{A}/\mathscr{A}^0$ on $S/\mathscr{A}^0$ must be a reflection, because it leaves fixed the image in $S/\mathscr{A}^0$ of any circle left fixed by a reflection of the action of $\mathscr{A}$ on S. Let f be this reflection, and let $p: S \rightarrow S/\mathscr{A}^0$ be the natural projection. We wish to construct a 2-vertex graph G cellularly imbedded in $S/\mathscr{A}^0$, such that G contains no branch points of the covering map p and no fixed points of f, such that each face of the imbedding contains at most one branch point, and such that $f(G) = G$. Then the preimage graph $p^{-1}(G)$ is invariant under the action of $\mathscr{A}$ on S, and the action of $\mathscr{A}$ on the vertex set of $p^{-1}(G)$ is transitive and free. It follows from Theorem 6.2.3 that $p^{-1}(G)$ is the natural action of a group on a Cayley graph. The imbedding of $p^{-1}(G)$ in S is cellular, since the imbedding of G in $S/\mathscr{A}^0$ is cellular and has at most one branch point in each face.

To construct the imbedded graph, we first construct half of G and then reflect it to get the other half. Let T be one-half of the reflection f. Suppose that T is a surface of genus g with r boundary components corresponding to the r dividing circles $C_1, \ldots, C_r$ of the reflection f. The case for $g = 1$ and $r = 2$ is illustrated in Figure 6.11. Begin the construction of G by imbedding a bouquet of $2g$ circles at a vertex v in the surface T so that the complement is an open disk with r holes (see the left of Figure 6.11). Next, for each branch point in the interior of T, imbed a loop at vertex v that encloses that branch point but no others (see the middle of Figure 6.11). Finally, for each dividing

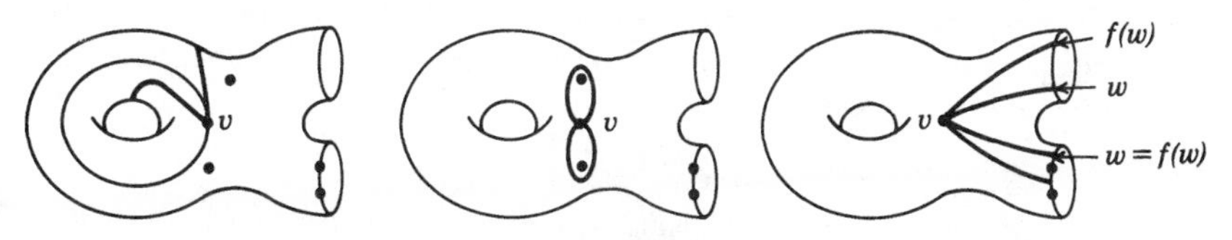

Figure 6.11. The imbedding of half of G in T (black dots are branch points).

circle C_i, choose a finite set W_i of points in C_i, such that each component of $C_i - W_i$ contains exactly one branch point; if C_i contains no branch points, let W_i consist of a single point in C_i. Then, for each point w in W_i, imbed an arc in T running from v to w, and, if $f(w) \neq w$, then imbed another arc from v to $f(w)$ (see the right of Figure 6.11). Let G_0 be the resulting collection of loops and arcs in T, and let $G = G_0 \cup f(G_0)$. Then G is a graph with two vertices v and $f(v)$ (the points in w are now just midpoints of edges). By construction, it follows that $f(G) = G$ and that the imbedding of G is cellular with at most one branch point inside each face. □

Corollary. *Let $\mathscr{A}$ be a finite group. Then the symmetric genus $\sigma(\mathscr{A})$ is the minimum genus over all surfaces on which $\mathscr{A}$ acts, and the strong symmetric genus $\sigma^0(\mathscr{A})$ is the minimum genus over all surfaces on which $\mathscr{A}$ acts preserving orientation.* □

The imbedding $G \to S/\mathscr{A}^0$ constructed in the proof of Theorem 6.2.4 is more important to later sections of this chapter than the theorem itself is. Let $C(\mathscr{A}, X)$ be the Cayley graph $p^{-1}(G)$. Then G is the quotient of $C(\mathscr{A}, X)$ by the index-2 subgroup $\mathscr{A}^0$. This implies that the imbedding $C(\mathscr{A}, X) \to S$ and the covering $p: S \to S/\mathscr{A}^0$ can be obtained by a voltage-graph construction. Moreover, by Theorem 6.1.1, the voltages can be assigned in $\mathscr{A}$, if desired, instead of $\mathscr{A}^0$: the loops in G correspond to elements of the generating set X that are contained in $\mathscr{A}^0$, while the arcs from v to $f(v)$ correspond to orientation-reversing elements of the generating set X. (Note that if the loop e at vertex v corresponds to the generator x, then so does the reflected loop $f(e)$, because the edges of $p^{-1}(e)$ and $p^{-1}(f(e))$ all belong to the same orbit of the action of $\mathscr{A}$.) The voltage-graph construction can be used to read off information about the presentation for $\mathscr{A}$ with generating set X. One example suffices here.

Example 6.2.4. *Let $\mathscr{A}$ be a finite group acting with reflections on the surface S such that the quotient surface $S/\mathscr{A}^0$ is the sphere and such that the covering $p: S \to S/\mathscr{A}^0$ has three branch points in $S/\mathscr{A}^0$, one of which lies on the dividing circle of the reflection f of the quotient action of $\mathscr{A}/\mathscr{A}^0$ on $S/\mathscr{A}^0$. Since f takes branch points to branch points, it follows that the remaining two branch points must be paired by f. In particular, they have the same order. The appearance of graph G is then as illustrated in Figure 6.12 The generator z corresponding to the edge e through the dividing circle of f must be an involution, by Theorem 6.2.3, because at least one edge of its preimage $p^{-1}(e)$ passes through a circle left pointwise-fixed by a reflection of $\mathscr{A}$. Since the order of the net voltage on each face equals the order of the branch point inside that face, the group $\mathscr{A}$ can be presented*

$$\langle x, z: z^2 = 1, x^q = 1, (xzx^{-1}z)^r = 1, \dots \rangle,$$

where r is the order of the branch point on the dividing circle and where q is the order of the other paired branch points.

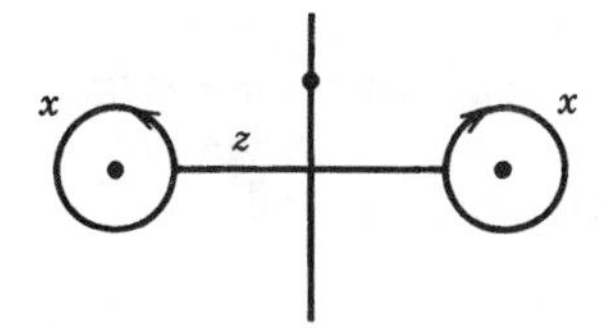

Figure 6.12. The imbedding $G \to S/\mathscr{A}^0$ with voltages in the group $\mathscr{A}$.

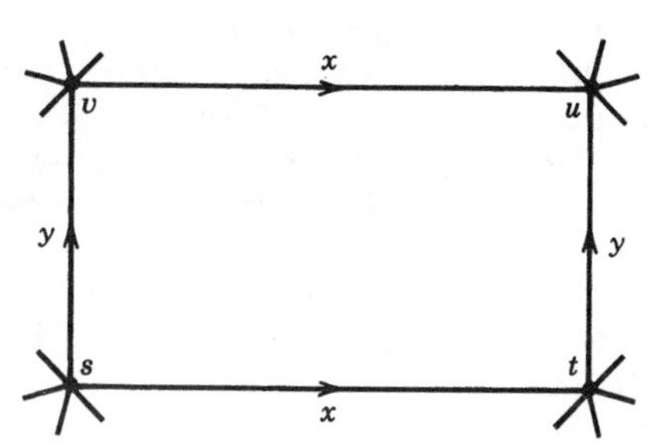

Figure 6.13. A quadrilateral in an imbedding of a Cayley graph $C(\mathscr{A}, X)$, where $\mathscr{A}$ is abelian.

6.2.6. Are Genus and Symmetric Genus the Same?

By definitions, $\gamma(\mathscr{A}) \leq \sigma(\mathscr{A})$ for any group $\mathscr{A}$. We shall show for most abelian groups $\mathscr{A}$, the inequality is strict, that is, $\gamma(\mathscr{A}) < \sigma(\mathscr{A})$.

Let $\mathscr{A} = \mathscr{Z}_{10} \times \mathscr{Z}_{10} \times \mathscr{Z}_{10}$, and let $X = \{x, y, z\}$ be the canonical generating set. By Section 6.1, the Cayley graph $C(\mathscr{A}, X)$ has a quadrilateral imbedding, and the genus of this imbedding is $\gamma(\mathscr{A})$. We claim, however, that no symmetric imbedding of $C(\mathscr{A}, X)$ can have any quadrilaterals at all. The only quadrilaterals in $C(\mathscr{A}, X)$ correspond to commutators in the generators. Suppose that the relator $xyx^{-1}y^{-1}$ bounds a face in a symmetric imbedding of $C(\mathscr{A}, X)$, as illustrated in Figure 6.13. Then the rotations at the vertices of this quadrilateral must have the form (starting at the lower left vertex):

$$s. \quad \ldots xy \ldots$$
$$t. \quad \ldots yx^{-1} \ldots$$
$$u. \quad \ldots x^{-1}y^{-1} \ldots$$
$$v. \quad \ldots y^{-1}x \ldots$$

Since the imbedding is symmetric, each of these rotations must be the same or reversed. Since the reverse of $\ldots xy \ldots$ is $\ldots yx \ldots$, the rotation at t is the same as the rotation at s. By similar reasoning, the rotations at u and v must be the same as those at t and u. From the information given at s, t, and u, this rotation must be $\ldots xyx^{-1}y^{-1} \ldots$. The information given at vertex v implies that x immediately follows y^{-1} in this same rotation, but that is impossible because z and z^{-1} must appear somewhere in the rotation. We conclude that no symmetric imbedding of $C(\mathscr{A}, X)$ can have a quadrilateral $xyx^{-1}y^{-1}$, and hence $C(\mathscr{A}, X)$ has no symmetric, quadrilateral imbedding. Since $\gamma(\mathscr{A})$ is strictly less than the genus of any Cayley graph for $\mathscr{A}$ corresponding to any noncanonical generating set, it follows that $\gamma(\mathscr{A}) < \sigma(\mathscr{A})$.

Of course, it should not be surprising that minimal-genus Cayley graph imbeddings can be highly nonsymmetric. Jungerman and White's constructions of quadrilateral imbeddings given in Section 6.1 are quite complicated partly because so many different rotations are needed. Even the more easily pictured tubing construction of Section 3.5, which suffices to obtain a quadrilateral imbedding for $\mathscr{A} = \mathscr{Z}_{10} \times \mathscr{Z}_{10} \times \mathscr{Z}_{10}$, is complicated if viewed in terms of rotations; in fact, eight different rotations are used (Exercise 15; see also Exercise 6.1.1).

The inequality $\sigma(\mathscr{A}) \leq \sigma^0(\mathscr{A})$ can also be strict. In particular, the example with $\mathscr{A} = \mathscr{Z}_2 \times \mathscr{Z}_4$ at the beginning of this section satisfies $\sigma(\mathscr{A}) = 0$ and $\sigma^0(\mathscr{A}) = 1$ (Exercise 14).

6.2.7. Euclidean Space Groups and the Torus

Group actions on a surface do not simply appear from nowhere; they arise from some geometric structure of the surface. To understand a particular group action, one often must search for some geometry that the group action respects. Perhaps the most revealing example is the relationship between the torus and plane Euclidean geometry.

Example 6.2.5. *Consider the pattern of bricks shown in Figure* 6.14, *which the reader has probably seen in a courtyard or patio. Imagine the pattern extending infinitely in all directions in the plane. What rigid motions, that is, Euclidean isometries of the plane, take this pattern to itself? First, one can simply "translate" the whole pattern, so that, for instance, the set of four outlined bricks around point u displaces those around point v. Second, one can "rotate" the pattern* 90° *about the point u (note that this rotation takes the entire global pattern onto itself, not just the four local bricks). Third, one can "reflect" the pattern about a horizontal or vertical line such as k or l. Finally, one can translate half the distance from u to v, at the same time reflecting about the line through u and v (this last type of motion is called a "glide"). The whole pattern can be generated as the orbit under rigid motions of a single half-brick like that shaded near the point w in Figure* 6.14.

The collection of all Euclidean isometries leaving the pattern in Figure 6.14 invariant forms a group $\mathscr{C}$, called the "symmetry group" of the pattern. This group has two important properties:

1. It contains translations in linearly independent (i.e., different) directions in the plane.
2. It is discrete; that is, for any point u, the orbit $\{c(u) \mid c \in \mathscr{C}\}$ has no accumulation points.

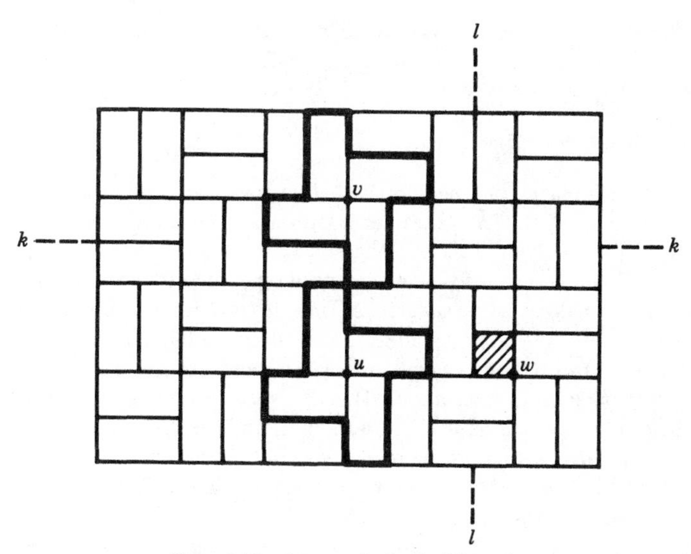

Figure 6.14. A pattern in the Euclidean plane.

Another way of stating property 2 is that for any point u, there is a minimum distance d such that every isometry in the group $\mathscr{C}$ either moves u at least the distance d or does not move u at all. Any group of isometries of the Euclidean plane satisfying properties 1 and 2 is called a (two-dimensional) "Euclidean space group" (or alternatively a Euclidean "crystallographic group"). Every pattern in the Euclidean plane having translational symmetry in more than one direction and having no infinitesimal symmetry (N.B.: a pattern of vertical stripes has infinitesimally small vertical translations as symmetries) has a Euclidean space group as its symmetry group. Conversely, every Euclidean space group has patterns for which it is the symmetry group: one first finds a "fundamental domain" like the half-brick in Figure 6.14, next draws any motif on this domain, and then replicates the motif throughout the plane using the given space group.

The history of space groups, from the art of the Alhambra and of M. C. Escher to the crystallography of Federov (1891) to the mathematics of Bieberbach (1911), is too rich to elaborate here. The reader is urged to read Weyl (1952) or Schattschneider (1978) for an informal account. Lyndon (1985) and Coxeter and Moser (1957) provide the mathematical details. The important facts we need about Euclidean space groups are summarized in the following theorem.

Theorem 6.2.5. *There are exactly* 17 *Euclidean space groups* (*up to group isomorphism*), 5 *of which contain only orientation-preserving isometries. The collection* $\mathscr{T}$ *of all translations contained in a given Euclidean space group* $\mathscr{C}$ *is a* *normal subgroup of* $\mathscr{C}$, *is isomorphic to* $\mathscr{Z} \times \mathscr{Z}$, *and has as quotient* $\mathscr{C}/\mathscr{T}$ *either the trivial group or one of the cyclic groups* $\mathscr{Z}_2, \mathscr{Z}_3, \mathscr{Z}_4, \mathscr{Z}_6$ *or one of the dihedral groups* $\mathscr{D}_2, \mathscr{D}_3, \mathscr{D}_4, \mathscr{D}_6$.

Proof. Omitted. □

The first statement of Theorem 6.2.5 is a lovely contribution to mathematics. Although it may well have been known to artists for centuries, it was not proved until Polya (1924) and Niggli (1924). Most of the facts about the subgroup $\mathscr{T}$ are easily verified. The composition of two translations is clearly a translation, making $\mathscr{T}$ a subgroup. That $\mathscr{T}$ is normal follows from the classification of isometries as either translations, rotations, or glides (the conjugate of a translation preserves orientation and has no fixed points and, hence, must be another translation). That $\mathscr{T}$ is isomorphic to $\mathscr{Z} \times \mathscr{Z}$ follows from the obvious commutativity of translations and the discreteness of $\mathscr{C}$ (see Exercise 17). That $\mathscr{C}/\mathscr{T}$ must be on the given list of groups follows from the "crystallographic restriction" that any rotation in a Euclidean space group must have order 2, 3, 4, or 6 (see Exercise 18).

The present reason we are interested in Euclidean space groups is that they give rise to group actions on the torus. For example, if we simply "roll up" the pattern in Figure 6.14 in both the vertical and horizontal directions, identifying in the process the top and bottom edges and the right and left edges of Figure 6.14, we obtain a similar pattern on the torus. Let t be the vertical translation from u to v, and let s be the horizontal translation from u to w. Then the "rolling up" is accomplished by taking the quotient of the Euclidean plane by the group $\mathscr{N}$ generated by s^3 and t^2. In particular, the finite quotient group $\mathscr{C}/\mathscr{N}$ acts on the torus as the symmetry group of the rolled-up pattern. The following theorem gives the general situation.

Theorem 6.2.6. *Let* $\mathscr{C}$ *be a Euclidean space group and* $\mathscr{N}$ *any normal subgroup of finite index in* $\mathscr{C}$. *Then the quotient group* $\mathscr{C}/\mathscr{N}$ *acts on the torus or on the sphere.*

Proof. Let $\mathscr{T}$ be the subgroup of translations in $\mathscr{C}$. Let $\mathscr{C}^0$ be the orientation-preserving subgroup of $\mathscr{C}$, and let $\mathscr{N}^0 = \mathscr{C}^0 \cap \mathscr{N}$. Since $\mathscr{N}$ has finite index in $\mathscr{C}$, the normal subgroup $\mathscr{N}^0 \cap \mathscr{T}$ must contain translations in two directions. Thus the quotient surface $S' = E/\mathscr{N}^0 \cap \mathscr{T}$ of the Euclidean plane E by the action of $\mathscr{N}^0 \cap \mathscr{T}$ is a torus (Exercise 17 shows that $\mathscr{N}^0 \cap \mathscr{T}$ is generated by two translations that roll up the plane in two directions). The natural action of the quotient group $\mathscr{N}^0/\mathscr{N}^0 \cap \mathscr{T}$ on S' has quotient $S = E/\mathscr{N}^0$. Since $\mathscr{N}^0$ preserves orientation, it follows that the surface S is orientable and that the projection $S' \to S$ is a regular branched covering. By the Riemann–Hurwitz equation, we know that $\chi(S') \le \chi(S)$. Since S' is the torus, it follows that S must be the torus or the sphere.

In order to complete the proof, we first observe that the group $\mathscr{C}/\mathscr{N}^0$ acts naturally on S. Thus, if $\mathscr{N}^0 = \mathscr{N}$, we already have the desired action.

Alternatively, if $\mathcal{N}^0 \neq \mathcal{N}$, then $\mathcal{N}\mathcal{C}^0 = \mathcal{C}$, since the subgroup $\mathcal{C}^0$ has index 2 in $\mathcal{C}$. By the fundamental isomorphism theorem for quotient groups, we know that $\mathcal{C}^0/\mathcal{N}^0$ is isomorphic to $\mathcal{N}\mathcal{C}^0/\mathcal{N}$. Since $\mathcal{C}^0/\mathcal{N}^0$ acts on S and $\mathcal{N}\mathcal{C}^0 = \mathcal{C}$, it follows that $\mathcal{C}/\mathcal{N}$ acts on the torus or on the sphere. □

Corollary. *If $\mathcal{A}$ is a finite quotient group of a Euclidean space group, then* $\sigma(\mathcal{A}) \leq 1$.

One can also define "spherical space groups" using spherical geometry and "hyperbolic space groups" using hyperbolic geometry. In both cases we require the group to be discrete. For the sphere, which is a compact surface, this automatically means the group is finite. For the hyperbolic plane, we require the quotient space under the group action to be compact, in analogy to condition 1. In either case, there is no subgroup analogous to the translation subgroup $\mathcal{T}$ for Euclidean space groups.

6.2.8. Triangle Groups

Triangle groups are a particularly important kind of space group, especially for non-Euclidean geometry. We give here a sketchy account of the salient features of these groups. For more details, pictures, history, and rigor, we highly recommend Magnus (1974). For geometry, both Euclidean and non-Euclidean, see Coxeter (1961) or Lyndon (1985).

Example 6.2.6. *We begin with a "fundamental" isosceles right triangle in the Euclidean plane, such as the one outlined in the pattern in Figure* 6.15. *Let x and y be the reflections about the lines along the legs of the triangle, and let z be reflection about the line along the hypotenuse. The group $\mathcal{C}$ of isometries generated by x, y, and z flops the triangle all over the plane, creating the pattern shown in Figure* 6.15. *Half the triangles in the pattern have the same orientation*

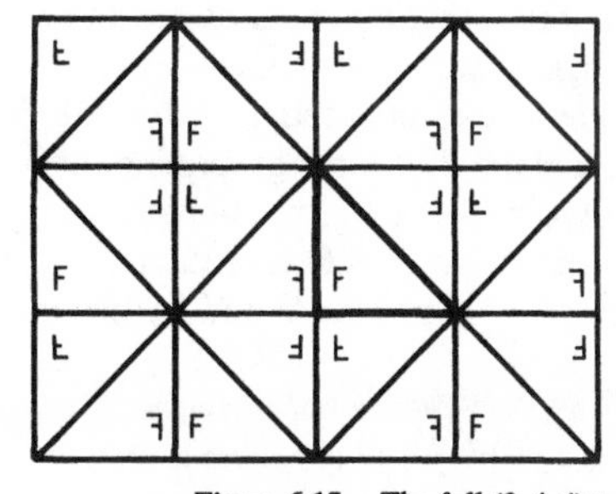

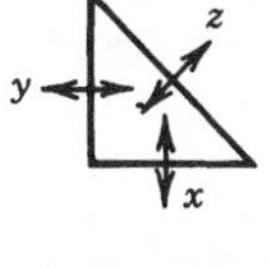

Figure 6.15. The full (2, 4, 4) triangle group.

as the fundamental triangle and half the opposite; the forward and backward letter Fs tell which is which. The group $\mathscr{C}$ is the symmetry group of the pattern and hence is a Euclidean space group.

This construction applies to certain other triangles. Since the triangles of the pattern must fit together around each vertex in a circle of paired "forward" and "backward" triangles, each angle α of the fundamental triangle must be of the form π/p, where p is an integer (so that 2α fits into 2π a whole number of times). A "(p, q, r) triangle" is one with angles $\pi/p, \pi/q, \pi/r$, where p, q, and r are positive integers. Because the sum of the angles of a Euclidean triangle is π, the only Euclidean (p, q, r) triangles are $(2,4,4)$, $(2,3,6)$, and $(3,3,3)$. There are, however, other kinds of geometry. In spherical geometry, where lines are great circles, triangles have angle sum greater than π. The possible spherical (p, q, r) triangles are $(1, r, r)$, $(2,2,r)$, for any $r \geq 2$, and $(2,3,3)$, $(2,3,4)$, $(2,3,5)$; although $(1, r, r)$ is a degenerate triangle, it still can be used. In hyperbolic geometry, which can be modeled by the interior of the unit disk, where "lines" are circles orthogonal to the boundary and "reflections" are inversions in those circles, triangles have angle sum less than π. There are hyperbolic (p, q, r) triangles for any p, q, r such that $1/p + 1/q + 1/r < 1$.

Given any (p, q, r) triangle, the group of isometries, either Euclidean or spherical or hyperbolic, generated by reflections x, y, z in the three sides of the triangle is called the "full (p, q, r) triangle group". The symmetry group of Figure 6.15 is the full $(2,4,4)$ triangle group. The number of elements of a triangle group equals the number of copies of the original triangle in the pattern, because there is a unique isometry in the group taking the fundamental triangle to any other triangle (isometries are determined by what they do to a single triangle and no isometry in the group takes the fundamental triangle to itself, other than the identity). Thus, Euclidean and hyperbolic triangle groups are infinite, because the Euclidean and hyperbolic plane are unbounded, but spherical triangle groups are finite. In fact, since the area of a triangle on the unit sphere is equal to the excess of its angle sum compared with π, the orders of the full triangle groups for the sphere are easily computed. For example, the area of a $(2,3,5)$ triangle on the unit sphere is $\pi/2 + \pi/3 + \pi/5 - \pi$, or $\pi/30$. Since the total area of the unit sphere is 4π, the order of the full $(2,3,5)$ triangle group must be 120.

A presentation in terms of generators and relations for the full triangle group (p, q, r) is not difficult to obtain. The three generating reflections x, y, and z clearly have order 2. The product of two reflections in lines meeting at an angle α is a rotation by 2α about the intersection of the two lines. Thus xy, yz, and xz are rotations by $2\pi/p$, $2\pi/q$, and $2\pi/r$ and hence have orders p, q, and r. So far, we have

$$x^2 = y^2 = z^2 = 1, (xy)^p = (yz)^q = (xz)^r = 1$$

These are in fact the only relators needed. We sketch a proof. First, note that the dual graph to the triangular pattern for a (p, q, r) triangle group $\mathscr{C}$ is a Cayley graph for $\mathscr{C}$, since $\mathscr{C}$ acts on the dual graph transitively, fixing no vertex (i.e., fixing no triangle in the original pattern). Any relator in the generators corresponds to a closed walk in this Cayley graph. Since the Cayley graph is imbedded in the plane, every cycle, and hence very closed walk, can be expressed in terms of the face boundary cycles, which simply correspond to the relators $(xy)^p = (yz)^q = (xz)^r = 1$. Thus, every relator can be expressed in terms of the already given relators. (We are being vague about what we mean by "expressed", but this can be made rigorous.)

Given a full (p, q, r) triangle group $\mathscr{C}$, the subgroup $\mathscr{B}$ generated by the rotations xy, yz, and xz at the three vertices of the fundamental triangle is called the "ordinary" (p, q, r) triangle group. Since $(xy)(yz) = xy^2z = xz$, this group is generated by $u = xy$ and $v = yz$ alone. It has the presentation

$$\langle u, v : u^p = v^q = (uv)^r = 1\rangle$$

Moreover, this subgroup has index 2 in the full triangle group [the map $f: \mathscr{C} \to \mathscr{Z}_2$ given by $f(x) = f(y) = f(z) = 1$ has kernel $\mathscr{B}$], and thus $\mathscr{B}$ coincides with the orientation-preserving subgroup $\mathscr{C}^0$.

We summarize these facts in the following theorem.

Theorem 6.2.7. *A full (p, q, r) triangle group $\mathscr{C}$ has the presentation*

$$\langle x, y, z : x^2 = y^2 = z^2 = 1, (xy)^p = (yz)^q = (xz)^r = 1\rangle$$

Its orientation-preserving subgroup $\mathscr{C}^0$ is the ordinary triangle group generated by the rotations $u = xy$, $v = yz$ and has the presentation

$$\langle u, v : u^p = v^q = (uv)^r = 1\rangle$$

If (p, q, r) is $(1, r, r)$, $(2, 2, r)$, $(2, 3, 3)$, $(2, 3, 4)$, or $(2, 3, 5)$, then the triangle is spherical and the group $\mathscr{C}$ is finite of order $4/(1/p + 1/q + 1/r - 1)$. If (p, q, r) is $(2, 3, 6)$, $(2, 4, 4)$, or $(3, 3, 3)$, then the triangle is Euclidean and $\mathscr{C}$ is an infinite Euclidean space group. Otherwise, $\mathscr{C}$ is a hyperbolic full triangle group.

An especially interesting application of triangle groups to graph imbeddings can be found in the work of Jones and Singerman (1978) and Bryant and Singerman (1985). By a clever combinatorial definition of graph imbedding or "map", they show that every map is the quotient of part of some $(2, q, r)$ triangle pattern in the sphere, Euclidean plane, or hyperbolic plane. Map automorphisms are then interpreted as quotients of subgroups of the ordinary $(2, q, r)$ triangle group in the orientable case and the full $(2, q, r)$ triangle group in the nonorientable case.

6.2.9. Exercises

1. Prove that an imbedding of a Cayley graph $C(\mathscr{A}, X)$ is weakly symmetric if and only if there is a subgroup $\mathscr{B}$ of index 2 in the group $\mathscr{A}$ such that every vertex of the Cayley graph labeled by an element of $\mathscr{B}$ has the same rotation and all other vertices have the reversed rotation.
2. Let S be a surface with r boundary components. Then $\chi(S)$ is defined to be $\chi(S) - r$, where S is the closed surface obtained by spanning each boundary component with a disk. Find all orientable surfaces S with boundary such that $\chi(S) = -5$.
3. Let S_0 be one-half of a reflection of the surface S. Prove that $\chi(S) = 2\chi(S_0)$.
4. Find all reflections of the surface of genus 2.
5. Given simplexes σ in $\mathbb{R}^m$ and τ in $\mathbb{R}^n$, a map $f\colon \sigma \to \tau$ is said to be "linear" if f takes vertices of σ to vertices of τ, and if f is the restriction to σ of an affine map $F\colon \mathbb{R}^m \to \mathbb{R}^n$ [i.e. $F(x) = L(x) + b$, where $L\colon \mathbb{R}^m \to \mathbb{R}^n$ is a linear transformation and $b \in \mathbb{R}^n$]. Find a linear map from the simplex σ in $\mathbb{R}^2$ having vertices $(0,0)$, $(1,0)$, $(0,1)$, to the simplex τ in $\mathbb{R}^3$ having vertices $(1,2,1)$, $(3,5,2)$, $(0,2,8)$. Prove that a linear map between simplexes is uniquely determined by what it does to vertices.
6. Given simplicial complexes K and L, a continuous map $f\colon |K| \to |L|$ is "piecewise linear" if there are subdivisions K', L' of K, L such that f restricted to each simplex of K' is linear (K' is a subdivision of K if $|K'| = |K|$ and every vertex of K is a vertex of K'). Let K be the simplicial complex in $\mathbb{R}^2$ consisting of the 2-simplex with vertices $(0,0)$, $(0,1)$, and $(1,0)$ (and all its faces). Find a piecewise-linear map $f\colon |K| \to |K|$ that leaves fixed the two legs of the right triangle $|K|$ and takes the point $(1/3, 2/3)$ to the point $(2/3, 1/3)$.
7. The set of fixed points of a piecewise-linear homeomorphism can be quite complicated. Let L be the two-dimensional simplicial complex in $\mathbb{R}^2$ formed by four right triangles with vertices at $(0,0)$, $(\pm 1, 0)$, $(0, \pm 1)$. Use the function f of Exercise 6 to construct a piecewise-linear homeomorphism $g\colon |L| \to |L|$ that leaves fixed all points of $|L|$ on the x axis, on the y axis, or in the second quadrant. Sketch $|L|$ and its set of fixed points. Does g have finite order?
8. Prove that if $h\colon \mathbb{R}^2 \to \mathbb{R}^2$ is a linear transformation of finite order leaving the x axis fixed, then either h is the identity or h is a reflection across the x axis.
9. Let g be an orientation-preserving, piecewise-linear homeomorphism $g\colon |L| \to |L|$, where $|L|$ is a surface. Use Exercise 8 to prove that if g has finite order and leaves some 1-simplex of L fixed, then g is the identity.
10. Use Exercise 9 to prove Theorem 6.2.2 for piecewise-linear homeomorphisms.

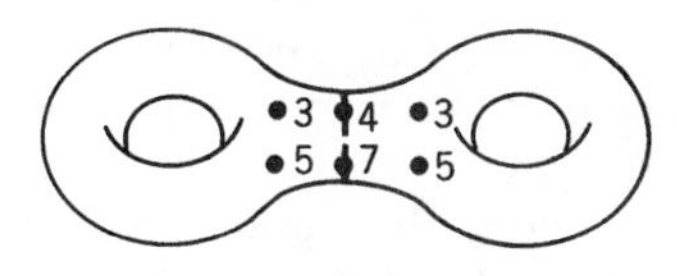

Figure 6.16. A quotient surface with branch points and quotient reflection.

11. Prove that $\psi_{\bar{a}\bar{b}} = \psi_{\bar{a}}\psi_{\bar{b}}$, where ψ defines the natural quotient action of the quotient group $\mathscr{A}/\mathscr{B}$ on the quotient surface $S/\mathscr{B}$.

12. Suppose that the group $\mathscr{A}$ acts on a surface S with reflections so that the quotient surface $S/\mathscr{A}^0$ is a surface of genus 2. Suppose that the quotient action of $\mathscr{A}/\mathscr{A}^0$ on $S/\mathscr{A}^0$ is a reflection with a single dividing circle and that the branch points of $p: S \to S/\mathscr{A}^0$ have orders and locations as given by Figure 6.16. Find a partial presentation for the group $\mathscr{A}$ by constructing a 2-vertex voltage graph as in the proof of Theorem 6.2.4.

13. Prove that a Cayley graph imbedding $C(\mathscr{A}, X) \to S$ is strongly symmetric if and only if it is the derived imbedding of a 1-vertex imbedded voltage graph.

14. Prove that $\sigma^0(\mathscr{Z}_2 \times \mathscr{Z}_4) = 1$ (see also Theorem 6.3.1).

15. List all the rotations used in the quadrilateral imbedding given in Section 3.5 for the Cayley graph $C(\mathscr{A}, X)$, where $\mathscr{A} = \mathscr{Z}_{10} \times \mathscr{Z}_{10} \times \mathscr{Z}_{10}$ and $X = \{x, y, z\}$, a canonical generating set for $\mathscr{A}$.

16. Give an example of an involution on the torus that is not a reflection.

17. Let $\mathscr{T}$ be a discrete group of translations of the Euclidean plane containing translations in independent directions. Choose any point P. Let Q be the closest point to P in the orbit of P, and let R be the closest point to P in the orbit of P but not on the line through P and Q. Prove that the translations by vectors $\overline{PQ}$ and $\overline{PR}$ generate $\mathscr{T}$.

18. Let $\mathscr{C}$ be a Euclidean space group, and let $r \in \mathscr{C}$ be a rotation about some point P. Let Q be the closest point to P in the orbit of P. Prove that if r is a rotation through an angle of less than $\pi/3$, then $r(Q)$ is closer to Q than Q is to P. Why does this contradict the choice of Q? Use a similar argument to show that r is not a rotation of order 5 either.

6.3. GROUPS OF SMALL SYMMETRIC GENUS

A first project in studying the genus of groups is to identify all the groups of genus 0, then maybe all the groups of genus 1. We shall see that the analogous problem for symmetric genus happens to be a quite tractable problem, and that preliminary consideration of the symmetric genus gives remarkably accurate insight into the general case, whose derivation is otherwise rather lengthy. Study of the symmetric genus is additionally rewarding, in that it also reveals rich connections with Euclidean and non-Euclidean geometry. Moreover, the moderately easy success in determining all the groups of symmetric genus 0 or

1 encourages the study of groups of given symmetric genus greater than 1, where we encounter Hurwitz's theorem that establishes a bound on the order of any group acting on a surface of given genus.

The geometric group actions introduced at the end of the previous section provide an abundance of groups acting on the torus and the sphere. An optimist might hope that all actions arise geometrically, that every finite group acting on the sphere is isomorphic to a spherical space group, and that every finite group acting on the torus is a quotient of a Euclidean space group. The main theorems of this section show that such optimism is justified. They also classify all such groups in terms of lists of partial presentations.

6.3.1. The Riemann–Hurwitz Equation Revisited

The main tool used to study symmetric genus is the Riemann–Hurwitz equation, which has appeared here as Theorem 4.2.3. Because the equation was first introduced in the context of general (not necessarily regular) branched coverings, it will be helpful to restate the equation in terms of group actions and regular branched coverings.

Let the group $\mathscr{A}$ act pseudofreely on the surface S, and let $p\colon S \to S/\mathscr{A}$ be the associated regular branched covering map. Let $Y \subset S/\mathscr{A}$ be the branch point set of p. Recall that, by Theorems 4.3.4 and 4.3.5, for any given branch point $y \in Y$, the order of every prebranch point in the covering surface S is the same. It follows that the deficiency of y is given by the equation

$$\operatorname{def}(y) = \#\mathscr{A} - \#\mathscr{A}/r_y$$

where r_y is the order of the branch point y. The Riemann–Hurwitz equation then assumes the second form

$$\chi(S) = \#\mathscr{A}\left[\chi(S/\mathscr{A}) - \sum_{y \in Y} (1 - 1/r_y)\right]$$

The key observation about the order of a branch point is that if the regular branched covering $p\colon S \to S/\mathscr{A}$ corresponds to an imbedded voltage graph, then the order of each branch point is the order of the net voltage on the face containing that branch point.

Example 6.2.4 Revisited. *Suppose that beyond the original circumstances, we further specify that $\chi(S) = 0$. Then, since $\chi(S/\mathscr{A}^0) = 2$, the second form of the Riemann–Hurwitz equation implies that*

$$0 = \#\mathscr{A}^0(2 - (1 - 1/q) - (1 - 1/q) - (1 - 1/r))$$

We conclude that $2/q + 1/r = 1$. It is easily verified by simple arithmetic that the only possibilities are $q = 4$, $r = 2$, and $q = 3$, $r = 3$. It follows that there are only two classes of groups that act on the torus in the manner described in the example itself, namely, the groups with these presentations:

$$\langle x, z\colon z^2 = 1, x^4 = 1, (xzx^{-1}z)^2 = 1, \ldots \rangle$$

$$\langle x, z\colon z^2 = 1, x^3 = 1, (xzx^{-1}z)^3 = 1, \ldots \rangle$$

6.3.2. Strong Symmetric Genus 0

Because dihedral groups are encountered frequently in classifying the groups that act on the sphere, a word or two about presentations for dihedral groups is in order. The dihedral group $\mathscr{D}_n$ is defined classically to be the group of symmetries of a regular n-sided polygon. Its order is $2n$, and it has two obvious presentations. If the generator x is rotation of the n-gon by $360°/n$, and if the generator y is any reflection of the n-gon, then we have the presentation

$$\mathscr{D}_n = \langle x, y\colon x^n = y^2 = 1,\ yxyx = 1 \rangle$$

If x and y are reflections of the n-sided polygon through different diameters making an angle of $180°/n$ with each other, then we have

$$\mathscr{D}_n = \langle x, y\colon x^2 = y^2 = (xy)^n = 1 \rangle.$$

In both presentations, it is easily verified that the permutations x and y generate the group $\mathscr{D}_n$ and that the given relations hold. No more relations are needed to define $\mathscr{D}_n$ because it can be shown that the order of the group presented is the same as the order of $\mathscr{D}_n$, namely, $2n$ (see Exercise 1). The second presentation given is particularly useful since it characterizes the dihedral groups as the groups generated by a pair of involutions.

We divide the classification of groups acting on the sphere into two cases. The first case is the subclass of orientation-preserving actions, the groups with strong symmetric genus 0.

Theorem 6.3.1. *For a finite group $\mathscr{A}$, the strong symmetric genus $\sigma^0(\mathscr{A})$ is 0 if and only if either the group $\mathscr{A}$ is isomorphic to one of the cyclic groups $\mathscr{Z}_n$ or to one of the dihedral groups $\mathscr{D}_n$, or else the group $\mathscr{A}$ has one of the following presentations and the given order*:

$$\text{(a)}\quad \langle x, y\colon x^3 = y^2 = (xy)^3 = 1, \ldots \rangle, \quad \#\mathscr{A} = 12$$

$$\text{(b)}\quad \langle x, y\colon x^3 = y^2 = (xy)^4 = 1, \ldots \rangle, \quad \#\mathscr{A} = 24$$

$$\text{(c)}\quad \langle x, y\colon x^3 = y^2 = (xy)^5 = 1, \ldots \rangle, \quad \#\mathscr{A} = 60$$

Proof. If $\sigma^0(\mathscr{A}) = 0$, then the group $\mathscr{A}$ acts on the sphere S preserving orientation. By the corollary to Theorem 6.2.2, the action is pseudofree. Let $p: S \to S/\mathscr{A}$ be the associated regular branched covering, with branch set B. From the second form of the Riemann–Hurwitz equation, we observe that

$$2 = \#\mathscr{A}\left[\chi(S/\mathscr{A}) - \sum_{y \in B} (1 - 1/r_y)\right]$$

We infer from this equation that the characteristic $\chi(S/\mathscr{A})$ must be positive. Since the quotient surface $S/\mathscr{A}$ is orientable, the only possibility is that $S/\mathscr{A}$ is itself the sphere and that $\chi(S/\mathscr{A}) = 2$. Moreover, because $1 - 1/r_y \geq \frac{1}{2}$ for each branch point y, it follows that there are at most three branch points. If there are fewer than two, then

$$2 = \#\mathscr{A}\left[2 - \sum(1 - 1/r_y)\right] > \#\mathscr{A}$$

from which it follows that $\mathscr{A}$ is the trivial group. We conclude that there are either two or three branch points.

Suppose, on one hand, that there are two branch points. Then imbed the 1-vertex, 1-loop graph B_1 in the sphere $S/\mathscr{A}$ so that the loop separates the branch points. By Theorem 4.3.5, the covering $p: S \to S/\mathscr{A}$ is the derived covering for some voltage assignments on the imbedded bouquet B_1 into the group $\mathscr{A}$. Since B_1 has only one edge, the group $\mathscr{A}$ must be generated by a single element, that is, the group $\mathscr{A}$ is cyclic.

Now suppose, on the other hand, that there are three branch points. Let $q \leq r \leq s$ be the orders of those branch points. The only values for the triple (q, r, s) that leave the right-hand side of the Riemann–Hurwitz equation positive are $(2, 2, n)$, $(2, 3, 3)$, $(2, 3, 4)$, $(2, 3, 5)$. Imbed the bouquet B_2 into the quotient surface $S/\mathscr{A}$ as in Figure 6.17 (in which heavy dots represent the branch points of the given orders). Again it follows from Theorem 4.3.5 that the covering $p: S \to S/\mathscr{A}$ is obtainable by assigning the two loops voltages $x, y \in \mathscr{A}$, as shown in Figure 6.17. Since the order of the net voltage on each face equals the order of the enclosed branch point, we infer that the group $\mathscr{A}$ may be presented

$$\mathscr{A} = \langle x, y \colon x^r = y^q = (xy)^s = 1, \ldots \rangle$$

If $(q, r, s) = (2, 2, n)$, then $\mathscr{A}$ is the dihedral group $\mathscr{D}_n$. The other three possibilities for (q, r, s) yield the three presentations given in the theorem.

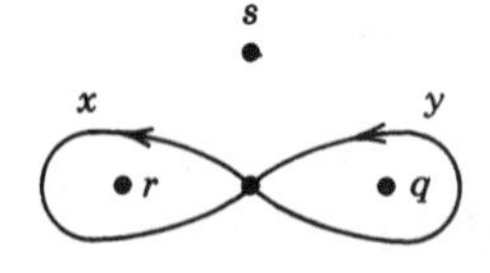

Figure 6.17. An imbedded voltage graph.

Moreover, in each of these cases the order of $\mathscr{A}$ is determined exactly by the Riemann–Hurwitz equation. For example, if $(q, r, s) = (2, 3, 5)$, then

$$2 = \#\mathscr{A}(2 - 1/2 - 2/3 - 4/5) = \#\mathscr{A}/30$$

so $\#\mathscr{A} = 60$.

We have shown that if $\sigma^0(\mathscr{A}) = 0$, then $\mathscr{A}$ is as described in the theorem. The converse, that if $\mathscr{A}$ is isomorphic to $\mathscr{Z}_n$ or $\mathscr{D}_n$ or has presentations (a), (b), or (c), then $\mathscr{A}$ acts on the sphere, follows directly from the voltage-graph constructions already given in this proof. □

The reader should recognize immediately the presentation $\langle x, y \colon x^q = y^r = (xy)^s = 1, \ldots \rangle$ as an ordinary spherical (q, r, s) triangle group with perhaps some extra relations. In particular, by Theorem 6.2.7, the given presentation yields a group of the required order without any extra relations. [For those uncomfortable with the geometry of the sphere, or maybe just to make the reader appreciate geometric arguments, Exercises 2–4 show how to use algebraic methods to establish that presentations (a)–(c) are automatically finite even without any extra relations.] Thus presentations (a)–(c) are precisely the ordinary $(2, 3, 3)$, $(2, 3, 4)$, and $(2, 3, 5)$ ordinary triangle groups.

To identify these spherical triangle groups as more familiar groups, all we need to do is find a group having two generators of the right order whose product has the right order. The alternating group $\mathscr{A}_4$ is generated by $x = (1\,2\,3)$, $y = (1\,2)(3\,4)$. Since $xy = (1)(2\,4\,3)$ has order 3, the group presented by (a) must be $\mathscr{A}_4$. Similarly, presentation (b) is the symmetric group $\mathscr{S}_4$ generated by $x = (1\,2\,3)$ and $y = (3\,4)$, and presentation (c) is the alternating group $\mathscr{A}_5$ generated by $x = (1\,2\,3)$ and $y = (2\,4)(3\,5)$.

Corollary (Maschke, 1896). *The finite group $\mathscr{A}$ has strong symmetric genus* 0 *if and only if it is isomorphic to one of the groups, $\mathscr{Z}_n$, $\mathscr{D}_n$, $\mathscr{A}_4$, $\mathscr{S}_4$, or $\mathscr{A}_5$. These groups correspond respectively to the ordinary* $(1, n, n)$, $(2, 2, n)$, $(2, 3, 3)$, $(2, 3, 4)$, $(2, 3, 5)$ *triangle groups.*

We have credited Maschke with the classification of the groups of symmetrics and strong symmetric genus 0, because his 1896 paper implies such a classification. However, Maschke was actually computing groups of genus (i.e., White genus) 0. In truth, the classification of spherical space groups is more a matter of folklore, and the general topological problem of finding which finite groups act on the sphere can be traced through a variety of papers, including Kerekjarto (1919), Scherrer (1929), and Babai et al. (1972).

The case of orientation-reversing actions on the sphere is also handled by the Riemann–Hurwitz equation. It is complicated, however, by reflections.

Theorem 6.3.2. *Suppose that the finite group $\mathscr{A}$ acts on the sphere, reversing orientation. The action has no reflections if and only if $\mathscr{A}$ is isomorphic to $\mathscr{Z}_{2n}$*

for some $n > 1$. The action has reflections if and only if either $\mathscr{A}$ is isomorphic to $\mathscr{D}_n$, $\mathscr{Z}_2 \times \mathscr{Z}_n$, $\mathscr{Z}_2 \times \mathscr{D}_n$ or $\mathscr{A}$ has one of the following presentations and the given order:

(a) $\langle x, y: x^3 = y^2 = (xyx^{-1}y)^2 = 1, \ldots \rangle$, $\#\mathscr{A} = 24$

(b) $\langle x, y, z: x^2 = y^2 = z^2 = (xy)^2 = (yz)^3 = (xz)^3 = 1, \ldots \rangle$, $\#\mathscr{A} = 24$

(c) $\langle x, y, z: x^2 = y^2 = z^2 = (xy)^2 = (yz)^3 = (xz)^4 = 1, \ldots \rangle$, $\#\mathscr{A} = 48$

(d) $\langle x, y, z: x^2 = y^2 = z^2 = (xy)^2 = (yz)^3 = (xz)^5 = 1, \ldots \rangle$, $\#\mathscr{A} = 120$

Moreover, in (a), *the orientation-preserving subgroup $\mathscr{A}^0$ is generated by x and yxy, and in* (b)–(d), *$\mathscr{A}^0$ is generated by xy and yz. In particular, these subgroups have index* 2 *in $\mathscr{A}$.*

Proof. Suppose $\mathscr{A}$ acts without reflections on the sphere S. Then, by Theorem 6.2.2, the group $\mathscr{A}$ acts pseudofreely with nonorientable quotient. By the Riemann–Hurwitz equation, the quotient surface must still have positive Euler characteristic. Hence, $\chi(S/\mathscr{A}) = 1$ and the quotient surface $S/\mathscr{A}$ is the projective plane. The Riemann–Hurwitz equation now reads

$$2 = \#\mathscr{A}\left[1 - \sum(1 - 1/r_y)\right]$$

and therefore we conclude there is at most one branch point, lest the right side become negative. If the order of the branch point is n, then $\#\mathscr{A} = 2n$. Moreover, with a single branch point in the projective plane, the covering $p: S \to S/\mathscr{A}$ can be obtained from an imbedded voltage group with a single type-1 loop. Therefore the group $\mathscr{A}$ is cyclic.

Suppose $\mathscr{A}$ acts with reflections. The orientation-preserving subgroup $\mathscr{A}^0$, however, still acts pseudofreely. As in Theorem 6.3.1, the quotient surface $S/\mathscr{A}^0$ is again the sphere, and the regular covering $p: S \to S/\mathscr{A}^0$ has two or three branch points. Suppose there are two branch points. As in Theorem 6.3.1, they must have the same order, call it n. The quotient group $\mathscr{A}/\mathscr{A}^0$ acts on $S/\mathscr{A}^0$ as a single reflection. Either both branch points lie on the dividing circle of this reflection or one lies on one side and one lies on the other side. The regular covering $p: S \to S/\mathscr{A}^0$ can then be obtained by a voltage graph as illustrated on the left of Figure 6.18 for the first case, and the right of Figure 6.18 for the second case. In both cases, edges crossing the reflection line correspond to reflections in the action of $\mathscr{A}$ on the sphere S, and hence the voltages on those edges have order 2. Presentations for the group $\mathscr{A}$ are given by setting the order of the net voltage on each face equal to the order of the enclosed branch point. In the first case, we get the presentation

$$\langle x, y: x^2 = y^2 = (xy)^n = 1, \ldots \rangle$$

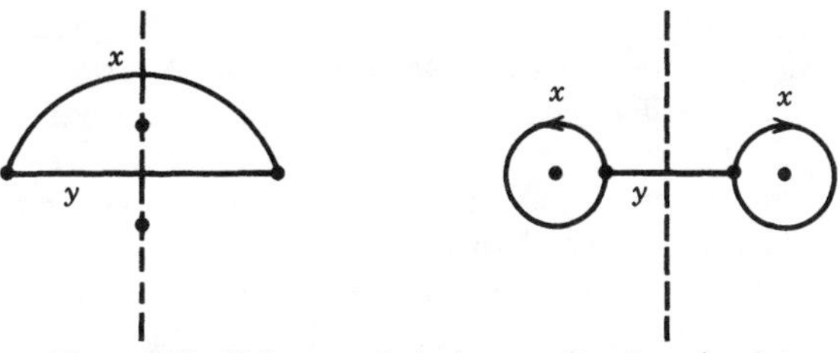

Figure 6.18. Voltage graphs in the case of two branch points.

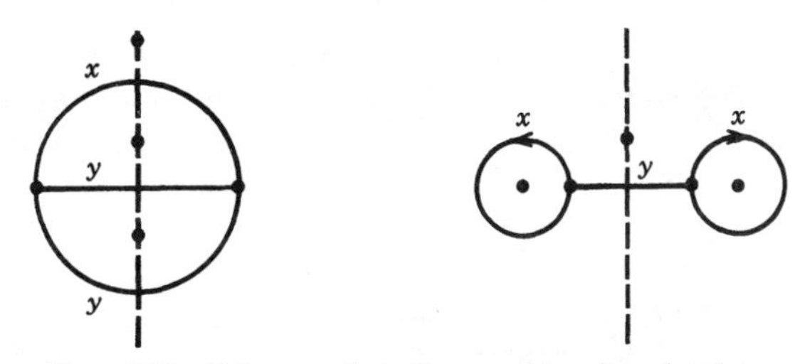

Figure 6.19. Voltage graphs in the case of three branch points.

and thus $\mathscr{A}$ is the dihedral group $\mathscr{D}_n$. In the second case, we get

$$\langle x, y: x^n = y^2 = yxyx^{-1} = 1, \dots \rangle$$

and thus $\mathscr{A}$ is the group $\mathscr{Z}_2 \times \mathscr{Z}_n$.

Suppose instead there are three branch points. Let their orders be $q \leq r \leq s$. As in Theorem 6.3.1, possible values for (q, r, s) are $(2, 2, n)$, $(2, 3, 3)$, $(2, 3, 4)$, $(2, 3, 5)$. There are again two ways that the branch points are located with respect to the dividing circle of the reflection for the action of $\mathscr{A}/\mathscr{A}^0$ on $S/\mathscr{A}^0$. All three branch points may be located on the dividing circle, as on the left of Figure 6.19. Or, if two of the branch points have the same order, those two may be located on either side of the dividing circle, with the third branch point lying on the circle, as on the right of Figure 6.19. The covering $p: S \to S/\mathscr{A}^0$ can then be obtained from an imbedded voltage graph, as in Figure 6.19. If (q, r, s) is $(2, 2, n)$, we obtain the presentations for $\mathscr{A}$:

$$\langle x, y, z: x^2 = y^2 = z^2 = (xy)^2 = (yz)^2 = (xz)^n = 1, \dots \rangle$$
$$\langle x, y: x^2 = y^2 = (xy)^{2n} = 1, \dots \rangle$$

The second presentation is, of course, the dihedral group $\mathscr{D}_n$. In the first presentation, since y commutes with both x and z, and since the subgroup generated by x and z is the dihedral group $\mathscr{D}_n$, the group $\mathscr{A}$ must be $\mathscr{Z}_2 \times \mathscr{D}_n$

(unless the subgroups $\langle y \rangle$ and $\langle x, z \rangle$ have a nontrivial intersection, in which case y is a redundant generator and $\mathscr{A}$ is simply the dihedral group $\mathscr{D}_n$).

For the other possible values of (q, r, s), we obtain the presentations (a)–(d) given in the statement of the theorem. Presentations (a) and (b) correspond to the two ways of locating branch points when $(q, r, s) = (2, 3, 3)$, whereas (c) and (d) correspond to the only possible branch-point locations from $(2, 3, 4)$ and $(2, 3, 5)$. The subgroup that must be of index two in each presentation is just the orientation-preserving subgroup, which is also the isotropy group for the given voltage graph. The order of $\mathscr{A}$ is determined as given in each case, because $\#\mathscr{A} = 2\#\mathscr{A}^0$ and $\#\mathscr{A}^0$ is determined by Theorem 6.3.1.

Since all the voltage graph constructions given in this proof depend only on $\mathscr{A}$ having the desired presentation, the conditions $\mathscr{A}$ must satisfy are not only necessary but also sufficient in order that $\mathscr{A}$ act on the sphere as described. □

It can be shown (Exercise 6) that the subgroup $\langle x, yxy \rangle$ in presentation (a) and the subgroup $\langle xy, yz \rangle$ in presentations (b)–(d) must have index at most two for purely algebraic reasons. Since these subgroups are just the groups (a)–(c) of Theorem 6.3.1, they are automatically finite without any extra relations. Therefore the presentations (a)–(b) of Theorem 6.3.2 are also automatically finite without any extra relations. This means again that only one group is defined by each presentation with the given subgroup restriction. In Exercise 5 the reader is asked to find the appropriate generating sets to present $\mathscr{Z}_2 \times \mathscr{A}_4$ by (a), $\mathscr{S}_4$ by (b), $\mathscr{Z}_2 \times \mathscr{S}_4$ by (c), and $\mathscr{Z}_2 \times \mathscr{A}_5$ by (d).

Again, the reader should recognize the full triangle groups appearing in the proof of the theorem and in presentations (b)–(d). On the other hand, the presentation

$$\langle x, y \colon x^m = y^2 = [x, y]^n = 1, \ldots \rangle$$

which yields either $\mathscr{Z}_2 \times \mathscr{Z}_m$, for $n = 1$, or presentation (a), for $m = n = 3$, appears to be new. In fact, it can be obtained from an ordinary (m, m, n) triangle group by adjoining a reflection in the perpendicular bisector to the side opposite the π/n angle, or as the index-2 subgroup of the full $(2, m, 2n)$ triangle group generated by rotation of order m at the π/m vertex and reflection in the side opposite the π/m vertex (see Exercise 20). Call this group a "hybrid (m, m, n) triangle" group, since there appears to be no standard terminology for this group. Thus we can summarize Theorem 6.3.2 as follows.

Corollary (Maschke, 1896). *$\sigma(\mathscr{A}) = 0$ and $\sigma^0(\mathscr{A}) \neq 0$ if and only if $\mathscr{A}$ is isomorphic to one of the groups $\mathscr{Z}_2 \times \mathscr{Z}_{2n}$, $\mathscr{Z}_2 \times \mathscr{D}_n$, $\mathscr{Z}_2 \times \mathscr{A}_4$, $\mathscr{Z}_2 \times \mathscr{S}_4$, $\mathscr{Z}_2 \times \mathscr{A}_5$. These groups correspond, respectively, to the hybrid $(2, 2, n)$, the full*

$(2,2,n)$, *the hybrid* $(2,3,3)$, *the full* $(2,3,4)$, *and the full* $(2,3,5)$ *triangle groups*.

6.3.3. Symmetric Genus 1

The classification of groups acting on the torus again uses the Riemann–Hurwitz equation, but not surprisingly, there are now more combinations of quotient surfaces, branch points, and reflections. The list of presentations looks bewildering at first, but the reader should know what to expect. There should be at least 17 partial presentations, none of which limits the order of the group without the addition of extra relations. These correspond to quotients of the 17 Euclidean space groups. If there are exactly 17 partial presentations, we might have struck gold—a quick check with the list of presentations given in Coxeter and Moser (1957) for the Euclidean space groups will settle the issue. The reader should also find full and ordinary $(2,3,6)$, $(2,4,4)$, and $(3,3,3)$ triangle groups and hybrid $(2,4,4)$ and $(3,3,3)$ triangle groups. One last hint before counting presentations in the statement of Theorem 6.3.3: presentations with the same letter but different subscripts are different presentations for the same underlying group. The only reason both presentations are listed is for convenience in later computations.

Theorem 6.3.3. *The group* $\mathscr{A}$ *acts on the torus and preserves orientation if and only if* $\mathscr{A}$ *has one of the following presentations*

(a) $\langle x, y \colon [x, y] = 1, \ldots \rangle$

(b_1) $\langle x, y, z \colon x^2 = y^2 = z^2 = (xyz)^2 = 1, \ldots \rangle$

(b_2) $\langle x, y, z \colon y^2 = z^2 = (xy)^2 = (xz)^2 = 1, \ldots \rangle$

(c) $\langle x, y \colon x^3 = y^3 = (xy)^3 = 1, \ldots \rangle$

(d) $\langle x, y \colon x^4 = y^2 = (xy)^4 = 1, \ldots \rangle$

(e) $\langle x, y \colon x^3 = y^2 = (xy)^6 = 1, \ldots \rangle$

The group $\mathscr{A}$ *acts on the torus and reverses orientation, but without reflections, if and only if* $\mathscr{A}$ *has one of the following presentations, with the given subgroup* $\mathscr{B}$ *having index* 2 *in* $\mathscr{A}$:

(f_1) $\langle x, y \colon xyx^{-1}y = 1, \ldots \rangle$, $\mathscr{B} = \langle x^2, y \rangle$

(f_2) $\langle x, y \colon x^2y^2 = 1, \ldots \rangle$, $\mathscr{B} = \langle x^2, xy \rangle$

(g_1) $\langle x, y \colon y^2 = (x^2y)^2 = 1, \ldots \rangle$, $\mathscr{B} = \langle x^2, y, xyx \rangle$

(g_2) $\langle x, y \colon (xy)^2 = (x^{-1}y)^2 = 1, \ldots \rangle$, $\mathscr{B} = \langle x^2, xy, y^2 \rangle$

The group $\mathscr{A}$ acts on the torus with reflections if and only if $\mathscr{A}$ has one of the following presentations, with the given subgroup $\mathscr{B}$ having index 2 in $\mathscr{A}$:

(h) $\langle x, y, z: y^2 = z^2 = [x, y] = [x, z] = 1, \dots \rangle, \quad \mathscr{B} = \langle x, yz \rangle$

(i) $\langle x, y: y^2 = [x^2, y] = 1, \dots \rangle, \quad \mathscr{B} = \langle x^2, xy \rangle$

(j) $\langle x, y, z, w: x^2 = y^2 = z^2 = w^2 = (xy)^2 = (yz)^2 = (zw)^2 = (wx)^2 = 1, \dots \rangle, \quad \mathscr{B} = \langle xy, yz, zw \rangle$

(k) $\langle x, y, z: x^2 = y^2 = z^2 = (yz)^2 = (xyxz)^2 = 1, \dots \rangle, \quad \mathscr{B} = \langle x, yz, yxy \rangle$

(l_1) $\langle x, y, z: x^2 = y^2 = z^2 = xyzyxz = 1, \dots \rangle, \quad \mathscr{B} = \langle x, y, zxz \rangle$

(l_2) $\langle x, y, z: y^2 = z^2 = (xy)^2 = [x, z] = 1, \dots \rangle, \quad \mathscr{B} = \langle x, y, zyz \rangle$

(m) $\langle x, y, z: x^2 = y^2 = z^2 = (xy)^3 = (yz)^3 = (xz)^3 = 1, \dots \rangle, \quad \mathscr{B} = \langle xy, yz \rangle$

(n) $\langle x, y: x^3 = y^2 = [x, y]^3 = 1, \dots \rangle, \quad \mathscr{B} = \langle x, yxy \rangle$

(o) $\langle x, y, z: x^2 = y^2 = z^2 = (xy)^2 = (yz)^4 = (xz)^4 = 1, \dots \rangle, \quad \mathscr{B} = \langle xy, yz \rangle$

(p) $\langle x, y: x^4 = y^2 = [x, y]^2 = 1, \dots \rangle, \quad \mathscr{B} = \langle x, yxy \rangle$

(q) $\langle x, y, z: x^2 = y^2 = z^2 = (xy)^2 = (yz)^3 = (xz)^6 = 1, \dots \rangle, \quad \mathscr{B} = \langle xy, yz \rangle$

Proof. Suppose $\mathscr{A}$ acts on the torus S preserving orientation. Then $\mathscr{A}$ acts pseudofreely. By the Riemann–Hurwitz equation

$$0 = \chi(S) = \#\mathscr{A}\left(\chi(S/\mathscr{A}) - \sum(1 - 1/r_y)\right)$$

It follows that $\chi(S/\mathscr{A}) \geq 0$ and that $S/\mathscr{A}$ is the torus or the sphere. If $S/\mathscr{A}$ is the torus, then there can be no branch points. Presentation (a) is thus obtained from a voltage assignment for a bouquet of two circles cellularly imbedded as usual in the torus $S/\mathscr{A}$. If $S/\mathscr{A}$ is the sphere, then, by the Riemann–Hurwitz equation, there are either four branch points all of order 2 or three branch points. Figure 6.20 illustrates two ways to imbed a bouquet of three circles in the sphere so that each face of the imbedding encloses one branch point of order two. The imbedded voltage graph on the left yields presentation (b_1), and the one on the right, (b_2). If there are three branch points of order $q \leq r \leq s$, then, from the Riemann–Hurwitz equation, the possible values for (q, r, s) are (3, 3, 3), (2, 4, 4), (2, 3, 6). Presentations (c), (d),

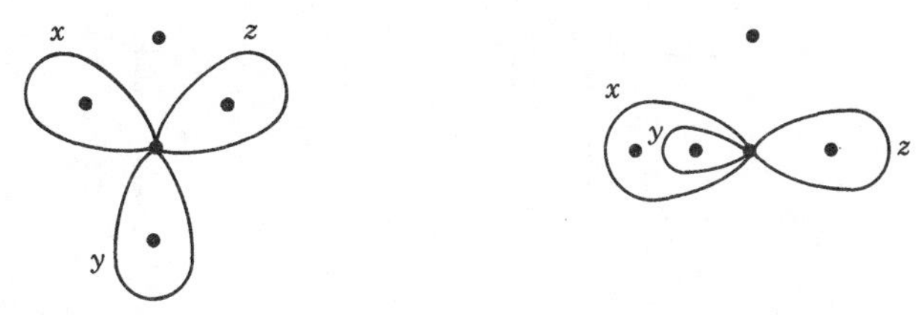

Figure 6.20. Two imbedded voltage graphs for the case of four branch points.

and (e), respectively, are obtained from these possibilities by an imbedded voltage graph of a bouquet of two circles, as in Theorem 6.3.1.

In all of the remaining cases, there is a subgroup $\mathscr{B}$ that is required to be of index 2. Given the action of the group $\mathscr{A}$ on the torus, the subgroup $\mathscr{B}$ is necessarily of index 2 because it corresponds to the orientation-preserving subgroup $\mathscr{A}^0$. Conversely, given a group $\mathscr{A}$ having one of the presentations (f_1)–(q), the index of the subgroup $\mathscr{B}$ must be 2 in order to construct the necessary voltage graph. In cases (f_1)–(g_2), the subgroup $\mathscr{B}$ must have index 2 so that Theorem 4.1.5 can be invoked to make the derived surface orientable; in cases (h)–(q), the subgroup $\mathscr{B}$ must have index 2 because the necessary voltage graph is a 2-vertex quotient graph. It should be noted that the generators given for the subgroup $\mathscr{B}$ are chosen to be irredundant. For example, from the voltage graph for presentation (l_1), one would expect to see zyz as one of the generators for $\mathscr{B}$. However, $zyz = zyxz \cdot zxz$, and since $xyzyxz = 1$ implies that $zyxz = yx$, we have $zyz = yx \cdot zxz \in \langle x, y, zxz \rangle = \mathscr{B}$.

Suppose that $\mathscr{A}$ acts on the torus, reversing orientation but without reflections. Then $\mathscr{A}$ acts pseudofreely, with nonorientable quotient surface $S/\mathscr{A}$. Since $\chi(S/\mathscr{A}) \geq 0$, from the Riemann–Hurwitz equation, $S/\mathscr{A}$ is either the Klein bottlê or the projective plane. If $S/\mathscr{A}$ is the Klein bottle, then, by the Riemann–Hurwitz equation, there are no branch points. Two ways of imbedding a bouquet of two circles in the Klein bottle are illustrated on the upper left and lower left of Figure 6.21. The corresponding voltage graphs yield presentations (f_1) and (f_2), respectively. If $S/\mathscr{A}$ is the projective plane, then there must be two branch points of order 2, by the Riemann–Hurwitz equation. Two ways of imbedding a bouquet of two circles enclosing the branch points are illustrated on the upper and lower right of Figure 6.21. The corresponding voltage graphs yield presentations (g_1) and (g_2), respectively.

Suppose that $\mathscr{A}$ acts on the torus with reflections. The orientation-preserving subgroup $\mathscr{A}^0$ acts pseudofreely. As with the presentations (a)–(e) already considered, the quotient surface $S/\mathscr{A}^0$ is either the torus or the sphere. Suppose it is the torus. By the Riemann–Hurwitz equation, the covering $p: S \to S/\mathscr{A}^0$ has no branch points. As usual, the quotient group $\mathscr{A}/\mathscr{A}^0$ acts on $S/\mathscr{A}^0$ by a single reflection. There are two possibilities for this reflection. One is the "obvious" reflection both of whose dividing circles are left point-

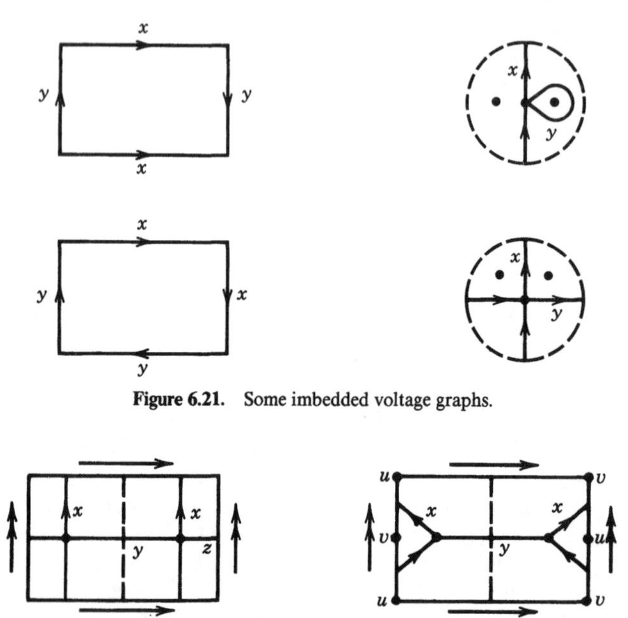

Figure 6.21. Some imbedded voltage graphs.

Figure 6.22. Two voltage graphs imbedded in the torus.

wise fixed; it is illustrated on the left in Figure 6.22. The other possible reflection is the one discussed in Example 6.2.2 and is illustrated on the right in Figure 6.22 (the dotted line down the middle of the rectangle is the dividing circle left pointwise fixed by the reflection, whereas the left and right sides of the rectangle form the dividing circle that is taken to itself with a half-rotation). In both cases, the figure shows a 2-vertex voltage graph imbedded in the torus, invariant under the given reflection, as in the proof of Theorem 6.2.4. Voltages on edges crossing a dividing circle left pointwise fixed are involutions. These two cases yield presentations (h) and (i).

Finally, suppose that the quotient surface $S/\mathscr{A}^0$ is the sphere. As in the orientation-preserving case, the covering $p\colon S \to S/\mathscr{A}^0$ has either four branch points of order 2 or else three branch points. With four branch points of order 2, there are three ways the points can be located with respect to the dividing circle of the quotient reflection: all four on the circle, two on and two off, all four off. Figure 6.23 illustrates the corresponding imbedded voltage graphs; two possible voltage graphs are shown for the case of no branch points on the dividing circle. These imbedded voltage graphs yield presentations (j), (k), (l_1) and (l_2), respectively. The last five presentations, (m)–(q), correspond to the case of three branch points, as in presentations (a)–(d) of Theorem 6.3.2, only here the branch-point orders are (3, 3, 3), (2, 4, 4), or (2, 3, 6).

The converse, that any group $\mathscr{A}$ having one of the presentations (a)–(q) with the given subgroup restrictions must act on the torus as described,

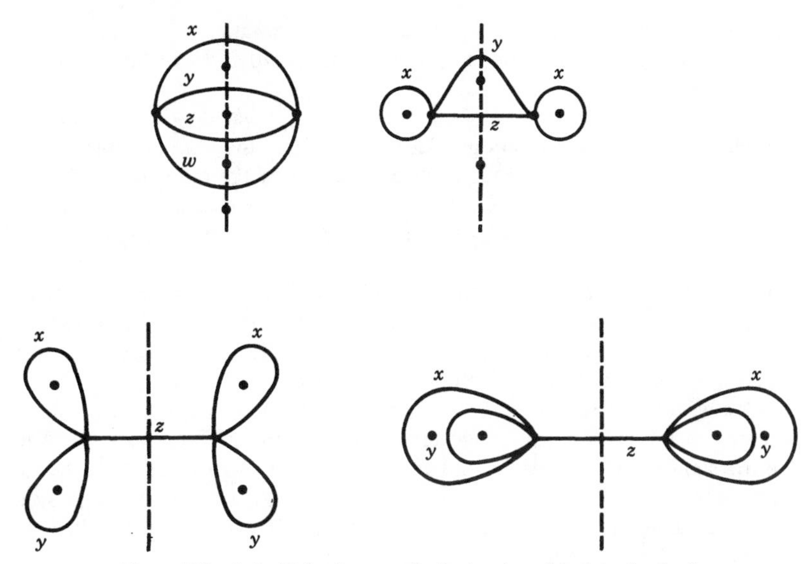

Figure 6.23. Imbedded voltage graphs for four branch points of order 2.

follows directly from the voltage-graph construction applied to the imbedded voltage graphs already listed. □

The number of different classes of groups given in Theorem 6.3.3 is 17. In the following corollary, $p1$, $p2$, and so on are the standard crystallographic notations for the Euclidean space groups.

Corollary. *If the finite group $\mathscr{A}$ acts on the torus, then $\mathscr{A}$ is a quotient of a Euclidean space group. The presentations a–q of Theorem* 6.3.3 *correspond respectively to the groups* $p1$, $p2$, $p3$, $p4$, $p6$, pg, pgg, pm, cm, pmm, cmm, pmg, $p3m1$, $p31m$, $p4m$, $p4g$, $p6m$.

Proof. See Coxeter and Moser (1957) and Exercise 21. □

6.3.4. The Geometry and Algebra of Groups of Symmetric Genus 1

Some further comments on Theorem 6.3.3 are necessary. The subgroup $\mathscr{B}$ corresponds to the quotient of the orientation-preserving subgroup of the underlying Euclidean space group. Thus $\mathscr{B}$ has index at most 2, for geometric reasons. Alternatively, one can also show this by algebraic methods. For example, in presentation g_2 the subgroup $\mathscr{B}$ is normal, since $xx^2x^{-1} = x^2$, $x(xy)x^{-1} = x^2y^2(xy)^{-1} \in \mathscr{B}$, and $xy^2x^{-1} = (xy)y^2(xy)^{-1} \in \mathscr{B}$; similarly $y\mathscr{B}y^{-1} = \mathscr{B}$. Given that $\mathscr{B}$ is normal, it is clear that $\mathscr{B}$ has index at most 2, because the quotient group $\mathscr{A}/\mathscr{B}$ has the presentation $\langle x, y: x^2 = y^2 = xy = 1\rangle$. The other presentations, $(f_1) - (q)$, are handled by similar arguments (Exercises 6–8).

What if $\mathscr{B}$ does not have index 2 in $\mathscr{A}$, that is, what if $\mathscr{B} = \mathscr{A}$? This can happen if $\mathscr{A}$ is a space group quotient $\mathscr{C}/\mathscr{N}$ where the normal subgroup $\mathscr{N}$ contains orientation reversing isometries. Since $\mathscr{B}$ is the quotient of $\mathscr{C}^0$, we expect that $\mathscr{A}$ must then have one of the orientation-preserving presentations (a)–(e). This can be derived algebraically. For example, in presentation (g_2) again, let $u = xy$, $v = y^{-1}x$, $w = x^{-1}y^{-1}$. Then $x^2 = uv$ and $y^2 = (uv)^{-1}$, from which it follows that u, v, w generate $\mathscr{A} = \mathscr{B}$. Moreover, from the given relations $(xy)^2 = (x^{-1}y)^2 = 1$, it follows that $u^2 = v^2 = w^2 = 1$ and $(uvw)^2 = (xy^{-1})^2 = 1$, thus yielding presentation (b_1). The other presentations, $(f_1) - (q)$, are considered in Exercises 9–11.

Finally, by Theorem 6.2.5, we know that every one of the presentations of Theorem 6.3.3 has a normal, 2-generator, abelian group $\mathscr{T}$ corresponding to the translation subgroup of the underlying Euclidean space group $\mathscr{C}$, such that $\mathscr{A}/\mathscr{T}$ is trivial or $\mathscr{Z}_2, \mathscr{Z}_3, \mathscr{Z}_4, \mathscr{Z}_6, \mathscr{D}_2, \mathscr{D}_3, \mathscr{D}_4, \mathscr{D}_6$. Again this can be derived by purely algebraic methods (see Exercises 12–15). The following theorem summarizes this discussion.

Theorem 6.3.4. *If the group $\mathscr{A}$ has one of the presentations* (f_1)–(q), *then the given subgroup $\mathscr{B}$ has index at most* 2, *and if $\mathscr{B} = \mathscr{A}$, then $\mathscr{A}$ also has one of the presentations* (a)–(e). *Thus, if $\mathscr{A}$ has any of the presentations* (a)–(q), *with or without subgroup restrictions, then $\sigma(\mathscr{A}) \le 1$. Finally, if $\mathscr{A}$ has one of the presentations of Theorem* 6.3.3, *then $\mathscr{A}$ has a normal, 2-generator, abelian subgroup $\mathscr{T}$ such that $\mathscr{A}/\mathscr{T}$ is trivial or one of the following groups*:

$$\begin{array}{ll} \mathscr{Z}_2(b, f, h, i) & \mathscr{D}_2(g, j, k, l) \\ \mathscr{Z}_3(c) & \mathscr{D}_3(m, n) \\ \mathscr{Z}_4(d) & \mathscr{D}_4(o, p) \\ \mathscr{Z}_6(e) & \mathscr{D}_6(q) \end{array}$$

6.3.5. Hurwitz's Theorem

The problem of determining all groups of a given symmetric genus $g > 1$ would appear to be hopelessly complicated, in light of the long list of presentations just given for groups of symmetric genus 1. In one way, however, the situation for $g > 1$ is easier, or at least quite different, from the case $g = 1$: the number of groups of given symmetric genus $g > 1$ is finite. The following theorem, first proved by Hurwitz (1892), for automorphisms of a Riemann surface, provides a bound upon the order of any group acting on a given surface of genus $g > 1$.

Theorem 6.3.5 (Hurwitz). *Suppose the group $\mathscr{A}$ acts on the surface of genus $g > 1$. Then $\#\mathscr{A} \le 168(g - 1)$.*

Proof. Let $p: S \to S/\mathscr{A}^0$ be the regular covering associated with the pseudofree action of the orientation-preserving subgroup $\mathscr{A}^0$. By the Riemann–Hurwitz equation,

$$2 - 2g = \chi(S) = \#\mathscr{A}^0\left[\chi(S/\mathscr{A}^0) - \sum(1 - 1/r_y)\right]$$

If $\chi(S/\mathscr{A}^0) < 0$, then $\chi(S/\mathscr{A}^0) \leq -2$. Therefore $2 - 2g \leq \#\mathscr{A}^0(-2)$, and hence $\#\mathscr{A} \leq 2\#\mathscr{A}^0 \leq 2(g-1)$. If $\chi(S/\mathscr{A}^0) = 0$, then there must be branch points; otherwise $2 - 2g = \#\mathscr{A}^0(0)$, contradicting the assumption that $g > 1$. Since $r_y \geq 2$ for any branch point y, $\Sigma(1 - 1/r_y) \leq \frac{1}{2}$. Thus $2 - 2g \leq \#\mathscr{A}^0(-\frac{1}{2})$, and hence $\#\mathscr{A} \leq 2\#\mathscr{A}^0 \leq 8(g-1)$.

If $\chi(S/\mathscr{A}^0) > 0$, then $\chi(S/\mathscr{A}^0) = 2$. Because $2 - 2g$ is negative, so is $2 - \Sigma(1 - 1/r_y)$. In particular, there must be more than two branch points. If there are five or more branch points, then $2 - \Sigma(1 - 1/r_y) \leq \frac{1}{2}$, and hence $\#\mathscr{A} \leq 8(g-1)$ again. Suppose there are four branch points. The negative value for $2 - \Sigma(1 - 1/r_y)$ nearest to 0 is $-\frac{1}{6}$, obtained with three branch points of order 2 and one of order 3. Thus $2 - 2g \leq \#\mathscr{A}^0(-\frac{1}{6})$, and hence $\#\mathscr{A} \leq 24(g-1)$. Finally, suppose there are three branch points of orders $q \leq r \leq s$. Then $2 - \Sigma(1 - 1/r_y) = 1/q + 1/r + 1/s - 1$, which has $-\frac{1}{42}$ for its negative value nearest to 0, occurring when $(q, r, s) = (2, 3, 7)$. Thus, in this case, $2 - 2g \leq \#\mathscr{A}^0(-\frac{1}{42})$, from which it follows that $\#\mathscr{A} \leq 168(g-1)$. □

The value of $1/q + 1/r + 1/s - 1$ next nearest to 0 is $-\frac{1}{24}$, occurring for $(q, r, s) = (2, 3, 8)$. Thus the proof of Theorem 6.3.5 leads to the following corollaries.

Corollary. *Suppose* $\sigma^0(\mathscr{A}) > 1$. *Then* $\#\mathscr{A} > 48(\sigma^0(\mathscr{A}) - 1)$ *if and only if* $\#\mathscr{A} = 84(\sigma^0(\mathscr{A}) - 1)$ *and* $\mathscr{A}$ *has the presentation*

$$\langle x, y: x^2 = y^3 = (xy)^7 = 1, \dots \rangle$$

Corollary. *Suppose* $\sigma(\mathscr{A}) > 1$. *Then* $\#\mathscr{A} > 96(\sigma(\mathscr{A}) - 1)$ *if and only if* $\#\mathscr{A} = 168(\sigma(\mathscr{A}) - 1)$ *and* $\mathscr{A}$ *has the presentation*

$$\langle x, y, z: x^2 = y^2 = z^2 = 1, (xy)^2 = (yz)^3 = (xz)^7 = 1, \dots \rangle$$

Conder (1980) has shown that the symmetric group $\mathscr{S}_n$ has a presentation of the form given in the second corollary for all $n > 167$. Thus $\sigma(\mathscr{S}_n) = n!/168 + 1$, $n > 167$. He has also shown that the alternating group $\mathscr{A}_n$ has a presentation of the form given in the first corollary for all $n > 167$. Thus $\sigma(\mathscr{A}_n) = \sigma^0(\mathscr{A}_n) = n!/168 + 1$, $n > 167$. Conder (1984) has also computed the symmetric genus of all the other symmetric and alternating groups. Glover and Sjerve (1986) have computed the strong symmetric genus of the simple groups PSL(2, p), where p is a prime.

A careful inspection of the proof of Hurwitz's theorem leads to the following refinement. We leave it to the reader to verify the details. First, we define the group $\mathscr{A}$ to be a "proper" quotient of the full triangle group $\mathscr{C}$ if the image of $\mathscr{C}^0$ is a proper subgroup of $\mathscr{A}$. That is, if

$$\mathscr{A} = \langle x, y, z : x^2 = y^2 = z^2 = 1, (xy)^q = (yz)^r = (xz)^s = 1, \ldots \rangle$$

then the subgroup generated by xy and yz is required to be a proper subgroup. Similarly, the presentation

$$\mathscr{A} = \langle x, y : x^q = y^2 = [x, y]^s = 1, \ldots \rangle$$

makes $\mathscr{A}$ a "proper" quotient of the hybrid (q, q, s) triangle group if the group generated by x and yxy is a proper subgroup of $\mathscr{A}$.

Theorem 6.3.6 (Refined Hurwitz Theorem). *Suppose* $\sigma(\mathscr{A}) > 1$ *and* $\#\mathscr{A} > 24(\sigma(\mathscr{A}) - 1)$. *Then one of the following holds:*

a. $\mathscr{A}$ *is an ordinary* (q, r, s) *triangle group quotient where* (q, r, s) *is either* $(2, 4, 5)$ *or* $(2, 3, s)$ *with* $7 \le s \le 11$.
b. $\mathscr{A}$ *is either a proper quotient of a full* (q, r, s) *triangle group quotient, where* (q, r, s) *is* $(2, r, s)$ *with* $3 \le r \le 5$ *or* $(3, 3, s)$ *with* $s = 4, 5$, *or* $\mathscr{A}$ *is a proper quotient of a hybrid* $(3, 3, 4)$, $(3, 3, 5)$, *or* $(5, 5, 2)$ *triangle group.*

Moreover, if $\#\mathscr{A} > 48(\sigma(\mathscr{A}) - 1)$, *then* $\mathscr{A}$ *is either a quotient of the ordinary* $(2, 3, 7)$ *triangle group or a proper quotient of the full* (q, r, s) *triangle group, where* (q, r, s) *is* $(2, 4, 5)$, *or* $(2, 3, s)$ *with* $7 \le s \le 11$.

6.3.6. Exercises

1. Prove that the group $\langle x, y : y^2 = x^n = yxyx = 1 \rangle$ has order $2n$ by proving that the subgroup $\langle x \rangle$ is normal of index 2. Prove that the group $\langle x, y : x^2 = y^2 = (xy)^n = 1 \rangle$ has order $2n$ by proving that the subgroup $\langle xy \rangle$ is normal of index 2.
2. Prove that the group $\langle x, y : x^3 = y^2 = (xy)^3 = 1 \rangle$ has order 12 by showing that $\mathscr{B}, \mathscr{B}y, \mathscr{B}yx^{-1}$ are all of the right cosets of the subgroup $\mathscr{B} = \langle x \rangle$. (*Hint:* Use $yxy = x^{-1}yx^{-1}$ and $yx^{-1}y = xyx$.)
3. Prove that the group $\langle x, y : x^4 = y^2 = (xy)^3 = 1 \rangle$ has order 24 by showing that $\mathscr{B}, \mathscr{B}yx^r, \mathscr{B}yx^2y$, $r = 0, 1, 2, 3$, are all of the right cosets of the subgroup $\mathscr{B} = \langle x \rangle$.
4. Prove that the group $\langle x, y : x^5 = y^2 = (xy)^3 = 1 \rangle$ has order 60 by showing that $\mathscr{B}, \mathscr{B}yx^r, \mathscr{B}yx^2yx^r, \mathscr{B}yx^2yx^{-2}y$, $r = 0, 1, 2, 3, 4$, are all of the right cosets of the subgroup $\mathscr{B} = \langle x \rangle$. (For example, $\mathscr{B}yx^{-2}y$ is in this list of cosets because $\mathscr{B}yx^2yx$ is in the list, and $\mathscr{B}yx^2yx = \mathscr{B}xyxxyx = \mathscr{B}yx^{-1}yyx^{-1}y = \mathscr{B}yx^{-2}y$.)

5. Find generators for $\mathscr{Z}_2 \times \mathscr{A}_4$, $\mathscr{S}_4$, $\mathscr{Z}_2 \times \mathscr{S}_4$, $\mathscr{Z}_2 \times \mathscr{A}_5$ that give, respectively, presentations (a)–(d) of Theorem 6.3.2.
6. Prove that the subgroup $\mathscr{B} = \langle xy, yz \rangle$ has index at most 2 in the group $\mathscr{A} = \langle x, y, z: x^2 = y^2 = z^2 = 1, \ldots \rangle$. Do the same for the subgroup $\mathscr{B} = \langle x, yxy \rangle$ in the group $\mathscr{A} = \langle x, y: y^2 = 1, \ldots \rangle$.
7. Prove for presentations (f_1)–(g_1) of Theorem 6.3.3 that the given subgroup $\mathscr{B}$ is of index at most 2.
8. Same as Exercise 7 for presentations (h)–(l_2) of Theorem 6.3.3.
9. Prove that if $\mathscr{B} = \mathscr{A}$ in presentations (f_1), (f_2), (h), or (i) of Theorem 6.3.3, then $\mathscr{A}$ has presentation (a).
10. Prove that in Theorem 6.3.3 if $\mathscr{B} = \mathscr{A}$ in presentations (g_1), (g_2), (j), (k), (l_1), then $\mathscr{A}$ has presentation (b_1), and if $\mathscr{B} = \mathscr{A}$ in presentation (l_2), then $\mathscr{A}$ has presentation (b_2).
11. Show that presentations (m)–(q) of Theorem 6.3.3 correspond, respectively, to presentations (c), (c), (d), (d), (e) when $\mathscr{B} = \mathscr{A}$.
12. Verify that the normal subgroup $\mathscr{T}$ of Theorem 6.3.4 is given by $\langle x^{-1}y, xy^{-1} \rangle$ for presentation (c), $\langle xyx, x^2y \rangle$ for (d), and $\langle xyx^{-1}y, x^{-1}yxy \rangle$ for (e).
13. Verify that the normal subgroup $\mathscr{T}$ of Theorem 6.3.4 is $\mathscr{B}$ itself in presentations (f_1), (h), (i).
14. Verify that the normal subgroup $\mathscr{T}$ of Theorem 6.3.4 is given by $\langle xyx^{-1}y^{-1}, x^{-1}yxy \rangle$ for presentation (g_2), $\langle xz, yw \rangle$ for (j), $\langle xyxy, yxz \rangle$ for (k), and $\langle x, yz \rangle$ for (l_2).
15. Verify that the normal subgroup $\mathscr{T}$ of Theorem 6.3.4 is given by $\langle yxyz, xyzy \rangle$ for presentations (m) and (o), $\langle x^{-1}yx^{-1}y, xyxy \rangle$ for (n) and (p), and $\langle [yz, xy], [zy, xy] \rangle$ for (q).
16. Determine $\sigma(\mathscr{Q})$ and $\sigma^0(\mathscr{Q})$, where $\mathscr{Q} = \langle i, j: i^4 = 1, i^2 = j^2 \rangle$, the quaternion group of order 8.
17. Find $\sigma(\mathscr{A})$ and $\sigma^0(\mathscr{A})$ for all groups of order $\#\mathscr{A} < 16$.
18.* Determine $\sigma^0(\mathscr{A})$ for all groups $\mathscr{A}$ such that $\sigma(\mathscr{A}) = 0$.
19.* Compute $\sigma(\mathscr{A})$ and $\sigma^0(\mathscr{A})$ as for many groups as you can in Coxeter and Moser's (1957) list of all groups of order $\#\mathscr{A} < 32$.
20. Prove that the group obtained by adjoining to the ordinary (m, m, n) triangle group the reflection in the perpendicular bisector to the side opposite the π/n angle, has presentation

$$\langle x, y: x^m = y^2 = [x, y]^n = 1, \ldots \rangle$$

(*Hint:* x is the rotation of order m at one π/m vertex, and y is the reflection in the bisector.)

21. Prove the corollary to Theorem 6.3.3 by comparing the presentations given in the theorem to the presentations in Table 3 of Coxeter and Moser (1957).

* denotes a problem whose solution is lengthy.

6.4. GROUPS OF SMALL GENUS

Now that the lists of all groups of symmetric genus 0 or 1 have been compiled in Section 6.3, it is time to ask about genus rather than symmetric genus. Is it possible that every group of genus 0 or 1 appears in the lists of Section 6.3? Since $\gamma(\mathscr{A}) \leq \sigma(\mathscr{A})$ for any finite group $\mathscr{A}$, every group in these lists has genus at most 1. On the other hand, from the example at the end of Section 6.2, the inequality $\gamma(\mathscr{A}) \leq \sigma(\mathscr{A})$ is strict for most abelian groups, so one would not expect these lists to be complete. In this section, perhaps surprisingly, it will be shown that if $\gamma(\mathscr{A}) \leq 1$, then $\gamma(\mathscr{A}) = \sigma(\mathscr{A})$, with three exceptions. Thus, with the addition of three extra presentations, the lists of Section 6.3 classify all groups of genus 0 or 1. At the same time, it is shown that Hurwitz's theorem applies to genus as well as to symmetric genus, that is, $\#\mathscr{A} \leq 168[\gamma(\mathscr{A}) - 1]$ if $\gamma(\mathscr{A}) > 1$. It follows that the number of groups of a given genus $g > 1$ is finite.

The classification of groups of genus 0 or 1 begins with Maschke (1896), who shows that if $\gamma(\mathscr{A}) = 0$, then $\mathscr{A}$ is isomorphic to one of the groups $\mathscr{Z}_n$, $\mathscr{D}_n$, $\mathscr{A}_4$, $\mathscr{S}_4$, or $\mathscr{A}_5$ or to the direct product of $\mathscr{Z}_2$ with any of these groups, from which it follows that $\sigma(\mathscr{A}) = \gamma(\mathscr{A})$ when $\gamma(\mathscr{A}) = 0$. Baker (1931) extended some of Maschke's work to the torus, but it was Proulx (1977, 1978) who did all the massive work needed to classify groups of genus 1. Tucker (1983, 1984a) clarified the connections of Proulx's results to group actions on surface and space groups. He used those connections to prove that three of the four group presentations not fitting into Proulx's list of general classes of Euclidean space group quotients are truly exceptional. The imbedded-Cayley-graph version of Hurwitz's theorem was proved first by Tucker (1980) and was later refined by Tucker (1984b). This last paper includes also a proof of Proulx's classification. Finally, this last paper opens the way to the determination by Tucker (1984c) of the only group of genus 2.

6.4.1. An Example

The basic idea behind the results of this section is fairly simple. Suppose that the group $\mathscr{A}$ has a Cayley graph imbedded in a surface whose genus is small compared with $\#\mathscr{A}$ (in particular, genus 0 or 1). Then this imbedding must have many faces. If it has many faces, there must be some faces of small size. A face of small size gives a short closed walk in the Cayley graph, which in turn corresponds to a short relator in the group $\mathscr{A}$. Upon inspection, the group presentations appearing in the lists in Section 6.3 all seem to have few generators and short relators. If the imbedding forces enough short relators in

$\mathscr{A}$, then $\mathscr{A}$ will have one of the listed presentations. As example will illustrate the details of this argument in a particular case.

Example 6.4.1. *Let $X = \{x, y\}$ be an irredundant generating set for the group $\mathscr{A}$, such that both x and y have order greater than 3. Suppose there is an imbedding $C(\mathscr{A}, X) \to S$ such that $\#\mathscr{A} > -20\chi(S)$. The number of vertices of the graph $C(\mathscr{A}, X)$ is $\#\mathscr{A}$, and the number of edges is $2\#\mathscr{A}$. Consider the faces of the imbedding. A boundary of a two-sided or three-sided face gives a relation such as $x^2 = 1$ or $xyx = 1$; clearly any such relation contradicts irredundancy or the assumed orders of x and y. Thus every face has four or more sides. Let F' be the set of four-sided faces and F'' the set of faces with more than four sides. By the Euler polyhedral equation,*

$$\chi(S) = \#\mathscr{A} - \#2\mathscr{A} + \#F' + \#F''$$

Also, since the sum of the face sizes taken over all faces is twice the number of edges, it follows that

$$4\#F' + 5\#F'' \leq 4\#\mathscr{A}$$

Suppose that $\#F' = 0$. Then

$$\chi(S) = -\#\mathscr{A} + \#F' + \#F'' \leq -\#\mathscr{A} + 4\#\mathscr{A}/5,$$

and hence $\#\mathscr{A} \leq -5\chi(S)$, contradicting our assumption that $\#\mathscr{A} > -20\chi(S)$. Thus there must be a four-sided face and, consequently, a relator of length 4. The relation $xyx^{-1}y^{-1} = 1$ gives presentation (a) *of Theorem* 6.3.3. *The relation $xyx^{-1}y = 1$ yields presentation* (f_1) *of Theorem* 6.3.3, *and $x^2y^2 = 1$ yields presentation* (f_2) *of Theorem* 6.3.3. *The only other relators of length* 4 *not contradicting irredundancy are $(xy)^2$, $(xy^{-1})^2$, x^4, y^4. Suppose four four-sided faces all meet at one vertex v. Assuming the only relators of length* 4 *are the four just given, then there are two cases depending on whether or not x and y edges alternate in the rotation at vertex v. If they alternate, as depicted on the left of Figure* 6.24, *then $(xy)^2$ and $(xy^{-1})^2$ are relators yielding presentation* (g_2) *of Theorem* 6.3.3. *If they do not alternate, as on the right of Figure* 6.24, *then x^4, y^4 and $(xy^{\pm 1})^2$ are relators, which gives presentation* (d) *of Theorem* 6.3.3 *(replace y by y^{-1} if necessary).*

Suppose instead that there is no vertex at which four four-sided faces meet. Then counting vertices by counting four-sided faces and multiplying by 4, *one obtains $4\#F' \leq 3\#\mathscr{A}$. Then since $\#F'' \leq 4(\#\mathscr{A} - \#F')/5$,*

$$\begin{aligned}\chi(S) &\leq -\#\mathscr{A} + \#F' + 4(\#\mathscr{A} - \#F')/5 = -\#\mathscr{A}/5 + \#F'/5 \\ &\leq -\#\mathscr{A}/5 + 3\#\mathscr{A}/20\end{aligned}$$

Therefore, $\#\mathscr{A} \leq -20\chi(S)$, again contradicting the assumption.

We conclude that if $\#\mathscr{A} > -20\chi(S)$, in particular if $\chi(S) \geq 0$, then the group $\mathscr{A}$ has a presentation in the lists in Section 6.3. Of course, there are many other cases to consider, depending on the number of elements in the given generating set and their orders. It is the consideration of these cases that occupies most of this section.

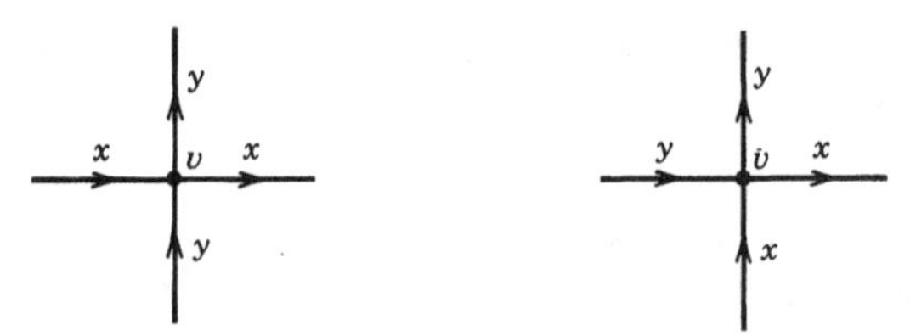

Figure 6.24. The rotation at vertex v.

6.4.2. A Face-Size Inequality

The determination of groups of low genus depends, as in Example 6.4.1, on some delicate computations for the number of faces of various sizes. The following inequality is the main tool in these computations.

Theorem 6.4.1. *Let $G \to S$ be a cellular imbedding, where G is a regular graph of valence d. Let f_i be the number of i-sided faces. Then for any integer $r > 1$,*

$$\left(\frac{rd}{2} - r - d\right)\#V + r\chi(S) \leq \sum_{i=1}^{r-1} (r-i) f_i$$

Proof. Since $\Sigma_{i=1}^{\infty} i f_i = 2\#E$, it follows that

$$\sum_{i=1}^{r-1} i f_i + r\left(\sum_{i=1}^{\infty} f_i - \sum_{i=1}^{r-1} f_i\right) \leq 2\#E$$

which can be rewritten

$$r\sum_{i=1}^{\infty} f_i \leq 2\#E + \sum_{i=1}^{r-i} (r-i) f_i$$

Multiplying the Euler polyhedral equation by r and rearranging, one obtains

$$r\#E - r\#V + r\chi(S) = r\sum_{i=1}^{\infty} f_i$$

Therefore,

$$r\#E - r\#V + r\chi(S) \leq 2\#E + \sum_{i=1}^{r-1} (r-i) f_i$$

The substitution $\#E = d\#V/2$ now yields the desired inequality. □

We shall put this inequality to immediate use. The following corollaries show that it is necessary to consider only Cayley graphs of valence 4 or less when studying groups of low genus.

Corollary 1. *Let $G \to S$ be an imbedding of a regular simple graph of valence $d > 6$. Then $\#V \leq -6\chi(S)$. In particular, there are only finitely many regular simple graphs of given valence $d > 6$ and genus $g > 1$, and there are no regular simple graphs of valence $d > 6$ and genus $g \leq 1$.*

Proof. By the inequality of Theorem 6.4.1, with $r = 3$,

$$(d/2 - 3)\#V + 3\chi(S) \leq 0$$

since $f_1 = f_2 = 0$. If $d > 6$, then $d/2 - 3 \geq 1/2$ so $\#V \leq -6\chi(S)$, as claimed. □

Corollary 2. *Let $C^1(\mathscr{A}, X) \to S$ be an imbedding of an alternative Cayley graph of valence $d > 4$, where X is an irredundant generating set for the group $\mathscr{A}$. Then $\#\mathscr{A} \leq -12\chi(S)$. In particular, if $d(\mathscr{A})$ denotes the minimal valence over all Cayley graphs for $\mathscr{A}$, then there are only finitely many groups $\mathscr{A}$ with given genus and given $d(\mathscr{A}) > 4$, and there are no groups $\mathscr{A}$ with $d(\mathscr{A}) > 4$ and $\gamma(\mathscr{A}) \leq 1$.*

Proof. By the inequality of Theorem 6.4.1, with $r = 4$,

$$(d - 4)\#\mathscr{A} + 4\chi(S) \leq f_3$$

since $f_1 = f_2 = 0$ by irredundancy. Also by irredundancy, the only three-sided faces correspond to a relation of the form $x^3 = 1$, $x \in X$. Thus there are at most $\lfloor d/2 \rfloor$ three-sided faces meeting at any vertex. Thus $3f_3 \leq \lfloor d/2 \rfloor \#V$. By the previous corollary, we need consider only the cases $d = 5$ and $d = 6$. For $d = 6$, $f_3 \leq \#V$, and hence

$$2\#V + 4\chi(S) \leq f_3 \leq \#V$$

which gives $\#V \leq -4\chi(S)$. For $d = 5$, $f_3 \leq 2\#V/3$ and hence

$$\#V + 4\chi(S) \leq \tfrac{2}{3}\#V$$

which gives $\#V \leq -12\chi(S)$. □

The second corollary leaves open the question of whether there are infinitely many regular graphs of valence $d \leq 6$ and given genus $g > 1$. In fact, there are infinitely many such graphs (see Exercises 1–3). These corollaries also raise other questions. Is the hypothesis of irredundancy necessary? The following example illustrates the difficulties here.

Example 6.4.2 (N. Wormald). *Let*

$$\mathscr{A} = \langle x, y, z, w\colon x^{2n} = y^2 = z^2 = w^2 = 1, z = x^n, w = xy \rangle, \qquad n > 2$$

Clearly z and w are redundant, and $\mathscr{A}$ is actually the dihedral group of order $4n$. Consider the Cayley graph $C(\mathscr{A}, X)$, where $X = \{x, y, z, w\}$. The subgraph of $C(\mathscr{A}, X)$ of edges colored x or z forms two disjoint $2n$-cycles with "diameters" (see left of Figure 6.25). In particular for $n \geq 3$, the Cayley graph $C(\mathscr{A}, X)$ contains disjoint copies of $K_{3,3}$ and therefore $\gamma(C(\mathscr{A}, X)) > 1$. On the other

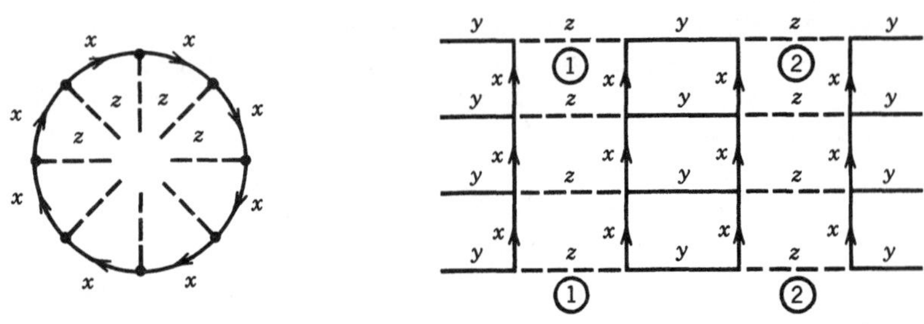

Figure 6.25. A subgraph of the Cayley graph $C(\mathscr{A}, X)$ and a partial imbedding of $C(\mathscr{A}, X)$.

hand, if a pair of antipodal x-edges is removed from each x-cycle, the resulting graph can be imbedded in the plane; in effect, the crossing of the z-diameters can be undone with a twist at the expense of two x-edges crossing (see the right of Figure 6.25). *The four remaining x-edges can be reattached using two handles (hitting the plane at spots marked* 1 *and* 2 *in Figure* 6.25). *Thus* $\gamma(C(\mathscr{A}, X)) = 2$.

This example provides infinitely many Cayley graphs of valence 5 and genus 2. Most evidence, however, indicates that this phenomenon is limited to genus 2. A reasonable conjecture is that the second corollary is true without the restriction of irredundancy, if $g > 1$ is replaced by $g > 2$. L. Babai has made the stronger conjecture that there are only finitely many graphs of given genus $g > 2$ whose automorphism groups act transitively on vertices.

6.4.3. Statement of Main Theorem

As was noted at the beginning of this chapter, one needs to consider only irredundant generating sets to compute the genus of a group. Since the condition of irredundancy is so frequently used, call a Cayley graph $C(\mathscr{A}, X)$ "irredundant" if the generating set X is irredundant.

Theorem 6.4.2 (Proulx, 1978; Tucker, 1980, 1984a). *Let* $C(\mathscr{A}, X) \to S$ *be an imbedding of an irredundant Cayley graph in a surface S. If* $\#\mathscr{A} > -84\chi(S)$, *then* $\mathscr{A}$ *has one of the presentations listed in Theorems* 6.3.1–6.3.3 *or one of these presentations*:

$$\text{(a)}\quad \langle x, y\colon x^3 = y^3 = xyxy^{-1}x^{-1}y^{-1} = 1\rangle, \quad \#\mathscr{A} = 24$$

$$\text{(b)}\quad \langle x, y\colon x^3 = y^2 = (xy)^3(x^{-1}y)^3 = 1\rangle, \quad \#\mathscr{A} = 48$$

$$\text{(c)}\quad \langle x, y\colon x^3 = y^2 = (xyxyx^{-1}y)^2 = 1\rangle, \quad \#\mathscr{A} = 48$$

To give some idea of the power of this theorem, we list three immediate corollaries.

Corollary 1. *If* $\gamma(\mathscr{A}) > 1$, *then* $\#\mathscr{A} \leq 168(\gamma(\mathscr{A}) - 1)$. *Thus there are only finitely many groups of a given genus* $g > 1$.

Proof. If $\#\mathscr{A} > 168(\gamma(\mathscr{A}) - 1)$, then an irredundant Cayley graph for $\mathscr{A}$ can be imbedded in a surface S such that $\#\mathscr{A} > -84\chi(S)$. But then, by Theorem 6.4.2, we have $\sigma(\mathscr{A}) \leq 1$ and hence $\gamma(\mathscr{A}) \leq 1$. □

Analogously to the inequality for symmetric genus in Hurwitz's theorem, the inequality of Corollary 1 becomes an equality if and only if the group $\mathscr{A}$ is a proper quotient of the full (2, 3, 7) triangle group. In particular, it follows from the work of Conder (1980) that

$$\gamma(\mathscr{S}_n) = \sigma(\mathscr{S}_n) = 1 + n!/168$$

for all $n > 167$.

Corollary 2. *If* $\gamma(\mathscr{A}) \leq 1$, *then either* $\mathscr{A}$ *is one of the three groups* (a), (b), (c) *of Theorem* 6.4.2 *or else* $\sigma(\mathscr{A}) \leq 1$. *In particular, the lists of Section* 6.3 *together with* (a), (b), (c) *classify all groups of genus* 0 *or* 1.

Proof. If $\gamma(\mathscr{A}) \leq 1$, then an irredundant Cayley graph for $\mathscr{A}$ imbeds in a surface S of Euler characteristic $\chi(S) \geq 0$. In particular, $\#\mathscr{A} > -84\chi(S)$. □

Corollary 3. *If the group* $\mathscr{A}$ *has a Cayley graph imbeddable in the Klein bottle, then it has a Cayley graph imbeddable in the torus.*

Proof. The surface S in Theorem 6.4.2 is allowed to be nonorientable. □

This last corollary does not say that any Cayley graph imbeddable in the Klein bottle is imbeddable in the torus. In fact, for redundant Cayley graphs this is not true.

Example 6.4.2 Revisited. *The two cycles with diameters formed by x-z edges imbed in a pair of projective planes so that the x-cycles bound faces* (*view the projective plane as a disk with antipodal points identified*). *If the two projective planes are then joined by a tube to form a Klein bottle, the remaining y- and w-edges can be added around the tube. Thus the Cayley graph* $C(\mathscr{A}, X)$ *imbeds in the Klein bottle but not in the torus.*

In general, Theorem 6.4.2 and its corollaries apply to the genus of a group, not the genus of a Cayley graph. Part of the difficulty is that in the proof of Theorem 6.4.2, the given generating set X occasionally must be changed to obtain one of the presentations in the Section-6.3 lists; thus one obtains a Cayley graph for $\mathscr{A}$ imbeddable in the torus, but it might not be the original Cayley graph $C(\mathscr{A}, X)$. In fact, this difficulty can be overcome, but it involves some intricate group theory that would take us far afield (Tucker, 1984b). Thus, in Theorem 6.4.2, without extra hypotheses, it can be shown that the given Cayley graph $C(\mathscr{A}, X)$ imbeds in the torus or sphere whenever $\#\mathscr{A} > -84\chi(S)$. It also follows that there are only finitely many irredundant Cayley graphs of a given genus $g > 1$, that an irredundant Cayley graph imbeddable in the Klein bottle is imbeddable in the torus, and that the lists of Section 6.3 together with presentations (a)–(c) of Theorem 6.4.2 classify all generating sets and groups for irredundant Cayley graphs of genus 0 or 1.

The other difficulty in applying the results of this section to the genus of a Cayley graph, rather than the genus of a group, is the restriction of irredundancy. Example 6.4.2 shows that this is a serious problem. It is conjectured that this example is, in effect, the only example that makes redundant Cayley graphs different.

The proof of Theorem 6.4.2 occupies the remainder of this chapter. By the second corollary to Theorem 6.4.1, only Cayley graphs of valence 3 or 4 need be considered. The proof is broken into two main arguments: valence 4 and valence 3. These are, in turn, broken into cases and subcases depending on the nature of the given generating set X.

The proof of Theorem 6.4.2 involves much manipulation of generators and relators. The study of groups given by generators and relators (usually called combinatorial group theory) has many famous logical pitfalls, but here the obvious intuitive approach is fine. Some agreement on terminology is, nevertheless, in order. A "word" in a generating set X for the group $\mathscr{A}$ is a finite sequence of elements of X and inverses of elements of X. The "length" of the word is the number of terms in the sequence. A "relator" is a word the product of whose terms is the identity; by abuse of language, we shall refer to the product itself as a relator. Any conjugate or inverse of a relator is a relator, as is any cyclic permutation. A relator is "reduced" if no cyclic permutation of it can be shortened by using another relator to replace one subword by another word. For example, if $xyxy$ is a relator, then $yzxzyx$ would not be a reduced relator, since $zxzyxy = zxzx^{-1}$. A relator of length k is called a k-relator.

6.4.4. Proof of Theorem 6.4.2: Valence $d = 4$

The inequality of Theorem 6.4.1 will be used repeatedly. It is given again here in the specific case $d = 4$ for easy reference:

$$(r-4)\#\mathscr{A} + r\chi(S) \leq \sum_{i=1}^{r-1} (r-i)f_i$$

Case Ia. $X = \{x, y\}$, both x and y have order greater than 3.
This case was handled in Example 6.4.1.

Case Ib. $X = \{x, y\}$, $x^3 = 1$, and y has order greater than 3.
By the inequality of Theorem 6.4.1, with $r = 5$,

$$\#\mathscr{A} + 5\chi(S) \leq 2f_3 + f_4$$

The only 3-relator, by irredundancy, is x^3, and hence $f_3 \leq \#\mathscr{A}/3$. Thus $f_4 > 0$ if $\#\mathscr{A} > -15\chi(S)$. The only reduced 4-relators that do not immediately give a Section-6.3 presentation are y^4 or $(xy^{\pm 1})^2$. If y^4 is the only 4-relator, then $f_4 \le \#\mathscr{A}/4$, which implies $\#\mathscr{A} \le -60\chi(S)$. Therefore we assume $(xy)^2$ is a relator (if both $(xy)^2$ and $(xy^{-1})^2$ are relators, we have presentation (f_1) of Theorem 6.3.3). We claim that if $\#\mathscr{A} > -42\chi(S)$, then y has order at most 6, in which case we have presentation (a) or (b) of Theorem 6.3.1 or (c) of Theorem 6.3.3. Suppose every k-relator, $k \le 6$, involves x. The total number of sides of edges labeled x available for building faces is $2\#\mathscr{A}$. By irredundancy, any relator using x once must use it twice. Thus by counting occurrences of x, we obtain

$$3f_3 + 2f_4 + 2f_5 + 2f_6 \le 2\#\mathscr{A}$$

Combining this inequality with Theorem 6.4.1, when $r = 7$, we get

$$3\#\mathscr{A} + 7\chi(S) \le 4f_3 + 3f_4 + 2f_5 + f_6 \le f_3 + f_4 + 2\#\mathscr{A}$$

Since $(xy)^2$ is the only 4-relator, there can be at most two faces of size 4 meeting at any vertex, and hence $4f_4 \le 2\#\mathscr{A}$. Since $f_3 \le \#\mathscr{A}/3$, it follows that $\#\mathscr{A} > -42\chi(S)$. Thus if $\#\mathscr{A} > -42\chi(S)$, there must be a k-relator, $k \le 6$, not involving x, that is, y has order at most 6, as claimed.

Case Ic. $X = \{x, y\}, x^3 = y^3 = 1$.
By Theorem 6.4.1, with $r = 7$,

$$3\#\mathscr{A} + 7\chi(S) \le 4f_3 + 3f_4 + 2f_5 + f_6$$

The only 3-relators are x^3 and y^3, so $3f_3 \le 2\#\mathscr{A}$. Therefore if $\#\mathscr{A} > -21\chi(S)$, at least one of f_4, f_5, or f_6 is nonzero. Any 4-relator yields one of the presentations of Section 6.3. Since $x^2 = x^{-1}$ and $y^2 = y^{-1}$, there is no reduced 5-relator. The only 6-relators, up to cyclic permutation, interchange of x and y, or replacement of x by x^{-1} or y by y^{-1}, are $(xy)^3$, $xyxyxy^{-1}$, $xyxyx^{-1}y^{-1}$, or $xyx^{-1}yxy^{-1}$. The relator $(xy)^3$ gives presentation (c) of Theorem 6.3.3. The relator $xyxyxy^{-1}$ implies that $(xy)^2 = yx^{-1}$. Thus yx^{-1} is in the subgroup $\langle xy \rangle$. So is $yx^{-1} \cdot xy = y^2 = y^{-1}$, and therefore so is x. Thus $\mathscr{A}$ is cyclic generated by xy, and obviously $\sigma(\mathscr{A}) = 0$. The relator $xyxyx^{-1}y^{-1}$ implies $xyx = yxy^{-1}$. Therefore xyx has order 3, which yields $(xyx)^3 = xyx^{-1}yx^{-1}yx = 1$. Cyclically permuting this last relator gives us $(x^{-1}y)^3$; this is just presentation (c) of Theorem 6.3.3 again when x is replaced by x^{-1}. Finally, the relator $xyx^{-1}yxy^{-1}$ becomes $xyxy^{-1}x^{-1}y^{-1}$ after replacing y by y^{-1} and cyclically permuting; this gives the first exceptional presentation (a) of this theorem. We conclude that if $\#\mathscr{A} > -21\chi(S)$, then the group $\mathscr{A}$ has one of the desired presentations.

Case IIa. $X = \{x, y, z\}$, $y^2 = z^2 = 1$, x has order greater than 3.
By irredundancy, $f_3 = 0$. If there is no vertex at which four faces of size 4 meet, then $4f_4 \le 3\#\mathscr{A}$. Thus by Theorem 6.4.1, with $r = 5$,

$$\#\mathscr{A} + 5\chi(S) \le 2f_3 + f_4 \le 3\#\mathscr{A}/4$$

and therefore $\#\mathscr{A} \leq -20\chi(S)$. On the other hand, if four faces of size 4 meet at a vertex, then it is easily verified that, no matter what the rotation at that vertex, the group $\mathscr{A}$ must have presentation (b_1), (h), or (l_2) of Theorem 6.3.3.

Case IIb. $X = \{x, y, z\}$, $y^2 = z^2 = 1$, $x^3 = 1$.

As in Case Ib, if $\#\mathscr{A} > -15\chi(S)$, there must be a 4-relator. If $xyx^{\pm 1}y$ and $xzx^{\pm 1}z$ are both relators, then one of the presentations (b_1), (h), or (l_2) of Theorem 6.3.3 is obtained, as in Case IIa. Therefore assume that $xzx^{\pm 1}z$ is not a relator. The only possible 4-relators are then $xyx^{\pm 1}y$ and $yzyz$. Let f_x be the number of faces of size 4 corresponding to $xyx^{\pm 1}y$ and f_z the number of faces of size 4 corresponding to $yzyz$. We claim that $f_3 + f_x \leq \#\mathscr{A}/2$. It would then follow from Theorem 6.4.1, with $r = 6$, that

$$2\#\mathscr{A} + 6\chi(S) \leq 3f_3 + 2(f_x + f_z) + f_5 \leq f_3 + \#\mathscr{A} + f_z + f_5$$

Since the cycles $yzyz$ give a 2-factor of the Cayley graph $C(\mathscr{A}, X)$, $f_z \leq \#\mathscr{A}/4$. Any reduced 5-relator contradicts irredundancy, so $f_5 = 0$. Since $f_3 \leq \#\mathscr{A}/3$, the above inequality then gives us $\#\mathscr{A} \leq -72\chi(S)/5 \leq -15\chi(S)$.

To establish the claim that $f_3 + f_x \leq \#\mathscr{A}/2$, consider the subgraph of the Cayley graph generated by x- and y-edges. With the relators x^3 and $xyx^{\pm 1}y$, the subgroup $\langle x, y\rangle$ is just $\mathscr{Z}_2 \times \mathscr{Z}_3$ or the dihedral group $\mathscr{D}_3$. Thus a component of the x-y subgraph is simply a 6-vertex prism with a triangular top and bottom (corresponding to x^3) and three quadrilateral sides (corresponding to $xyx^{\pm 1}y$). As soon as a single z-edge is inserted at a vertex in the prism, one of the five faces of the prism is destroyed (as a face). At least two of the prism faces are destroyed when all six z-edges are inserted. Thus each prism can contribute at most three faces to the imbedding. Since there are $\#\mathscr{A}/6$ prisms, it follows that $f_3 + f_x \leq 3\#\mathscr{A}/6$, as claimed.

Case III. $X = \{x, y, z, w\}$, $x^2 = y^2 = z^2 = w^2 = 1$.

As in Case Ia, if $\#\mathscr{A} > -20\chi(S)$ there must be a vertex at which four faces of sizes 4 meet. By irredundancy, each reduced 4-relator involves two generators. Thus there is some cyclic ordering of the generators such that $(xy)^2, (yz)^2, (zw)^2$, and $(wx)^2$ are all relators, which gives us presentation (j) of Theorem 6.3.3.

6.4.5. Proof of Theorem 6.4.2: Valence $d = 3$

There are two main cases to consider here: two generators, one of order 2, and three generators, all of order 2. The form of Theorem 6.4.1 used here is

$$(r/2 - 3)\#\mathscr{A} + r\chi(S) \leq \sum_{i=1}^{r-1} (r - i)f_i$$

Case Ia. $X = \{x, y\}$, $y^2 = 1$, x has order greater than 5.

There can be no 3-relator, by irredundancy. The only possible reduced 4-relators are $xyx^{\pm 1}y$; this gives presentation (a) or (f_1) of Theorem 6.3.3 since $y = y^{-1}$. The only reduced 5-relators are $yxyx^{\pm 2}$. The relator $yxyx^{-2}$ implies $yxy = x^2$. Thus, $x = yx^2y = yxy \cdot yxy = x^2 \cdot x^2$, which means x has order 3, contradicting the assumed order of x (in any case $x^2 = x^{-1}$ so that the relator $yxy = x^2$ makes $\mathscr{A}$ the dihedral group $\mathscr{D}_3$). A similar argument handles $yxyx^2$. We conclude that $f_3 = f_4 = f_5 = 0$. By Theorem 6.4.1, with $r = 7$,

$$\#\mathscr{A}/2 + 7\chi(S) \leq f_6$$

Therefore if $\#\mathscr{A} > -14\chi(S)$, there must be a 6-relator.

Possible reduced 6-relators are, up to cyclic permutations and replacement of x by x^{-1},

$$x^2yx^{\pm 2}y, \quad yxyx^{\pm 3}, \quad xyxyx^{-1}y, \quad (xy)^3, \quad x^6$$

The relator x^2yx^2y gives presentation (g_1) and $x^2yx^{-2}y$ gives (i) of Theorem 6.3.3. If $yxyx^{-3}$ is a relator, then $yxyx = x^4$. Thus $(yx)^2$ commutes with x. Since $(yx)^2$ also commutes with yx, it must commute with y. Let $u = yx$. Then u and y generate $\mathscr{A}$, and $u^2yu^{-2}y$ is a relator yielding presentation (g_1) of Theorem 6.3.3 again. The same argument applies to the relator $yxyx^3$. The relator $xyxyx^{-1}y$ implies $x = yxyx^{-1}y$, which means x is conjugate to y and hence x has order 2, a contradiction. Finally, the remaining relators, $(xy)^3$ and x^6, together give us presentation (e) of Theorem 6.3.3 with generators y and $u = xy$. If three six-sided faces meet at a single vertex, then both $(xy)^3$ and x^6 are relators, or one of the other relators is present; in either case, we have a desired presentation. If there is no vertex at which three six-sided faces meet, then $6f_6 \leq 2\#\mathscr{A}$ and, by Theorem 6.4.1 again, $\#\mathscr{A} \leq -42\chi(S)$.

Case Ib. $X = \{x, y\}$, $y^2 = 1$, $x^5 = 1$.

As in Case Ia, $f_3 = f_4 = 0$. If $f_6 = 0$, then, by Theorem 6.4.1, with $r = 7$,

$$\#\mathscr{A}/2 + 7\chi(S) \leq 2f_5 \leq 2\#\mathscr{A}/5$$

and hence $\#\mathscr{A} \leq -70\chi(S)$. If $f_6 \neq 0$, then the previous analysis of 6-relators leads to a presentation from Section 6.3.

Case Ic. $X = \{x, y\}$, $y^2 = 1$, $x^4 = 1$.

As in Case Ia, $f_3 = f_5 = 0$. We can assume $f_6 = 0$, by the analysis of 6-relators in Case Ia. Since $x^3 = x^{-1}$, there are no reduced 7-relators involving two y's. Thus the only possible reduced 7-relators are of the form $x^2yxyx^{\pm 1}y$. If x^2yxyxy is a relator, then $xyxyxy = x^{-1}$; thus $x \in \langle xy \rangle$, which makes $\mathscr{A}$ cyclic, and $\sigma(\mathscr{A}) = 0$. If $x^2yxyx^{-1}y$ is a relator, then $xyxyx^{-1} = x^{-1}y$. Thus $x^{-1}y$ is conjugate to x, and hence $x^{-1}y$ has order 4. This gives us presentation (e) of Theorem 6.3.3, generated by x and $u = x^{-1}y$. We assume therefore that $f_7 = 0$. Theorem 6.4.1, with $r = 9$, then yields

$$3\#\mathscr{A}/2 + 9\chi(S) \leq 5f_4 + f_8$$

Since $f_4 \leq \#\mathscr{A}/4$, it follows that $f_8 > 0$ when $\#\mathscr{A} > -36\chi(S)$.

Consider then the possible reduced 8-relators. Since $x^3 = x^{-1}$, there are no reduced 8-relators involving two y's. The only 8-relators involving three y's are of the form $yx^2yx^2yx^{\pm 1}$, but then $yx^2yx^2y = x^{\pm 1}$, which implies that $x^{\pm 1}$ is conjugate to y and has order 2, a contradiction. Therefore we need only consider 8-relators involving four y's. Up to cyclic permutation and replacement of x by x^{-1}, the possibilities are

$$(xy)^4, \quad (xy)^3x^{-1}y, \quad (xy)^2(x^{-1}y)^2, \quad (xyx^{-1}y)^2$$

The relator $(xy)^4$ gives us presentation (e) of Theorem 6.3.3 again. The relator $(xy)^3x^{-1}y$ implies that $xyx = yx^{-1}yxy$, that is, xyx is conjugate to y and therefore has order 2. Hence $xyxxyx$ is a relator, yielding presentation (g_1) of Theorem 6.3.3, as in Case Ia. The relator $(xy)^2(x^{-1}y)^2$ implies that $(xy)^2 = (yx)^2$. Thus $y(xy)^2y = y(yx)^2y = (xy)^2$, and this gives us presentation (i) of Theorem 6.3.3, generated by y and $u = xy$. Finally, the relator $(xyx^{-1}y)^2$ together with x^4 yields presentation (p) of Theorem 6.3.3.

Case Id. $X = \{x, y\}$, $y^2 = 1$, $x^3 = 1$.

Since $x^2 = x^{-1}$, there are no reduced relators of odd length. By the analysis of the previous cases, we can assume that $f_4 = f_6 = f_8 = 0$. Up to cyclic permutation and replacement of x by x^{-1}, the possible 10-relators are

$$(xy)^5, \quad (xy)^4x^{-1}y, \quad (xy)^3(x^{-1}y)^2, \quad (xy)^2x^{-1}yxyx^{-1}y$$

The relator $(xy)^5$ gives us presentation (c) of Theorem 6.3.1. Since $xy \cdot yx = x^2 = x^{-1}$, the elements xy and yx together generate the group $\mathscr{A}$. Therefore the relator $(xy)^4x^{-1}y$, which makes $yx = (xy)^4 \in \langle xy \rangle$, implies that $\mathscr{A}$ is cyclic. The relator $(xy)^3(x^{-1}y)^2$ implies that $xyxyxyx^{-1}yx^{-1} = y$, but then y is conjugate to x, contradicting the assumed orders of x and y. The relator $(xy)^2x^{-1}yxyx^{-1}y$ implies that $yxyx^{-1}yxyx^{-1}y = x^{-1}$, and again x^{-1} is conjugate to y, a contradiction. Therefore we can assume that $f_{10} = 0$. Then by Theorem 6.4.1, with $r = 14$ (recall that $f_{11} = f_{13} = 0$, since there are no odd-length relators),

$$4\#\mathscr{A} + 14\chi(S) \leq 11f_3 + 2f_{12}$$

Since $f_3 \leq \#\mathscr{A}/3$, it follows that $f_{12} \neq 0$ when $\#\mathscr{A} > -42\chi(S)$.

Up to cyclic permutation and replacement of x by x^{-1}, the possible 6-relators are

$$(xy)^6, \quad (xy)^5x^{-1}y, \quad (xy)^4(x^{-1}y)^2, \quad (xy)^3x^{-1}yxyx^{-1}y, \quad (xyxyx^{-1}y)^2$$
$$(xy)^3(x^{-1}y)^3, \quad (xy)^2x^{-1}yxy(x^{-1}y)^2, \quad (xyx^{-1}y)^3$$

The relator $(xy)^6$ gives us presentation (e) of Theorem 6.3.3. The relator $(xy)^5x^{-1}y$ implies that $\mathscr{A}$ is cyclic, as before, for 10-relators. The relator $(xy)^4(x^{-1}y)^2$ implies that $yxyxyx^{-1}yx^{-1}y = x^{-1}yx^{-1}$; that is, $x^{-1}yx^{-1}$ is conjugate to y. Therefore $(x^{-1}yx^{-1})^2$ is a relator, which in turn gives the relator $(xy)^2$, making $\mathscr{A}$ the dihedral group $\mathscr{D}_3$. Similarly, the relator $(xy)^3x^{-1}yxyx^{-1}y$ implies that $yxyx^{-1}yxyx^{-1}y = x^{-1}yx^{-1}$, from which it follows again that $x^{-1}yx^{-1}$ is conjugate to y and $\mathscr{A} = \mathscr{D}_3$. The relators $(xyxyx^{-1}y)^2$ and $(xy)^3(x^{-1}y)^3$ give us, respectively, the exceptional presentations (c) and (b). The relator $(xy)^2x^{-1}yxy(x^{-1}y)^2$ can be rewritten $(xy)^2x^{-1}yx^{-1} \cdot x^{-1}y(x^{-1}y)^2$, from which we obtain $xyxyx^{-1}yx^{-1} = (yx)^3$. Then $(yx)^3$ is conjugate to y, which means that x has order 6 and that $\mathscr{A}$

has presentation (e) of Theorem 6.3.3 again. Finally, the relator $(xyx^{-1}y)^3$ gives us presentation (n) of Theorem 6.3.3.

Case IIa. $X = \{x, y, z\}$, $x^2 = y^2 = z^2 = 1$, xy, yz, xz all have order greater than 2.

By irredundancy, there are no reduced relators of length 3 or 5, and the only relators of length 4 are of the form $(xy)^2$, contradicting the assumed orders of xy, yz, and xz. Therefore $f_3 = f_4 = f_5 = 0$. As in Case Ia, if $\#\mathscr{A} > -14\chi(S)$, then $f_6 \neq 0$. Up to cyclic permutation and interchanging the names x, y, z, the possible 6-relators are

$$(xyz)^2, \quad xyzyxz, \quad (xy)^3, \quad (yz)^3, \quad (xz)^3$$

Relators $(xyz)^2$ and $xyzyxz$ yield, respectively, presentations (b_1) and (l_1) of Theorem 6.3.3. The remaining three relators together give presentation (m) of Theorem 6.3.3. Thus if three faces of size 6 meet at a vertex, then one of the desired presentations is obtained. As in Case Ia, if $\#\mathscr{A} > -42\chi(S)$, then three faces of size 6 meet at a vertex.

Case IIb. $X = \{x, y, z\}$, $x^2 = y^2 = z^2 = 1, (xy)^2 = 1$, yz and xz have order greater than 3.

Observe that if, say, $(xy)^2 = (yz)^2 = 1$, then y commutes with both x and z, which means that $\mathscr{A} = \mathscr{Z}_2 \times \mathscr{D}_n$ and $\sigma(\mathscr{A}) = 0$. Therefore at most one of xy, yz, and xz can have order 2. By the analysis in Case IIa, we can assume that $f_3 = f_5 = f_6 = 0$. The only possible reduced 7-relator, up to cyclic permutation or interchange of x and y, is $zxzyzyx$ (recall that x and y commute, since $(xy)^2 = 1$). Thus $zyzy = xzx$ and $xzxz = yzy$, which imply $(zy)^2$ and $(xz)^2$ have order 2. This gives us presentation (o) of Theorem 6.3.3. Therefore we assume that $f_7 = 0$. As in Case Ic, $f_8 \neq 0$ if $\#\mathscr{A} > -36\chi(S)$.

Up to cyclic permutation or interchange of x and y, the possible reduced 8-relators are

$$zxzxyzxy, \quad (zxzy)^2, \quad (zx)^2(zy)^2, \quad (xz)^4, \quad (yz)^4$$

The relator $zxzxyzxy$ implies that $xzxyzx = zy$, which means that zy is conjugate to xy and has order 2, a contradiction of the assumed order of yz. The relator $(zxzy)^2$ gives us presentation (k) of Theorem 6.3.3 (interchange x and z). Given the relator $(zx)^2(zy)^2$, let $u = zxz$ and $v = xy$. Then $(uv)^2 = (zxzxy)^2 = (zx)^2y(zx)^2y = (zx)^2y(yz)^2y = (zx)^2(zy)^2 = 1$, and $(zuzv)^2 = (xxy)^2 = 1$. Thus we have presentation (k) of Theorem 6.3.3 again, generated by u, v, and z. Finally, if both $(xz)^4$ and $(yz)^4$ are relators, then we have presentation (o) of Theorem 6.3.3. Thus if two 8-sided faces meet at a vertex, we have one of the desired presentations. On the other hand, if there is at most one 8-sided face at any vertex, then by Theorem 6.4.1, with $r = 9$,

$$3\#\mathscr{A}/2 = 9\chi(S) \leq 5f_4 + f_8 \leq 5\#\mathscr{A}/4 + \#\mathscr{A}/8$$

and hence $\#\mathscr{A} \leq -72\chi(S)$.

Case IIc. $X = \{x, y, z\}$, $x^2 = y^2 = z^2 = 1, (xy)^2 = (yz)^3 = 1$.

We claim that if $\#\mathscr{A} > -84\chi(S)$, then xz must have order 6 or less, and hence $\mathscr{A}$ has presentation (b), (c), or (d) of Theorem 6.3.2 or (q) of Theorem 6.3.3. The claim is equivalent to showing that there is a relator of length 13 or less not involving y. Indeed, suppose that every relator of length 13 or less involved y. By the analysis of previous cases, we can assume that $f_3 = f_5 = f_7 = f_8 = 0$ and that the only 4-relator is $(xy)^2$ and only 6-relator is $(yz)^3$. Suppose we could show that $f_9 = 0$, that every 10-relator uses at least four y's, and that every 11-relator uses at least three y's (by irredundancy, any relator using y must use it twice). There are only $\#\mathscr{A}$ sides of edges labeled y available for building faces. Counting occurrences of y, we would have

$$2f_4 + 3f_6 + 4f_{10} + 3f_{11} + 2f_{12} + 2f_{13} \leq \#\mathscr{A}$$

Then by Theorem 6.4.1, with $r = 14$,

$$4\#\mathscr{A} + 14\chi(S) \leq 10f_4 + 8f_6 + 4f_{10} + 3f_{11} + 2f_{12} + f_{13} \leq 8f_4 + 5f_6 + \#\mathscr{A}$$

Since $f_4 \leq \#\mathscr{A}/4$ and $f_6 \leq \#\mathscr{A}/6$, it follows that $\#\mathscr{A} \leq -84\chi(S)$.

Therefore, to establish our claim, we must eliminate all 9-relators, all 10-relators using three y's or fewer, and all 11-relators using two y's. In the following computation, it should be recalled that because $(xy)^2 = (yz)^3 = 1$, x and y commute and $yzy = zyz$. There is no reduced 9-relator using only two z's or five or more z's. The only reduced 9-relator using three z's is $(zxy)^3$. If we let $u = xy$, then $(uy)^2$, $(yz)^3$, and $(zu)^3$ are relators, thus giving us presentation (b) of Theorem 6.3.2, generated by u, y, and z. Any reduced 9-relator using four z's must contain an occurrence of zyz, and hence can be turned into a 9-relator with three z's, since $zyz = yzy$. It is easily verified, using the relations $xy = yx$ and $yzy = zyz$, that the only 10-relator involving two y's that needs to be considered is $(yxzxz)^2$. Let $u = yx$; then $(uzxz)^2$ and $(ux)^2$ are relators, yielding presentation (k) of Theorem 6.3.3, generated by u, x, and z. In any reduced 10-relator involving three y's, there must occur the word $yxzy$, $yzyx$, or yzy. In each case, the relations $xy = yx$ and $yzy = zyz$ can be used to make a 10-relator involving two y's. Finally, by similar arguments, the only 11-relators involving two y's that one needs to consider are of the form $yzxzy(zx)^3$ and $yzxzxzy(zx)^2$. The first implies that $(zx)^3$ is conjugate to x and that hence zx has order 6, giving us presentation (q) of Theorem 6.3.3. The second implies that $(zx)^2$ is conjugate to x and that hence zx has order 4, giving us presentation (c) of Theorem 6.3.2. □

6.4.6. Remarks about Theorem 6.4.2

We have not yet addressed the exceptional groups (a)–(c) of Theorem 6.4.2. These groups do have genus 1; toroidal imbeddings are given in Figures 6.26–6.28. The groups are related; group (a) is contained in both (b) and (c) as the normal index-2 subgroup corresponding to the normal closure of x,

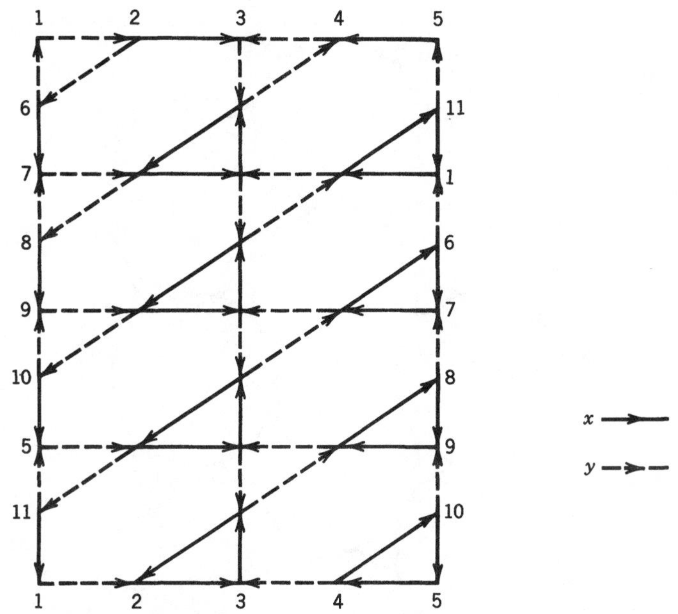

Figure 6.26. A toroidal imbedding of group (a): $\langle x, y: x^3 = y^3 = xyxy^{-1}x^{-1}y^{-1} = 1\rangle$. Vertices with the same number are to be identified.

namely, $\langle x, yxy\rangle$ (see Exercises 4 and 5). The orders of the groups (a)–(c) are then easily computed from the order of any one of them; the order of (c) is computed in Exercise 6. None of these groups has symmetric genus 1; Theorem 6.3.4 is used in Exercise 7 to show that the group (a) does not act on the torus, and therefore (b) and (c) cannot either, since (a) is contained in both (b) and (c). Also, as a perhaps curious coincidence, group (c) is the classical group GL(2, 3) of 2×2 invertible matrices over the field with three elements, which means that group (a) must be the subgroup SL(2, 3) of matrices with determinant 1 (see Exercise 8). In summary, we have the following result.

Theorem 6.4.3 (Tucker, 1984a). *The groups* (a)–(c) *of Theorem* 6.4.2 *are the only groups* $\mathscr{A}$ *such that* $\gamma(\mathscr{A}) = 1$ *and* $\sigma(\mathscr{A}) > 1$. *In addition, group* (a) *is a subgroup of both* (b) *and* (c), *and* (c) *is the group* GL(2, 3).

As it stands, Theorem 6.4.2 does not quite classify groups of genus 0. However, a reconsideration of the proof of Theorem 6.4.2 in Exercises 9–13 leads to the following theorem that, in fact, does imply $\sigma(\mathscr{A}) = 0$ whenever $\gamma(\mathscr{A}) = 0$. An interesting consequence of this theorem is that, although $\gamma(\mathscr{Z}_2 \times \mathscr{A}_5) = 0$, there is no two-element generating set X for the group $\mathscr{Z}_2 \times \mathscr{A}_5$ such that $\gamma(C(\mathscr{Z}_2 \times \mathscr{A}_5, X)) = 0$ (see Exercise 16).

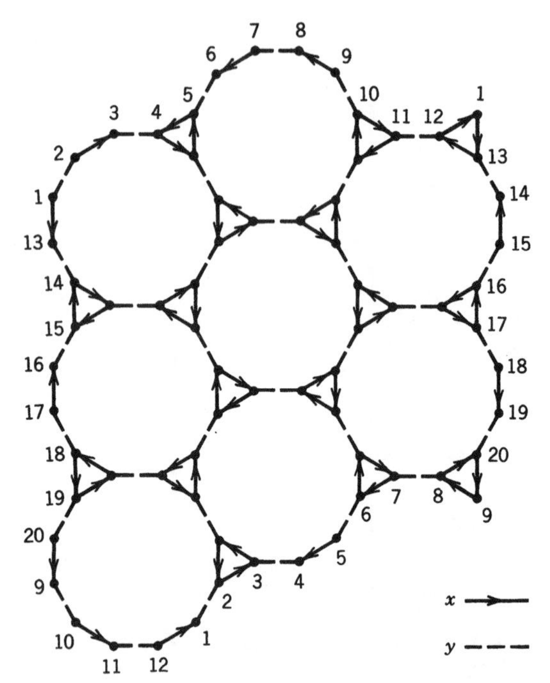

Figure 6.27. A toroidal imbedding of group (b): $\langle x, y\colon x^3 = y^2 = (xy)^3(x^{-1}y)^2 = 1\rangle$. Vertices with the same number are to be identified.

Theorem 6.4.4 (Maschke, 1896). *If the irredundant Cayley graph $C(\mathscr{A}, X)$ imbeds in a surface S such that $\chi(S) > 0$, then either $\mathscr{A} = \mathscr{Z}_3 \times \mathscr{Z}_3$ or $\sigma(\mathscr{A}) = 0$. Moreover, $\gamma(\mathscr{Z}_3 \times \mathscr{Z}_3) = 1$, although there is a projective-plane imbedding of a Cayley graph for $\mathscr{Z}_3 \times \mathscr{Z}_3$.*

The actual proof of Theorem 6.4.2 is probably as important as the final result, at least for the computation of groups of low genus. Many of the bounds on $\#\mathscr{A}$ that appear in the proof can be improved (Exercises 17–20), and additional information about the structure of the given group $\mathscr{A}$ can be determined when $\#\mathscr{A}$ is near one of these bounds. The computation of $\gamma(\mathscr{S}_5)$ in Exercises 21–23 gives one example of how the proof of Theorem 6.4.2 can be refined in a specific case. As another example, we state without proof the following Cayley-graph version of the refined Hurwitz theorem (Theorem 6.3.6) given at the end of the previous section.

Theorem 6.4.5 (Tucker, 1984b). *Suppose that $\gamma(\mathscr{A}) > 1$ and that $\#\mathscr{A} > 24(\gamma(\mathscr{A}) - 1)$. Then either the group $\mathscr{A}$ is as described in Theorem* 6.3.6, *with the word "proper" omitted throughout, or else $\mathscr{A}$ has a presentation*

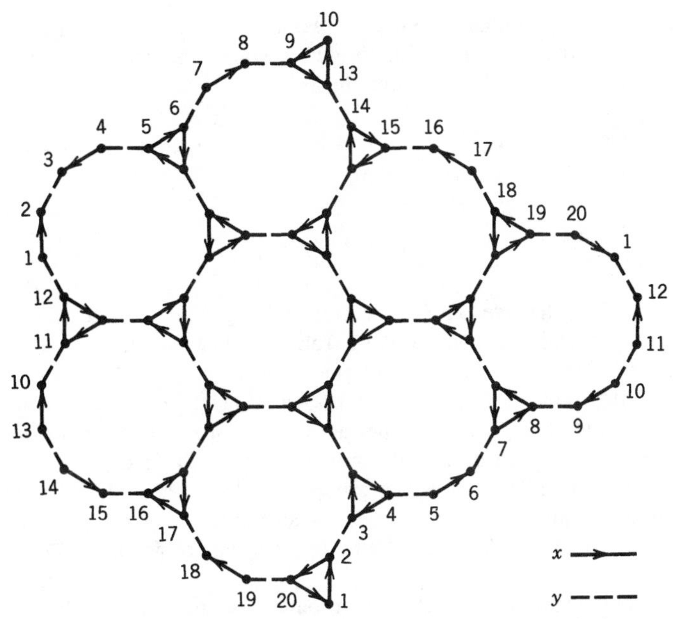

Figure 6.28. A toroidal imbedding of group (c): $\langle x, y: x^3 = y^2 = (xyxyx^{-1}y)^2 = 1\rangle$. Vertices with the same number are to be identified.

of the form

$$\langle x, y: x^3 = y^2 = 1, w = 1, \dots \rangle$$

where w *is a word in* x, x^{-1}, *and* y *of length* 14, 16, 18, 20, *or* 22. *If* $\#\mathscr{A} > 48(\gamma(\mathscr{A}) - 1)$, *then* $\mathscr{A}$ *is as described in Theorem* 6.3.6, *without the word "proper".*

Proof. Omitted. □

Conder (1987) has considered all possible words w. For $w = (xy)^4(x^{-1}y)^4$, Conder (1986) has shown that $\#\mathscr{A} > 24(\gamma(\mathscr{A}) - 1)$ and $\mathscr{A}$ is not described in Theorem 6.3.6. Thus the exceptional presentations of this type really do provide examples of conflicts between $\gamma(\mathscr{A})$ and $\sigma(\mathscr{A})$, even when the order of $\mathscr{A}$ is large compared with $\gamma(\mathscr{A})$. The role of the word "proper" in all of this is not understood. It is known that improper quotients have Cayley-graph imbeddings in the surface of "correct" Euler characteristic, but that surface must be nonorientable. The genus of such groups is unknown; the alternating groups $\mathscr{A}_n$, $n > 167$, are an example of such groups, due to Conder (1980).

Nevertheless, Theorem 6.4.5 greatly restricts the possibilities for a group of small genus greater than 1. The reader is encouraged to try to use it to complete the proof of the following theorem.

Theorem 6.4.6 (Tucker, 1984c). *There is exactly one group of genus* 2. *It has order* 96, *and it has the presentation*

$$\langle x, y, z : x^2 = y^2 = z^2 = (xy)^2 = (yz)^3 = (xz)^8 = 1, \quad [y, (xz)^4] = 1 \rangle$$

Proof. That the given group has an imbedding in the surface of genus 2 (a symmetric imbedding at that!) follows from the given presentation. The reader can verify that the subgroup generated by $y' = xy$ and $x' = yz$ is the exceptional group of genus 1 with presentation (c) in Theorem 6.4.2. In particular, since this group does not act on the torus, neither does the given group. Since the given group does not have order 24 or order 48, it is not exceptional either. Thus the given group must have genus 2. An exhaustive analysis of groups of order between 24 and 168, using the restrictions of Theorem 6.4.5, shows that there are no other groups of genus 2. □

The group given in Theorem 6.4.6 probably does not look familiar. Coxeter (1982) provides a full account of its origins in geometry and combinatorics; it is, for example, the automorphism group of the generalized Petersen graph $G(8,3)$ (see Frucht et al., 1971).

Finally, it is possible to adapt the methods used in the proof of Theorem 6.4.2 to give a partial answer to another fundamental question about the genus of a group. Recall Babai's theorem that if $\mathscr{B}$ is a subgroup of $\mathscr{A}$, then $\gamma(\mathscr{B}) \leq \gamma(\mathscr{A})$. One would expect that the same holds true for quotients: to wit, that if $\mathscr{Q}$ is a quotient of $\mathscr{A}$, then $\gamma(\mathscr{Q}) \leq \gamma(\mathscr{A})$. This is clearly true for symmetric and strong symmetric genus, by the natural action of the quotient group on the quotient surface. Unfortunately, taking the quotient of a Cayley graph can increase its genus. For instance, in Example 6.4.2 (due to Wormald), the genus-2 Cayley graph for the presentation

$$\langle x, y, z : x^{2n} = y^2 = z^2 = 1, [x, y] = [x, z] = 1, z = x^n \rangle$$

is a quotient of the genus-1 Cayley graph for

$$\langle x, y, z : x^{2n} = y^2 = z^2 = 1, [x, y] = [x, z] = 1 \rangle.$$

Of course, the first presentation is redundant, and in fact the underlying group $\mathscr{Z}_2 \times \mathscr{Z}_{2n}$ has genus 0, but this example does indicate that some subtle questions about generating sets are involved, which is not the case for the corresponding subgroup problem.

6.4.7. Exercises

1. Given graphs G and H and edges d in G and e in H, define $G_d * H_e$ to be the graph obtained by deleting edges d and e and adding a pair of edges joining the endpoints of d to the endpoints of e. Note that if G and H are regular of valence n, then so is $G_d * H_e$. Prove that if $\gamma(G-d)=\gamma(G)$ and $\gamma(H-e)=\gamma(H)$, then $\gamma(G_d * H_e)=\gamma(G)+\gamma(H)$.
2. Use Exercise 1 to construct infinitely many regular graphs of valence n, $3 \le n \le 5$, and genus $g \ge 0$ (Proulx, 1983).
3. Construct infinitely many regular graphs of valence 6 and genus $g > 0$ using Exercise 1 (Proulx, 1983). (*Hint:* First construct infinitely many valence-6 triangulations of the torus.)
4. Prove that the subgroup $\langle x, yxy\rangle$ of group (b) of Theorem 6.4.2 is isomorphic to group (a).
5. Prove that the subgroup $\langle x, yxy\rangle$ of group (c) of Theorem 6.4.2 is isomorphic to group (a) (let $u=x$, $v=yx^{-1}y$, and consider $uvuv^{-1}u^{-1}v^{-1}$).
6. In group (c) of Theorem 6.4.2, prove that $(xy)^4y(xy)^4y=1$. Show then that xy has order 8 and $(xy)^4$ is central. Conclude that group (c) has order 48, by considering the quotient by the normal subgroup $\langle (xy)^4\rangle$.
7. Let $\mathscr{A}$ be the group (a) of Theorem 6.4.2 and suppose that $\sigma(\mathscr{A})=1$. Use Theorem 6.3.4 to prove that then $\mathscr{A}$ must have a normal subgroup $\mathscr{N}$ with quotient $\mathscr{Z}_3$ and that both xy^{-1} and $y^{-1}x$ are in $\mathscr{N}$. Then show that $[xy^{-1}, y^{-1}x]=1$ implies that $x^{-1}y$, has order 2 and that therefore $\mathscr{A}$ has order 12, a contradiction.
8. Show that

$$x=\begin{pmatrix}0 & 1\\ -1 & -1\end{pmatrix}$$

and

$$y=\begin{pmatrix}1 & 0\\ 0 & -1\end{pmatrix}$$

generate GL(2, 3) and satisfy $x^3=y^2=(xyxyx^{-1}y)^2=1$. Then compute the order of GL(2, 3) and compare with the order of group (c) to conclude that (c) is isomorphic to GL(2, 3).
9. Prove that if $\chi(S)>0$ in Theorem 6.4.2, then $f_3 \ne 0$ for valence 4 and $f_6 \ne 0$ for valence 3, which implies that Cases Ia, IIa, III for valence 4 and Ia, IIa for valence 3 do not occur.
10. Show by counting x-edges, that if $\chi(S)>0$ in Theorem 6.4.2, then y has order at most 5 in Case Ib for valence 4, and yz has order 2 in Case IIb for valence 4. Conclude that if $\mathscr{A}$ has a valence-4,

irredundant Cayley graph, imbeddable in a surface S, $\chi(S) > 0$, then $\mathscr{A}$ is one of the groups $\mathscr{Z}_3 \times \mathscr{Z}_3$, $\mathscr{Z}_{15}$, $\mathscr{Z}_{12}$, $\mathscr{Z}_2 \times \mathscr{Z}_6$, $\mathscr{Z}_2 \times \mathscr{D}_3$ or else $\mathscr{A} = \langle x, y\colon x^3 = y^r = (xy)^2 = 1 \rangle$, $r = 3, 4, 5$.

11. Prove that if $\chi(S) > 0$ in Theorem 6.4.2, then in Case Ib and Ic for valence 3, xy has order 3. (For Case Ic you will need to show that $f_7 = 0$.)

12. Prove that if $\chi(S) > 0$ in Theorem 6.4.2, then in Case Id for valence 3, either $(xy)^4$ or $[x, y]^2$ or $(xy)^5$ is a relator.

13. Prove that if $\chi(S) > 0$ in Theorem 6.4.2, then Case IIb for valence 3 cannot occur, and in Case IIc, the element xz has order 3 or 4.

14. Use Theorem 6.4.1, with $r = 7$, to prove that $\gamma(\mathscr{Z}_3 \times \mathscr{Z}_3) \neq 0$.

15. Give a projective-plane imbedding of a Cayley graph for $\mathscr{Z}_3 \times \mathscr{Z}_3$.

16. Show that the group $\mathscr{A} = \mathscr{Z}_2 \times \mathscr{A}_5$ has a two-element generating set. Prove, however, that although $\sigma(\mathscr{A}) = 0$, $\gamma(C(\mathscr{A}, X)) > 0$ for any two-element generating set X. Thus for a given group $\mathscr{A}$, it is not necessarily true that $\gamma(C(\mathscr{A}, X))$ is minimized by choosing a generating set of minimal size (Sit, 1980).

17. By considering possible 5-relators in Case Ib, valence 4, of Theorem 6.4.2, prove that if $\#\mathscr{A} > -12\chi(S)$ (rather than $-60\chi(S)$), then xy has order 2. Thus, in this case, if $\#\mathscr{A} > -42(S)$, then $\gamma(\mathscr{A}) \leq 1$.

18. By considering possible 7-relators in Case Ib, valence 3, show that $\gamma(\mathscr{A}) \leq 1$ if $\#\mathscr{A} > -20\chi(S)$ (rather than $-70\chi(S)$).

19. In Case IIb, valence 3, of Theorem 6.4.2, if yz has order 4, prove by counting y-edges and by using Theorem 6.4.1, that if $\#\mathscr{A} > -36\chi(S)$, then xz also has order 4 and $\gamma(\mathscr{A}) \leq 1$.

20. Use Exercises 17–19 to prove that if $\#\mathscr{A} > -42\chi(S)$ in Theorem 6.4.2, then either $\gamma(\mathscr{A}) \leq 1$ or $\mathscr{A} = \langle x, y, z\colon x^2 = y^2 = z^2 = 1, (xy)^2 = (yz)^3 = 1 \ldots \rangle$.

21. Prove that the symmetric group $\mathscr{S}_5$ has no normal subgroup as in Theorem 6.3.4 and hence that $\gamma(\mathscr{S}_5) > 1$. Then show that therefore $\mathscr{S}_5$ has no presentation of the form $\langle x, y\colon x^r = y^s = (xy)^t = 1, \ldots \rangle$ or $\langle x, y, z\colon x^2 = y^2 = z^2 = (xy)^r = (yz)^s = (xz)^t, \ldots \rangle$ such that $\{2, 3\} \subset \{r, s, t\}$. (*Hint:* Every element of $\mathscr{S}_5$ has order at most 6.)

22. Suppose that $\gamma(\mathscr{S}_5) \leq 3$. Use Exercises 17–20 to show that all valence-4 cases and Case Ib, Id, IIc for valence 3 of Theorem 6.4.2 do not occur. Eliminate the other valence-3 cases using Theorem 6.4.1 and $\#\mathscr{S}_5 \geq -30\chi(S)$. (Difficult cases are Ic, where you must show $f_8 = f_9 = 0$, IIa, where 7-relators must be considered, and IIb, where you must show that yz has order 4 and xz order 5, which is impossible for $\mathscr{S}_5$.)

23. Prove that $\gamma(\mathscr{S}_5) = 4$ (Proulx, 1981).

24. Let $\mathscr{A}$ be the K-metacyclic group $\langle x, y\colon x^p = y^{p-1} = yxy^{-1}x^r = 1 \rangle$, p a prime, $r \neq \pm 1$. Prove that $\gamma(\mathscr{A}) \leq 1$ if and only if $p = 5$ or 7 (Gross and Lomonaco, 1980).

References

Additional works of interest are listed in the Supplementary Bibliography. Numbers in angle brackets indicate the page citations of the references.

Mohammed Abu-Sbeih

—— and T. D. Parsons (1983). Embeddings of bipartite graphs, *J. Graph Theory 7*, 325–334. ⟨210⟩

Jin Akiyama

—— and M. Kano (1985). Factors and factorization of graphs—a survey, *J. Graph Theory 9*, 1–42. (This entire issue of the journal is devoted to factorizations of graphs.) ⟨79⟩

James W. Alexander

(1920). Note on Riemann spaces, *Bull. Amer. Math. Soc. 26*, 370–372. ⟨180⟩

(1924). An example of a simply connected surface bounding a region which is not simply connected, *Proc. Nat. Acad. Sci. USA 10*, 8–10. ⟨100⟩

Seth R. Alpert

(1973). The genera of amalgamations of graphs, *Trans. Amer. Math. Soc. 178*, 1–39. ⟨154⟩

(1975). Two-fold triple systems and graph imbeddings, *J. Combin. Theory Ser. A 18*, 101–107. ⟨105⟩

—— and J. L. Gross (1976). Components of branched coverings of current graphs, *J. Combin. Theory Ser. B 20*, 283–303. ⟨87⟩

see also J. L. Gross

Dana Angluin

—— and A. Gardiner (1981). Finite common coverings of pairs of regular graphs, *J. Combin. Theory Ser. B 30*, 184–187. ⟨80⟩

Kenneth Appel

—— and W. Haken (1976). Every planar map is four-colorable, *Bull. Amer. Math. Soc. 82*, 771–712. ⟨38, 215⟩

Dan Archdeacon

(1981). A Kuratowski theorem for the projective plane, *J. Graph Theory 5*, 243–246. ⟨53⟩

(1984). Face colorings of embedded graphs, *J. Graph Theory 8*, 387–398. ⟨46⟩

(1986). The orientable genus is nonadditive, *J. Graph Theory 10*, 385–401. ⟨154⟩

—— and P. Huneke (1985). On cubic graphs which are irreducible for non-orientable surfaces, *J. Combin. Theory Ser. B 39*, 233–264. ⟨53⟩

Louis Auslander

—— with T. A. Brown and J. W. T. Youngs (1963). The imbedding of graphs in manifolds, *J. Math. and Mech. 12*, 629–634. ⟨140, 141⟩

Laszlo Babai

(1976). Embedding graphs in Cayley graphs, *Probl. Comb. Theorie des Graphes*, Proc. Conf. Paris-Orsay (J.-C. Bermond et al., eds.), CNRS Paris, 13–15. ⟨17⟩

(1977). Some applications of graph contractions, *J. Graph Theory 1*, 125–130. ⟨92ff⟩

—— with W. Imrich and L. Lovasz (1972). Finite homeomorphism groups of the 2-sphere, *Colloq. Math. Janos Bolyai 8, Topics in Topology*, Keszthely (Hungary). ⟨287⟩

R. P. Baker

(1931). Cayley diagrams on the anchor ring, *Amer. J. Math. 53*, 645–669. ⟨300⟩

Hyman Bass

see J.-P. Serre

Joseph Battle

—— with F. Harary, Y. Kodama, and J. W. T. Youngs (1962). Additivity of the genus of a graph, *Bull. Amer. Math. Soc. 68*, 565–568. ⟨152⟩

Lowell W. Beineke

see S. Stahl

K. N. Bhattacharya

(1943). A note on two-fold triple systems, *Sankhyā 6*, 313–314. ⟨105⟩

Ludwig Bieberbach

(1911). Uber die Bewegungsgruppen der euklidische Raume I, *Math. Ann. 70*, 297–336; II, *72* (1912), 400–412. ⟨277⟩

Norman L. Biggs

—— with E. K. Lloyd and R. J. Wilson (1976). *Graph Theory, 1736–1936*, Clarendon Press, Oxford. ⟨35, 79⟩

R H Bing

(1954). Locally tame sets are tame, *Ann. of Math. 59*, 145–158. ⟨18⟩

Garrett Birkhoff

see S. MacLane

Donald W. Blackett

(1967). *Elementary Topology: A Combinatorial and Algebraic Approach*, Academic Press, Orlando. ⟨110⟩

Bela Bollobas

—— and P. Catlin (1980). Hadwiger's conjecture is true for almost every graph, *European J. Combin. 1*, 195–199. ⟨40⟩

W. W. Boone

(1954). Certain simple unsolvable problems of group theory, *Nederl. Akad. Wetensch. Indag. Math. 16*, 231–237 and 492–497; *17*, 252–256 and 571–577; *19*, 22–27 and 227–232. ⟨14⟩

André Bouchet

(1982a). Constructions of covering triangulations with folds, *J. Graph Theory 6*, 57–74. ⟨175, 210⟩

(1982b). Constructing a covering triangulation by means of a nowhere-zero dual flow, *J. Combin. Theory Ser. B 32*, 316–325. ⟨210⟩

(1983). A construction of a covering map with faces of even lengths, *Ann. Discrete Math. 17*, 119–130. ⟨210⟩

Matthew G. Brin

—— and C. C. Squier (1986). On the genus of $Z_3 \times Z_3 \times Z_3$, *Europ. J. Combin.*, to appear. ⟨262⟩

Edward M. Brown

(1969). The Hauptvermutung for 3-complexes, *Trans. Amer. Math. Soc. 144*, 173–196. ⟨18, 98⟩

T. A. Brown

see L. Auslander

Robin P. Bryant

—— and D. Singerman (1985). Foundations of the theory of maps on surfaces with boundary, *Quart. J. Math. Oxford Ser. (2) 36*, 17–41. ⟨281⟩

W. Burnside

(1911). *Theory of Groups of Finite Order* (1st ed., 1897). Cambridge Univ. Press, Cambridge. ⟨14, 265⟩

James W. Cannon

(1984). The combinatorial structure of cocompact discrete hyperbolic groups, *Geom. Dedicata 16*, 123–148. ⟨14⟩

Paul Catlin

(1979). Hajos graph-coloring conjecture: Variations and counterexamples, *J. Combin. Theory Ser. B 26*, 268–274. ⟨41⟩

see also B. Bollobas

Augustin-Louis Cauchy

(1813). Recherches sur les polyedres—premier memoire, *J. de l'Ecole Poly. 9* (Cah. 16), 68–86. [Extract in Biggs, Lloyd, and Wilson (1976).] ⟨27⟩

Arthur Cayley

(1878). The theory of groups: Graphical representations, *Amer. J. Math. 1*, 174–176. ⟨10⟩

Marston D. E. Conder

(1980). Generators of the alternating and symmetric groups, *J. London Math. Soc. (2) 22*, 75–86. ⟨305, 315⟩

(1984). Some results on quotients of triangle groups, *Bull. Austral. Math. Soc. 30*, 73–90. ⟨297⟩

(1986). A note on Cayley graphs, *J. Combin. Theory Ser. B 40*, 362–368. ⟨315⟩

(1987). Three-relator quotients of the modular group, *Quart. J. Math. Oxford*, to appear. ⟨315⟩

Harold Scott MacDonald Coxeter

(1961). *Introduction to Geometry*, Wiley, New York. ⟨279⟩

(1982). The group of genus two, *Rend. Sem. Mat. Brescia 7*, 219–248. ⟨316⟩

—— and W. O. J. Moser (1957). *Generators and Relations for Discrete Groups* (4th ed., 1980). Springer-Verlag, Berlin and New York. ⟨14, 277, 291, 295⟩

Richard W. Decker

—— with H. H. Glover and J. P. Huneke (1981). The genus of the 2-amalgamations of graphs, *J. Graph Theory 5*, 95–102. ⟨154⟩

—— with H. H. Glover and J. P. Huneke (1985). Computing the genus of the 2-amalgamations of graphs, *Combinatorica 5*, 271–282. ⟨154⟩

Max Dehn

(1911). Uber unendliche diskontinuierliche Gruppen, *Math. Ann. 72*, 413–421. ⟨10, 14⟩

G. Demoucron

—— with Y. Malgrange and R. Pertuiset (1964). Graphs planaires: Reconnaissance et construction de representations planaires topologiques, *Rev. Française Recherche Operationnelle 8*, 33–47. ⟨52⟩

Gabriel A. Dirac

—— and S. Schuster (1954). A theorem of Kuratowski, *Nederl. Akad. Wetensch. Indag. Math. 16*, 343–348. ⟨264⟩

Richard A. Duke

(1966). The genus, regional number, and Betti number of a graph, *Canad. J. Math 18*, 817–822. ⟨134⟩

(1971). How is a graph's Betti number related to its genus?, *Amer. Math. Monthly 78*, 386–388. ⟨136⟩

W. Dyck

(1882). Gruppentheoretische Studien, *Math. Ann. 20*, 1–45. ⟨265⟩

Jack R. Edmonds

(1960). A combinatorial representation for polyhedral surfaces (abstract), *Notices Amer. Math. Soc. 7*, 646. ⟨113, 117, 192⟩

(1965). On the surface duality of linear graphs, *J. Res. Nat. Bur. Standards Sect. B*, 121–123. ⟨119, 135⟩

Samuel Eilenberg

(1934). Sur les transformations periodique de la surface de sphere, *Fund. Math. 22*, 28–41. ⟨269⟩

Arnold Emch

(1929). Triple and multiple systems, their geometric configurations and groups, *Trans. Amer. Math. Soc. 31*, 25–42. ⟨105⟩

Paul Erdős

—— and S. Fajtlowicz (1981). On the conjecture of Hajos, *Combinatorica 1*, 141–143. ⟨41⟩

Leonhard Euler

(1736). Solution problematis ad geometriam situs pertinentis, *Comment. Acad. Sci. I. Petropolitanae 8*, 128–140. ⟨34⟩

(1750). Letter to C. Goldbach [excerpted in Biggs, Lloyd, and Wilson (1976)]. ⟨27⟩

Cloyd L. Ezell

(1979). Observations on the construction of covers using permutation voltage assignments, *Discrete Math. 28*, 7–20. ⟨88⟩

Siemion Fajtlowicz

see P. Erdős

Istvan Fary

(1948). On straight line representation of planar graphs, *Acta Sci. Math. (Szeged) 11*, 229–233. ⟨48⟩

E. S. Federov

(1891). *Zapiski Mineralogischeskogo Imperatorskogo S. Petersburgskogo Obstrechestva (2) 28*, 345–390. ⟨277⟩

Ion S. Filotti

—— with G. Miller and J. Reif (1979). On determining the genus of a graph in $O(v^{O\langle G\rangle})$ steps, *Proc. 11th Annual ACM Symp. Theory of Computing*, 27–37. ⟨53⟩

Ralph H. Fox

(1957). Covering spaces with Singularities, *Algebraic Geometry and Topology: A Symposium in Honor of S. Lefschetz*, Princeton University Press, Princeton, 243–257. ⟨175⟩

Philip Franklin

(1934). A six color problem, *J. Math. Phys. 16*, 363–369. ⟨137, 218⟩

Roberto Frucht

(1938). Herstellung von Graphen mit vorgegebener abstrakten Gruppe, *Compositio Math. 6*, 239–250. ⟨70⟩

—— with J. E. Graver and M. Watkins (1971). The groups of generalized Petersen graphs, *Proc. Camb. Philos. Soc. 70*, 211–218. ⟨59, 316⟩

Merrick L. Furst

—— with J. L. Gross and L. McGeoch (1985a). Finding a maximum-genus graph imbedding, *J. Assoc. Comp. Mach.*, to appear. ⟨133, 146⟩

—— with J. L. Gross and R. Statman (1985b). Genus distributions for two classes of graphs, *J. Combin. Theory Ser. B*, to appear. ⟨147⟩

see also J. L. Gross

A. Gardiner

see D. Angluin

Brian L. Garman

(1979). Voltage graph embeddings and the associated block designs, *J. Graph Theory 2*, 181–187. ⟨105⟩

Henry H. Glover

—— with P. Huneke and C. Wang (1979). 103 graphs that are irreducible for the projective plane, *J. Combin. Theory Ser. B 27*, 332–370. ⟨53⟩

—— and D. Sjerve (1985). Representing $PSL_2(p)$ on a Riemann surface of least genus, *Enseignement Math. 31*, 305–325. ⟨297⟩

see also R. W. Decker

Chris Godsil

—— and W. Imrich (1985). Embedding graphs in Cayley graphs, *Graphs and Combinatorics*, to appear. ⟨17⟩

Jack E. Graver

see R. Frucht

Hubert Brian Griffiths

(1976). *Surfaces* (2nd ed., 1981). Cambridge University Press, Cambridge. ⟨110⟩

Jonathan L. Gross

(1974). Voltage graphs, *Discrete Math. 9*, 239–246. ⟨57, 165, 176⟩

(1977). Every connected regular graph of even degree is a Schreier coset graph, *J. Combin. Theory Ser. B 22*, 227–232. ⟨78⟩

(1978). An infinite family of octahedral crossing numbers, *J. Graph Theory 2*, 171–178. ⟨209, 214⟩

—— and S. R. Alpert (1973). Branched coverings of graph imbeddings, *Bull. Amer. Math. Soc. 79*, 942–945. ⟨194⟩

—— and S. R. Alpert (1974). The topological theory of current graphs, *J. Combin. Theory Ser. B 17*, 218–233. ⟨117, 165, 194, 196, 205⟩

—— and M. Furst (1987). Hierarchy for imbedding-distribution invariants of a graph, *J. Graph Theory 11*, to appear. ⟨147⟩

—— and F. Harary (1980). Some problems in topological graph theory, *J. Graph Theory 4*, 253–263. ⟨250⟩

—— with B. Jackson, T. D. Parsons, and T. Pisanski (1982). Wrapped coverings and their realizations, *Congress. Numer. 35*, 333–347. ⟨207⟩

—— and E. W. Klein (1986). A Kuratowski-type theorem for the average genus of a graph, in preparation. ⟨147⟩

—— and S. J. Lomonaco (1980). A determination of the toroidal k-metacyclic groups, *J. Graph Theory 4*, 165–172. ⟨318⟩

—— with D. P. Robbins and T. W. Tucker (1986). Genus distributions for bouquets of circles, preprint. ⟨147⟩

—— and R. H. Rosen (1979). A linear-time planarity algorithm for 2-complexes, *J. Assoc. Comput. Mach. 20*, 611–617. ⟨117⟩

—— and R. H. Rosen (1981). A combinatorial characterization of planar 2-complexes, *Colloq. Math. 44*, 241–247. ⟨117⟩

—— and T. W. Tucker (1974). Quotients of complete graphs: Revisiting the Heawood map-coloring problem, *Pacific J. Math. 55*, 391–402. ⟨57, 67, 230⟩

—— and T. W. Tucker (1977). Generating all graph coverings by permutation voltage assignments, *Discrete Math. 18*, 273–283. ⟨74, 81, 84, 176, 177⟩

—— and T. W. Tucker (1979a). Fast computations in voltage graph theory, *Ann. New York Acad. Sci. 319*, 254–257. ⟨91, 170⟩

—— and T. W. Tucker (1979b). Local maxima in graded graphs of imbeddings, *Ann. New York Acad. Sci. 319*, 254–257. ⟨150⟩

see also S. R. Alpert and M. L. Furst

William Gustin

(1963). Orientable embedding of Cayley graphs, *Bull. Amer. Math. Soc. 69*, 272–275. ⟨193ff⟩

H. Hadwiger

(1943). Uber eine Klassification der Streckenkomplexe, *Vierteljschr. Naturforsch. Ges. Zurich 88*, 133–142. ⟨40⟩

Gary Haggard

(1980). Loops in duals, *Amer. Math. Monthly 87*, 654–656. ⟨46⟩

Wolfgang R. Haken

see K. Appel

Haim Hanani

(1961). The existence and construction of balanced incomplete block designs, *Ann. Math. Stat. 32*, 361–386. ⟨105⟩

Frank Harary

(1969). *Graph Theory*, Addison-Wesley, Reading, Massachusetts. ⟨2⟩

—— and R. H. Rosen (1976). On the planarity of 2-complexes, *Colloq. Math. 36*, 101–108. ⟨117⟩

see also J. Battle and J. L. Gross

Philip J. Heawood

(1890). Map-colour theorem, *Quart. J. Math. 24*, 332–338. ⟨215ff⟩

Lothar Heffter

(1891). Uber das Problem der Nachbargebiete, *Math. Ann. 38*, 477–580. ⟨113, 117, 137, 191, 217ff⟩

(1897). Uber Triplesysteme, *Math. Ann. 49*, 101–112. ⟨105⟩

Hugh M. Hilden

(1976). Three-fold branched coverings of S^3, *Amer. J. Math. 98*, 989–997. ⟨180⟩

John E. Hopcroft

—— and R. E. Tarjan (1974). Efficient planarity testing, *J. Assoc. Comput. Mach. 21*, 271–283. ⟨52⟩

John Philip Huneke

see D. Archdeacon, R. W. Decker, and H. H. Glover

A. Hurwitz

(1892). Uber algebraische Gebilde mit eindeutigen Transformationen in sich, *Math. Ann. 41*, 403–442. ⟨296⟩

Wilfried Imrich

see L. Babai and C. Godsil

Bradley W. Jackson

—— with T. D. Parsons and T. Pisanski (1981). A duality theorem for graph embeddings, *J. Graph Theory 5*, 55–77. ⟨207, 210⟩

see also J. L. Gross and T. D. Parsons

David M. Jackson

(1987). Counting cycles in permutations by group characters, with an application to a topological problem, *Trans. Amer. Math. Soc. 299*, 785–801. ⟨147⟩

Alain Jacques

(1969). *Constellations et Proprietes Algebrigues des Graphs*, Ph.D. thesis, University of Paris. ⟨194⟩

Francoise Jaeger

(1976). Thesis, Université de Grenoble. ⟨145⟩

Gareth A. Jones

—— and D. Singerman (1978). Theory of maps on orientable surfaces, *Proc. London Math. Soc. (3) 37*, 273–307. ⟨281⟩

Mark Jungerman

(1974). Ph.D. thesis, University of California, Santa Cruz. ⟨245⟩

(1975). The genus of the symmetric quadripartite graph, *J. Combin. Theory Ser. B 19*, 181–187. ⟨195, 201⟩

(1978). The non-orientable genus of the n-cube, *Pacific J. Math. 76*, 443–451. ⟨159⟩

—— and A. T. White (1980). On the genus of finite abelian groups, *European J. Combin. 1*, 243–251. ⟨249ff⟩

Mikio Kano

see J. Akiyama

Richard M. Karp

(1972). Reducibility among combinatorial problems, in *Complexity of Computer Computation* (R. Miller and J. Thatcher, eds.), Plenum, New York. ⟨35⟩

Abraham Karrass

see W. Magnus

A. B. Kempe

(1879). On the geographical problem of four colors, *Amer. J. Math. 2*, 193–204. ⟨219⟩

B. Kerekjarto

(1919). Uber die endlichen topologische Gruppen der Kugelflache, *Amsterdam Akad. Versl. 28*, 555–556. ⟨287⟩

(1921). Uber die periodische Transformationen der Kreischeibe und der Kugelflache, *Math. Ann. 80*, 36–38. ⟨269⟩

E. Ward Klein

see J. L. Gross

Felix Klein

(1911). *Elementar-Mathematik vom hoheren Standpunkte aus, Vol. 2*, Teubner, Leipzig. ⟨26⟩

Donald Erwin Knuth

(1969). *The Art of Computer Programming, Vol. 2*, Addison-Wesley, Reading, Massachusetts. ⟨161⟩

Yukihiro Kodama

see J. Battle

Denes König

(1916). Uber graphen und ihre Anwendung auf Determinantheorie und Mengenlehre, *Math. Ann. 77*, 453–465. ⟨80⟩

S. Kundu

(1974). Bounds on the number of disjoint spanning trees, *J. Combin. Theory Ser. B 17*, 199–203. ⟨145⟩

Kasimierz Kuratowski

(1930). Sur le probleme des courbes gauches en topologie, *Fund. Math. 15*, 271–283. ⟨28⟩

Henry M. Levinson

(1970). On the genera of graphs of group presentations, *Ann. New York Acad. Sci. 175*, 277–284. ⟨249, 264⟩

—— and B. Maskit (1975). Special embeddings of Cayley diagrams, *J. Combin. Theory Ser. B 18*, 12–17. ⟨265⟩

S.-A.-J. Lhuilier

(1811). Memoire sur la polyedrometrie, *Ann. de Mathematiques 3*, 169–189. [Extract in Biggs, Lloyd, and Wilson (1976).] ⟨27⟩

J. B. Listing

(1861). Der Census raumlicher Complexe oder Verallgemeinerung des Euler'schen Satzes von den Polyedern, *Abh. K. Ges. Wiss. Göttingen Math. Cl 10*, 97–182. ⟨27⟩

E. Keith Lloyd

see N. L. Biggs

Samuel J. Lomonaco

see J. L. Gross

Laszlo Lovasz

see L. Babai

Roger C. Lyndon

(1985). *Groups and Geometry* (London Math. Soc. Lecture Note Series, No. 101), London Math. Soc., Cambridge. ⟨277, 279⟩

—— and P. E. Schupp (1977). *Combinatorial Group Theory*, Springer-Verlag, Berlin and New York. ⟨14⟩

C. Maclachlin

(1965). Abelian groups of automorphisms of compact Riemann surfaces, *Proc. London Math. Soc. 15*, 699–712. ⟨265⟩

Saunders MacLane

(1937). A structural characterization of planar combinatorial graphs, *Duke Math. J. 3*, 460–472. ⟨54⟩

—— and G. Birkhoff (1967), *Algebra*, Macmillan, New York. ⟨254⟩

Wilhelm Magnus

(1974). *Non-Euclidean Tessalations and Their Groups*, Academic Press, Orlando. ⟨279⟩

—— with A. Karrass and D. Solitar (1966). *Combinatorial Group Theory*, Wiley-Interscience, New York. ⟨14⟩

Y. Malgrange

see G. Demoucron

H. Maschke

(1896). The representation of finite groups, *Amer. J. Math. 18*, 156–194. ⟨287, 290, 300, 314⟩

Bernard Maskit

see H. Levinson

Lyle McGeoch

see M. L. Furst

Martin Milgram

—— and P. Ungar (1977). Bounds for the genus of graphs with given Betti number, *J. Combin. Theory Ser. B 23*, 227–233. ⟨136⟩

Gary Miller

see I. S. Filotti

John Milnor

(1961). Two complexes which are homeomorphic but combinatorially distinct, *Ann. of Math. 74*, 575–590. ⟨18, 98⟩

(1965a). *Lectures on the h-Cobordism Theorem* (Princeton Mathematical Notes), Princeton University Press, Princeton. ⟨109⟩

(1965b). *Topology from the Differentiable Viewpoint*, University Press of Virginia, Charlottesville. ⟨185⟩

August F. Möbius

(1865). Uber die Bestimmung des Inhaltes eines Polyeders, *Ber. K. Sachs. Ges. Wiss. Leipzig Math.-Phys. Cl 17*. ⟨25⟩

Bojan Mohar

—— with T. Pisanski, M. Skoviera, and A. T. White (1985). The cartesian product of three triangles can be imbedded into a surface of genus 7, *Discrete Math. 56*, 87–89. ⟨262⟩

Edwin E. Moise

(1952). Affine structures in 3-manifolds, V, *Ann. of Math. 56*, 96–114. ⟨18⟩

J. Montesinos

(1976). Three-manifolds as 3-fold branched covers of S^3, *Quart. J. Math. Oxford 27*, 85–94. ⟨180⟩

William O. J. Moser

see H. S. M. Coxeter

James R. Munkres

(1975). *Topology: A First Course*, Prentice-Hall, Englewood Cliffs, New Jersey. ⟨95⟩

Seiya Negami

(1983). Uniqueness and faithfulness of embedding of toroidal graphs, *Discrete Math. 44*, 161–180. ⟨147⟩

(1985). Unique and faithful embeddings of projective-planar graphs, *J. Graph Theory 9*, 235–243. ⟨147⟩

M. H. A. Newman

(1954). *Elements of the Topology of Plane Sets of Points*, Cambridge Univ. Press, Cambridge. ⟨100⟩

P. Niggli

(1924). Die Flachensymmetrien homogener Diskontinuen, *Z. Kristallogr. Mineralog. Petrog. Abt. A 60*, 283–298. ⟨278⟩

Edward A. Nordhaus

—— with B. M. Stewart and A. T. White (1971). On the maximum genus of a graph, *J. Combin. Theory Ser. B 11*, 258–267. ⟨132⟩

—— with R. D. Ringeisen, B. M. Stewart, and A. T. White (1972). A Kuratowski-type theorem for the maximum genus of a graph, *J. Combin. Theory Ser. B 12*, 260–267. ⟨132, 148⟩

P. S. Novikov

(1955). On the algorithmic unsolvability of the word problem in groups (in Russian), *Trudy Mat. Inst. Steklov 44*, Izdat. Akad. Nauk. SSSR, Moscow. ⟨14⟩

Christos D. Papakyriakopoulos

(1943). A new proof of the invariance of the homology groups of a complex, *Bull. Soc. Math. Grèce 22*, 1–154. ⟨18, 98⟩

Torrence D. Parsons

—— with T. Pisanski and B. Jackson (1980). Dual imbeddings and wrapped quasi-coverings of graphs, *Discrete Math. 31*, 43–52. ⟨207, 210⟩

see also M. Abu-Sbeih, J. L. Gross, and B. W. Jackson

David Pengelley

(1975). Self-dual orientable embeddings of K_n, *J. Combin. Theory Ser. B 18*, 46–52. ⟨209⟩

R. Pertuisset

see G. Demoucron

Julius Petersen

(1891). Die Theorie der regularen Graphen, *Acta Math. 15*, 193–220. ⟨35, 36⟩

Tomaz Pisanski

(1980). Genus of cartesian products of regular bipartite graphs, *J. Graph Theory 4*, 31–42. ⟨155, 159⟩

(1982). Nonorientable genus of cartesian products of regular graphs, *J. Graph Theory 6*, 31–42. ⟨156⟩

see also J. L. Gross, B. W. Jackson, B. Mohar, and T. D. Parsons

Georg Polya

(1924). Uber die Analogie der Kristallsymmetrie in der Ebene, *Z.Kristallogr. Mineralog. Petrog. Abt. A 60*, 278–282. ⟨278⟩

Viera Krnanova Proulx

(1978). Classification of the toroidal groups, *J. Graph Theory 2*, 269–273. [Extraction of results from Ph.D. thesis (1977), Columbia University.] ⟨300, 304⟩

(1981). On the genus of symmetric groups, *Trans. Amer. Math. Soc. 266*, 531–538. ⟨318⟩

(1983). On the genus of five- and six-regular graphs, *J. Graph Theory 7*, 231–234. ⟨317⟩

Tibor Rado

(1925). Uber den Begriff der Riemannschen Flache, *Acta Litt. Sci. Szeged 2*, 101–121. ⟨95⟩

John Reif

see I. S. Filotti

Richard D. Ringeisen

(1972). Determining all compact orientable 2-manifolds upon which $K_{m,n}$ has 2-cell imbeddings, *J. Combin. Theory Ser. B 12*, 101–104. ⟨148⟩

(1978). Survey of results on the maximum genus of a graph, *J. Graph Theory 3*, 1–13. ⟨142⟩

see also E. A. Nordhaus

Gerhard Ringel

(1955). Uber drei kombinatorische Probleme am *n*-dimensionalen Wurfel und Wurfelgitter, *Abh. Math. Sem. Univ. Hamburg 20*, 10–19. ⟨30, 159⟩

(1959). *Farbungsprobleme auf Flachen und Graphen*, Springer-Verlag, Berlin and New York. ⟨30, 236ff⟩

(1961). Uber das Problem der Nachbargebiete auf orientierbaren Flachen, *Abh. Math. Sem. Univ. Hamburg 25*, 105–127. ⟨192, 230, 234⟩

(1965). Das Geschlecht des vollstandigen paaren Graphen, *Abh. Math. Sem. Univ. Hamburg 28*, 139–150. ⟨30⟩

(1974). *Map Color Theorem*, Springer-Verlag, Berlin and New York. ⟨113, 216, 240, 245⟩

—— and J. W. T. Youngs (1968). Solution of the Heawood map-coloring problem, *Proc. Nat. Acad. Sci. USA 60*, 438–445. ⟨38, 215⟩

Neil Robertson

—— and P. D. Seymour (1983). Graph minors, I: excluding a forest, *J. Combin. Theory Ser. B 35*, 39–61. ⟨53⟩

—— and P. D. Seymour (1984). Graph minors, III: planar tree-width, *J. Combin. Theory Ser. B 36*, 49–64. ⟨53⟩

—— and P. D. Seymour (1986a). Graph minors, V: excluding a planar graph, *J. Combin. Theory Ser. B 41*, 92–114. ⟨53⟩

—— and P. D. Seymour (1986b). Graph minors, VI: disjoint paths across a disc, *J. Combin. Theory Ser. B 41*, 115–138. ⟨53⟩

—— and P. D. Seymour (1987). Graph minors, II: algorithmic aspects of tree width, *J. Algorithms*, to appear. ⟨53⟩

Ronald H. Rosen

see J. L. Gross and F. Harary

Gert Sabidussi

(1958). On a class of fixed-point free graphs, *Proc. Amer. Math. Soc. 9*, 800–804. ⟨69, 272⟩

Doris Schattschneider

(1978). The plane symmetry groups: their recognition and notation. *Amer. Math. Monthly 85*, 439–450. ⟨277⟩

W. Scherrer

(1929). Zur Theorie der endlichen Gruppen topologischer Abbildungen von geschlossenen Flachen in sich, *Comment. Math. Helv. 1*, 69–119. ⟨287⟩

Otto Schreier

(1927). Die Untergruppen der freien Gruppen, *Abh. Math. Sem. Univ. Hamburg 5*, 161–183. ⟨10⟩

Paul E. Schupp

see R. C. Lyndon

Seymour Schuster

see G. A. Dirac

Jean-Pierre Serre

—— with H. Bass (1977). Groups discretes: arbres, amalgams et SL_2, Cours redige avec la collaboration of H. Bass, *Astérisque 46*. ⟨14⟩

Paul D. Seymour

see N. Robertson

David Singerman

see R. P. Bryant and G. A. Jones

Cho Wei Sit

(1976). Quotients of complete multipartite graphs, *Pacific J. Math. 63*, 531–538. ⟨236⟩

(1980). Ph.D. thesis, Columbia University. ⟨250, 318⟩

Denis Sjerve

see H. H. Glover

Martin Skoviera

see B. Mohar

Stephen Smale

(1961). Generalized Poincaré's conjecture in dimensions greater than four, *Ann. of Math. 74*, 391–406. ⟨109⟩

Donald Solitar

see W. Magnus

Craig C. Squier

see M. G. Brin

Saul Stahl

(1978). Generalized embedding schemes, *J. Graph Theory 2*, 41–52. ⟨113, 136⟩

(1979). Self-dual embeddings of Cayley graphs, *J. Combin. Theory Ser. B 27*, 92–107. ⟨209⟩

(1980). Permutation-partition pairs: a combinatorial generalization of graph embeddings, *Trans. Amer. Math. Soc. 259*, 129–145. ⟨154⟩

(1981). Permutation-partition pairs II. Bounds on the genus of amalgamations of graphs, *Trans. Amer. Math. Soc. 271*, 175–182.

—— and L. W. Beineke (1977). Blocks and the nonorientable genus of graphs, *J. Graph Theory 1*, 75–78. ⟨155⟩

—— and A. T. White (1976). Genus embeddings for some complete tripartite graphs, *Discrete Math. 14*, 279–296. ⟨87, 170⟩

John Stallings

(1971). *Group Theory and Three-Dimensional Manifolds*, Yale University Press, New Haven. ⟨14⟩

Richard Statman

see M. L. Furst

Bonnie M. Stewart

see E. A. Nordhaus

Robert E. Tarjan

see J. E. Hopcroft

Charles M. Terry

—— with L. R. Welch and J. W. T. Youngs (1967). The genus of K_{12s}, *J. Combin. Theory 2*, 43–60. ⟨235⟩

—— with L. R. Welch and J. W. T. Youngs (1970). Solution of the Heawood map-coloring problem—case 4, *J. Combin. Theory 8*, 170–174. ⟨233⟩

Carsten Thomassen

(1980). Planarity and duality of finite and infinite graphs, *J. Combin. Theory Ser. B 29*, 244–271. ⟨42⟩

H. Tietze

(1910). Einige Bemerkerungen uber das Problem des Kartenfarbens auf einseitigen Flachen, *Jahresber. Deutsch. Math.-Verein. 19*, 155–159. ⟨217⟩

Marvin D. Tretkoff

(1975). Covering space proofs in combinatorial group theory, *Comm. Algebra 3*, 429–457. ⟨14⟩

(1980). A topological approach to the theory of groups acting on trees, *J. Pure Appl. Algebra 16*, 323–333. ⟨14⟩

Albert W. Tucker

(1936). Branched and folded coverings, *Bull. Amer. Math. Soc. 42*, 859–862. ⟨175⟩

Thomas W. Tucker

(1980). The number of groups of a given genus, *Trans. Amer. Math. Soc. 258*, 167–179. ⟨300, 304⟩

(1983). Finite groups acting on surfaces and the genus of a group, *J. Combin. Theory Ser. B 34*, 82–98. ⟨265, 273, 300⟩

(1984a) On Proulx's four exceptional toroidal groups, *J. Graph Theory 8*, 29–33. ⟨300, 304, 313⟩

(1984b). A refined Hurwitz theorem for imbeddings of irredundant Cayley graphs, *J. Combin. Theory Ser. B 36*, 244–268. ⟨300, 305, 314⟩

(1984c). There is one group of genus two, *J. Combin. Theory B 36*, 269–275. ⟨300, 316⟩

see also J. L. Gross

William Tutte

(1947). The factorizations of linear graphs, *J. London Math. Soc. 22*, 107–111. ⟨79⟩

(1949). On the imbedding of linear graphs in surfaces, *Proc. London Math. Soc. (2) 51*, 474–483. ⟨46⟩

(1960). Convex representation of graphs, *Proc. London Math. Soc. 10*, 304–320. ⟨48⟩

Peter Ungar

see M. Milgram

V. G. Vizing

(1964). On an estimate of the chromatic class of a p-graph (in Russian), *Diskret. Analiz 3*, 25–30. ⟨160⟩

Chin San Wang

see H. H. Glover

Mark E. Watkins

see R. Frucht

Lloyd R. Welch

see C. M. Terry

Hermann Weyl

(1952). *Symmetry*, Princeton University Press, Princeton. ⟨277⟩

Arthur T. White

(1970). The genus of repeated cartesian products of bipartite graphs, *Trans. Amer. Math. Soc. 151*, 393–404. ⟨155ff⟩

(1972). On the genus of a group, *Trans. Amer. Math. Soc. 173*, 203–214. ⟨249, 256⟩

(1974). Orientable imbeddings of Cayley graphs, *Duke. Math. J. 41*, 353–371. ⟨207⟩

(1978). Block designs and graph imbeddings, *J. Combin. Theory Ser. B 25*, 166–183. ⟨105⟩

(1973). *Graphs, Groups and Surfaces* (rev. ed., 1984), North-Holland, Amsterdam. ⟨93⟩

see also M. Jungerman, B. Mohar, E. A. Nordhaus, and S. Stahl

Hassler Whitney

(1932). Non-separable and planar graphs, *Trans. Amer. Math. Soc. 34*, 339–362. ⟨54⟩

(1933). A set of topological invariants for graphs, *Amer. J. Math. 55*, 231–235. ⟨49⟩

Robin J. Wilson

see N. L. Biggs

Nguyen Huy Xuong

(1979). How to determine the maximum genus of a graph, *J. Combin. Theory Ser. B 26*, 217–225. ⟨133, 144ff⟩

J. W. T. Youngs

(1963). Minimal imbeddings and the genus of a graph, *J. Math. and Mech. 12*, 303–315. ⟨117⟩

(1967). The Heawood map coloring conjecture, *Graph Theory and Theoretical Physics* (F. Harary, ed.), Academic Press, Orlando, 313–354. ⟨194⟩

(1968a). Remarks on the Heawood conjecture (nonorientable case), *Bull. Amer. Math. Soc. 74*, 347–353. ⟨199, 237⟩

(1968b). The nonorientable genus of K_n, *Bull. Amer. Math. Soc. 74*, 354–358. ⟨237⟩

see also L. Auslander, G. Ringel, and C. M. Terry

Bibliography

Pointers are given to further results on topics discussed in the text and also to several major research areas mentioned scarcely or not at all. A few useful monographs are cited. We have favored papers with topological content and tended to exclude papers not yet published.

Michael Albertson

—— and J. P. Hutchinson (1977). The independence ratio and genus of a graph, *Trans. Amer. Math. Soc. 226*, 161–173.

—— and J. P. Hutchinson (1978). On the independence ratio of a graph, *J. Graph Theory 2*, 1–8.

—— and J. P. Hutchinson (1980). On six-chromatic toroidal graphs, *Proc. London Math. Soc. (3) 41*, 533–556.

—— and J. P. Hutchinson (1980). Hadwiger's conjecture for graphs on the Klein bottle, *Discrete Math. 29*, 1–11.

V. B. Alekseev

—— and V. S. Gonchakov (1976). Thickness of arbitrary complete graphs, *Mat. Sb. 101*, 212–230.

Amos Altschuler

(1973). Construction and enumeration of regular maps on the torus, *Discrete Math. 4*, 201–217.

I. Anderson

—— and A. T. White (1978). Current graphs and bi-embeddings, *J. Graph Theory 2*, 231–239.

Dan Archdeacon

(1986). The non-orientable genus is additive, *J. Graph Theory 10*, 363–383.

Konhei Asano

(1986). The crossing number of $K_{1,3,n}$ and $K_{2,3,n}$, *J. Graph Theory 10*, 1–8.

M. Behzad

—— with G. Chartrand and L. Lesniak-Foster. *Graphs and Digraphs*, Prindle, Weber, and Schmidt, Boston.

Lowell W. Beineke

—— and F. Harary (1969). The thickness of the complete graph, *Canad. J. Math. 21*, 850–859.

—— with F. Harary and J. W. Moon (1964). On the thickness of the complete bipartite graph, *Proc. Cambridge Philos. Soc. 60*, 1–5.

—— and R. D. Ringeisen (1978). On crossing numbers of certain products of graphs, *J. Combin. Theory. Ser. B 24*, 134–136.

—— and R. D. Ringeisen (1980). On the crossing numbers of products of cycles and graphs of order four, *J. Graph Theory 4*, 145–155.

see also R. K. Guy

D. Benard

(1978). Orientable imbedding of line graphs, *J. Combin. Theory Ser. B. 24*, 34–43.

Edward A. Bender

—— and E. R. Canfield (1986). The asymptotic number of rooted maps on a surface, *J. Combin. Theory Ser. A 43*, 244–257.

—— and L. B. Richmond (1986). A survey of the asymptotic behavior of maps, *J. Combin. Theory Ser. B 40*, 297–329.

Claude Berge

(1973). *Graphs and Hypergraphs*, North-Holland, Amsterdam.

Frank Bernhart

—— and P. C. Kainen (1979). The book thickness of a graph, *J. Combin. Theory Ser. B 27*, 320–331.

Norman L. Biggs

(1971). Automorphisms of imbedded graphs. *J. Combin. Theory 11*, 132–138.

(1974). *Algebraic Graph Theory*, Cambridge University Press, Cambridge.

—— and A. T. White (1979). *Permutation Groups and Combinatorial Structures*, Cambridge University Press, Cambridge.

Gary S. Bloom

—— see E. F. Schmeichel

J. Ch. Boland

(1967). Embedding of graphs into orientable surfaces, *Nederl. Akad. Wetensch. Indag. Math. 29*, 33–44.

J. Adrian Bondy

—— and U. S. R. Murty (1976). *Graph Theory with Applications*, American Elsevier, New York.

O. V. Borodin

—— and L. S. Melnikov (1974). The chromatic number of a pseudosurface, *Diskret. Analiz 24*, 8–20.

Morton Brown

—— and R. Connelly (1975). On graphs with a constant link—II, *Discrete Math. 11*, 199–232.

William G. Brown

(1966). On the enumeration of non-planar maps, *Mem. Amer. Math. Soc. 65*.

E. Rodney Canfield

see E. A. Bender

Gary Chartrand

see M. Behzad

F. W. Clarke

—— with A. D. Thomas and D. A. Waller (1980). Embeddings of covering projections of graphs, *J. Combin. Theory. Ser. B. 28*, 10–17.

David S. Cochran

—— and R. H. Crowell (1970). $H_2(G')$ for tamely embedded graphs, *Quart. J. Math. Oxford Ser. (2) 21*, 25–27.

H. Coldeway

see H. Zieschang

R. Connelly

see M. Brown

John H. Conway

—— and C. McA. Gordon (1983). Knots and links in spatial graphs, *J. Graph Theory 7*, 445–453.

Richard H. Crowell

see D. S. Cochran

A. K. Dewdney

(1972). The chromatic number of a class of pseudo-manifolds, *Manuscripta Math. 6*, 311–320.

Gabriel A. Dirac

(1953). The coloring of maps, *J. London Math. Soc. 28*, 476–480.

(1957). Short proof of a map colour theorem, *Canad. J. Math. 9*, 225–226.

Dragomir Z. Djokovic

(1974). Automorphisms of graphs and coverings, *J. Combin. Theory Ser. B 16*, 243–247.

W. Dorfler

(1978). Double covers of hypergraphs and their properties, *Ars Combin. 6*, 293–313.

Allan L. Edmonds

—— with J. H. Ewing and R. S. Kulkarni (1982). Regular tesselations of surfaces and $(P, Q, 2)$ triangle groups, *Ann. of Math. 116*, 113–132.

John H. Ewing

see A. L. Edmonds

M. Farzan

—— and D. A. Waller (1977). Antipodal embeddings of graphs, *J. London Math. Soc. (2) 15*, 377–383.

Massimo Ferri

(1979). Crystallisation of 2-fold branched coverings of S^3, *Proc. Amer. Math. Soc. 73*, 271–276.

—— with C. Gagliardi and L. Graselli (1986). A graph-theoretical representation of *PL*-manifolds—a survey on crystallizations, *Aequationes Math. 31*, 121–141.

Steve Fisk

(1977). Geometric coloring theory, *Adv. in Math. 24*, 298–340.

Carlo Gagliardi

(1981). Extending the concept of genus to dimension n, *Proc. Amer. Math. Soc. 81*, 473–481.

see also M. Ferri

V. S. Gonchakov

see V. B. Alekseev

Cameron McA. Gordon

see J. H. Conway

L. Grasselli

see M. Ferri

Jack Graver

—— and M. Watkins (1977). *Combinatorics with Emphasis on the Theory of Graphs*, Springer-Verlag, New York.

Jonathan L. Gross

(1975). The genus of nearly complete graphs—case 6, *Aequationes Math. 13*, 243–249.

Branko Grunbaum

(1970). Polytopes, graphs, and complexes, *Bull. Amer. Math. Soc. 76*, 1131–1201.

Richard K. Guy

(1969). The decline and fall of Zarankiewicz's theorem, *Proof Techniques in Graph Theory* (F. Harary, ed.), Academic Press, Orlando.

(1972). Crossing number of graphs, *Graph Theory and Applications*, Springer-Verlag, New York, 111–124.

—— and L. W. Beineke (1966). The coarseness of the complete graph, *Canad. J. Math. 20*, 888–894.

Frank Harary

—— and W. Tutte (1965). A dual form of Kuratowski's theorem, *Canad. Math. Bull. 8*, 17–20.

see also L. W. Beineke

Nora Hartsfield

(1986). The toroidal splitting number of the complete graph K_n, *Discrete Math. 62*, 35–47.

Peter Hoffman

—— and B. Richter (1984). Embedding graphs in surfaces, *J. Combin. Theory Ser. B 36*, 65–84.

Joan P. Hutchinson

—— and George F. McNulty (1983). Connected graphs of genus g with complementary orbits, *Discrete Math. 45*, 255–275.

see also M. O. Albertson

T. Ito

(1981). On a graph of O'Keefe and Wong, *J. Graph Theory 5*, 87–94.

Bradley W. Jackson

—— and G. Ringel (1984). Coloring island maps, *Bull. Austral. Math. Soc. 29*, 151–165.

—— and G. Ringel (1984). The splitting number of complete bipartite graphs, *Arch. Math. 42*, 178–184.

—— and G. Ringel (1984). Heawood's empire problem on the plane, *J. Reine Angew. Math. 347*, 148–153.

—— and G. Ringel (1985). Heawood's empire problem, *J. Combin. Theory Ser. B 38*, 168–178.

Lynne D. James

—— and G. A. Jones (1985). Regular orientable imbeddings of complete graphs, *J. Combin. Theory Ser. B 39*, 353–367.

Gareth A. Jones

see L. D. James

Paul C. Kainen

—— and A. T. White (1978). On stable crossing numbers, *J. Graph Theory 2*, 181–187.

see also F. Bernhart

Daniel J. Kleitman

(1970). The crossing number of $K_{5,n}$, *J. Combin. Theory 9*, 315–323.

Hudson V. Kronk

(1969). An analogue to the Heawood map-coloring problem, *J. London Math. Soc. (2) 1*, 550–552.

Ravi S. Kulkarni

see A. L. Edmonds

Gerard Lallament

—— and D. Perrin (1981). A graph covering construction of all finite complete biprefix codes, *Discrete Math. 36*, 261–272.

Frank Thomson Leighton

(1982). Finite common coverings, *J. Combin. Theory Ser. B 33*, 231–238.

Linda Lesniak-Foster

see M. Behzad

Sostenes Lins

—— and A. Mandel (1985). Graph-encoded 3-manifolds, *Discrete Math. 57*, 261–284.

Bernard Liouville

(1978). Sur le genre d'une somme cartesienne de graphes, *Problemes combinatoires et theorie des * *2* * graphes*, Colloques Internat. CNRS 260, Paris, 275–277.

Richard J. Lipton

—— and R. E. Tarjan (1979). A separator theorem for planar graphs, *SIAM J. Appl. Math. 36*, 177–189.

—— and R. E. Tarjan (1980). Applications of a planar separator theorem, *SIAM J. Comput. 9*, 615–627.

Heidi Mahnke

(1972). The necessity of non-abelian groups in the case 0 of the Heawood map-coloring problem, *J. Combin. Theory B 13*, 263–265.

Joseph Malkevitch

(1970). Properties of planar graphs with uniform vertex and face structure, *Mem. Amer. Math. Soc. 99*.

Arnaldo Mandel

see S. Lins

George F. McNulty

see J. P. Hutchinson

L. S. Melnikov

see O. V. Borodin

Bojan Mohar

(1986). A common cover of graphs and 2-cell embeddings, *J. Combin. Theory Ser. B 40*, 94–106.

John W. Moon

see L. W. Beineke

Bruce P. Mull

—— and R. G. Rieper and A. T. White (1987). Enumerating 2-cell imbeddings of complete graphs, *Proc. Amer. Math. Soc.*, to appear.

Ronald C. Mullin

(1965). On counting rooted triangular maps, *Canad. J. Math. 17*, 373–382.

L. Nebesky

(1981). Every connected, locally connected graph is upper embeddable, *J. Graph Theory 5*, 197–199.

Seiya Negami

(1987). Enumeration of projective-planar embeddings of graphs, *Discrete Math.*, to appear.

Lee P. Neuwirth

(1966). Imbedding in low dimension, *Illinois J. Math. 10*, 470–478.

Torrence D. Parsons

(1980). Circulant graph imbeddings, *J. Combin. Theory Ser B 29*, 310–320.

Dominique Perrin

see G. Lallament

Giustina Pica

—— with T. Pisanski and A. G. S. Ventre. The genera of amalgamations of cube graphs, *Glasnik Mat. 19*, 21–26.

Tomaz Pisanski

—— with J. Shawe-Taylor and J. Vrabec (1983). Edge-colorability of graph bundles, *J. Combin. Theory Ser. B 35*, 12–19.

see also G. Pica

L. Bruce Richmond

see E. A. Bender

Bruce Richter

—— and H. Shank (1984). The cycle space of an embedded graph, *J. Graph Theory 8*, 365–369.

see also P. Hoffman

Robert G. Rieper

see B. P. Mull

Richard D. Ringeisen

see L. W. Beineke

Gerhard Ringel

see B. Jackson

Thomas L. Saaty

(1967). Two theorems on the minimum number of intersections for complete graphs, *J. Combin. Theory 2*, 571–584.

Edward F. Schmeichel

—— and G. S. Bloom (1979). Connectivity, genus, and the number of components in vertex-deleted subgraphs, *J. Combin. Theory Ser. B 27*, 198–201.

Herbert Shank

see B. Richter

John Shawe-Taylor

see T. Pisanski

Jozef Siran

—— and M. Skoviera (1985). Quotients of connected regular graphs of even degree, *J. Combin. Theory Ser. B 38*, 214–225.

Martin Skoviera

(1986). A contribution to the theory of voltage graphs, *Discrete Math. 61*, 281–292.

see also J. Siran

Z. Skupien

(1966). Locally Hamiltonian and planar graphs, *Fund. Math. 58*, 193–200.

Saul Stahl

(1983). The average genus of classes of graph embeddings, *Congr. Number. 40*, 375–388.

Richard P. Stanley

(1981). Factorization of permutations into *n*-cycles, *Discrete Math. 37*, 255–262.

H. Joseph Straight

(1979). Cochromatic number and the genus of a graph, *J. Graph Theory 3*, 43–51.

Robert E. Tarjan

see R. J. Lipton

William T. Tutte

(1973). What is a map? *New Directions in the Theory of Graphs* (F. Harary, ed.), Academic Press, Orlando, 309–325.

(1986). From topology to algebra, *J. Graph Theory 10*, 331–338.

see also F. Harary

Brian Ummel

(1978). The product of nonplanar complexes does not imbed in 4-space, *Trans. Amer. Math. Soc. 242*, 319–328.

Aldo G. S. Ventre

see G. Pica

Andrew Vince

(1983). Combinatorial maps, *J. Combin. Theory Ser. B 34*, 1–21.

(1983). Regular combinatorial maps, *J. Combin. Theory Ser. B 35*, 256–277.

E. Vogt

see H. Zeischang

Joze Vrabec

see T. Pisanski

Derek A. Waller

(1976). Double covers of graphs, *Bull. Austral. Math. Soc. 14*, 233–248.

see also F. W. Clarke and M. Farzan

Mark Watkins

see J. C. Graver

Arthur T. White

(1983). Ringing the changes, *Math. Proc. Camb. Philos. Soc. 94*, 203–215.

see also I. Anderson and B. G. Mull

Nick C. Wormald

(1981). Counting unrooted planar maps, *Discrete Math.* *36*, 205–225.

Joseph Zaks

(1974). The maximum genus of cartesian products of graphs, *Canad. J. Math.* *26*, 1025–1035.

Heiner Zieschang

—— with E. Vogt and H. Coldeway (1980). *Surfaces and Planar Discontinuous Groups* (Lecture Notes in Mathematics, Vol. 835), Springer-Verlag, New York.

Supplementary Bibliography

The following are among the more significant papers that have been published in topological graph theory since the first edition went to press. We regret inadvertent omissions. A few of these papers were cited in preliminary form in the first edition. We have attempted to include survey articles or books that might contain extensive bibliographies in a particular area.

M. Anderson and R. B. Richter, Self-dual Cayley maps, *European J. Combin.* 21 (2000), 419–430.

Dan Archdeacon, Calculations on the average genus and genus distribution of graphs, Nineteenth Southeastern Conference on Combinatorics, Graph Theory, and Computing (Baton Rouge, LA, 1988), *Congr. Numer.* 67 (1988), 114–124.

———, Densely embedded graphs, *J. Combin. Theory Ser. B* 54 (1992), 13–36.

———, The medial graph and voltage-current duality, *Discrete Math.* 104 (1992), 111–141.

———, Problems in topological graph theory—-questions I can't answer, *Proceedings of the 10th Workshop on Topological Graph Theory* (Yokohama, 1998), *Yokohama Math. J.* 47 (1999), Special Issue, 89–92.

———, Self-dual embeddings of complete multipartite graphs, *J. Graph Theory* 18 (1994), 735–749.

———, Topological graph theory: a survey, Surveys in graph theory (San Francisco, CA, 1995), *Congr. Numer.* 115 (1996), 5–54.

Dan Archdeacon, C. Paul Bonnington, and Charles H. C. Little, An algebraic characterization of planar graphs, *J. Graph Theory* 19 (1995), 237–250.

Dan Archdeacon, C. Paul Bonnington, and Jozef A. Siran, Nebesky-type characterization for relative maximum genus, *J. Combin. Theory Ser. B* 73 (1998), 77–98.

Dan Archdeacon and David A. Grable, The genus of a random graph, *Discrete Math.* 142 (1995), 21–37.

Dan Archdeacon, Pavol Gvozdjak, and Jozef Siran, Constructing and forbidding automorphisms in lifted maps, *Math. Slovaca* 47 (1997), 113–129.

Dan Archdeacon and R. Bruce Richter, The construction and classification of self-dual spherical polyhedra, *J. Combin. Theory Ser. B* 54 (1992), 37–63.

———, On the parity of planar covers, *J. Graph Theory* 14 (1990), 199–204.

Dan Archdeacon, R. Bruce Richter, Jozef Siran, and Martin Skoviera, Branched coverings of maps and lifts of map homomorphisms, *Australas. J. Combin.* 9 (1994), 109–121.

D. Archdeacon, J. Siran, and M. Skoviera, Self-dual regular maps from medial graphs, *Acta Math. Univ. Comenian.* (N.S.) 61 (1992), 57–64.

Laszlo Babai, Vertex-transitive graphs and vertex-transitive maps, *J. Graph Theory* 15 (1991), 587–627.

D. W. Barnette, Decomposition theorems for the torus, projective plane and Klein bottle, *Discrete Math.* 70 (1988), 1–16.

———, Generating projective plane polyhedral maps, *J. Combin. Theory Ser. B* 51 (1991), 277–291.

D. W. Barnette and Allan L. Edelson, All 2-manifolds have finitely many minimal triangulations, *Israel J. Math.* 67 (1989), 123–128.

V. Batagelj, D. Marusic, B. Mohar and T. Pisanski (eds.), *Algebraic and Topological Methods in Graph Theory*, Papers from the Second Slovenian Conference (Lake Bled, 1991), *Discrete Math.* 134 (1994), no. 1–3, North-Holland Publishing Co., Amsterdam.

C. P. Bonnington, M. J. Grannell, T. S. Griggs, and J. Siran, Exponential families of non-isomorphic triangulations of complete graphs, to appear, *J. Comb. Theory Ser. B.*

C. Paul Bonnington, Wilfried Imrich, and Mark E. Watkins, Separating double rays in locally finite planar graphs, *Discrete Math.* 145 (1995), 61–72.

C. P. Bonnington and C. H. C. Little, The classification of combinatorial surfaces using 3-graphs, *Australas. J. Combin.* 5 (1992), 87–102.

C. Paul Bonnington and Charles H. C. Little, *The Foundations of Topological Graph Theory*, Springer-Verlag, New York, 1995.

Ljiljana Brankovic, Mirka Miller, Jan Plesnmk, Joe Ryan, and Jozef Siran, Large graphs with small degree and diameter: a voltage assignment approach, *Australas. J. Combin.* 18 (1998), 65–76.

Matthew G. Brin, David E. Rauschenberg, and Craig C. Squier, On the genus of the semi-direct product of Z_9 by Z_3, *J. Graph Theory* 13 (1989), 49–61.

Matthew G. Brin and Craig C. Squier, On the genus of $Z_3 \times Z_3 \times Z_3$, *European J. Combin.* 9 (1988), 431–443.

Richard Brunet, Bojan Mohar, and R. Bruce Richter, Separating and nonseparating disjoint homotopic cycles in graph embeddings, *J. Combin. Theory Ser. B* 66 (1996), 201–231.

J. Chen and J. L. Gross, Kuratowski-type theorems for average genus, *J. Combin. Theory Ser. B* 57 (1993), 100–121.

———, Limit points for average genus, (I): 3-connected and 2-connected simplicial graphs, *J. Combin. Theory Ser. B* 55 (1992), 83–103.

———, Limit points for average genus (II): 2-connected non-simplicial graphs (with J. Chen), *J. Combin. Theory Ser. B* 56 (1992), 108–129.

———, No lower limit points for average genus, *Graph Theory, Combinatorics, and Algorithms, Vol. 1* (Kalamazoo, MI, 1992), 183–194, Wiley-Interscience, New York, 1995.

J. Chen, J. L. Gross, and D. Archdeacon, Maximum genus and connectivity, *Discrete Math.* 149 (1996), 19–30.

J. Chen, J. L. Gross, and R. G. Rieper, Lower bounds for the average genus, *J. Graph Theory* 19 (1995), 281–296.

———, Overlap matrices and total imbedding distributions, *Discrete Math.* 128 (1994), 73–94.

J. Chen, S. P. Kanchi, and J. L. Gross, A tight lower bound on the maximum genus of a simplicial graph, *Discrete Math.* 156 (1996), 83–102.

Marston (D. E.) Conder, Asymmetric combinatorially-regular maps, *J. Algebraic Combin.* 5 (1996), 323–328.

———, Groups of minimal genus including C_2 extensions of PSL(2, q) for certain q, *Quart. J. Math. Oxford Ser. (2)* 38 (1987), 449–460.

———, Hurwitz groups: a brief survey, *Bull. Amer. Math. Soc.* (N.S.) 23 (1990), 359–370.

———, The symmetric genus of alternating and symmetric groups, *J. Combin. Theory Ser. B* 39 (1985), 179–186.

———, The symmetric genus of the Mathieu groups, *Bull. London Math. Soc.* 23 (1991), 445–453.

Marston Conder and Peter Dobcsanyi, Determination of all regular maps of small genus, to appear, *J. Comb. Theory Ser. B*.

Marston Conder and Brent Everitt, Regular maps on non-orientable surfaces, *Geom. Dedicata* 56 (1995), 209–219.

M. D. E. Conder, R. A. Wilson, and A. J. Woldar, The symmetric genus of sporadic groups, *Proc. Amer. Math. Soc.* 116 (1992), 653–663.

Tamara Dakic, Roman Nedela, and Tomas Pisanski, Embeddings of tensor product graphs, *Graph Theory, Combinatorics, and Algorithms, Vol. 2* (Kalamazoo, MI, 1992), 893–904, Wiley, New York, 1995.

Alice M. Dean and Joan P. Hutchinson, Relations among embedding parameters for graphs, *Graph Theory, Combinatorics, and Applications, Vol. 1* (Kalamazoo, MI, 1988), 287–296, Wiley, New York, 1991.

J. R. Fiedler, J. P. Huneke, R. B. Richter, and N. Robertson, Computing the orientable genus of projective graphs, *J. Graph Theory* 20 (1995), 297–308.

Steve Fisk, Bojan Mohar, and Roman Nedela, Minimal locally cyclic triangulations of the projective plane, *J. Graph Theory* 18 (1994), 25–35.

M. L. Furst, J. L. Gross, and L. A. McGeoch, Finding a maximum-genus graph imbedding, *J. ACM* 35 (1988), 523–534.

———, Genus distribution for two classes of graphs, *J. Combin. Theory Ser. B* 46 (1989), 22–36.

Z. Gao, R. B. Richter, and P. D. Seymour, Irreducible triangulations of surfaces, *J. Combin. Theory Ser. B* 68 (1996), 206–217.

A. Gardiner, R. Nedela, J. Siran, and M. Skoviera, Characterisation of graphs which underlie regular maps on closed surfaces, *J. London Math. Soc. (2)* 59 (1999), 100–108.

John Gimbel and Carsten Thomassen, Coloring graphs with fixed genus and girth, *Trans. Amer. Math. Soc.* 349 (1997), 4555–4564.

Maurits de Graaf and Alexander Schrijver, Making curves minimally crossing by Reidemeister moves, *J. Combin. Theory Ser. B* 70 (1997), 134–156.

M. J. Grannell, T. S. Griggs, J. Siran, Surface embeddings of Steiner triple systems, *J. Comb. Designs* 6 (1998), 325–336.

J. E. Graver and M. E. Watkins, *Locally Finite, Planar, Edge-transitive Graphs*, Memoirs of the AMS Vol. 126, No. 601 (1997).

Jonathan L. Gross and Jianer Chen, Algebraic specification of interconnection networks by permutation voltage graph morphisms, *Mathematical Systems Theory* 29 (1996), 451–470.

Jonathan L. Gross and Merrick L. Furst, Hierarchy for imbedding-distribution invariants of a graph, *J. Graph Theory* 11 (1987), 205–220.

J. L. Gross, E. W. Klein and R. G. Rieper, On the average genus of a graph, *Graphs and Combinatorics* 9 (1993), 153–162.

J. L. Gross and R. G. Rieper, Local extrema in genus-stratified graphs, *J. Graph Theory* 15 (1991), 159–171.

J. L. Gross, D. P. Robbins, and T. W. Tucker, Genus distributions for bouquets of circles, *J. Combin. Theory Ser. B* 47 (1989), 292–306.

Jonathan L. Gross and Thomas W. Tucker, Stratified graphs for imbedding systems, *Discrete Math.* 143 (1995), 71–86.

Jonathan L. Gross and Jay Yellen, *Graph-Theory and Its Applications*, CRC Press, 1999.

B. Grunbaum and G. C. Shephard, Edge-transitive planar graphs, *J. Graph Theory* 11 (1978), 141–156.

———, *Tilings and Patterns*, W. H. Freeman and Co., New York, 1989.

Pavol Gvozdjak and Jozef Siran, Regular maps from voltage assignments, *Graph structure theory* (Seattle, WA, 1991), 441–454, Contemp. Math., 147, Amer. Math. Soc., Providence, RI, 1993.

N. Hartsfield and G. Ringel, Minimal quadrangulations of orientable surfaces, *J. Comb. Theory Ser. B* 46 (1989), 84–95.

Joan P. Hutchinson, Coloring ordinary maps, maps of empires and maps of the moon, *Math. Mag.* 66 (1993), 211–226.

———, On short noncontractible cycles in embedded graphs, *SIAM J. Discrete Math.* 1 (1988), 185–192.

———, Three-coloring graphs embedded on surfaces with all faces even-sided, *J. Combin. Theory Ser. B* 65 (1995), 139–155.

Joan P. Hutchinson and Gary L. Miller, On deleting vertices to make a graph of positive genus planar, *Discrete Algorithms and Complexity* (Kyoto, 1986), 81–98, Perspect. Comput., 15, Academic Press, Boston, Mass., 1987.

R. Jajcay, Automorphism Groups of Cayley Maps, *J. Combin. Theory Ser. B* 59 (1993), 297–310.

S. Jendrol, R. Nedela, and M. Skoviera, Constructing regular maps and graphs from planar quotients, *Math. Slovaca* 47 (1997), 155–170.

G. A. Jones, Maps on surfaces and Galois groups, *Math. Slovaca* 47 (1997), 1–33.

E. C. Kirby and Tomaz Pisanski, Aspects of topology, genus and isomerism in closed 3-valent networks, Girona 97 special issue, *J. Math. Chem.* 23 (1998), 151–167.

Hideo Komuro, Atsuhiro Nakamoto, and Seiya Negami, Diagonal flips in triangulations on closed surfaces with minimum degree at least 4, *J. Combin. Theory Ser. B* 76 (1999), 68–92.

V. P. Korzhik and H.-J. Voss, On the number of non-isomorphic orientable regular embeddings of complete graphs, to appear, *J. Comb Theory Ser. B.*

Alex Kumjian and David Pask, C*-algebras of directed graph and group actions, *Ergod. Thy. and Dyn. Sys.* 19 (1999), 1503–1519.

Jin Ho Kwak, Jang-Ho Chun, and Jaeun Lee, Enumeration of regular graph coverings having finite abelian covering transformation groups, *SIAM J. Discrete Math.* 11 (1998), 273–285 (electronic).

Jin Ho Kwak, Se-Goo Kim, and Jaeun Lee, Distributions of regular branched prime-fold coverings of surfaces, *Discrete Math.* 156 (1996), 141–170.

Jin Ho Kwak and Jaeun Lee, Distribution of branched D_p-coverings of surfaces, *Discrete Math.* 183 (1998), 193–212.

———, Enumeration of graph embeddings, *Discrete Math.* 135 (1994), 129–151.

Serge Lawrencenko and Seiya Negami, Constructing the graphs that triangulate both the torus and the Klein bottle, *J. Combin. Theory Ser. B* 77 (1999), 211–218.

———, Irreducible triangulations of the Klein bottle, *J. Combin. Theory Ser. B* 70 (1997), 265–291.

Jaeun Lee, Enumeration of graph coverings, *Proceedings of the 10th Workshop on Topological Graph Theory* (Yokohama, 1998), *Yokohama Math. J.* 47 (1999), Special Issue, 101–111.

D. Marusic and B. Mohar, *Graph theory*, Papers from the 3rd Slovenian Conference (Lake Bled, 1995), *Discrete Math.* 182 (1998), no. 1–3. North-Holland Publishing Co., Amsterdam, 1998, pp. iii–x and 1–336.

Dragan Marusic and Tomaz Pisanski, The remarkable generalized Petersen graph $G(8,3)$, *Math. Slovaca* 50 (2000), 117–121.

C. L. May and J. Zimmerman, The group of symmetric Euler characteristic −3, to appear, *Houston J. Math.*

———, Groups of small strong symmetric genus, to appear, *J. Group Theory*.

———, Groups of small symmetric genus, *Glasgow Math J.* 37 (1995), 101–115.

———, The groups of symmetric genus 3, *Houston J. Math* 23 (1997), 573–590.

———, The symmetric genus of finite abelian groups, *Illinois J. Math.* 37 (1993), 400–423.

Bojan Mohar, Face covers and the genus of apex graphs, to appear, *J. Comb. Theory Ser. B.*

———, A linear time algorithm for embedding graphs in an arbitrary surface, *SIAM J. Discrete Math.* 12 (1999), 6–26 (electronic).

———, An obstruction to embedding graphs in surfaces. *Discrete Math.* 78 (1989), 135–142.

———, On the orientable genus of graphs with bounded nonorientable genus, *Graph Theory* (Lake Bled, 1995), *Discrete Math.* 182 (1998), 245–253.

Bojan Mohar, Tomaz Pisanski, and Martin Skoviera, The maximum genus of graph bundles, *European J. Combin.* 9 (1988), 215–224.

Bojan Mohar, Tomaz Pisanski, and Arthur T. White, Embeddings of Cartesian products of nearly bipartite graphs, *J. Graph Theory* 14 (1990), 301–310.

Bojan Mohar, Neil Robertson, and Richard P. Vitray, Planar graphs on the projective plane, *Discrete Math.* 149 (1996), 141–157.

Bojan Mohar and Carsten Thomassen, *Graphs on Surfaces*, Johns Hopkins University Press, scheduled for publication in 2001.

Bruce P. Mull, Robert G. Rieper, and Arthur T. White, Enumerating 2-cell imbeddings of connected graphs, *Proc. Amer. Math. Soc.* 103 (1988), 321–330.

Roman Nedela and Martin Skoviera, Exponents of orientable maps, *Proc. London Math. Soc.* (3) 75 (1997), 1–31.

———, Regular embeddings of canonical double coverings of graphs, *J. Combin. Theory Ser. B* 67 (1996), 249–277.

———, Regular maps from voltage assignments and exponent groups, *European J. Combin.* 18 (1997), 807–823.

Tomaz Pisanski and Thomas W. Tucker, The genus of low rank hamiltonian groups, *Discrete Math* 78 (1989), 157–167.

———, The genus of a product of a group with an abelian group, *Europ. J. Combin.* 10 (1989), 469–475.

Tomaz Pisanski, Thomas W. Tucker, and David Witte, The nonorientable genus of some metacyclic groups, *Combinatorica* 12 (1992), 1–11.

Tomaz Pisanski and Arthur T. White, Nonorientable embeddings of groups, *European J. Combin.* 9 (1988), 445–461.

R. B. Richter, J. Siran, R. Jajcay, T. W. Tucker, and M. E. Watkins, Cayley maps, preprint (2000), 80 pages.

R. Bruce Richter and Carsten Thomassen, Relations between crossing numbers of complete and complete bipartite graphs, *Amer. Math. Monthly* 104 (1997), 131–137.

Neil Robertson, Daniel Sanders, Paul Seymour, and Robin Thomas, The four-colour theorem, *J. Combin. Theory Ser. B* 70 (1997), 2–44.

Neil Robertson and P. D. Seymour, Graph minors, VIII, A Kuratowski theorem for general surfaces, *J. Combin. Theory Ser. B* 48 (1990), 255–288.

Neil Robertson and P. D. Seymour, eds., *Graph Structure Theory* (Seattle, Washington, 1991) Contemp. Math. 147, Amer. Math Soc., Providence, RI, 1993.

Neil Robertson, P. D. Seymour, and Robin Thomas, Linkless embeddings of graphs in 3-space, *Bull. Amer. Math. Soc.* (N.S.) 28 (1993), 84–89.

Neil Robertson and Robin Thomas, On the orientable genus of graphs embedded in the Klein bottle, *J. Graph Theory* 15 (1991), 407–419.

Neil Robertson and Richard Vitray, Representativity of surface embeddings, *Paths, flows, and VLSI-layout* (Bonn, 1988), 293–328, *Algorithms Combin. 9*, Springer, Berlin, 1990.

Vojtech Rodl and Robin Thomas, On the genus of a random graph, *Random Structures Algorithms* 6 (1995), 1–12.

A. Schrijver, Disjoint circuits of prescribed homotopies in a graph on a compact surface, *J. Combin. Theory Ser. B* 51 (1991), 127–159.

———, On the uniqueness of kernels, *J. Combin. Theory Ser. B* 55 (1992), 146–160.

John Shawe-Taylor and Tomaz Pisanski, Homeomorphism of 2-complexes is graph isomorphism complete, *SIAM J. Comput.* 23 (1994), 120–132.

Jozef Siran and Martin Skoviera, Characterization of the maximum genus of a signed graph, *J. Combin. Theory Ser. B* 52 (1991), 124–146.

———, Quotients of connected regular graphs of even degree, *J. Combin. Theory Ser. B* 38 (1985), 214–225.

J. Siran and M. Skoviera, Regular maps from Cayley graphs, Part 2: Antibalanced Cayley maps, *Discrete Math.* 124 (1994) 179–191.

Jozef Siran, Martin Skoviera, and Heinz-Jurgen Voss, Sachs triangulations and regular maps, *Algebraic and Topological Methods in Graph Theory* (Lake Bled, 1991). *Discrete Math.* 134 (1994), 161–175.

J. Siran, T. W. Tucker, and M. E. Watkins, Realizing finite edge-transitive orientable maps, to appear, *J. Graph Theory.*

Martin Skoviera and Roman Nedela, The maximum genus of vertex-transitive graphs, *Discrete Math.* 78 (1989), no. 1–2, 179–186.

M. Skoviera and J. Siran, Regular maps from Cayley graphs, Part 1: Balanced Cayley maps, *Discrete Math.* 109 (1992), 265–276.

Robin Thomas, An update on the four-color theorem, *Notices Amer. Math. Soc.* 45 (1998), 848–859.

Carsten Thomassen, Five-coloring maps on surfaces, *J. Combin. Theory Ser. B* 59 (1993), 89–105.

———, The graph genus problem is NP-complete, *J. Algorithms* 10 (1989), 568–576.

———, The Jordan-Schonflies theorem and the classification of surfaces, *Amer. Math. Monthly* 99 (1992), 116–130.

———, Tilings of the torus and the Klein bottle and vertex-transitive graphs on a fixed surface, *Trans. Amer. Math. Soc.* 323 (1991), 605–635.

Thomas W. Tucker, Some topological graph theory for topologists: a sampler of covering space techniques, *Topology and Combinatorial Group Theory*, ed. P. Latiolais, Springer-Verlag, New York, 1990.

———, Symmetric embeddings of Cayley graphs in nonorientable surfaces, *Graph Theory, Combinatorics and Applications* (Kalamazoo, 1988), 1105–1120, Wiley-Interscience, New York, 1991.

Andrew Vince, Graphs and finitely presented groups, *Graphs and algorithms* (Boulder, CO, 1987), 167–185, Contemp. Math., 89, Amer. Math. Soc., Providence, RI, 1989.

———, Map duality and generalizations, *Ars Combin.* 39 (1995), 211–229.

Andrew Vince and C. H. C. Little, Discrete Jordan curve theorems, *J. Combin. Theory Ser. B* 47 (1989), 251–261.

R. P. Vitray, Representativity and flexibility on the projective plane, *Graph Structure Theory* (Seattle, WA, 1991), 341–347, Contemp. Math., 147, Amer. Math. Soc., Providence, RI, 1993.

Arthur T. White, Cayley maps, *Surveys in graph theory* (San Francisco, CA, 1995), *Congr. Numer.* 115 (1996), 105–130.

———, Efficient imbeddings of finite projective planes, *Proc. London Math. Soc.* (3) 70 (1995), 33–55.

———, An introduction to random topological graph theory, *Combin. Probab. Comput.* 3 (1994), 545–555.

Steve Wilson, Families of regular graphs in regular maps, to appear, *J. Combin. Theory Ser. B.*

Thomas Zaslavsky, The largest parity demigenus of a simple graph, *J. Combin. Theory Ser. B* 70 (1997), 325–345.

———, A mathematical bibliography of signed and gain graphs and allied areas, Manuscript prepared with Marge Pratt. *Electron. J. Combin.* 5 (1998), Dynamic Surveys 8, 124 pp.

———, Matroids determine the embeddability of graphs in surfaces, *Proc. Amer. Math. Soc.* 106 (1989), 1131–1135.

———, Orientation embedding of signed graphs, *J. Graph Theory* 16 (1992), 399–422.

———, The projective-planar signed graphs, *Discrete Math.* 113 (1993), 223–247.

Table of Notations

GRAPHS

$G, H, \ldots$	arbitrary graphs, 1
E_G, E	the edge set of graph G, 1
V_G, V	the vertex set of graph G, 1
$d, e, \ldots$	edges, 1
$u, v, \ldots$	vertices, 2
$V_G(e), V(e)$	the endpoint set of edge e, 1
e^+, e^-	the plus, minus directions of edge e, 5
$G - e$	the graph G with edge e deleted, 33
G/e	the graph G with edge e contracted, 39
G^c	the edge complement of graph G, 19
$G \times H$	the cartesian product of graphs G and H, 19
$G + H$	the suspension (join) of graphs G and H, 20
$G *_f H$	the amalgamation of graphs G and H by map f, 20
$G *_v H$	the amalgamation of graphs G and H at vertex v, 151
G^{simp}	the maximal simplicial subgraph of G, 38
G^{smooth}	the smoothing of graph G, 40
$C(\mathscr{A}, X)$	the Cayley color graph for group $\mathscr{A}$ and generating set X, 11
$C(\mathscr{A}, X)^0$	the Cayley graph for group $\mathscr{A}$ and generating set X, 11
$C(\mathscr{A}, X)^1$	the alternative Cayley graph for group $\mathscr{A}$ and generating set X, 12
$S(\mathscr{A}: \mathscr{B}, X)$	the Schreier color graph for the cosets of subgroup $\mathscr{B}$ and generating set X, 73
$S(\mathscr{A}: \mathscr{B}, X)^0$	the Schreier graph for the cosets of subgroup $\mathscr{B}$ and generating set X, 73
$S(\mathscr{A}: \mathscr{B}, X)^1$	the alternative Schreier graph for the cosets of subgroup $\mathscr{B}$ and generating set X, 73
T	a tree, 8
W	a walk, 8
C_n	the cycle of length n, 8

P_n	the path with n vertices (and $n-1$ edges), 8
K_n	the complete graph on n vertices, 10
$K_{m,n}$	the complete bipartite graph with partsizes m, n, 15
$K_{m,n,r,\ldots}$	the complete multipartite graph with partsizes $m, n, r, \ldots$, 196
B_n	the bouquet of n circles, 15
O_n	the n-octahedron graph, 20
Q_n	the n-cube graph, 19
$G(n, k)$	a generalized Petersen graph, 59
$H_{(n)}$	the composition of graph H by a set of n independent vertices, 210
$K(S)$	the set of Kuratowski graphs for the surface S, 53

GROUPS

$\mathscr{A}, \mathscr{B}, \ldots$	arbitrary groups, 11
$a, b, \ldots$	group elements, 11
X	a generating set, 11
$x, y, \ldots$	generators, 11
$1_{\mathscr{A}}, 1$	the identity element of group $\mathscr{A}$, 12
$\langle x, \ldots : R = 1, \ldots \rangle$	the group with generators $x, \ldots$ and relators $R, \ldots$, 172
$\langle x, y, \ldots \rangle$	the subgroup generated by $x, y, \ldots$, 290
$[x, y]$	the commutator $xyx^{-1}y^{-1}$, 290
$\mathscr{A}: \mathscr{B}$	the collection of cosets of the subgroup $\mathscr{B}$ in the group $\mathscr{A}$, 73
$\mathscr{A}/\mathscr{B}$	the quotient of group $\mathscr{A}$ by the normal subgroup $\mathscr{B}$, 80
$\mathscr{A} \times \mathscr{B}$	the direct product of groups $\mathscr{A}$ and $\mathscr{B}$, 15
$\mathscr{Z}_n$	the cyclic group of order n, 10
$\mathscr{D}_n$	the dihedral group of order $2n$, 22
$\mathscr{S}_n$	the full symmetric group on n symbols, 13
$\mathscr{A}_n$	the alternating group on n symbols, 15
π, ρ	permutations, 83
$(c_1, \ldots, c_n)$	the cycle structure of a permutation having c_i cycles of length i, 83
$\mathrm{GL}(2, p)$	the general linear group of 2×2 matrices over the field of p elements, 313
$\mathrm{SL}(2, p)$	the special linear group of 2×2 matrices of determinant 1, 313
$\mathrm{PSL}(2, p)$	the projective special linear group, 297
$\mathrm{Aut}(G)$	the automorphism group of graph G, 6
$\pi_1(G, u)$	the fundamental group of graph G based at vertex u, 92

SURFACES AND COMPLEXES

S_g	the orientable surface of genus g, 24
N_k	the nonorientable surface with k crosscaps, 26
K	a (simplicial) complex, 96
σ	a simplex, 96
$\langle v_0, \ldots, v_k \rangle$	the k-simplex having vertices $v_0, \ldots, v_k$, 96
$\|K\|$	the carrier of complex K, 96
$K^{(r)}$	the r-skeleton of complex K, 96
(V, C)	an abstract complex with vertex set V, 97
star(u)	the star of a vertex, 106
link(v)	the link of a vertex, 107
$S \circ T$	the disk sum of surfaces S and T, 131

GROUP ACTIONS AND COVERINGS

ϕ_a, ψ_a	graph automorphisms or surface homeomorphisms corresponding to group element a, 21, 182
$[v]$, $[e]$, $[x]$	the orbit of a vertex, edge, or point under a group action, 21, 186
$G/\mathscr{A}$, $S/\mathscr{A}$	the graph, surface quotient under the action of group $\mathscr{A}$, 22, 186
$p: G \to G/\mathscr{A}$, $p: S \to S/\mathscr{A}$	the quotient map for the action of group $\mathscr{A}$, 22, 186
$p: \tilde{G} \to G$	a covering projection for graphs, 75
$p: \tilde{S} \to S$	a covering projection, possibly branched, for surfaces, 174
Y	the branchpoint set of a branched covering, 174
$\mathscr{A}^0$	the orientation-preserving subgroup of a surface action of group $\mathscr{A}$, 269
def(y)	the deficiency of branchpoint y, 175
r_y	the order of the branchpoint y for a regular covering, 284
$\mathscr{C}$	a symmetry group, 276
$\mathscr{T}$	the translation subgroup of a Euclidean space group, 277

IMBEDDINGS AND ROTATIONS

$i: G \to S, G \to S$	an imbedding of graph G in surface S, 26
F_G, F	the set of faces for an imbedding of graph G, 26
f_i	the number of faces of size i, 302
G^*	the graph dual to an imbedding of G, 31

e^*, v^*, f^*	dual edges, vertices, faces, 31
e^{+*}	the dual plus direction of edge e, 202
$v.e_1 \ldots e_n$	a rotation at vertex v, 112
e^1	a type-1 edge, 112
ρ	a rotation as a permutation of the directed edge set, 161
$\ldots 2\bar{3} \ldots$	a rotation $\ldots x_2 x_3^{-1} \ldots$ for a Cayley graph with generators x_2, x_3, 257

VOLTAGE AND CURRENT GRAPHS

$\langle G, \alpha \rangle$	an (ordinary) voltage graph with voltage assignment α, 57
$\langle G, \alpha/\mathscr{B} \rangle$	a relative voltage graph, 74
$\langle G, \alpha \rangle_n$	a permutation voltage graph with assignment in the symmetric group $\mathscr{S}_n$, 81
G^α	the derived graph for voltage assignment α, 57
$p: G^\alpha \to G$	the natural projection of the derived graph onto the base graph, 61
$\langle G \to S, \alpha \rangle$	an imbedded (ordinary) voltage graph, 162
$\langle G \to S, \alpha \rangle_n$	an imbedded permutation voltage graph, 162
S^α	the derived surface for an imbedded voltage graph, 162
$G^\alpha \to S^\alpha$	the derived imbedding for an imbedded voltage graph, 162
$p: S^\alpha \to S$	a natural projection of the derived surface onto the base surface, 176
$\langle G \to S, \beta \rangle$	a current graph with current assignment β, 194
$G_\beta \to S_\beta$	the derived imbedding for current assignment β, 194
$(e, a), e_a$	edges of a derived graph, 57
$(v, a), v_a$	vertices of a derived graph, 57
$(e, b)^-, (e^-, b)$	minus directions of derived edges for a current graph, 194
$\langle G^* \to S, \beta^* \rangle$	the voltage graph dual to the current graph $\langle G \to S, \beta \rangle$, 204
$H_\beta^* \to S_\beta$	the dual derived imbedding for the current graph $\langle H \to S, \beta \rangle$, 206
$p: H_\beta^* \to H$	the natural projection of the dual derived graph onto the base current graph, 207
$\mathscr{A}(u)$	the local (voltage) group at vertex u, 87
$\mathscr{A}^0(u)$	the type-0 local (voltage) group, 169
$\alpha(v, T)$	the T-potential of vertex v, 89
$\alpha_T(e)$	the T-voltage for edge e, 90

GENUS AND RELATED PARAMETERS

$\gamma(G)$	the (minimum) genus of graph G, 29
$\gamma_M(G)$	the maximum genus of graph G, 132
$\bar{\gamma}(G)$	the (minimum) crosscap number of graph G, 29
$\bar{\gamma}_M(G)$	the maximum crosscap number of graph G, 133
$\beta(G)$	the Betti number (cycle rank) of graph G, 135
$\xi(G, T)$	the deficiency of graph G with respect to the spanning tree T, 143
$\xi(G)$	the deficiency of graph G, 143
$\text{girth}(G)$	the girth of graph G, 29
$\text{chr}(G)$	the chromatic number of graph G, 37
$\text{GR}(G)$	the genus range of graph G, 132
$\text{CR}(G)$	the crosscap range of graph G, 133
$\text{ER}(G)$	the Euler characteristic range of graph G, 148
$\gamma(\mathscr{A})$	the genus of group $\mathscr{A}$, 249
$\sigma(\mathscr{A})$	the symmetric genus of group $\mathscr{A}$, 264
$\sigma^0(\mathscr{A})$	the strong symmetric genus of group $\mathscr{A}$, 264
$\chi(G \to S)$	the Euler characteristic of the imbedding $G \to S$, 121
$\chi(S)$	the Euler characteristic of surface S, 130
$H(S)$	the Heawood number of surface S, 215
$\text{chr}(S)$	the chromatic number of surface S, 37
$I(n)$	the integer $\lceil (n-3)(n-4)/12 \rceil$, 221
$\bar{I}(n)$	the integer $\lceil (n-3)(n-4)/6 \rceil$, 221

GENERAL NOTATION

$\{x \mid P\}$	the set of all x such that P holds, 1
$v \in V$, $v \notin V$	v is, is not, an element of set V, 1
$A \subset B$	set A is contained in set B, 12
$A \cup B$	the union of sets A and B, 18
$A - B$	the set of elements in A but not in B, 18
$A \times B$	the cartesian product of sets A and B, 11
$\#V$	the number of elements in the set V, 3
$<$, $\leq$	strictly less than, less than or equal, 29
$>$, $\geq$	strictly greater than, greater than or equal, 19
$a \equiv b \bmod n$	the integers a and b are congruent modulo n, 146
$\lfloor x \rfloor$	the "floor of x", i.e., the greatest integer less than or equal to x, 38
$\lceil x \rceil$	the "ceiling of x", i.e., the least integer greater than or equal to x, 38

NOTATION USED IN FIGURES

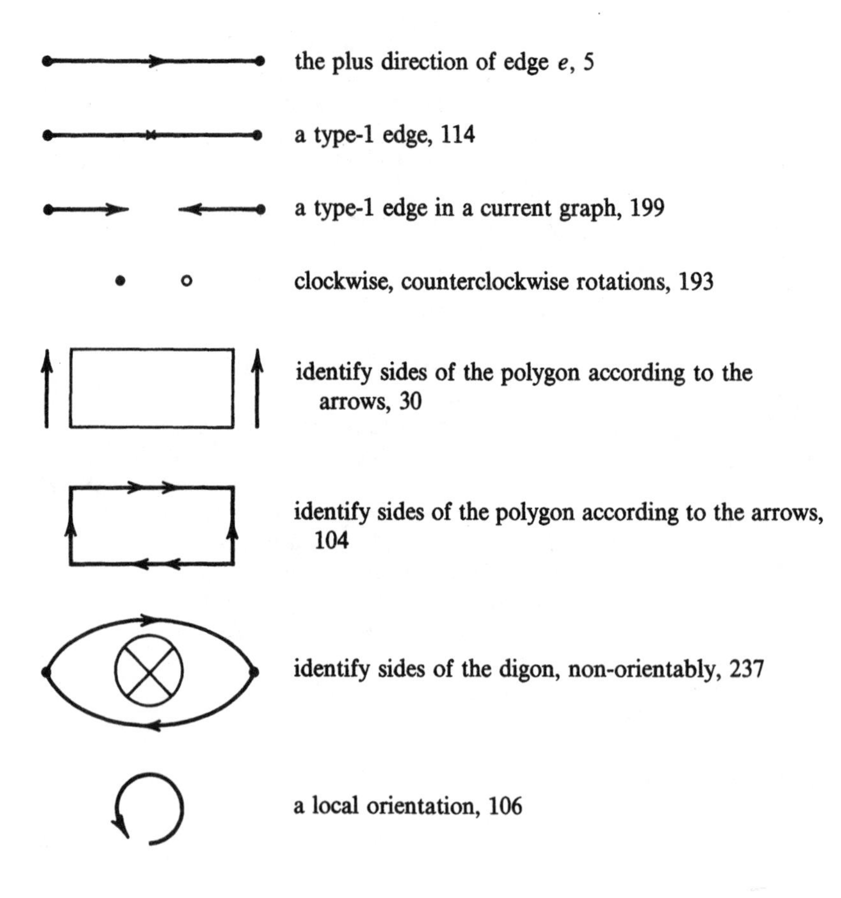

the plus direction of edge e, 5

a type-1 edge, 114

a type-1 edge in a current graph, 199

clockwise, counterclockwise rotations, 193

identify sides of the polygon according to the arrows, 30

identify sides of the polygon according to the arrows, 104

identify sides of the digon, non-orientably, 237

a local orientation, 106

Subject Index

A CATALOG OF SELECTED

DOVER BOOKS

IN SCIENCE AND MATHEMATICS

Astronomy

BURNHAM'S CELESTIAL HANDBOOK, Robert Burnham, Jr. Exhaustive guide to the stars beyond our solar system. Andromeda to Cetus in Vol. 1; Chamaeleon to Orion in Vol. 2; and Pavo to Vulpecula in Vol. 3. Hundreds of illustrations. Index in Vol. 3. 2,000pp. 6⅛ x 9¼. 23567-X, 23568-8, 23673-0 Pa., Three-vol. set $46.85

THE EXTRATERRESTRIAL LIFE DEBATE, 1750–1900, Michael J. Crowe. First detailed, scholarly study in English of early ideas about existence of intelligent extraterrestrial life. Examines ideas of Kant, Herschel, Voltaire, Percival Lowell, many others. 16 illustrations. 704pp. 5⅜ x 8½. 40675-X Pa. $19.95

A HISTORY OF ASTRONOMY, A. Pannekoek. Well-balanced study covers Ptolemaic theory, work of Copernicus, Kepler, Newton, Eddington's work on stars, much more. Illustrated. References. 521pp. 5⅜ x 8½. 65994-1 Pa. $15.95

AMATEUR ASTRONOMER'S HANDBOOK, J. B. Sidgwick. Timeless, comprehensive coverage of telescopes, mirrors, lenses, mountings, telescope drives, micrometers, spectroscopes, more. 189 illustrations. 576pp. 5⅝ x 8¼. (Available in U.S. only.) 24034-7 Pa. $13.95

STARS AND RELATIVITY, Ya. B. Zel'dovich and I. D. Novikov. Vol. 1 of *Relativistic Astrophysics* by famed Russian scientists. General relativity, properties of matter under astrophysical conditions, stars and stellar systems. Deep physical insights, clear presentation. 1971 edition. 544pp. 5⅜ x 8½. 69424-0 Pa. $14.95

Chemistry

CHEMICAL MAGIC, Leonard A. Ford. Second Edition, Revised by E. Winston Grundmeier. Over 100 unusual stunts demonstrating cold fire, dust explosions, much more. Text explains scientific principles and stresses safety precautions. 128pp. 5⅜ x 8½. 67628-5 Pa. $5.95

THE DEVELOPMENT OF MODERN CHEMISTRY, Aaron J. Ihde. Authoritative history of chemistry from ancient Greek theory to 20th-century innovation. Covers major chemists and their discoveries. 209 illustrations. 14 tables. Bibliographies. Indices. Appendices. 851pp. 5⅜ x 8½. 64235-6 Pa. $24.95

CATALYSIS IN CHEMISTRY AND ENZYMOLOGY, William P. Jencks. Exceptionally clear coverage of mechanisms for catalysis, forces in aqueous solution, carbonyl- and acyl-group reactions, practical kinetics, more. 864pp. 5⅜ x 8½. 65460-5 Pa. $19.95

THE HISTORICAL BACKGROUND OF CHEMISTRY, Henry M. Leicester. Evolution of ideas, not individual biography. Concentrates on formulation of a coherent set of chemical laws. 260pp. 5⅜ x 8½. 61053-5 Pa. $8.95

A SHORT HISTORY OF CHEMISTRY (3rd edition), J. R. Partington. Classic exposition explores origins of chemistry, alchemy, early medical chemistry, nature of atmosphere, theory of valency, laws and structure of atomic theory, much more. 428pp. 5⅜ x 8½. (Available in U.S. only.) 65977-1 Pa. $12.95

Mathematics

COMPLEX VARIABLES, Francis J. Flanigan. Unusual approach, delaying complex algebra till harmonic functions have been analyzed from real variable viewpoint. Includes problems with answers. 364pp. 5⅜ x 8½. 61388-7 Pa. $10.95

AN INTRODUCTION TO THE CALCULUS OF VARIATIONS, Charles Fox. Graduate-level text covers variations of an integral, isoperimetrical problems, least action, special relativity, approximations, more. References. 279pp. 5⅜ x 8½. 65499-0 Pa. $10.95

CATASTROPHE THEORY FOR SCIENTISTS AND ENGINEERS, Robert Gilmore. Advanced treatment describes mathematics of theory grounded in the work of Poincaré, R. Thom, other mathematicians. Also important applications to problems in mathematics, physics, chemistry, and engineering. 1981 edition. xvii + 666pp. 6⅛ x 9¼. 67539-4 Pa. $17.95

INTRODUCTION TO DIFFERENCE EQUATIONS, Samuel Goldberg. Exceptionally clear exposition of important discipline with applications to sociology, psychology, economics. Many illustrative examples; over 250 problems. 260pp. 5⅜ x 8½. 65084-7 Pa. $10.95

UNBOUNDED LINEAR OPERATORS: Theory and Applications, Seymour Goldberg. Classic presents systematic treatment of the theory of unbounded linear operators in normed linear spaces with applications to differential equations. Bibliography. 199pp. 5⅜ x 8½. 64830-3 Pa. $7.95

NUMERICAL METHODS FOR SCIENTISTS AND ENGINEERS, Richard Hamming. Classic text stresses frequency approach in coverage of algorithms, polynomial approximation, Fourier approximation, exponential approximation, other topics. Revised and enlarged 2nd edition. 721pp. 5⅜ x 8½. 65241-6 Pa. $17.95

POPULAR LECTURES ON MATHEMATICAL LOGIC, Hao Wang. Noted logician's lucid treatment of historical developments, set theory, model theory, recursion theory and constructivism, proof theory, more. 3 appendixes. Bibliography. 1981 edition. ix + 283pp. 5⅜ x 8½. 67632-3 Pa. $10.95

INTRODUCTION TO NUMERICAL ANALYSIS (2nd Edition), F. B. Hildebrand. Classic, fundamental treatment covers computation, approximation, interpolation, numerical differentiation and integration, other topics. 150 new problems. 669pp. 5⅜ x 8½. 65363-3 Pa. $16.95

THE FUNCTIONS OF MATHEMATICAL PHYSICS, Harry Hochstadt. Comprehensive treatment of orthogonal polynomials, hypergeometric functions, Hill's equation, much more. Bibliography. Index. 322pp. 5⅜ x 8½. 65214-9 Pa. $12.95

THREE PEARLS OF NUMBER THEORY, A. Y. Khinchin. Three compelling puzzles require proof of a basic law governing the world of numbers. Challenges concern van der Waerden's theorem, the Landau-Schnirelmann hypothesis and Mann's theorem, and a solution to Waring's problem. Solutions included. 64pp. 5⅜ x 8½. 40026-3 Pa. $4.95

Math–Decision Theory, Statistics, Probability

LINEAR PROGRAMMING AND ECONOMIC ANALYSIS, Robert Dorfman, Paul A. Samuelson, and Robert M. Solow. First comprehensive treatment of linear programming in standard economic analysis. Game theory, modern welfare economics, Leontief input-output, more. 525pp. 5⅜ x 8½. 65491-5 Pa. $17.95

DICTIONARY/OUTLINE OF BASIC STATISTICS, John E. Freund and Frank J. Williams. A clear concise dictionary of over 1,000 statistical terms and an outline of statistical formulas covering probability, nonparametric tests, much more. 208pp. 5⅜ x 8½. 66796-0 Pa. $8.95

GAMES AND DECISIONS: Introduction and Critical Survey, R. Duncan Luce, and Howard Raiffa. Superb nontechnical introduction to game theory, primarily applied to social sciences. Utility theory, zero-sum games, n-person games, decision-making, much more. Bibliography. 509pp. 5⅜ x 8½. 65943-7 Pa. $14.95

FIFTY CHALLENGING PROBLEMS IN PROBABILITY WITH SOLUTIONS, Frederick Mosteller. Remarkable puzzlers, graded in difficulty, illustrate elementary and advanced aspects of probability. Detailed solutions. 88pp. 5⅜ x 8½. 65355-2 Pa. $5.95

OPTIMIZATION THEORY WITH APPLICATIONS, Donald A. Pierre. Broad spectrum approach to important topic. Classical theory of minima and maxima, calculus of variations, simplex technique and linear programming, more. Many problems, examples. 640pp. 5⅜ x 8½. 65205-X Pa. $17.95

PROBABILITY THEORY: A Concise Course, Y. A. Rozanov. Highly readable, self-contained introduction covers combination of events, dependent events, Bernoulli trials, etc. 148pp. 5⅜ x 8¼. 63544-9 Pa. $8.95

STATISTICAL METHOD FROM THE VIEWPOINT OF QUALITY CONTROL, Walter A. Shewhart. Important text explains regulation of variables, uses of statistical control to achieve quality control in industry, agriculture, other areas. 192pp. 5⅜ x 8½. 65232-7 Pa. $8.95

THE COMPLEAT STRATEGYST: Being a Primer on the Theory of Games of Strategy, J. D. Williams. Highly entertaining classic describes, with many examples, how to select best strategies in conflict situations. 268pp. 5⅜ x 8½. 25101-2 Pa. $9.95

Math–Geometry and Topology

ELEMENTARY CONCEPTS OF TOPOLOGY, Paul Alexandroff. Elegant, intuitive approach to topology from set-theoretic topology to Betti groups; how concepts of topology are useful in math and physics. 57pp. 5⅜ x 8½. 60747-X Pa. $4.95

COMBINATORIAL TOPOLOGY, P. S. Alexandrov. Clearly written, well-organized, three-part text begins by dealing with certain classic problems without using the formal techniques of homology theory and advances to the central concept, the Betti groups. Numerous detailed examples. 654pp. 5⅜ x 8½. 40179-0 Pa. $18.95

Physics

INTRODUCTION TO QUANTUM MECHANICS With Applications to Chemistry, Linus Pauling and E. Bright Wilson Jr. Classic undergraduate text by Nobel Prize winner applies quantum mechanics to chemical and physical problems. Numerous tables and figures enhance the text. Chapter bibliographies. Appendices. Index. 468pp. 5⅜ x 8½. 64871-0 Pa. $13.95

TENSOR ANALYSIS FOR PHYSICISTS, J. A. Schouten. Concise exposition of the mathematical basis of tensor analysis, integrated with well-chosen physical examples of the theory. Exercises. Index. Bibliography. 289pp. 5⅜ x 8½. 65582-2 Pa. $13.95

RELATIVITY IN ILLUSTRATIONS, Jacob T. Schwartz. Clear nontechnical treatment makes relativity more accessible than ever before. Over 60 drawings illustrate concepts more clearly than text alone. Only high school geometry needed. Bibliography. 128pp. 6⅛ x 9¼. 25965-X Pa. $7.95

THE ELECTROMAGNETIC FIELD, Albert Shadowitz. Comprehensive undergraduate text covers basics of electric and magnetic fields, builds up to electromagnetic theory. Also related topics, including relativity. Over 900 problems. 768pp. 5⅝ x 8¼. 65660-8 Pa. $19.95

GREAT EXPERIMENTS IN PHYSICS: Firsthand Accounts from Galileo to Einstein, edited by Morris H. Shamos. 25 crucial discoveries: Newton's laws of motion, Chadwick's study of the neutron, Hertz on electromagnetic waves, more. Original accounts clearly annotated. 370pp. 5⅜ x 8½. 25346-5 Pa. $12.95

RELATIVITY, THERMODYNAMICS AND COSMOLOGY, Richard C. Tolman. Landmark study extends thermodynamics to special, general relativity; alsoapplications of relativistic mechanics, thermodynamics to cosmological models. 501pp. 5⅜ x 8½. 65383-8 Pa. $15.95

LIGHT SCATTERING BY SMALL PARTICLES, H. C. van de Hulst. Comprehensive treatment including full range of useful approximation methods for researchers in chemistry, meteorology and astronomy. 44 illustrations. 470pp. 5⅜ x 8½. 64228-3 Pa. $14.95

STATISTICAL PHYSICS, Gregory H. Wannier. Classic text combines thermodynamics, statistical mechanics and kinetic theory in one unified presentation of thermal physics. Problems with solutions. Bibliography. 532pp. 5⅜ x 8½. 65401-X Pa. $14.95

Prices subject to change without notice.